Ruilin Long

Martingale Spaces and Inequalities

Ruilin Long

Martingale Spaces and Inequalities

Springer Fachmedien Wiesbaden GmbH

vieweg

Mathematical Subject Classification: 60 Gxx, 60Hxx

Originally published by Friedr. Vieweg & Sohn Verlagsgesellschaft mbH, Braunschweig/Wiesbaden, 1993

Typesetting: Peking University Press

ISBN 978-3-322-99268-0 ISBN 978-3-322-99266-6 (eBook)
DOI 10.1007/978-3-322-99266-6

Preface

In the past twenty years, the H_p-BMO Theory on $\mathbf{R}^n$ has undergone a flourishing development, which should partly give the credit to the application of some martingale idea and methods. It would be valuable to exhibit some examples concerning this point. As one of the key parts of Calderón-Zygmund's real method which first appeared in the 50's, Calderón-Zygmund Decomposition is exactly the so-called stopping time argument in nature which already existed in the Probability Theory early in the 30's, although such a close relationship between Calderón-Zygmund Decomposition and the stopping time argument perhaps was not realized consciously at that time. But after the 70's we actually used the stopping time argument intentionally as a method of thinking in Analysis. Later, when classical H_p Theory had undergone an evolution from one chapter in the Complex Variable Theory to an independent branch (the key step to accelerate this evolution was D.Burkholder-R.Gundy-M.Silverstein's well-known work in the early 70's on the maximal function characterization of H_p), Martingale H_p-BMO Theory soon appeared as a counterpart of the classical H_p-BMO Theory. Owing to the simplicity of the structure in martingale setting, many new ideas and methods might be produced easier on this stage. These new things have shown a great effect on the classical H_p-BMO Theory. For example, the concept of atomic decomposition of H_p was first germinated in martingale setting; the good λ-inequality, which is a powerful tool to compare the integrability of two related measurable functions, was also originally found in obtaining martingale inequalities; the constructive proof of the Fefferman-Stein Decomposition of a BMO function was first got in ternary martingale situation. In addition, there are also many applications of Martingale Theory to Harmonic Function Theory and many recent applications to Analysis and especially to Harmonic Analysis. Among them, two examples are worth to be mentioned. One is that D. Burkholder described an important kind of Banach spaces (called UMD spaces) by using martingales, another is that by using martingales as a tool, a much more simplified proof of the important $T(b)$ Theorem in Calderón-Zygmund Singular Integral Theory was given. From the above-cited examples we can see what an important role Martingale Theory has played in the development of Analysis, especially of Harmonic Analysis.

As we mentioned above, since the early 70's, accompanying the development of H_p-BMO Theory on $\mathbf{R}^n$, a corresponding branch in Martingale Theory was born, which may be called the Martingale Spaces and Inequalities. Until now, most of the important facts in H_p-BMO Theory on $\mathbf{R}^n$ have been found to have their satisfactory counterparts in the martingale setting. Some of them, such as the duality theory of H_1-BMO; L^p inequalities concerning the maximal function operators, the square function operators and the conditional square function operators can be found in A.Garsia's famous monograph "Martingale Inequalities". This is so far the sole book dealing with martingale spaces and inequalities systematically which gives us a very good summary on the advances obtained before the early 70's in this field. Since its publication this branch of Martingale Theory has become more and more mature. For instance, martingale inequalities between various operators (even weighted inequalities) can be discussed in more detail; H_1 can be expanded to the regular H_p ($p < 1$); BMO space establishes its relation with Carleson measures, A_p weights etc.; moreover, because of D.Burkholder's contribution great success has been scored in martingale transforms, which are considered to be an analogue to classical singular integral operators. It seems that nobody has ever compiled all of the above-mentioned materials into one systematic book. In order to help setting up the link between the two fields, Probability and Analysis, the book "Martingale Spaces and Inequalities" is presented to the readers.

Finally, some words are to be made concerning the materials and arrangement of the book. Generally, martingales could be considered with respect to discrete indices or continuous indices. This book will be focused only on the situation of discrete indices, because this part is the most mature and the easiest part. But we don't think that this will form a real limitation, because those who are major in Probability can make their own judgement which subjects of the book can and which cannot be transformed from the discrete to the continuous index situation and how this is transformed. On the other hand, they may also consult C.Dellacherie-P.Meyer's book "Probabilities et Potentials" as a reference. As for those who are major in Analysis, it seems that only the discrete index part will be sufficient at present. For the discrete index part, this book will contain eight chapters, which are concerning: (1) Preparatory knowledge from Probability; (2) H_p and some related spaces (which includes main points in Garsia's book except the part concerning BMO); (3) Martingale Φ-inequalities; (4) BMO Martingales; (5) Weights and weighted inequalities; (6) Martingale transforms (which will deal with the results concerning the best constants in martingale transform inequalities obtained by D. Burkholder in the early and middle 80's); (7) Regular martingales; (8) Some applications of martingales in Harmonic Analysis (the simplified proof of the $T(b)$ Theorem will be included). The book is rewritten from author's former text book in Chinese which was used several times for graduate students at the Department of Mathematics, Peking University in the early 1980's, and was published by Peking University Press in 1985. In the present English version, many supplements to the Chinese edition have been embodied. Here the author would give his hearty thanks to his teachers Professor Cheng Min-Teh and Professor J.P.Kahane, who introduced him to Harmonic Analysis and Martingale Theory. He is also indebted to the editor Qiu Shuqing of Peking University Press, and the editors of Vieweg, who helped him to improve the writing of the book.

Contents

1 Probabilistic Preliminaries

This chapter will be devoted to the probabilistic preliminaries, such as conditional expectations, stopping times, martingales and its convergences and decompositions, etc., which are needed in what follows. It is probably useful for those who are not familiar with Probability Theory.

1.1 Conditional expectations

Let $(\Omega, \mathcal{F}, \mu)$ be a complete probability space, $\mathcal{F}_1 \subset \mathcal{F}$ be a complete sub-σ-field. For all $f \in L^1(\Omega, \mathcal{F}, \mu)$,

$$\nu(F) = \int_F f d\mu, \quad \forall F \in \mathcal{F}_1, \tag{1.1.1}$$

defines a complex measure on $\mathcal{F}_1$, which is absolutely continuous with respect to $\mu|_{\mathcal{F}_1}$. By means of Radon-Nikodým's theorem, there is a unique function denoted by $E_{\mathcal{F}_1}(f)$ or $E(f|\mathcal{F}_1)$, which is measurable with respect to $\mathcal{F}_1$ and integrable with respect to $\mu|_{\mathcal{F}_1}$, and such that

$$\int_F E(f|\mathcal{F}_1) d\mu = \int_F f d\mu, \quad \forall F \in \mathcal{F}_1. \tag{1.1.2}$$

Now we give our definition.

Definition 1.1.1 *Let $f \in L^1(\Omega, \mathcal{F}, \mu)$. Then $E(f|\mathcal{F}_1)$ defined as above is called the conditional expectation of f with respect to $\mathcal{F}_1$. For any $F \in \mathcal{F}_1$, $E_{\mathcal{F}_1}(\chi_F)$ is called F's conditional probability, where χ_F denotes the characteristic (or indicator) function of F.*

Examples

1. Let $\mathcal{F}_1$ be a sub-σ-field generated by atoms $\{F_k\}_1^n$. (An atom of a measure space is a set A of which there is no subset B satisfying $0 < |B| < |A|$. Here $|\cdot|$ denotes measure in the underlying space as usual). Then for any $f \in L^1$,

$$E_{\mathcal{F}_1}(f) = \sum_{i=1}^{n} \frac{1}{|F_i|} \int_{F_i} f d\mu \chi_{F_i}. \tag{1.1.3}$$

This can be seen as follows. Since $E_{\mathcal{F}_1}(f)$ is measurable with respect to $\mathcal{F}_1$, so it is constant on each atom, and hence $E_{\mathcal{F}_1}(f) = \sum_1^n c_i \chi_{F_i}$. Now integrating two sides of the expression, and making use of (1.1.2), we get $c_i = \frac{1}{|F_i|} \int_{F_i} f d\mu$, and (1.1.3) follows.

2. In particular, let $F \in \mathcal{F}$, and $\mathcal{F}_1$ be generated by atoms $\{F, F^c\}$ (c denotes the complementary set as usual). Let $f = \chi_G$. Then

$$E_{\mathcal{F}_1}(\chi_G) = \frac{|G \cap F|}{|F|}\chi_F + \frac{|G \cap F^c|}{|F^c|}\chi_{F^c}.$$

Here $\frac{|G \cap F|}{|F|}$ is nothing but the classical concept of conditional probability of event G when the event F has occurred. This verifies the reason of the name of "conditional expectation".

We now give some basic properties of conditional expectation.

(a) $E_{\mathcal{F}_1}$ is (complex) linear, and commutable with the complex conjugate, i.e. $E_{\mathcal{F}_1}(\overline{f}) = \overline{E}_{\mathcal{F}_1}(f)$.

(b) $E_{\mathcal{F}_1}(1) = 1$.

(c) $E_{\mathcal{F}_1}$ is positive. That is to say $f \geq 0$ implies $E_{\mathcal{F}_1} f \geq 0$. (In fact, consider the set $F = \{E_{\mathcal{F}_1}(f) < 0\}$. Then $\int_F E_{\mathcal{F}_1}(f) d\mu = \int_F f d\mu \geq 0$ implies $|F| = 0$.)

(d) Martingale property. Let $\mathcal{F}_1$, $\mathcal{F}_2$ be two sub-σ-fields, and $\mathcal{F}_1 \subset \mathcal{F}_2$, then

$$E_{\mathcal{F}_1}(E_{\mathcal{F}_2}(f)) = E_{\mathcal{F}_1}(f), \ \forall f \in L^1. \tag{1.1.4}$$

We only have to verify (1.1.2). Let $F \in \mathcal{F}_1 \subset \mathcal{F}_2$, we have

$$\int_F E_{\mathcal{F}_1}(E_{\mathcal{F}_2}(f)) d\mu = \int_F E_{\mathcal{F}_2}(f) d\mu = \int_F f d\mu = \int_F E_{\mathcal{F}_1}(f) d\mu.$$

(e) We have

$$\|E_{\mathcal{F}_1}(f)\|_p \leq \|f\|_p, \quad 1 \leq p \leq \infty. \tag{1.1.5}$$

Here $\|\cdot\|_p$ denoteds the L^p-norm. This follows from (denoting the conjugate index of p by p', i.e. $\frac{1}{p} + \frac{1}{p'} = 1$, and making use of (d))

$$\|E_{\mathcal{F}_1}(f)\|_p = \sup_{g:\|g\|_{p'} \leq 1} |E(E_{\mathcal{F}_1}(f) g)| = \sup_{g:\|g\|_{p'} \leq 1} |E(fg)| \leq \|f\|_p,$$

where "sup" is taken over all g of the form

$$g = \sum_{i=1}^{n} c_i \chi_{F_i}, \quad \forall n, \ \forall \{c_i\}_1^n \subset \mathbf{C}, \ \forall \{F_i\}_1^n, \ F_i \in \mathcal{F}_1. \tag{1.1.6}$$

(f) For $f \in L^p$, $g \in L^{p'}(\mathcal{F}_1)$ (means g being $\mathcal{F}_1$-measurable as well) and $1 \le p \le \infty$, we have

$$E_{\mathcal{F}_1}(fg) = gE_{\mathcal{F}_1}(f). \tag{1.1.7}$$

This follows by taking limits. For $\mathcal{F}_1$-simple g like (1.1.6), (1.1.7) holds obviously. In fact, say $g = \chi_F$, $F \in \mathcal{F}_1$, then both of $E_{\mathcal{F}_1}(f\chi_F)$ and $E_{\mathcal{F}_1}(f)\chi_F$ are $\mathcal{F}_1$-measurable, and for all $G \in \mathcal{F}_1$,

$$\int_G E_{\mathcal{F}_1}(f\chi_F)d\mu = \int_{F\cap G} f d\mu = \int_G E_{\mathcal{F}_1}(f)\chi_F d\mu,$$

so $E_{\mathcal{F}_1}(f\chi_F) = E_{\mathcal{F}_1}(f)\chi_F$. Now let $\{g^{(n)}\}$ be a $\mathcal{F}_1$-simple sequence such that $g^{(n)} \to g$ in $L^{p'}$, then both sides of (1.1.7) converge in L^1, so their limits are the same.

(g) We have

$$|E_{\mathcal{F}_1}(f)| \le E_{\mathcal{F}_1}(|f|), \quad \text{a.e.} \quad \forall f \in L^1, \tag{1.1.8}$$

here a.e. means almost everywhere. First, assume f is real valued. Consider $F = \{E_{\mathcal{F}_1}(f) > E_{\mathcal{F}_1}(|f|)\}$. Then $F \in \mathcal{F}_1$, and

$$\int_F E_{\mathcal{F}_1}(f)d\mu = \int_F f d\mu \le \int_F |f| d\mu = \int_F E_{\mathcal{F}_1}(|f|)d\mu,$$

and hence $|F| = 0$. Analogously, $\{E_{\mathcal{F}_1}(f) < -E_{\mathcal{F}_1}|f|\}$ is also of measure 0. For complex valued f, there is $\mathcal{F}_1$-measurable $\theta(\omega)$ such that $E_{\mathcal{F}_1}(f)e^{i\theta(\omega)} = |E_{\mathcal{F}_1}(f)|$. By making use of (f), we have $E_{\mathcal{F}_1}(fe^{i\theta(\omega)}) = |E_{\mathcal{F}_1}(f)|$. And hence

$$|E_{\mathcal{F}_1}(f)| = E_{\mathcal{F}_1}(\operatorname{Re}(fe^{i\theta(\omega)})) \le E_{\mathcal{F}_1}(|\operatorname{Re}(fe^{i\theta(\omega)})|) \le E_{\mathcal{F}_1}(|f|).$$

(h) Parseval's equality. $f \in L^p$, $g \in L^{p'}$, $1 \le p \le \infty$. We have

$$E(E_{\mathcal{F}_1}(f)g) = E(fE_{\mathcal{F}_1}(g)). \tag{1.1.9}$$

In fact, each of them equals to $E(E_{\mathcal{F}_1}(f)E_{\mathcal{F}_1}(g))$. This can be seen as follows. Denote the σ-field $\{\emptyset, \Omega\}$ by $\mathcal{F}_0$, then $E_{\mathcal{F}_0}(f) = E(f)$. So, from (d) and (f), we have

$$E(E_{\mathcal{F}_1}(f)g) = E(E_{\mathcal{F}_1}(E_{\mathcal{F}_1}(f)g)) = E(E_{\mathcal{F}_1}(f)E_{\mathcal{F}_1}(g)).$$

(i) Hölder's inequality. For all $f \in L^p$, $g \in L^{p'}$, $1 \le p \le \infty$, we have

$$|E_{\mathcal{F}_1}(fg)| \le E_{\mathcal{F}_1}(|f|^p)^{\frac{1}{p}} E_{\mathcal{F}_1}(|g|^{p'})^{\frac{1}{p'}}. \tag{1.1.10}$$

Denote $F = \{E_{\mathcal{F}_1}(|f|^p) > 0,\ E_{\mathcal{F}_1}(|g|^{p'}) > 0\}$. Then on F, we have

$$E_{\mathcal{F}_1}(|f|^p)^{-\frac{1}{p}}|f|E_{\mathcal{F}_1}(|g|^{p'})^{-\frac{1}{p'}}|g| \le \frac{1}{p}E_{\mathcal{F}_1}(|f|^p)^{-1}|f|^p + \frac{1}{p'}E_{\mathcal{F}_1}(|g|^{p'})^{-1}|g|^{p'}.$$

Since $F \in \mathcal{F}_1$, we get

$$\begin{aligned}
&E_{\mathcal{F}_1}(E_{\mathcal{F}_1}(|f|^p)^{-\frac{1}{p}}E_{\mathcal{F}_1}(|g|^{p'})^{-\frac{1}{p'}}|f||g|)\chi_F \\
&\le E_{\mathcal{F}_1}\left(\frac{1}{p}E_{\mathcal{F}_1}(|f|^p)^{-1}|f|^p\chi_F\right) + E_{\mathcal{F}_1}\left(\frac{1}{p'}E_{\mathcal{F}_1}(|g|^{p'})^{-1}|g|^{p'}\chi_F\right) \\
&\le \frac{1}{p} + \frac{1}{p'} = 1,
\end{aligned}$$

$$E_{\mathcal{F}_1}(|fg|\chi_F) \leq E_{\mathcal{F}_1}(|f|^p)^{\frac{1}{p}} E_{\mathcal{F}_1}(|g|^{p'})^{\frac{1}{p'}}.$$

But on $F_f = \{E_{\mathcal{F}_1}(|f|^p) = 0\}$, $f = 0$ a.e. because of

$$\int_{F_f} |f|^p d\mu = \int_{F_f} E_{\mathcal{F}_1}(|f|^p) d\mu = 0.$$

Analogously $g = 0$, a.e. on F_g. So $fg = 0$ on $F^c = F_f \bigcup F_g$. This completes the proof of (1.1.10).

Before giving some other properties of conditional expectations, a slight generalization of the concept itself is worth to be formulated.

Definition 1.1.2 *Let f be nonnegative and measurable. Consider*

$$f^{(N)} = \begin{cases} f, & f \leq N, \\ N, & f > N. \end{cases} \tag{1.1.11}$$

Because of the positivity and hence the monotonity, $\lim\limits_{N\to\infty} E_{\mathcal{F}_1}(f^{(N)})$ exists almost everywhere (a.e.). We denote it by $E_{\mathcal{F}_1}(f)$.

Remark Such definition of $E_{\mathcal{F}_1}$ keeps its two characteristic properties, i.e. the $\mathcal{F}_1$-measurability and the integral equality $\int_F f d\mu = \int_F E_{\mathcal{F}_1}(f) d\mu$, for all f nonnegative and measurable, for all $F \in \mathcal{F}_1$. The former is obvious, and the latter follows from the monotone convergence theorem of integrals. Soon we will show the monotone convergence theorem for such defined conditional expectation, from which we see that the choice of $f^{(N)}$ in the definition is not essential. Basing on these facts, the generalized definition could be extended to real $f = f^+ - f^-$, provided $E_{\mathcal{F}_1}(f^+) < \infty$, a.e., or $E_{\mathcal{F}_1}(f^-) < \infty$, a.e., to complex f provided $E_{\mathcal{F}_1}(\operatorname{Re} f)$ and $E_{\mathcal{F}_1}(\operatorname{Im} f)$ could be defined, without loss of usual properties of originally defined $E_{\mathcal{F}_1}$. But we need $E_{\mathcal{F}_1}$ defined only for $f \in L^1$, and f nonnegative and measurable.

(j) Monotone convergence theorem. Let $\{f^{(n)}\}$ be an increasing (means nondecreasing) sequence of nonnegative measurable functions, such that $\lim\limits_{n\to\infty} f^{(n)}(\omega) = f(\omega)$, exists a.e. Then

$$\lim_{n\to\infty} E_{\mathcal{F}_1}(f^{(n)}) = E_{\mathcal{F}_1}(f), \quad \text{a.e.} \tag{1.1.12}$$

Now we prove it. Because of the positivity and monotonity of $E_{\mathcal{F}_1}$, $\lim\limits_{n\to\infty} E_{\mathcal{F}_1}(f^{(n)}) = h$ exists a.e., with h a $\mathcal{F}_1$-measurable function. Notice that $h \leq E_{\mathcal{F}_1}(f)$, a.e. But, for all $F \in \mathcal{F}_1$, we have

$$\int_F h d\mu = \int_F \lim_{n\to\infty} E_{\mathcal{F}_1}(f^{(n)}) d\mu = \int_F f d\mu = \int_F E_{\mathcal{F}_1}(f) d\mu,$$

and hence $h = E_{\mathcal{F}_1}(f)$, a.e. The proof is finished.

(k) Fatou's lemma. For any sequence $\{f^{(n)}\}$ of nonnegative measurable functions, we have

$$E_{\mathcal{F}_1}(\underline{\lim}\, f^{(n)}) \leq \underline{\lim}\, E_{\mathcal{F}_1}(f^{(n)}), \quad \text{a.e.} \tag{1.1.13}$$

This follows from (j). In fact, denote $g^{(n)} = \inf_{m \geq n} f^{(m)}$, then $\lim_{n\to\infty} g^{(n)} = \underline{\lim}\, f^{(n)}$, a.e. monotonically, so

$$E_{\mathcal{F}_1}(\underline{\lim}\, f^{(n)}) = E_{\mathcal{F}_1}(\lim g^{(n)}) = \lim E_{\mathcal{F}_1}(g^{(n)}) \leq \underline{\lim}\, E_{\mathcal{F}_1}(f^{(n)}).$$

(l) Dominated convergence theorem. Let $\{f^{(n)}\} \subset L^1$, $\lim_{n\to\infty} f^{(n)} = f$, a.e , and $|f^{(n)}| \leq g \in L^1$. Then

$$\lim_{n\to\infty} E_{\mathcal{F}_1}(f^{(n)}) = E_{\mathcal{F}_1}(f), \quad \text{a.e.} \tag{1.1.14}$$

This follows from (k). In fact, we have

$$\begin{aligned} E_{\mathcal{F}_1}(2g) - \overline{\lim_{n\to\infty}}\, E_{\mathcal{F}_1}(|f - f^{(n)}|) &= \underline{\lim_{n\to\infty}}\, E_{\mathcal{F}_1}(2g - |f - f^{(n)}|) \\ &\geq E_{\mathcal{F}_1}\left(\underline{\lim_{n\to\infty}}\,(2g - |f - f^{(n)}|)\right) = E_{\mathcal{F}_1}(2g), \end{aligned}$$

$$\overline{\lim_{n\to\infty}}\, E_{\mathcal{F}_1}(|f - f^{(n)}|) \leq 0.$$

Remark The condition $0 \leq f^{(n)} \leq g \in L^1$ in (l) could be replaced by $h \leq f^{(n)} \leq g$, $h, g \in L^1$, the nonnegativity in (j), (k), could be replaced by $f^{(n)} \geq h$, $h \in L^1$.

The last property of $E_{\mathcal{F}_1}$ we want to state is Jensen's inequality. Let $\varphi(u)$ be a convex function defined on (a, b). Notice that for all $u \in (a, b)$, for all those $\lambda \in (a, b)$ such that $\varphi'(\lambda)$ exists (only countable λ's may be exceptional), we have

$$\varphi(u) - \varphi(\lambda) \geq \varphi'(\lambda)(u - \lambda). \tag{1.1.15}$$

Let $f \in L^1$ be real with its values in (a, b) for a.e. ω, and such that $\varphi(f) \in L^1$ or $\varphi(f)$ being nonnegative. Then we have

(m) Let $\varphi(u)$ and f be as above. Then

$$\varphi(E_{\mathcal{F}_1}(f)) \leq E_{\mathcal{F}_1}(\varphi(f)), \quad \text{a.e.} \tag{1.1.16}$$

Now we prove it. Consider those λ and ω such that $\varphi'(\lambda)$ exists, and

$$a < u(= f(\omega)), \quad v(= E_{\mathcal{F}_1}(f)(\omega)) < b.$$

For such λ and u apply (1.1.15) and take the conditional expectation on both sides, we get

$$E_{\mathcal{F}_1}(\varphi(f)) \geq \varphi'(\lambda)(E_{\mathcal{F}_1}(f) - \lambda) + \varphi(\lambda). \tag{1.1.17}$$

Taking a sequence of λ's tending to v, the right-hand side of (1.1.17) tends to $\varphi(E_{\mathcal{F}_1}(f)$. This proves (1.1.16).

Remark The conditional expectation operator could be characterized by several simple conditions. Neveu [1] pointed that: Let T be an operator defined on $L^p(\Omega, \mathcal{F}, \mu)$, $1 \leq p < \infty$, being linear, positive, idempotent, norm-decreasing and such that $T(1) = 1$, then there exists a complete sub-σ-field $\mathcal{F}_1$ such that $T = E_{\mathcal{F}_1}$; in addition, when $p = 2$, the positivity assumption in the conditions could be taken off. For such kind of discussion, see also Rao [3]. We do not give details.

1.2 Stopping times

The real method in Harmonic Analysis was born in the 1950's. Its main content is so-called Calderón-Zygmund's decomposition. This is such a decomposition which for a given $f \in L^1_{\text{loc}}(\mathbf{R}^n)$, and a level $\lambda > 0$, divides $\mathbf{R}^n$ into two parts $O \bigcup F$, where $F = O^c$ and $O = \bigcup_k Q_k$, each Q_k is the biggest dyadic cube which makes $\frac{1}{|Q_k|}\int_{Q_k} |f| dy > \lambda$. Here the idea "the biggest" coincides with the idea "the earliest" in Probability Theory surprisingly. The latter is the so-called stopping time. It played a crucial role in whole Probability Theory as well as in this book just like Calderón-Zygmund's decomposition did in Analysis.

Definition 1.2.1 *Let $(\Omega, \mathcal{F}, \mu)$ be a probability space, $\{\mathcal{F}_n\}_{n \geq 0}$ be a nondecreasing (briefly, say increasing) family of complete sub-σ-fields such that $\mathcal{F} = \bigvee_n \mathcal{F}_n$ (that means $\bigcup_n \mathcal{F}_n$ generates $\mathcal{F}$). Denote the set of all nonnegative integers by $\mathbf{Z}^+$, and $\mathbf{Z}^+ \bigcup \{\infty\}$ by $\overline{\mathbf{Z}}^+$. A mapping T from Ω to $\overline{\mathbf{Z}}^+$ is called a stopping time with respect to $\{\mathcal{F}_n\}_{n \geq 0}$, if $\{\omega : T(\omega) = n\} \in \mathcal{F}_n$, for all n, or equivalently $\{\omega : T(\omega) \leq n\} \in \mathcal{F}_n$, for all n.*

Remark In what follows, we always consider $(\Omega, \mathcal{F}, \mu)$ endowed with a family $\{\mathcal{F}_n\}$ satisfying the preceding usual conditions. Many concepts are related to such a family such as stopping times, adaptation, predictability etc., without explicit indication unless otherwise stated.

Example Let $\{f_n\}_{n \geq 0}$ be any adapted process, that means f_n being $\mathcal{F}_n$-measurable for all n. Let B be any Borel set in $\mathbf{C}$. Then $T = \inf\{n : f_n \in B\}$ † is a stopping time. This is owing to $\{\omega : T = n\} = \{\omega : f_j \notin B, \forall j < n, f_n \in B\} \in \mathcal{F}_n$. This is a typical example of stopping times. So-called stopping time is just the earliest time when the underlying process stops at some place.

For each stopping time T, we can associate a sub-σ-field $\mathcal{F}_T$ called the field prior to T.

Definition 1.2.2 *Let T be a stopping time. Denote*

$$\mathcal{F}_T = \{F \in \mathcal{F} : F \bigcap \{T \leq n\} \in \mathcal{F}_n, \ \forall n\}. \tag{1.2.1}$$

† According to the usual convention, $\inf\{\emptyset\} = \infty$.

Remark In the definition, $\{T \le n\}$ could be replaced by $\{T = n\}$. And, $\mathcal{F}_T$ is a σ-field obviously. What is the meaning of $\mathcal{F}_T$? If we think of $\mathcal{F}_n$ as the collection of all events observed up to time n,then $\mathcal{F}_T$ could be thought of as a collection of all events observed at the random time T. Notice that according to the definition, all subsets of $\{T = \infty\}$ belong to $\mathcal{F}_T$.

Now give some elementary properties of stopping times and its related σ-field $\mathcal{F}_T$.

(a) Let T be any stopping time. Then $A \in \mathcal{F}_T \Leftrightarrow A \in \mathcal{F}$, and

$$T_A = \begin{cases} T, & \omega \in A, \\ \infty, & \omega \notin A, \end{cases}$$

is a stopping time. This is because of $\{T_A \le n\} = A \bigcap \{T \le n\}$.

(b) Let S be a stopping time, T be a $\overline{\mathbf{Z}}^+$-valued mapping which is $\mathcal{F}_S$-measurable and such that $S \le T$. Then T is also a stopping time. This follows from the fact $\{T \le n\} = \{S \le n\} \bigcap \{T \le n\}$. In particular, for any $n \in \mathbf{Z}^+$, $S+n$ are all stopping times but not $S - n$, except $n = 0$.

(c) Let T, S be two stopping times. Then $T + S$ and $T \vee S = \max(T, S)$, and $T \wedge S = \min(T, S)$ are stopping times. Furthermore, let $\{T_k\}$ be a sequence of stopping times. Then $T = \sup\limits_k T_k$ and $S = \inf\limits_k T_k$ are stopping times. These are because of

$$\{T + S \le n\} = \bigcup_{k=0}^{n} \{T = k\} \bigcap \{S \le n - k\},$$

$$\{T \le n\} = \bigcap_k \{T_k \le n\},$$

$$\{S \ge n + 1\} = \bigcap_k \{T_k \ge n + 1\}.$$

(d) Let T, S be two stopping times. Then all of $\{T \le S\}$, $\{S < T\}$, $\{S \le T\}$, $\{S > T\}$, $\{T = S\}$ belong to $\mathcal{F}_T \bigcap \mathcal{F}_S$, and $\mathcal{F}_T \bigcap \mathcal{F}_S = \mathcal{F}_{T \wedge S}$. To see these, it is enough to consider $\{T \le S\}$, and prove $\mathcal{F}_T \bigcap \mathcal{F}_S \subset \mathcal{F}_{T \wedge S}$. We have

$$\{T \le S\} \bigcap \{S = n\} = \bigcup_{m \le n} \{T = m\} \bigcap \{S = n\} \in \mathcal{F}_n,$$

$$\{T \le S\} \bigcap \{T = n\} = \{S \ge n\} \bigcap \{T = n\} \in \mathcal{F}_n;$$

and for all $A \in \mathcal{F}_T \bigcap \mathcal{F}_S$,

$$A \bigcap \{T \wedge S \le n\} = (A \bigcap \{T \le n\}) \bigcup (A \bigcap \{S \le n\}) \in \mathcal{F}_n.$$

(e) Let T, S be two stopping times, $A \subset \{T \le S\}$ and $A \in \mathcal{F}_S$, (or $A \subset \{T = S\}$ and $A \in \mathcal{F}_T \bigcap \mathcal{F}_S$), then $A \bigcap \mathcal{F}_T \subset A \bigcap \mathcal{F}_S$ (or $A \bigcap \mathcal{F}_T = A \bigcap \mathcal{F}_S$). In particular,

$T \leq S$ implies $\mathcal{F}_T \subset \mathcal{F}_S$. This can be seen as follows. Let $B \in \mathcal{F}_T$, we want to show $A \bigcap B \in \mathcal{F}_S$. We have

$$A \bigcap B \bigcap \{S \leq n\} = A \bigcap \{S \leq n\} \bigcap B \bigcap \{T \leq n\} \in \mathcal{F}_n,$$

and hence $A \bigcap B \in \mathcal{F}_S$, $A \bigcap B \in A \bigcap \mathcal{F}_S$. This proves $A \bigcap \mathcal{F}_T \subset A \bigcap \mathcal{F}_S$. For the second case, the preceding argument can be inverted.

(f) Let $A \subset \{T = S\}$, $A \in \mathcal{F}_S \bigcap \mathcal{F}_T$. Then for all $f \in L^1$, we have

$$E(f|\mathcal{F}_S)\chi_A = E(f|\mathcal{F}_T)\chi_A. \tag{1.2.2}$$

This follows from

$$E(f|\mathcal{F}_S)\chi_A = E(f\chi_A|\mathcal{F}_S)\chi_A = E(f\chi_A|\mathcal{F}_T)\chi_A = E(f|\mathcal{F}_T)\chi_A.$$

(g) Let $A \subset \{T \leq S\}$, $A \in \mathcal{F}_S$. Then for all $f \in L^1$, we have

$$E(f\chi_A|\mathcal{F}_T) = E(E(f\chi_A|\mathcal{F}_S)|\mathcal{F}_T). \tag{1.2.3}$$

Since both-hand sides are $\mathcal{F}_T$-measurable, so it remains to verify the identity (1.1.2). We have, for all $B \in \mathcal{F}_T$, noticing $A \bigcap B \in \mathcal{F}_S$,

$$\begin{aligned}\int_B E(f\chi_A|\mathcal{F}_T)d\mu &= \int_{B\cap A} f d\mu = \int_{B\cap A} E(f|\mathcal{F}_S)d\mu \\ &= \int_B E(f\chi_A|\mathcal{F}_S)d\mu = \int_B E(E(f\chi_A|\mathcal{F}_S)|\mathcal{F}_T)d\mu.\end{aligned}$$

(h) Let T be any stopping time, $f \in L^1$. Then f_T defined as $f_{T(\omega)}(\omega)$, where $f_n = E(f|\mathcal{F}_n)$, $n \geq 0$, $f_\infty = f$, satisfies

$$f_T = E(f|\mathcal{F}_T), \tag{1.2.4}$$

$$E(|f_T|) \leq E(|f|). \tag{1.2.5}$$

First we show that f_T is $\mathcal{F}_T$-measurable. In fact, we have

$$f_{T(\omega)}(\omega) = \sum_{n=0}^{\infty} f_n(\omega)\chi_{\{T=n\}} + f(\omega)\chi_{\{T=\infty\}}.$$

For any Borel set B in $\mathbf{C}$, we have

$$f_T^{-1}(B) \bigcap \{T \leq m\} = \bigcup_{n\leq m} (f_n^{-1}(B) \bigcap \{T = n\}) \in \mathcal{F}_m, \quad \forall m,$$

this proves that f_T is $\mathcal{F}_T$-measurable. Since $f_n = E(f|\mathcal{F}_n)$, we have

$$|f_n| \leq E(|f|\,|\mathcal{F}_n), \text{ and } \forall F \in \mathcal{F}_n, \ \int_F |f_n| d\mu \leq \int_F |f| d\mu,$$

and hence

$$\int_\Omega |f_{T(\omega)}(\omega)| d\mu = \sum_{n=0}^{\infty} \int_{\{T=n\}} |f_n(\omega)| d\mu + \int_{\{T=\infty\}} |f| d\mu \leq \int_\Omega |f| d\mu.$$

This gives (1.2.5). It remains to prove (1.2.4). We have, for all $F \in \mathcal{F}_T$,

$$\int_F f_T d\mu = \sum_{n=0}^{\infty} \int_{F\cap\{T=n\}} f_n d\mu + \int_{F\cap\{T=\infty\}} f d\mu$$
$$= \sum_{n=0}^{\infty} \int_{F\cap\{T=n\}} f d\mu + \int_{F\cap\{T=\infty\}} f d\mu = \int_F f d\mu.$$

Together with the $\mathcal{F}_T$-measurablility of f_T, we prove (1.2.4).

(i) Let T, S be two stopping times. Then for all $f \in L^1$, we have

$$(f_T)_S = f_{T\wedge S}. \tag{1.2.6}$$

In fact, denoting $A = \{T \leq S\}$, $A^c = \{T > S\}$, then A, $A^c \in \mathcal{F}_T \bigcap \mathcal{F}_S$, and

$$\begin{aligned}(f_T)_S &= E(E(f|\mathcal{F}_T)|\mathcal{F}_S)(\chi_A + \chi_{A^c}) \\ &= E(E(f\chi_A|\mathcal{F}_T)|\mathcal{F}_S) + E(E(f\chi_{A^c}|\mathcal{F}_T)|\mathcal{F}_S) \\ &= E(f\chi_A|\mathcal{F}_T) + E(f\chi_{A^c}|\mathcal{F}_S) = f_T\chi_A + f_S\chi_{A^c} = f_{T\wedge S}.\end{aligned}$$

We have shown that the difference of two stopping times is no longer a stopping time in general. But if we introduce a new enlarged family, then the difference may be a stopping time with respect to this new family.

(j) Let S be a stopping time. Denote $\mathcal{B}_n = \mathcal{F}_{S+n}$, $n \geq 0$. Let T be a $\mathcal{F}$-measurable mapping from Ω to $\overline{\mathbf{Z}}^+$. Then $R = T+S$ is a stopping time with respect to $\{\mathcal{F}_n\}_{n\geq 0}$, if and only if T is a stopping time with respect to $\{\mathcal{B}\}_{n\geq 0}$. In fact, assuming T is, and noticing $S+k$ being a stopping time, then we have

$$\{T+S \leq n\} = \bigcup_{k=0}^{n} (\{T=k\}\bigcap\{S+k \leq n\}) \in \mathcal{F}_n.$$

Assuming R is, then we have

$$\{T \leq n\} = \{T+S \leq S+n\} \in \mathcal{F}_{S+n} = \mathcal{B}_n.$$

The assertion is proved.

Remark This assertion shows that in order to $R = T+S$ being a stopping time, both of T and S being stopping times are slightly stronger. Furthermore, this assertion shows also that $T = R-S$ is a stopping time with respect to a new family $\{\mathcal{B}_n\} = \{\mathcal{F}_{S+n}\}$, provided R, S being stopping times with respect to $\{\mathcal{F}_n\}$.

As the end of the section, we devote some words to the continuous version of stopping times and related concepts, although, in this book, the continuous time case is not our concerned object. Let $\{\mathcal{F}_t\}_{t\geq 0}$ be a family of sub-σ-fields satisfying the usual conditions, and being right continuous in addition. Then the concepts of stopping time T and $\mathcal{F}_T$ are defined in the same way, and have same properties. But this time, a new σ-field should be introduced which has some finer properties.

Definition 1.2.3 *Let T be a stopping time. Define $\mathcal{F}_{T-}$ be the σ-field generated by the set family* $\{$*all* $A \in \mathcal{F}_0$ *and* $A = B\bigcap\{T > t\}$, $\forall t$, $\forall B \in \mathcal{F}_t\}$.

Remark Intuitively, $\mathcal{F}_{T-}$ consists of all those $\mathcal{F}_t$-sets strictly prior to T. So it can be thought that $\mathcal{F}_{T-} \subset \mathcal{F}_T$. It is indeed true. In fact, for all $B \in \mathcal{F}_t$, for all t, we have

$$B\bigcap\{T > t\}\bigcap\{T \leq r\} = \begin{cases} \emptyset, & r \leq t, \\ B\bigcap\{t < T \leq r\}, & t < r, \end{cases} \in \mathcal{F}_r.$$

This prove $\mathcal{F}_{T-} \subset \mathcal{F}_T$. Thus the $\mathcal{F}_{T-}$-measurability is stronger than $\mathcal{F}_T$-measurability. Many sets or functions related to a stopping time T could be shown to be not only $\mathcal{F}_T$- but also $\mathcal{F}_{T-}$-measurable. For example, T is $\mathcal{F}_{T-}$-measurable, since $\{T > t\} = \Omega\bigcap\{T > t\} \in \mathcal{F}_{T-}$; for any two stopping times T and S, $\{T < S\} \in \mathcal{F}_{S-}$, since $\{T < S\} = \bigcup_r(\{T < r\}\bigcap\{S > r\}) \in \mathcal{F}_{S-}$($r$ runs through rational numbers); in addition, any $\mathcal{F}$-measurable subset A of $\{T = \infty\}$ is in $\mathcal{F}_{T-}$, the proof of which is as follows. For all t, for all $F \in \mathcal{F}_t$,

$$F\bigcap\{T = \infty\} = \bigcap_{n \geq t} F\bigcap\{T > n\} \in \mathcal{F}_{T-},$$

this means $\mathcal{F}_t\bigcap\{T = \infty\} \subset \mathcal{F}_{T-}$, so $\mathcal{F}\bigcap\{T = \infty\} \subset \mathcal{F}_{T-}$. It is just what was to be proved.

1.3 Martingales, super-(or sub-)martingales

In this section we want to introduce the concepts of martingales and the decompositions and the convergences of martingales. Let $(\Omega, \mathcal{F}, \mu)$ be a complete probability space with a family $\{\mathcal{F}_n\}_{n\geq 0}$ of sub-σ-fields satisfying the usual conditions, i.e. $\{\mathcal{F}_n\}$ is increasing, each $(\Omega, \mathcal{F}_n, \mu)$ is complete, and $\mathcal{F} = \bigvee_n \mathcal{F}_n$.

Definition 1.3.1 *Let $Q = (Q_n)_{n\geq 0}$ be a process. Q is said to be adapted, if Q_n is $\mathcal{F}_n$ measurable, for all n; is said to be (strictly) predictable, if Q_n is $\mathcal{F}_{(n-1)\vee 0}$-measurable, for all n.*

Definition 1.3.2 *Let $f = (f_n)_{n\geq 0}$ be an adapted process. f is said to be a martingale (with respect to $\{\mathcal{F}_n\}_{n\geq 0}$) if each $f_n \in L^1$, and*

$$E(f_{n+1}|\mathcal{F}_n) = f_n, \quad n = 0, 1, 2, \cdots; \tag{1.3.1}$$

is said to be a super-(or sub-)martingale, if the equality in (1.3.1) is replaced by $\leq$ (or $\geq$).

Remark Obviously, $f = (f_n)_{n\geq 0}$ is a supermartingale, if and only if $-f$ is a submartingale.

Examples

1. Let $f \in L^1$, and $f_n = E(f|\mathcal{F}_n)$, $n \geq 0$. Then $f = (f_n)_{n\geq 0}$ is a martingale. Let $(d_n)_{n\geq 0}$ be an adapted process such that $d_n \in L^1$ and $E(d_n|\mathcal{F}_{n-1}) = 0$, $n \geq 1$, then $f = (f_n)_{n\geq 0}$ with $f_n = \sum_0^n d_k$, is a martingale. Inversely, each martingale

$f = (f_n)_{n\geq 0}$ could be generated in such way. In fact, denote $d_n = \Delta_n f = f_n - f_{n-1}$, $n \geq 0$, here f_{-1} is meant 0 like we will always do for any process, then $(d_n)_{n\geq 0}$ is an adapted process such that $E(d_n|\mathcal{F}_{n-1}) = 0$, $n \geq 1$, and $f_n = \sum_0^n d_k$, $n \geq 0$.

Remark When $f \in L^2$, then $(d_n)_{n\geq 0} = (\Delta_n f)_{n\geq 0}$ is an orthogonal system in L^2, since (say $k > l$)

$$E(d_k\bar{d}_l) = E(E(d_k\bar{d}_l|\mathcal{F}_{k-1})) = E(\bar{d}_l E(d_k|\mathcal{F}_{k-1})) = 0.$$

So, each martingale in L^2, is an orthogonal series.

2. About super-(or sub-)martingale. Let $f = (f_n)_{n\geq 0}$ be a martingale, $1 \leq p < \infty$. Then $(|f_n|^p)_{n\geq 0}$ is a submartingale, since

$$|f_n|^p \leq |E(f_{n+1}|\mathcal{F}_n)|^p \leq E(|f_{n+1}|^p|\mathcal{F}_n), \quad n \geq 0.$$

Let $f = (f_n)_{n\geq 0}$ be a submartingale, $\varphi(u)$ be an increasing convex function, then $(\varphi(f_n))_{n\geq 0}$ is still a submartingale, this comes from the Jensen's inequality

$$\varphi(f_n) \leq \varphi(E(f_{n+1}|\mathcal{F}_n)) \leq E(\varphi(f_{n+1})|\mathcal{F}_n), \quad n \geq 0.$$

Similarly, let $f = (f_n)_{n\geq 0}$ be a supermartingale, $\varphi(u)$ be increasing concave, then $(\varphi(f_n))_{n\geq 0}$ is still a supermartingale, since denoting the inverse function of $\varphi(u)$ by $\psi(v)$, which is increasing convex, we get

$$E(f_{n+1}|\mathcal{F}_n) = E(\psi(\varphi(f_{n+1}))|\mathcal{F}_n) \geq \psi(E(\varphi(f_{n+1})|\mathcal{F}_n)),$$
$$\varphi(f_n) \geq \varphi(E(f_{n+1}|\mathcal{F}_n)) \geq E(\varphi(f_{n+1})|\mathcal{F}_n), \quad n \geq 0.$$

Remark The class of sub-(or super-) martingales is closed under the action of any increasing convex (or concave) function. Here "increasing" could not be taken off as shown by following simple example. Let $(f_n)_{n\geq 0}$ be a nonnegative supermartingale but not a submartingale. Then $(-f_n)_{n\geq 0}$ is a submartingale, but $(\varphi(-f_n))_{n\geq 0}$ with $\varphi(u) = |u|$ a convex function, is not submartingale. But for real martingale $f = (f_n)_{n\geq 0}$, and any convex function $\varphi(u)$, concave function $\psi(u)$, then $(\varphi(f_n))_{n\geq 0}$ is a submartingale, and $(\psi(f_n))_{n\geq 0}$ is a supermartingale by making use of Jensen's inequality.

3. The so-called dyadic martingale is a typical and important example of martingales. Let $([0,1), \mathcal{B}, dx)$ be Lebesgue's probability space, with the family $\{\mathcal{F}_n\}_{n\geq 0}$ generated as follows

$$\mathcal{F}_n = \sigma\text{-field generated by atoms } F_j^{(n)} = \left[\frac{j}{2^n}, \frac{j+1}{2^n}\right), \quad j = 0, \cdots, 2^n - 1.$$

Then all martingales with respect to such $(\Omega, \mathcal{F}, \mu, \{\mathcal{F}_n\}_{n\geq})$ are called dyadic martingales. They are nothing but the 2^n-partial sum sequences of Walsh expansions. In fact, let $f \in L^1(0,1)$, its 2^n-partial sum of Walsh expansion is

$$S_{2^n}(f,x) = \int_0^1 f(x\dot{+}t)D_{2^n}(t)dt,$$

where the dyadic expressions $x\dot{+}t = (y_1, y_2, \cdots)$, $x = (x_1, x_2, \cdots)$, $t = (t_1, t_2, \cdots)$ satisfy $y_j = x_j \dot{+} t_j$ for all j, with $\dot{+}$ addition module 2, and

$$D_n(t) = 2^n \chi_{[0,2^{-n})}(t), \quad n \geq 0.$$

Thus, we have

$$\begin{aligned} S_{2^n}(f,x)\chi_{F_j^{(n)}} &= 2^n \int_0^{2^{-n}} f(x\dot{+}t)dt\chi_{F_j^{(n)}} \\ &= \frac{1}{|F_j^{(n)}|}\int_{F_j^{(n)}} f(t)dt\chi_{F_j^{(n)}} = E(f|\mathcal{F}_n)\chi_{F_j^{(n)}}. \end{aligned}$$

We have connected the dyadic martingales with Walsh expansions. This connection remains to hold in more general case. Let X be a locally compact Abelian group, $(X_j)_{-\infty}^{\infty}$ be an increasing sequence of compact and open subgroups satisfying

$$\text{index of } X_j \text{ in } X_{j+1} \text{ is finite}, \ \bigcup X_j = X, \ \bigcap X_j = \{0\}.$$

Let G_j be the annihinator of X_j in G, that is $G_j = \{t \in G : (x,t) = 1, \ \forall x \in X_j\}$. Then $(G)_{-\infty}^{\infty}$ is a decreasing sequence of compact and open subgroups satisfying

$$\text{index of } G_{j+1} \text{ in } G_j \text{ is finite}, \ \bigcup G_j = G, \ \bigcap G_j = \{0\}.$$

Let $\Omega = G$, $\mathcal{F}$ be the Borel family of G, μ be the Haar measure on G, and $\mathcal{F}_j$ be generated by all cosets of G_j. Then $\{\mathcal{F}_j\}_{j=-\infty}^{\infty}$ is an increasing family of sub-σ-fields of $\mathcal{F}$, such that $\bigcap_j \mathcal{F}_j$ is trivial and $\bigvee_j \mathcal{F}_j = \mathcal{F}$. Notice that each $(G, \mathcal{F}_j, \mu)$ is atomic, and each atom, i.e. each coset $t + G_j$ makes $0 < |t + G_j| < \infty$ since G_j is compact and open. For all $f \in L^1(G)$, $E(f|\mathcal{F}_j)$ is constant on each $t + G_j$, i.e.

$$\begin{aligned} E(f|\mathcal{F}_j)(t) &= \frac{1}{|t+G_j|}\int_{t+G_j} f(y)d\mu(y) \\ &= \frac{1}{|G_j|}\int_{G_j} f(t-y)d\mu(y) \\ &= \frac{1}{|G_j|}\chi_{G_j} * f(t), \quad \forall t \in t + G_j, \end{aligned}$$

here $*$ denotes the convolution operator. Now consider the Fourier transform. Since G_j is compact and open, we have

$$\left(\frac{1}{|G_j|}\chi_{G_j}\right)^{\wedge}(x) = \chi_{X_j}(x).$$

And hence

$$(E(f|\mathcal{F}_j))^{\wedge}(x) = \chi_{X_j}(x)\hat{f}(x) = (S_{X_j}f)^{\wedge}(x).$$

This means that the conditional expectation $E_{\mathcal{F}_j}$ is nothing but the Fourier partial sum S_{X_j} over the subgroup X_j. In particular, when $G = D_2$ is the dyadic group, and

$$G_j = \{x = (x_i)_1^{\infty} : x_i = 0, \ i = 1, 2, \cdots, j\}, \quad j = 1, 2, \cdots,$$

$\mathcal{F}_j$ is generated by all cosets of G_j, then we reduces it to the Walsh expansion case.

First, we define the simplest classes of martingales, or super-, or submartingales.

Definition 1.3.3 *Let $1 \le p \le \infty$. For any martingale, or super-, or submartingale $f = (f_n)_{n\ge 0}$, denote $\|f\|_p = \sup_n \|f_n\|_p$. When $\|f\|_p < \infty$, we say that f is a L^p-martingale, L^p-supermartingale or L^p-submartingale (we also say that f is a L^p-bounded martingale, or super-(sub-)martingale), in symbols $f \in L^p$. When $f_n = E(f|\mathcal{F}_n)$, for all n, for some $f \in L^p$, we say that f is a L^p_u-martingale.*

Remark There is a slight confusion for the symbol L^p which denotes the spaces of martingales (or super-, or sub-martingales), as well as the usual Lebesgue spaces. The confusion is not essential in the case $1 < p < \infty$. We will see in 1.3.2 that when $1 < p < \infty$, $L^p = L^p_u$, and $\|f\|_p = \|f_\infty\|_p$, but $L^1_u \subsetneqq L^1$ in general. For all $f \in L^1$, we have that f_∞ exists pointwise, and $f_\infty \in L^1$, and $\|f_\infty\|_1 \le \|f\|_1$. Furthermore, the equality holds if and only if $f \in L^1_u$.

Sometimes, when a martingale, or super-, or submartingale has a limit f_∞, we want to know if $(f_n)_{0\le n\le\infty}$ is still a martingale , or super-(sub-) martingale in following sense.

Definition 1.3.4 *$f = (f_n)_{0\le n\le\infty}$ is called a martingale, or super-, or submartingale, if $(f_n)_{n\ge 0}$ is so, and*

$$f_n = (or\ \ge, or\ \le)\ E(f_\infty|\mathcal{F}_n),\ \forall n \ge 0.$$

Remark In the martingale case, $f_n = E(f_\infty|\mathcal{F}_n)$ implies $E(f_m|\mathcal{F}_n) = f_n$, for all $m \ge n$, in fact $E(f_\infty|\mathcal{F}_n) = E(E(f_\infty|\mathcal{F}_m)|\mathcal{F}_n) = E(f_m|\mathcal{F}_n)$, for all $m \ge n$.

1.3.1 Decompositions of martingales, super-(or sub-)martingales

First, consider Doob's decomposition of super-(or sub-) martingales.

Theorem 1.3.1.1 *Each supermartingale $f = (f_n)_{n\ge 0}$ can be expressed as a difference of a martingale $M = (M_n)_{n\ge 0}$, and an increasing, nonnegative and predictable process $A = (A_n)_{n\ge 0}$. When $A_0 = 0$, the decomposition is unique. Furthermore when $f \in L^1$, then $A_\infty \in L^1$ and $M \in L^1$. Same assertions hold for submartingales with "difference" replaced by "sum".*

Proof. Let $Q = (Q_n)_{n\ge 0}$ be a supermartingale, that is $(Q_n)_{n\ge 0}$ is adapted and $Q_n - E(Q_{n+1}|\mathcal{F}_n) \ge 0$, a.e., $n = 0, 1, \cdots$. We have

$$\begin{aligned} Q_n &= Q_0 + \sum_1^n \{Q_k - E(Q_k|\mathcal{F}_{k-1})\} - \sum_{k=0}^{n-1} E(Q_k - Q_{k+1}|\mathcal{F}_k) \\ &= M_n - A_n, \end{aligned} \tag{1.3.1.1}$$

$$M_n = Q_0 + \sum_{k=1}^{n}\{Q_k - E(Q_k|\mathcal{F}_{k-1})\}, \quad n \geq 1,\ M_0 = Q_0,$$

$$A_n = \sum_{k=0}^{n-1} E(Q_k - Q_{k+1}|\mathcal{F}_k),\ A_0 = 0.$$

Obviously, $M = (M_n)_{n\geq 0}$ is a martingale, $A = (A_n)_{n\geq 0}$ is an increasing nonnegative and predictable process. We have got the desired decomposition of Q. The same definition of M and A gives the decomposition of Q, $Q = M + (-A)$, for the submartingale $Q = (Q_n)_{n\geq 0}$.

We want to prove the uniqueness of the decomposition. Assume Q has two decompositions

$$Q_n = M_n - A_n = N_n - B_n, \quad n = 0, 1, \cdots,\ A_0 = B_0 = 0.$$

Then $M_n - N_n = A_n - B_n$. Thus $M - N = (M_n - N_n)_{n\geq 0}$ is a martingale such that $M_n - N_n$ is $\mathcal{F}_{n-1}$-measurable. Such martingale must make $M_n - N_n = M_0 - N_0 = 0$. And hence $A_n = B_n$. The uniqueness has been proved.

Now assume $Q = (Q_n)_{n\geq 0} \in L^1$. We have

$$\begin{aligned} E(A_n) &= \sum_{k=0}^{n-1} E(Q_k - E(Q_{k+1}|\mathcal{F}_k)) \\ &= E(Q_0) - E(Q_n) \leq 2\sup_n \|Q_n\|_1 \leq \infty. \end{aligned}$$

Since A is nonnegative and increasing, $A_\infty \in L^1$. At last $E(|M_n|) \leq E(A_n) + E(|Q_n|) \leq c\sup_n \|Q_n\|_1 < \infty$. The proof is finished. □

Now for a special kind of supermartingale, called potential, we give an interesting expression by means of Doob's decomposition.

Definition 1.3.1.2 *A potential is an nonnegative supermartingale $Q = (Q_n)_{n\geq 0}$, which is uniformly integrable (see §1.3.2), and has the pointwise limit 0.*

Theorem 1.3.1.3 *For each potential $Q = (Q_n)_{n\geq 0}$, there is a unique associated $A = (A_n)_{n\geq 0}$, $A_0 = 0$, which is nonnegative, increasing and predictable, such that*

$$Q_n = E(A_\infty - A_n|\mathcal{F}_n), \quad \forall n \geq 0. \tag{1.3.1.2}$$

Proof. For $Q = (Q_n)_{n\geq 0}$, make Doob's decomposition $Q = M - A$, $M = Q + A$ is a nonnegative martingale, and is uniformly integrable, and hence from following Theorem 1.3.2.9 we know that $M_n = E(M_\infty|\mathcal{F}_n)$. We have (noticing $Q_\infty = M_\infty - A_\infty$)

$$\begin{aligned} Q_n &= M_n - A_n \\ &= E(Q_\infty|\mathcal{F}_n) + M_n - E(M_\infty|\mathcal{F}_n) + E(A_\infty - A_n|\mathcal{F}_n) \\ &= E(A_\infty - A_n|\mathcal{F}_n). \end{aligned}$$

This proves (1.3.1.2). Now we prove the uniqueness of A. We have

$$Q_{n+1} = E(A_\infty|\mathcal{F}_{n+1}) - A_{n+1},$$

$$E(Q_{n+1}|\mathcal{F}_n) = E(A_\infty|\mathcal{F}_n) - A_{n+1},$$

$$Q_n - E(Q_{n+1}|\mathcal{F}_n) = A_{n+1} - A_n.$$

Thus, $A_n = \sum_{k=0}^{n-1}\{Q_k - E(Q_{k+1}|\mathcal{F}_k)\}$ is determined uniquely. The proof is finished. □

Remark For a potential $Q = (Q_n)_{n\geq 0}$, A in (1.3.1.2) is called the canonical increasing process associated with it.

Now consider Krickeberg's decomposition for all L^1-martingales.

Theorem 1.3.1.4 *Let $f = (f_n)_{n\geq 0}$ be a real martingale, then f can be decomposed as $f = f^{(1)} - f^{(2)}$, $f^{(i)} = (f_n^{(i)})_{n\geq 0}$, $i = 1, 2$, being nonnegative martingales, if and only if $f \in L^1$. And in this case $f^{(i)}$ could be chosen such that*

$$\|f\|_1 = \sup_n \|f_n\|_1 = \|f^{(1)}\|_1 + \|f^{(2)}\|_1. \tag{1.3.1.3}$$

Furthermore, the decomposition satisfying (1.3.1.3) *is unique.*

Proof. Assume $f = f^{(1)} - f^{(2)}$, $f^{(i)}$ being nonnegative martingales, then obviously $f \in L^1$. On the contrary, assume $f = (f_n)_{n\geq 0} \in L^1$. Notice that $(f_n^+)_{n\geq 0}$, $(f_n^-)_{n\geq 0}$ are both nonnegative submartingale (since $\varphi(u) = u^+$ is convex and $(f_n^-)_{n\geq 0} = ((-f_n)^+)_{n\geq 0}$), and $\sup_n \|f_n^\pm\|_1 \leq \sup_n \|f_n\|_1 < \infty$. For fixed n, consider

$$f_{n,m} = E(f_{n+m}^+|\mathcal{F}_n), \quad m \geq 0,$$

$$g_{n,m} = E(f_{n+m}^-|\mathcal{F}_n), \quad m \geq 0.$$

Obviously, $(f_{n,m})_{m\geq 0}$, $(g_{n,m})_{m\geq 0}$ are increasing, since

$$\begin{aligned} f_{n,m+1} &= E(f_{n+m+1}^+|\mathcal{F}_n) = E(E(f_{n+m+1}^+|\mathcal{F}_{n+m})|\mathcal{F}_n) \\ &\geq E(f_{n+m}^+|\mathcal{F}_n) = f_{n,m}, \end{aligned}$$

analogously replacing (f_n) by $(-f_n)$, $g_{n,m+1} \geq g_{n,m}$. So we have

$$f_{n,m} \to f_n^{(1)}, \quad g_{n,m} \to f_n^{(2)}, \quad \text{a.e., increasingly.}$$

Obviously, $f_n^{(i)}$ is measurable with respect to $\mathcal{F}_n$, and $f^{(i)} \in L^1$, since

$$\begin{aligned} \|f_n^{(1)}\|_1 + \|f_n^{(2)}\|_1 &= \lim_{m\to\infty} E(f_{n,m} + g_{n,m}) = \lim_{m\to\infty} E(E(|f_{n+m}|\,|\mathcal{F}_n)) \\ &\leq \lim_{m\to\infty} \|f_{n+m}\|_1 \leq \sup_n \|f_n\|_1. \end{aligned} \tag{1.3.1.4}$$

In addition, we have

$$f_{n,m} - g_{n,m} = E((f_{n+m}^+ - f_{n+m}^-)|\mathcal{F}_n) = E(f_{n+m}|\mathcal{F}_n) = f_n,$$

and hence $f_n = f_n^{(1)} - f_n^{(2)}$. It remains to prove that $(f_n^{(i)})_{n\geq 0}$ are martingales. We have

$$f_{n,m+1} = E((f_{n+m+1}^+|\mathcal{F}_n) = E(E(f_{n+m+1}^+|\mathcal{F}_{n+1})|\mathcal{F}_n) = E(f_{n+1,m}|\mathcal{F}_n),$$

letting $m \to \infty$, we get $f_n^{(1)} = E(f_{n+1}^{(1)}|\mathcal{F}_n)$. Analogously, we have also $f_n^{(2)} = E(f_{n+1}^{(2)}|\mathcal{F}_n)$.

Thus we have got f's decomposition as a difference of two nonnegative martingales. In addition, (1.3.1.4) together with the obvious inequality $\sup_n \|f_n\|_1 \leq \sup_n \|f_n^{(1)}\|_1 + \sup_n \|f_n^{(2)}\|_1$ gives (1.3.1.3). It remains the proof of the uniqueness. Suppose that besides the preceding decomposition $f = f^{(1)} - f^{(2)}$, there were another decomposition $f = g^{(1)} - g^{(2)}$ such that $\|f\|_1 = \|g^{(1)}\|_1 + \|g^{(2)}\|_1$. Then we would have $f_n^+ \leq g_n^{(1)}$, and hence

$$f_n^{(1)} = \lim_{m\to\infty} E(f_{n+m}^+|\mathcal{F}_n) \leq \lim_{m\to\infty} E(g_{n+m}^{(1)}|\mathcal{F}_n) = g_n^{(1)}.$$

Analogously, replacing $(f_n)_{n\geq 0}$ by $(-f_n)_{n\geq 0}$, we get $f_n^{(2)} \leq g_n^{(2)}$. But we have

$$\begin{aligned} E(f_n^{(1)} + f_n^{(2)}) &= \|f^{(1)}\|_1 + \|f^{(2)}\|_1 = \|f\|_1 \\ &= \|g^{(1)}\|_1 + \|g^{(2)}\|_1 = E(g_n^{(1)} + g_n^{(2)}). \end{aligned}$$

Finally we get $f_n^{(1)} = g_n^{(1)}$, $i = 1, 2$. The proof is finished. □

Corollary 1.3.1.5 *Let $f = (f_n)_{n\geq 0}$ be a L^1-submartingale. Then there exists a nonnegative martingale $g = (g_n)_{n\geq 0}$ such that*

$$f_n^+ \leq g_n, \quad \textit{and} \ \sup_n E(f_n^+) = \sup_n E(g_n).$$

Proof. In the proof of Theorem 1.3.1.4, we have only used the fact that $(f_n^+)_{n\geq 0}$ is a L^1-nonnegative submartingale, which holds for L^1-submartingales $f = (f_n)_{n\geq 0}$, too. Defining $g_n = \lim_{m\to\infty} g_{n,m}$, with $g_{n,m} = E(f_{n+m}^+|\mathcal{F}_n)$, we have

$$\|g_n\|_1 = \lim_{m\to\infty} \|g_{n,m}\|_1 = \lim_{m\to\infty} \|f_{n+m}^+\|_1 = \sup_n \|f_n^+\|_1.$$

In addition, we also have

$$f_n^+ \leq E(f_{n+m}^+|\mathcal{F}_n) = g_{n,m}, \quad \forall m, \ f_n^+ \leq g_n.$$

The proof is finished. □

For L^1-martingales, for a given level $\lambda > 0$, there is a decomposition of Calderón-Zygmund's type, which is called Gundy's decomposition. We introduce it here. We mentioned in the Preface of the book, that for $f = (f_n)_{n\geq 0} \in L^1$ and τ a stopping time like $\tau = \inf\{n : |f_n| > \lambda\}$, $f = f^{(\tau)} + (f - f^{(\tau)})$ is a decomposition of Calderón-Zygmund's type provided $\|f^{(\tau)}\|_\infty \leq C\lambda$. But in general, $\|f^{(\tau)}\|_\infty \leq C\lambda$ is not true. Gundy's decomposition is a decomposition of Calderón-Zygmund's type in general case.

We need following two elementary facts.

Lemma 1.3.1.6 *Let $f = (f_n)_{n\geq 0}$ be a L^1-martingale, $\lambda > 0$, and $\tau = \inf\{n : |f_n| > \lambda\}$. Then*

$$|\{\tau < \infty\}| \leq \frac{1}{\lambda}\|f\|_1. \tag{1.3.1.5}$$

Proof. First, we consider the finite martingale $f^{(n)} = (f_{n\wedge m})_{m\geq 0}$ (called the f's stopped martingale at n), and

$$\tau_n = \inf\{m \leq n, |f_m| > \lambda\} = \begin{cases} \tau, & \tau \leq n; \\ \infty, & \tau > n. \end{cases}$$

Then $|f_{\tau_n}| > \lambda$ on $\{\tau_n < \infty\}$. And hence

$$\begin{aligned} |\{\tau_n < \infty\}| &\leq \frac{1}{\lambda}\int_{\{\tau_n<\infty\}} |f_{\tau_n}|d\mu = \frac{1}{\lambda}\sum_{m=0}^{n}\int_{\{\tau_n=m\}} |f_m|d\mu \\ &\leq \frac{1}{\lambda}\int_{\{\tau_n<\infty\}} |f_n|d\mu . \end{aligned}$$

Since $|\{\tau_n < \infty\}| \to |\{\tau < \infty\}|$, we get

$$|\{\tau < \infty\}| \leq \frac{1}{\lambda}\sup_n \|f_n\|_1 = \frac{1}{\lambda}\|f\|_1.$$

The proof is finished. □

Lemma 1.3.1.7 *Let $f = (f_n)_{n\geq 0}$ be a L^1-martingale, $\tau = \inf\{n : |f_n| > \lambda\}$. Then (denoting $\Delta_n f = f_n - f_{n-1}$, $n \geq 0$, $f_{-1} = 0$)*

$$\sum_0^\infty E(|\Delta_k f|^2 \chi_{\{\tau>k\}}) \leq 2\lambda\|f\|_1. \tag{1.3.1.6}$$

Proof. First we want to show that for any finite martingale $f = (f_n)_{n\geq 0}$ and any stopping time τ, we have the identity

$$\begin{aligned} \sum_{k=0}^{\tau-1} |\Delta_k f|^2 + |f_{\tau-1}|^2 &= \sum_{k=1}^{\tau} \bar{f}_{k-1}(f_{k-1} - f_k) \\ &+ \sum_{k=1}^{\tau} f_{k-1}(\bar{f}_{k-1} - \bar{f}_k) + f_\tau \bar{f}_{\tau-1} + \bar{f}_\tau f_{\tau-1}. \end{aligned} \tag{1.3.1.7}$$

In fact, at those points where $\tau < \infty$, we have

$$\sum_{k=0}^{\tau-1} |\Delta_k f|^2 + |f_{\tau-1}|^2 = \sum_{k=0}^{\tau-1} (f_k - f_{k-1})(\bar{f}_k - \bar{f}_{k-1}) + f_{\tau-1}\bar{f}_{\tau-1}$$

$$= 2\sum_{k=1}^{\tau} f_{k-1}\bar{f}_{k-1} - f_{\tau-1}\bar{f}_{\tau-1} - \sum_{k=1}^{\tau} f_k\bar{f}_{k-1} + f_\tau\bar{f}_{\tau-1}$$

$$- \sum_{k=0}^{\tau} \bar{f}_k f_{k-1} + \bar{f}_\tau f_{\tau-1} + f_{\tau-1}\bar{f}_{\tau-1}$$

$$= \sum_{k=1}^{\tau} \bar{f}_{k-1}(f_{k-1} - f_k) + \sum_{k=1}^{\tau} f_{k-1}(\bar{f}_{k-1} - \bar{f}_k) + f_\tau\bar{f}_{\tau-1} + \bar{f}_\tau f_{\tau-1}.$$

At those points where $\tau = \infty$, $\sum_0^{\tau-1}$, $f_{\tau-1}$, f_τ and $\sum_1^\tau$ are meant as $\sum_0^n$, f_n, f_n, $\sum_1^n$ respectively, (1.3.1.7) holds obviously.

Now for the L^1-martingale $f = (f_n)_{n\geq 0}$ and $\tau = \inf\{n : |f_n| > \lambda\}$, considering $f^{(n)} = (f_{n\wedge m})_{m\geq 0}$ and $\tau_n = \tau$, $\tau \leq n$; $= \infty$, $\tau > n$, and writing them as f and τ still, we have

$$E\left(\sum_{k=0}^{\tau-1} |\Delta_k f|^2\right) + E(|f_{\tau-1})|^2 = E(f_\tau\bar{f}_{\tau-1}) + E(\bar{f}_\tau f_{\tau-1})$$

$$\leq 2\lambda E(|f_\tau|) \leq 2\lambda\|f\|_1.$$

Letting $n \to \infty$, we get (1.3.1.6). The proof is finished. □

Theorem 1.3.1.8 *Let $f = (f_n)_{n\geq 0}$ with $f_n = \sum_0^n d_k$ be a martingale in L^1, and $\lambda > 0$. There exist martingales $X = (X_n)_{n\geq 0}$, $Y = (Y_n)_{n\geq 0}$, $Z = (Z_n)_{n\geq 0}$, with $X_n = \sum_0^n x_k$, $Y_n = \sum_0^n y_k$, $Z_n = \sum_0^n z_k$ such that $f = X + Y + Z$ and*

$$\|X\|_2^2 \leq 2\lambda\|f\|_1, \tag{1.3.1.8}$$

$$\left\|\sum_0^n |y_k|\right\|_1 \leq 4\|f\|_1, \tag{1.3.1.9}$$

$$|\{\sup_n |z_n| > 0\}| \leq \frac{\|f\|_1}{\lambda}. \tag{1.3.1.10}$$

Proof. Let $\tau = \inf\{n : |f_n| > \lambda\}$. Then τ is a stopping time. Now we make the decomposition by defining

$$x_0 = d_0\chi_{\{\tau>0\}}, \quad x_k = d_k\chi_{\{\tau>k\}} - E(d_k\chi_{\{\tau>k\}}|\mathcal{F}_{k-1}), \quad k \geq 1,$$

$$y_0 = d_0\chi_{\{\tau=0\}}, \quad y_k = d_k\chi_{\{\tau=k\}} - E(d_k\chi_{\{\tau=k\}}|\mathcal{F}_{k-1}), \quad k \geq 1,$$

$$z_0 = 0, \quad z_k = d_k\chi_{\{\tau<k\}}, \quad k \geq 1.$$

Because $\chi_{\{\tau\geq k\}}$ is $\mathcal{F}_{k-1}$-measurable, we have $E(d_k\chi_{\{\tau\geq k\}}|\mathcal{F}_{k-1}) = 0$ and hence $f = X + Y + Z$. In addition,

$$E(x_k|\mathcal{F}_{k-1}) = E(y_k|\mathcal{F}_{k-1}) = E(z_k|\mathcal{F}_{k-1}) = 0, \quad \forall k \geq 1,$$

so, X, Y, Z are martingales. Now the verifications of (1.3.1.8)–(1.3.1.10) are easy. It is obvious, $\sup_n |z_n|\chi_{\{\tau=\infty\}} = 0$, which gives (1.3.1.10). We have (noticing $|f_{\tau-1}| \leq \lambda$)

$$\begin{aligned}
E\Big(\sum_0^n |y_k|\Big) &\leq 2\sum_0^n E(|d_k|\chi_{\{\tau=k\}}) \\
&\leq 2\sum_0^n E(|f_k|\chi_{\{\tau=k\}}) + 2\lambda|\{\tau<\infty\}| \\
&\leq 2\sum_0^n E(E(|f_n|\chi_{\{\tau=k\}}|\mathcal{F}_k)) + 2\lambda|\{\tau<\infty\}| \\
&\leq 2\sup_n \|f_n\|_1 + 2\|f\|_1 = 4\|f\|_1,
\end{aligned}$$

which gives (1.3.1.9). As for (1.3.1.8), we have

$$\begin{aligned}
E(|x_k|^2) &= E([d_k\chi_{\{\tau>k\}} - E(d_k\chi_{\{\tau>k\}}|\mathcal{F}_{k-1})] \\
&\qquad \times [\bar{d}_k\chi_{\{\tau>k\}} - E(\bar{d}_k\chi_{\{\tau>k\}}|\mathcal{F}_{k-1})]) \\
&= E(|d_k|^2\chi_{\{\tau>k\}}) - E(\big|E(d_k\chi_{\{\tau>k\}}|\mathcal{F}_{k-1})\big|^2) \leq E(|d_k|^2\chi_{\{\tau>k\}}),
\end{aligned}$$

$$\Big\|\sum_0^n x_k\Big\|_2^2 = \sum_0^n E(|x_k|^2) \leq \sum_0^n E(|d_k|^2\chi_{\{\tau>k\}}) \leq 2\lambda\|f\|_1.$$

The proof of the theorem is finished. □

Remark The classical C-Z decomposition says that each $f \in L^1(R^n)$ can be decomposed, according to a given level $\lambda > 0$, as $g + h$ with g being good in the sense $\|g\|_\infty \leq C\lambda$, h being bad but controllable owing to $|\mathrm{supp} h| \leq \frac{C}{\lambda}\|f\|_1$. In martingale setting, the bad part of $f \in L^1$ is Z, the good part consists of $X + Y$. Here X is corresponding to g, but Y is a new one which does not occur in the classical case.

Now consider the Riesz' decomposition for supermartingales.

Theorem 1.3.1.9 *Let $f = (f_n)_{n\geq 0}$ be a supermartingale. Then the following assertions are equivalent.*

(a) *There exists a submartingale $g = (g_n)_{n\geq 0}$ such that $f_n \geq g_n$;*

(b) *There exists a unique decomposition $f = f^{(1)} + f^{(2)}$, with $f^{(1)}$ a martingale and $f^{(2)}$ a potential.*

Proof. (b) $\Rightarrow$(a) obviously, since $f_n \geq f_n^{(1)}$, for all n. Assume (a). Define $f_{n,m} = E(f_{n+m}|\mathcal{F}_n)$. Then $\{f_{n,m}\}_{m\geq 0}$ is decreasing. In fact

$$\begin{aligned} f_{n,m+1} &= E(f_{n+m+1}|\mathcal{F}_n) = E(E(f_{n+m+1}|\mathcal{F}_{n+m})|\mathcal{F}_n) \\ &\leq E(f_{n+m}|\mathcal{F}_n) = f_{n,m}. \end{aligned}$$

Since g is submartingale, meanwhile f is supermartingale, then

$$g_n \leq E(g_{n+m}|\mathcal{F}_n) \leq E(f_{n+m}|\mathcal{F}_n) = f_{n,m} \leq f_n.$$

This implies that $f_n^{(1)} = \lim_{m\to\infty} f_{n,m}$ exists monotonically and dominatedly. Now prove that $(f_n^{(1)})_{n\geq 0}$ is a martingale. This follows from

$$f_{n,m+1} = E(f_{n+m+1}|\mathcal{F}_n) = E(E(f_{n+m+1}|\mathcal{F}_{n+1})|\mathcal{F}_n) = E(f_{n+1,m}|\mathcal{F}_n),$$

and the dominated convergence theorem. It remains to show $f^{(2)} = f - f^{(1)}$ is a potential. Since $g_n \leq f_n^{(1)} \leq f_n$, $f^{(2)}$ is nonnegative. $f^{(2)}$ is a supermartingale obviously, since f, $-f^{(1)}$ are so. It remains to show that $E(f_n^{(2)}) \to 0$ (it is enough to imply: $(f_n^{(2)})$ is uniformly integrable and $f_n^{(2)} \to 0$, a.e., since $\lim_{n\to\infty} f_n^{(2)}$ exists pointwise by Theorem 1.3.2.10.). We have

$$f_0^{(1)} = \lim_{m\to\infty} E(f_{0+m}|\mathcal{F}_0),$$

$$E(f_0^{(1)}) = \lim_{m\to\infty} E(E(f_m|\mathcal{F}_0)) = \lim_{m\to\infty} E(f_m),$$

and hence

$$E(f_n^{(2)}) = E(f_n - f_n^{(1)}) = E(f_n) - E(f_0^{(1)}) \to 0.$$

Thus we get f's desired decomposition provided (a) holds.

At last prove the uniqueness. Suppose $f = f^{(1)} + f^{(2)} = g^{(1)} + g^{(2)}$. Then

$$f_{n,m} = E(f_{n+m}|\mathcal{F}_n) = E(g_{n+m}^{(1)} + g_{n+m}^{(2)}|\mathcal{F}_n) = g_n^{(1)} + E(g_{n+m}^{(2)}|\mathcal{F}_n).$$

Letting $m \to \infty$, the left-hand side is $f_n^{(1)}$, and the right-hand side will be $g_n^{(1)}$ once $\lim_{m\to\infty} E(g_{n+m}^{(2)}|\mathcal{F}_n) = 0$, a.e., to be shown. This is a fact holding for any potential $Q = (Q_n)_{n\geq 0}$. From

$$E(Q_{n+m}|\mathcal{F}_{n+m-1}) \leq Q_{n+m-1}, \quad \forall m,$$

we see that $\{E(Q_{n+m}|\mathcal{F}_{n+m-1})\}_{m\geq 1}$ is a uniformly integrable family and $\lim_{m\to\infty} E(Q_{n+m}|\mathcal{F}_{n+m-1}) = 0$, a.e. So

$$\lim_{m\to\infty} E(Q_{n+m}|\mathcal{F}_n) = \lim_{m\to\infty} E(E(Q_{n+m}|\mathcal{F}_{n+m-1})|\mathcal{F}_n) = 0, \text{ a.e.}$$

The proof of the theorem is finished. □

Corollary 1.3.1.10 *Each nonnegative supermartingale has a unique Riesz' decomposition.*

Corollary 1.3.1.11 *Each L^1-supermartingale has a unique Riesz' decomposition.*

Proof. Let $f = (f_n)_{n\geq 0}$ be a L^1-supermartingale. Then $-f = (-f_n)_{n\geq 0}$ is a L^1-submartingale. From Corollary 1.3.1.5, there exists nonnegative martingale $g = (g_n)_{n\geq 0}$ such that $-f_n \leq g_n$, so $f_n \geq -g_n$. The Theorem 1.3.1.9 gives the assertion. □

Remark Doob's decomposition says that each supermartingale could be controlled by a martingale, but Riesz' decomposition says that each supermartingale controls a martingale.

1.3.2 Convergences of martingales, super-(or sub-)martingales

First we want to discuss the uniform integrability of a process.

Definition 1.3.2.1 *Let B be a subset of $L^1(\Omega, \mathcal{F}, \mu)$*[†]*. B is said to be uniformly integrable, if*

$$\lim_{C\to\infty} \int_{\{|f|\geq C\}} |f| d\mu = 0, \quad \forall f \in B \text{ uniformly}. \tag{1.3.2.1}$$

Lemma 1.3.2.2 *A subset B of L^1 is uniformly integrable, if and only if*

$$\sup_{f\in B} E(|f|) < \infty, \tag{1.3.2.2}$$

and for all $\varepsilon > 0$, there exists $\delta > 0$ such that

$$\sup_{f\in B} \int_F |f| d\mu \leq \varepsilon, \quad \forall F \in \mathcal{F}, \text{ provided } \ |F| \leq \delta. \tag{1.3.2.3}$$

Proof. Assume (1.3.2.1). For all $F \in B$, we have

$$E(|f|) = \int_{\{|f|\leq C\}} |f| d\mu + \int_{\{|f|>C\}} |f| d\mu \leq C + 1.$$

In addition, for all $F \in \mathcal{F}$, we have

$$\int_F |f| d\mu = \int_{F\cap\{|f|\leq C\}} |f| d\mu + \int_{F\cap\{|f|>C\}} |f| d\mu \leq C|F| + \frac{\varepsilon}{2} \leq \varepsilon,$$

provided C is large enough and $|F| \leq \delta = \frac{\varepsilon}{2C}$. So, (1.3.2.2), (1.3.2.3) hold.

† We do not want to discuss the uniform integrability on general measure spaces but on probability spaces.

mentioned in the Preface of the book, that for $f = (f_n)_{n\geq 0} \in L^1$ and τ a stopping time like $\tau = \inf\{n : |f_n| > \lambda\}$, $f = f^{(\tau)} + (f - f^{(\tau)})$ is a decomposition of Calderón-Zygmund's type provided $\|f^{(\tau)}\|_\infty \leq C\lambda$. But in general, $\|f^{(\tau)}\|_\infty \leq C\lambda$ is not true. Gundy's decomposition is a decomposition of Calderón-Zygmund's type in general case.

We need following two elementary facts.

Lemma 1.3.1.6 *Let $f = (f_n)_{n\geq 0}$ be a L^1-martingale, $\lambda > 0$, and $\tau = \inf\{n : |f_n| > \lambda\}$. Then*

$$|\{\tau < \infty\}| \leq \frac{1}{\lambda}\|f\|_1. \tag{1.3.1.5}$$

Proof. First, we consider the finite martingale $f^{(n)} = (f_{n\wedge m})_{m\geq 0}$ (called the f's stopped martingale at n), and

$$\tau_n = \inf\{m \leq n, |f_m| > \lambda\} = \begin{cases} \tau, & \tau \leq n; \\ \infty, & \tau > n. \end{cases}$$

Then $|f_{\tau_n}| > \lambda$ on $\{\tau_n < \infty\}$. And hence

$$\begin{aligned} |\{\tau_n < \infty\}| &\leq \frac{1}{\lambda}\int_{\{\tau_n<\infty\}} |f_{\tau_n}| d\mu = \frac{1}{\lambda}\sum_{m=0}^{n}\int_{\{\tau_n = m\}} |f_m| d\mu \\ &\leq \frac{1}{\lambda}\int_{\{\tau_n<\infty\}} |f_n| d\mu . \end{aligned}$$

Since $|\{\tau_n < \infty\}| \to |\{\tau < \infty\}|$, we get

$$|\{\tau < \infty\}| \leq \frac{1}{\lambda}\sup_n \|f_n\|_1 = \frac{1}{\lambda}\|f\|_1.$$

The proof is finished. □

Lemma 1.3.1.7 *Let $f = (f_n)_{n\geq 0}$ be a L^1-martingale, $\tau = \inf\{n : |f_n| > \lambda\}$. Then (denoting $\Delta_n f = f_n - f_{n-1}$, $n \geq 0$, $f_{-1} = 0$)*

$$\sum_0^\infty E(|\Delta_k f|^2 \chi_{\{\tau > k\}}) \leq 2\lambda\|f\|_1. \tag{1.3.1.6}$$

Proof. First we want to show that for any finite martingale $f = (f_n)_{n\geq 0}$ and any stopping time τ, we have the identity

$$\begin{aligned} \sum_{k=0}^{\tau-1} |\Delta_k f|^2 + |f_{\tau-1}|^2 &= \sum_{k=1}^{\tau} \bar{f}_{k-1}(f_{k-1} - f_k) \\ &+ \sum_{k=1}^{\tau} f_{k-1}(\bar{f}_{k-1} - \bar{f}_k) + f_\tau \bar{f}_{\tau-1} + \bar{f}_\tau f_{\tau-1}. \end{aligned} \tag{1.3.1.7}$$

$$= \sum_{k=1}^{\infty}\sum_{n=k}^{\infty} E(|f|\chi_{\{|f|\in A_n\}}) = E(|f|) + \sum_{k=2}^{\infty} \int_{\{|f|\geq C_{k-1}\}} |f| d\mu$$
$$\leq E(|f|) + \sum_{k=2}^{\infty} \frac{1}{2^{k-1}} = E(|f|) + 1.$$

The proof is finished. □

Lemma 1.3.2.4 *Let $\{f^{(n)}\} \subset L^1$, $f^{(n)} \to f$, a.e. Then $f^{(n)} \to f$ in L^1, if and only if $\{f^{(n)}\}$ is uniformly integrable.*

Proof. Assume $f^{(n)} \to f$ in L^1. Then $f \in L^1$. And for all $F \in \mathcal{F}$,

$$\int_F |f^{(n)}| d\mu \leq \int_F |f| d\mu + \|f^{(n)} - f\|_1.$$

This implies (1.3.2.2) and (1.3.2.3). Conversely, assuming the uniform integrability of $\{f^{(n)}\}$, we have, for any given $\varepsilon \geq 0$,

$$E(|f^{(n)} - f|) \leq \int_\Omega |f^{(n)}\chi_{\{|f^{(n)}\leq C\}} - f\chi_{\{|f|\leq C\}}| d\mu$$
$$+ \int_{\{|f^{(n)}\geq C\}} |f^{(n)}| d\mu + \int_{\{|f|\geq C\}} |f| d\mu \leq \varepsilon,$$

provied n, C are large enough, since $\lim_{n\to\infty} (f^{(n)}\chi_{\{|f^{(n)}|\leq C\}} - f\chi_{\{|f|\leq C\}}) = 0$, a.e. boundedly. The proof is finished. □

Lemma 1.3.2.5 *Let $\{f^{(n)}\} \subset L^1$, $f^{(n)} \to f$, a.e. Then $f^{(n)} \to f$ in L^1, if and only if $\lim_n \|f^{(n)}\|_1 = \|f\|_1$.*

Proof. One half of the assertion is obvious. Now assume $f^{(n)} \to f$ a.e., and $\lim_{n\to\infty} \|f^{(n)}\|_1 = \|f\|_1$. First we want to prove $|f^{(n)}| \to |f|$ in L^1. We have

$$|f^{(n)}| + |f| = \min(|f^{(n)}|, |f|) + \max(|f^{(n)}|, |f|),$$
$$\min(|f^{(n)}|, |f|) \to |f|, \text{ a.e, and dominatedly by } |f|,$$

and hence,

$$E(\max(|f^{(n)}|, |f|) \to \|f\|_1,$$
$$E(\big||f^{(n)}| - |f|\big|) = E(\max(|f^{(n)}|,|f|) - \min(|f^{(n)}|,|f|)) \to 0.$$

This proves $|f^{(n)}| \to |f|$ in L^1. Now the uniform integrability of $\{f^{(n)}\}$ follows immediately. In fact, for all $F \in \mathcal{F}$, we have

$$\int_F |f^{(n)}| d\mu \leq \int_F |f| d\mu + \|\, |f^{(n)}| - |f|\, \|_1,$$

which implies the uniform integrability. The proof of the lemma is thus finished. □

Lemma 1.3.2.6 *Let $B \subset L^1$. Then the following assertions are equivalent.*
(a) *B is uniformly integrable;*
(b) *For any set sequence $\{F_k\}_k$, such that $F_k \downarrow \emptyset$, $\lim_{k\to\infty} \sup_{f\in B} \int_{F_k} |f| d\mu = 0$;*
(c) *For any disjoint set sequence $\{E_k\}_k$ satisfying $|E_k| \to 0$,*

$$\lim_{k\to\infty} \sup_{f\in B} \int_{E_k} |f| d\mu = 0.$$

Proof. Obviously, (a) implies (b), (b) implies (c). We want to show that "not (a)" implies "not (b)" and "not (b)" implies "not (c)". Assume B is not uniformly integrable. Then there exist $\varepsilon > 0$ and a set sequence $\{A_n\}$ such that $|A_n| \le 2^{-n}$ but $\sup_{f\in B} \int_{A_n} |f| d\mu \ge \varepsilon$. Set $F_n = \bigcup_{j\ge n} A_j$. Then $\sup_{f\in B} \int_{F_n} |f| d\mu \ge \varepsilon$, but $|F_n| \le 2^{-n+1}$, it is just "not (b)". Now assume that (b) is not true. Let $\{F_n\}$ be a set sequence such that $F_n \downarrow \emptyset$ but $\sup_{f\in B} \int_{F_n} |f| d\mu > 2\varepsilon$, for all n. Choose $f_0 \in B$ such that $\int_{F_0} |f_0| d\mu > 2\varepsilon$ and m_0 such that $\int_{F_0 - F_{m_0}} |f_0| d\mu > \varepsilon$. Then choose $f_1 \in B$ such that $\int_{F_{m_0+1}} |f_1| d\mu > 2\varepsilon$ and $m_1 > m_0 + 1$ such that $\int_{F_{m_0+1} - F_{m_1}} |f_1| d\mu > \varepsilon$, and so on. Set $E_0 = F_0 - F_{m_0}$, $E_1 = F_{m_0+1} - F_{m_1}, \cdots$. Thus we get a disjoint set sequence $\{E_k\}_k$ satisfying $|E_k| \to 0$ and $\sup_{f\in B} \int_{E_k} |f| d\mu > \varepsilon$, for all k. This completes the proof of the lemma. □

Before discussing the convergence of martingales and super- (or sub-) martingales, we need some weak type inequality for them.

Lemma 1.3.2.7 *Let $f = (f_n)_{0\le n\le\infty}$ be a submartingale with respect to $\{\mathcal{F}_n\}_{0\le n\le\infty}$, $\mathcal{F}_\infty = \mathcal{F} = \bigvee_n \mathcal{F}_n$. Then for all $\lambda \in \mathbf{R}$, we have*

$$\lambda \left| \left\{ \sup_n f_n > \lambda \right\} \right| \le \int_{\{\sup_n f_n > \lambda\}} f_\infty d\mu, \tag{1.3.2.6}$$

$$\lambda \left| \left\{ \inf_n f_n < \lambda \right\} \right| \ge \int_\Omega f_0 d\mu - \int_\Omega |f_\infty| d\mu. \tag{1.3.2.7}$$

Proof. For $\lambda \in \mathbf{R}$ given, define stopping times

$$\tau_1 = \inf\{n : f_n > \lambda\}, \quad \tau_2 = \inf\{n : f_n < \lambda\}.$$

Then

$$\{\tau_1 < \infty\} = \left\{ \sup_n f_n > \lambda \right\}, \quad \{\tau_2 < \infty\} = \left\{ \inf_n f_n < \lambda \right\},$$

$$f_{\tau_1} > \lambda, \text{ on } \{\tau_1 < \infty\}, \quad f_{\tau_2} < \lambda, \text{ on } \{\tau_2 < \infty\}.$$

We have

$$\lambda \left| \left\{ \sup_n f_n > \lambda \right\} \right| \le \int_{\{\sup_n f_n > \lambda\}} f_{\tau_1} d\mu = \sum_{n=0}^\infty \int_{\{\tau_1 = n\}} f_n d\mu$$

$$\leq \sum_0^\infty \int_{\{\tau_1=n\}} f_\infty d\mu = \int_{\{\sup_n f_n>\lambda\}} f_\infty d\mu.$$

And for (1.3.2.7), we have (denoting τ_2 by τ)

$$\begin{aligned} E(f_0) &= E(f_0\chi_{\{\tau=0\}}) + E(f_0\chi_{\{\tau>0\}}) \leq \lambda|\{\tau=0\}| + E(E(f_1|\mathcal{F}_0)\chi_{\{\tau>0\}}) \\ &= \lambda|\{\tau=0\}| + E(f_1\chi_{\{\tau=1\}}) + E(f_1\chi_{\{\tau>1\}}) \\ &\leq \lambda|\{\tau\leq 1\}| + E(f_1\chi_{\{\tau>1\}}) \leq \lambda|\{\tau\leq n\}| + E(f_n\chi_{\{\tau>n\}}), \\ E(f_0) &\leq \lambda|\{\tau<\infty\}| + E(f_\infty\chi_{\{\tau=\infty\}}). \end{aligned}$$

This proves (1.3.2.7). The proof is finished. □

Here is one of the main results about convergences.

Theorem 1.3.2.8 *Let $f=(f_n)_{n\geq 0}$ be a L^1-martingale. Then there exists f_∞ such that $\lim_{n\to\infty} f_n = f_\infty$, a.e., and $E(|f_\infty|) \leq \|f\|_1$.*

Proof. Without loss of generality, f can be assumed to be positive. Thus $g = (e^{-f_n})_{n\geq 0}$ is a nonnegative bounded submartingale. Notice

$$e^{-f_n} \text{ converges , a.e.} \Longleftrightarrow f_n \text{ converges , a.e.,}$$

which can be seen as follows. The part $\Leftarrow$ is obvious. Assume $e^{-f_0} \to g_\infty$, a.e. We claim that g_∞ could not be vanishing on a set of positive measure. Otherwise, we would have $f_n \to \infty$ a.e., on a set of positive measure, this would contracict to the fact $f=(f_n)_{n\geq 0}$ being a nonnegative martingale for which we should have

$$E(\underline{\lim}\, f_n) \leq \underline{\lim}\, E(f_n) = E(f_0) < \infty.$$

So $g_\infty > 0$, a.e., and $f_n \to \log g_\infty$, a.e. So our problem is reduced to the nonnegative bounded submartingale case.

We consider the case a little more general, i.e., $f=(f_n)_{n\geq 0}$ is a nonnegative L^2-submartingale. We want to show that f has its limit f_∞ both in L^2 and pointwise. Notice that $(f_n^2)_{n\geq 0}$ is also a submartingale, so $E(f_n^2) \uparrow a < \infty$. But for $n \geq m$, we have

$$0 \leq E(f_n^2 - f_m^2) = E((f_n-f_m)^2) + E(2f_m(f_n-f_m)),$$

$$E(f_m(f_n-f_m)) = E(f_m E(f_n - f_m|\mathcal{F}_m)) \geq 0,$$

so $E((f_n-f_m)^2) \to 0$ follows from $E(f_n^2 - f_m^2) \to 0$. This proves the L^2-convergence for a nonnegative L^2-submartingale.

Now consider the submartingale $(f_n - f_k)_{k\leq n\leq m}$ with respect to $\{\mathcal{F}_n\}_{k\leq n\leq m}$. For any $\varepsilon > 0$, applying Lemma 1.3.2.7, we get

$$\begin{aligned} &\left|\left\{\max_{k\leq n\leq m} |f_n - f_k| > \varepsilon\right\}\right| \\ &\leq \left|\left\{\max_{k\leq n\leq m} (f_n - f_k) > \varepsilon\right\}\right| + \left|\left\{\min_{k\leq n\leq m} (f_n - f_k) < -\varepsilon\right\}\right| \\ &\leq \frac{2}{\varepsilon} E(|f_m - f_k|) \leq \frac{2}{\varepsilon} E(|f_m - f_k|^2)^{\frac{1}{2}}. \end{aligned}$$

Thus, we have

$$|\{\overline{\lim} f_n - \underline{\lim} f_n > 2\varepsilon\}| \leq \left|\bigcap_{k=1}^{\infty}\left\{\sup_{n\geq k}|f_n - f_k| \geq \varepsilon\right\}\right|$$
$$\leq \lim_{k\to\infty}\left|\left\{\sup_{n\geq k}|f_n - f_k| \geq \varepsilon\right\}\right|$$
$$\leq \frac{2}{\varepsilon}\lim_{m\geq k\to\infty} E(|f_m - f_k|^2)^{\frac{1}{2}} = 0.$$

Since ε is arbitrary, the proof of the pointwise convergence of f is complete. Now Fatou's lemma gives $E(|f_\infty|) \leq \|f\|_1$. The proof of the theorem is finished. □

Theorem 1.3.2.9 *Let $f = (f_n)_{n\geq 0}$ be a martingale. Then the following assertions are equivalent:*

(a) *$(f_n)_{n\geq 0} \in L^1_u$, that is to say $(f_n)_{0\leq n\leq\infty}$ is a martingale;*

(b) *There exists $\varphi(u)$ like in Lemma 1.3.2.3, such that* $\sup_n$, $E(\varphi(|f_n|)) < \infty$;

(c) *$(f_n)_{n\geq 0}$ is uniformly integrable;*

(d) *$(f_n)_{n\geq 0}$ is convergent in L^1;*

(e) *$(f_n)_{n\geq 0}$ is a Cauchy sequence in L^1;*

(f) *$\lim_{n\to\infty} f_n = f_\infty$, a.e., and $\lim_{n\to\infty} E(|f_n|) = E(|f_\infty|) < \infty$.*

Proof. (a) $\Longrightarrow$ (b). $f_\infty \in L^1$, and hence there exists $\varphi(u)$ like as in Lemma 1.3.2.3 such that $E(\varphi(|f_\infty|)) < \infty$. Since $(|f_n|)_{0\leq n\leq\infty}$ is a submartingale, so $(\varphi(|f_n|))_{0\leq n\leq\infty}$ is. Thus, we have

$$\varphi(|f_n|) \leq E(\varphi(|f_\infty|)\,|\,\mathcal{F}_n), \quad E(\varphi(|f_n|) \leq E(\varphi(|f_\infty|)).$$

(b) $\Leftrightarrow$ (c). See Lemma 1.3.2.3.

(c) $\Leftrightarrow$ (d). See Lemma 1.3.2.4, since both of (c), and (d) implies $\lim_{n\to\infty} f_n = f_\infty$, a.e.

(d) $\Leftrightarrow$ (e), and (d) $\Leftrightarrow$ (a). Obvious.

(f) $\Leftrightarrow$ (d). See Lemma 1.3.2.5, since (d) implies $\lim_{n\to\infty} f_n = f_\infty$, a.e.

The proof is thus complete. □

Theorem 1.3.2.10 *Let $f = (f_n)_{n\geq 0}$ be a L^1-super-(or sub-)martingale. Then $\lim_{n\to\infty} f_n = f_\infty$, a.e., and $E(|f_\infty|) \leq \lim_{n\to\infty}\|f_n\|_1$.*

Proof. Doob's decomposition reduces the proof to the L^1-martingale case. □

Theorem 1.3.2.11 *Let $f = (f_n)_{n\geq 0}$ be a L^1-super-(or sub-)martingale. Then the asertions* (b)–(f) *in Theorem* 1.3.2.9, *are equivalent.*

Proof. All the assertions (b)–(f) imply $(f_n)_{n\geq 0} \in L^1$, and hence $\lim_{n\to\infty} f_n = f_\infty$, a.e. Then the Lemmas 1.3.2.3–1.3.2.5 give their equivalence. □

Remark For a super-(or sub-) martingale $f = (f_n)_{n\geq 0}$, the condition $(f_n)_{0\leq n\leq\infty}$ is a super-(or sub-)martingale can not imply the others in Theorem 1.3.2.9, even can not imply $\lim_{n\to\infty} f_n = f_\infty$ a.e. For example, let $f = (f_n)_{n\geq 0}$ be a nonpositive submartingale, and $f_\infty = 1$, then $(f_n)_{0\leq n\leq\infty}$ is still a submartingale but with $\lim_{n\to\infty} f_n \leq 0 \neq 1$. We will give some examples to show that "$(f_n)_{0\leq n\leq\infty}$ is a submartingale" does not imply the uniform integrability of $\{f_n\}$, see the example after Theorem 1.3.2.13.

For nonpositive submartingales (or nonnegative supermartingales), we have the following convergence result.

Theorem 1.3.2.12 *Let $f = (f_n)_{n\geq 0}$ be a nonnegative supermartingale (or nonpositive submartingale). Then $\lim_{n\to\infty} f_n = f_\infty$ a.e, and $(f_n)_{0\leq n\leq\infty}$ is still a super-(or sub-)martingale.*

Proof. Any nonnegative supermartingale $f = (f_n)$ is in L^1, since

$$E(|f_n|) = E(f_n) \leq E(f_0).$$

So $\lim_{n\to\infty} f_n = f_\infty$ a.e. Now for all $F \in \mathcal{F}_n$, we have

$$\int_F f_\infty d\mu \leq \lim_{m\to\infty} \int_F f_m d\mu = \lim_{m\to\infty} \int_F E(f_m|\mathcal{F}_n) d\mu \leq \int_F f_n d\mu.$$

This proves that $(f_n)_{0\leq n\leq\infty}$ is still a supermartingale. The proof is finished. □

Now consider the L^φ-convergence.

Theorem 1.3.2.13 *Let $(f_n)_{n\geq 0}$ be a martingale such that $\sup_n E(\varphi(|f_n|)) < \infty$, where $\varphi(u)$ is a convex function increasing and satisfying*

$$\varphi(0) = 0, \quad \lim_{u\to\infty} \frac{\varphi(u)}{u} = \infty. \tag{1.3.2.8}$$

Then $\lim_{n\to\infty} f_n = f_\infty$ a.e., and

$$E\Big(\varphi\Big(\frac{1}{3}|f_n - f_\infty|\Big)\Big) \to 0. \tag{1.3.2.9}$$

Proof. We have known that $(f_n)_{n\geq 0} \in L^1_u$, and $(\varphi(|f_n|))_{0\leq n\leq\infty}$ is a nonnegative submartingale, for any convex function φ satisfying (1.3.2.8). Thus

$$\sup_n E(\varphi(|f_n|)) \leq E(\varphi(|f_\infty|)) \leq \underline{\lim}\, E(\varphi(|f_n|)).$$

This shows that the pointwise limit f_∞ of $(f_n)_{n\geq 0}$ satisfies

$$\varphi(|f_n|) \to \varphi(|f_\infty|), \text{ a.e., and } \lim_{n\to\infty} E(\varphi(|f_n|)) = E(\varphi(|f_\infty|)). \tag{1.3.2.10}$$

From Lemma 1.3.2.5, we see that $(\varphi(|f_n|))_{n\geq 0}$ is uniformly integerable. Now we have

$$\begin{aligned} E\Big(\varphi\Big(\frac{1}{3}|f_n - f_\infty|\Big)\Big) &\leq E\Big(\varphi\Big(\frac{1}{3}|f_n\chi_{\{|f_n|\leq C\}} - f_\infty\chi_{\{|f_\infty|\leq C\}}| \\ &\quad + \frac{1}{3}|f_n|\chi_{\{|f_n|>C\}} + \frac{1}{3}|f_\infty|\chi_{\{|f_\infty|>C\}}\Big)\Big) \\ &\leq \frac{1}{3}E(\varphi(|f_n\chi_{\{|f_n|\leq C\}} - f_\infty\chi_{\{|f_\infty|\leq C\}}|)) \\ &\quad + \frac{1}{3}E(\varphi(|f_n|\chi_{\{|f_n|>C\}})) + \frac{1}{3}E(\varphi(|f_\infty|\chi_{\{|f_\infty|>C\}})), \end{aligned}$$

which tends to 0 owing to the bounded dominated convergence, and the uniform integrability of $\varphi(|f_n|)$. The proof is finished. □

Remark In particular, for $\varphi(u) = u^p$, $1 < p < \infty$, we see that $L^p = L^p_u$ with the same norm, since $\|f_n - f_\infty\|_p \to 0$ implies $\|f_\infty\|_p = \lim_{n\to\infty} \|f_n\|_p = \|f\|_p$.

Now we take an example to show $L^1_u \subsetneq L^1$.

Example Consider the dyadic martingale. Set $f_n = 2^n\chi_{[0,2^{-n})}$, then $(f_n)_{n\geq 0}$ is a martingale, since

$$E(f_{n+1}|\mathcal{F}_n) = 2^n \int_0^{2^{-n}} 2^{n+1}\chi_{[0,2^{-n-1})}dx\chi_{[0,2^{-n})} = f_n.$$

It is a nonnegative martingale and $E(f_n) = 1$, so it is in L^1. Its pointwise limit is $f_\infty = 0$. So $f \notin L^1_u$. The same example shows as well that for a L^1-super-(or sub-) martingale g, even g_∞ exists and makes $(g_n)_{0\leq n\leq\infty}$ being still a super-(or sub-) martingale, $(g_n)_{n\geq 0}$ is not necessarily uniformly integrable. In fact the preceding $f = (f_n)_{n\geq 0}$ with $f_\infty = 0$, is an example of supermartingale for which $(f_n)_{0\leq n\leq\infty}$ is a supermartingale meanwhile $(f_n)_{n\geq 0}$ is not uniformly integrable.

As the end of the section, we go on to discuss the conditions which characterize a L^1-martingale being a L^1_u-martingale.

Theorem 1.3.2.14 *Let $f = (f_n)_{n\geq 0}$ be a L^1-martingale. Denote $R_n = \inf\{m : |f_m| > n\}$. Then $\{f_n\}$ is uniformly integrable, if and only if*

$$\lim_{n\to\infty} \int_{\{R_n<\infty\}} |f_{R_n}|d\mu = 0. \tag{1.3.2.11}$$

Proof. Assume $\{f_n\}$ being uniformly integrable. Then there exists a convex function $\varphi(u)$ satisfying (1.3.2.8) such that $C = \sup_n E(\varphi(|f_n|)) < \infty$. Noticing $u \leq \varepsilon_n\varphi(u)$, when $u \geq n$, with $\varepsilon_n \to 0$, this implies

$$\int_{\{R_n<\infty\}} |f_{R_n}|d\mu \leq \varepsilon_n \int_{\{R_n<\infty\}} \varphi(|f_{R_n}|)d\mu = \varepsilon_n \sum_{k=0}^{\infty} \int_{\{R_n=k\}} \varphi(|f_k|)d\mu$$

$$\leq \varepsilon_n \sum_{k=0}^{\infty} \int_{\{R_n=k\}} \varphi(|f_\infty|)d\mu \leq C\varepsilon_n \to 0.$$

On the contrary, assume (1.3.2.11) holding. Denote $M_m f = \sup_{n\leq m} |f_n|$, $Mf = \sup_n |f_n|$, then $\{M_m f > n\} = \{R_n \leq m\}$, $\{Mf > n\} = \{R_n < \infty\}$. Making Doob's decomposition for $(|f_n|)_{n\geq 0}$, we get $|f_n| = g_n + A_n$, $f, g \in L^1$, $A_\infty \in L^1$. We have

$$\begin{aligned}
&\int_{\{|f_m|>n\}} |f_m|d\mu \\
&\leq \int_{\{R_n\leq m\}} |f_{R_n}|d\mu + \int_{\{R_n\leq m\}} (|f_m| - |f_{R_n}|)d\mu \\
&= \int_{\{R_n\leq m\}} |f_{R_n}|d\mu + E((g_m - g_{R_n})\chi_{\{R_n\leq m\}}) + E((A_m - A_{R_n})\chi_{\{R_n\leq m\}}) \\
&\leq \int_{\{R_n<\infty\}} |f_{R_n}|d\mu + E(E((g_m - g_{R_n})\chi_{\{R_n\leq m\}}|\mathcal{F}_{R_n})) + E(A_\infty \chi_{\{R_n<\infty\}}) \\
&= \sum_{i=1}^{3} I_i.
\end{aligned}$$

$I_1 = o(1)$, obviously, so I_3 is owing to $|\{R_n < \infty\}| = o(1)$ and $A_\infty \in L^1$ $\Big($In fact $|\{R_n < \infty\}| = o(\frac{1}{n})$, since

$$n|\{R_n < \infty\}| \leq \int_{\{R_n<\infty\}} |f_{R_n}|d\mu = o(1).\Big)$$

As for $I_2 = o(1)$, this follows from

$$I_2 = E((g_{m\wedge R_n} - g_{R_n})\chi_{\{R_n\leq m\}}) = 0.$$

The proof is finished. □

In some cases, the characterization condition (1.3.2.11) could be simplified as, for example, $\lim_{\lambda\to\infty} \lambda|\{Mf > \lambda\}| = 0$ or $\lim_{\lambda\to\infty} \lambda|\{S(f) > \lambda\}| = 0$, where Mf (we have known) and $S(f)$ are maximal function and square function associated to the martingale $f = (f_n)_{n\geq 0}$ studied in §2.1. Here we want to show that the so-called "regular" $(\Omega, \mathcal{F}, \mu, \{\mathcal{F}_n\}_{n\geq 0})$ studied in §7.1 is of the case. Before we formulate the theorem, we devote a few words to such regularity. $(\Omega, \mathcal{F}, \mu, \{\mathcal{F}_n\}_{n\geq 0})$ is called regular, if for all $n = 1, 2, \cdots$, for all $F_n \in \mathcal{F}_n$, there exists at least one $G_n \in \mathcal{F}_{n-1}$, such that $F_n \subset G_n$, and $|G_n| \leq d|F_n|$, $d \geq 1$ a constant. For such regular case, for any nonnegative adapted process $(\gamma_n)_{n\geq 0}$ with $\gamma_0 = 0$, and $\lambda > 0$, we could define a stopping time τ such that (see §7.1)

$$\begin{aligned}
&\gamma_\tau \leq \lambda, \quad \{M_m\gamma > \lambda\} \subset \{\tau \leq m-1\}, \\
&|\{\tau \leq m-1\}| \leq d|\{M_m\gamma > \lambda\}|, \quad \forall m,
\end{aligned} \tag{1.3.2.12}$$

where $M_m\gamma = \sup_{k\leq m} \gamma_k$. In §7.3, we will show that for such regular case, the ordered pairs $(Mf,\ S(f))$ and $(S(f),\ Mf)$ satisfy so-called good λ-inequality, i.e.

$$|\{Mf > \alpha\lambda\}| \leq \varepsilon_{\alpha,\beta}|\{Mf > \lambda\}| + d_{\alpha,\beta}|\{S(f) > \beta\lambda\}|, \quad \lambda > 0, \tag{1.3.2.13}$$

$$|\{S(f) > \alpha\lambda\}| \leq \varepsilon_{\alpha,\beta}|\{S(f) > \lambda\}| + d_{\alpha,\beta}|\{Mf > \beta\lambda\}|, \quad \lambda > 0, \tag{1.3.2.14}$$

where $\alpha > 1$, $\beta > 0$ small enough, $\lim_{\beta\to 0} \varepsilon_{\alpha,\beta} = 0$. For any nonnegative measurable f define

$$\theta_1(f) = \sup_{\lambda>0} \lambda|\{f > \lambda\}|, \quad \theta_2(f) = \lim_{\lambda\to\infty} \lambda|\{f > \lambda\}|. \tag{1.3.2.15}$$

For a pair (f, g) satisfying good λ-inequality, $\theta_i(f)$ could be controlled by $\theta_i(g)$, $i = 1, 2$, as shown by the following lemma.

Lemma 1.3.2.15 *Let (f, g) be a pair satisfying good λ-inequality. Then*

$$\theta_i(f) \leq \alpha\beta^{-1}d_{\alpha,\beta}(1 - \alpha\varepsilon_{\alpha,\beta})^{-1}\theta_i(g), \quad i = 1, 2. \tag{1.3.2.16}$$

Proof. Denote $\theta(\lambda) = \lambda|\{f > \lambda\}|$. From the good λ-inequality, we get

$$\theta(\alpha\lambda) \leq \alpha\varepsilon_{\alpha,\beta}\theta(\lambda) + \alpha\beta^{-1}d_{\alpha,\beta}\theta_1(g),$$

$$\begin{aligned}\theta(\lambda) &\leq \alpha\varepsilon_{\alpha,\beta}\theta\Big(\frac{\lambda}{\alpha}\Big) + C_{\alpha,\beta}\theta_1(g)\\ &\leq (\alpha\varepsilon_{\alpha,\beta})^n\theta\Big(\frac{\lambda}{\alpha^n}\Big) + C_{\alpha,\beta}(1 + \cdots + (\alpha\varepsilon_{\alpha,\beta})^{n-1})\theta_1(g).\end{aligned}$$

Let β be small enough such that $\alpha\varepsilon_{\alpha,\beta} < 1$. Letting $n \to \infty$, we get

$$\theta_1(f) \leq C_{\alpha,\beta}(1 - \alpha\varepsilon_{\alpha,\beta})^{-1}\theta_1(g).$$

When $\theta_2(g) < \infty$, then $\theta_1(g) < \infty$, and hence $\theta_2(f) \leq \theta_1(f) < \infty$. From the good λ-inequality, we get

$$\theta_2(f) \leq \alpha\varepsilon_{\alpha,\beta}\theta_2(f) + \alpha\beta^{-1}d_{\alpha,\beta}\theta_2(g).$$

Since $\theta_2(f) < \infty$, this gives the assertion about $\theta_2(f)$. the proof is finished. □

Now we can give a simple characterization for a L^1-martingale being in L^1_u in regular case.

Theorem 1.3.2.16 *Let $(\Omega, \mathcal{F}, \mu, \{\mathcal{F}_n\}_{n\geq 0})$ be regular, and $f = (f_n)_{n\geq 0} \in L^1$. Then $f \in L^1_u$, if and only if*

$$\lim_{\lambda\to\infty} \lambda|\{Mf > \lambda\}| = 0, \tag{1.3.2.17}$$

or equivalently,

$$\lim_{\lambda\to\infty} \lambda|\{S(f) > \lambda\}| = 0. \tag{1.3.2.18}$$

Proof. From Lemma 1.3.2.15, we see that (1.3.2.17) and (1.3.2.18) are equivalent. Consider only Mf. The uniform integrability of $\{f_n\}$ implies (1.3.2.11), and hence

$$n|\{Mf > n\}| = n|\{R_n < \infty\}| \leq \int_{\{R_n<\infty\}} |f_{R_n}|d\mu = o(1).$$

On the contrary, assume $\lambda|\{Mf > \lambda\}| = o(1)$. Assume $f_0 = 0$ without loss of generality. For the process $(|f_n|)_{n\geq 0}$ and $\lambda = n$, define a stopping time $\widetilde{R}_n$ satisfying (1.3.2.12). Then

$$\int_{\{\widetilde{R}_n<\infty\}} |f_{\widetilde{R}_n}|d\mu \leq n|\{\widetilde{R}_n < \infty\}| \leq dn|\{Mf > n\}| = o(1).$$

This inequality, just like (1.3.2.11), could imply the uniform integrability of $\{f_n\}$. In fact, from Doob's decomposition $|f_m| = g_m + A_m$, we have

$$\begin{aligned}\int_{\{|f_m|>n\}} |f_m|d\mu &\leq \int_{\{M_m f>n\}} |f_m|d\mu \leq \int_{\{\widetilde{R}_n\leq m-1\}} |f_m|d\mu \\ &\leq \int_{\{\widetilde{R}_n\leq m-1\}} |f_{\widetilde{R}_n}|d\mu + E(E(g_m - g_{\widetilde{R}_n})\chi_{\{\widetilde{R}_n\leq m-1\}}|\mathcal{F}_{\widetilde{R}_n}) \\ &\quad +E((A_m - A_{R_n})\chi_{\{\widetilde{R}_n\leq m-1\}}) = o(1).\end{aligned}$$

The proof of the theorem is finished. □

Notes to Chapter 1

The various concepts and related results that occur in this chapter are basic and elementary, which are well known and could be referred to in many related textbooks, for example, see Dellacherie-Meyer [1,2], Doob [1], Rao [3], and Yen [1] etc. Here we have only done some collections and organizations. The various decompositions in §1.3.1 are named according to their discoverers as usual. The proof of Gundy's decomposition(Theorem 1.3.1.8) is taken from Burkholder [2]. Lemma 1.3.2.6 is due to Dellacherie-Meyer-Yor [1]. The proof of the main convergence theorem (Theorem 1.3.2.8) is essentially due to R. Isaac, we take it from Rao [3]. The uniform integrability problem we discussed in §1.3.2 is related to the classical Riesz' brother's theorem. Martingales and super-(or sub-)martingales could be corresponded with the harmonic functions and super-(or sub-)harmonic functions. The L^1-boundedness of martingales is like the Poisson integral expression by bounded measure for harmonic functions. and L^1_u-property is like the Poisson integral expression by absolutely continuous measure. The classical Riesz' theorem says that if harmonic u and its conjugate harmonic v are both Poisson integrals of bounded measures, then the two associated measures must be absolutely continuous. Along this way there have been some related results in martingale setting, see for example Varopoulos [3] . Theorem 1.3.2.14 is a version of a known result, see K.M. Rao [1]. Azema-Gundy-Yor [1] established Theorem 1.3.2.16 in continuous case for continuous martingales. Here is its discrete version for regular martingales.

2 $H_p(p \geq 1)$ Martingales

In Chapter 1, we have introduced the concept of martingale and one kind of martingale spaces, i.e. L^p, and just mentioned two important operators defined on martingales, i.e. maximal operator M and square function operator S. In this chapter, we will study several other martingale spaces, among which Hardy space H_p is the most important one. Among other things, in the chapter, we will establish the $L^p(1 \leq p \leq \infty)$ equivalence between M and S (i.e. Davis' inequality and Burkholder-Gundy's inequality) establish the Fefferman's H_1-BMO duality by two different proofs (one of which is via atomic decomposition); discuss the weak compactness of subsets and the convergence of sequences in H_1; and as a comparison, we will introduce two versions of H_p, i.e. h_p and $\mathcal{P}_p$.

In what follows, let $(\Omega, \mathcal{F}, \mu, \{\mathcal{F}_n\}_{n \geq 0})$ be a complete probability space endowed with an increasing (means nondecreasing) family of sub-σ-fields satisfying the usual conditions, i.e. each $(\Omega, \mathcal{F}_n, \mu)$ is complete, and $\mathcal{F} = \bigvee_n \mathcal{F}_n$ (it means that $\mathcal{F}$ is generated by $\bigcup_n \mathcal{F}_n$).

2.1 Boundedness of M and S, equivalence between M and S

We have defined the maximal operator M and the square function operator S on martingales as follows: Let $f = (f_n)_{n \geq 0}$ be a martingale (for any process, the following definitions go well, but we prefer for martingales), denote

$$M_n f = \sup_{k \leq n} |f_k|, \quad Mf = M_\infty f = \sup_n |f_n|, \tag{2.1.1}$$

$$S_n(f) = \left(\sum_0^n |\Delta_k f|^2 \right)^{\frac{1}{2}}, \quad S(f) = S_\infty(f) = \left(\sum_0^\infty |\Delta_k f|^2 \right)^{\frac{1}{2}}, \tag{2.1.2}$$

$$\Delta_n f = f_n - f_{n-1}, \quad n \geq 0. \tag{2.1.3}$$

Here and in what follows, we keep the convention: for any process $(\gamma_n)_{n\geq 0}$, γ_{-1} is meant as 0 except when otherwise stated. According to this, $\Delta_0 f = f_0$. In order to show what the definitions mean, we take the dyadic martingale as an example. We have seen in §1.3 that, for $f = (f_n)_{n\geq 0}$ a dyadic martingale in $L^1_u([0,1])$, we have

$$f_n = \sum_I \frac{1}{|I|}\int_I f(y)dy\chi_I = \sum_0^{2^n} c_k\varphi_k(x),$$

where I are dyadic intervals of length 2^{-n}, and $\{\varphi_k\}$ is Walsh functions system. So in this case

$$Mf(x) = \sup_{I\ni x}\left|\frac{1}{|I|}\int_I f(y)dy\right|, \quad I \text{ are dyadic intervals},$$

$$S(f)(x) = \left(\sum_n \left|\sum_{2^{n-1}+1}^{2^n} c_k\varphi_k(x)\right|^2\right)^{\frac{1}{2}}.$$

They are nothing but the versions of Hardy-Littlewood maximal function and Littlewood-Paley function in the classical case. We are interested in the behavior of M and S on L^p, since our main object H_p space will be defined by them. In what follows, in most cases, we can deal with finite martingales (i.e. those martingales $f = (f_n)_{n\geq 0}$ such that $f_n = f_N$, for all $n \geq N$, for some N. Notice that for any martingale $f = (f_n)_{n\geq 0}$, and any N, $f^{(N)} = (f_{n\wedge N})_{n\geq 0}$ is a finite martingale, called stopped martingale of f at N.) at first, and then pass to the general martingales by taking limits. Thus some trouble concerning the convergence and integrability etc. can be avoided.

Now consider the behavior of M and S on L^1 martingales.

Theorem 2.1.1 *The maximal operator M is an operator of weak type* $(1,1)$, *that is for all $f \in L^1$, we have*

$$|\{Mf > \lambda\}| \leq \frac{1}{\lambda}\|f\|_1, \quad \forall\lambda > 0. \tag{2.1.4}$$

This follows from Lemma 1.3.1.6, since $\{Mf > \lambda\} = \{\tau < \infty\}$ with $\tau = \inf\{n : |f_n| > \lambda\}$.

Remarks

1. For $f \in L^1_u$, we could get more. Since $f_n = E(f_\infty|\mathcal{F}_n)$, denoting f_∞ by f (as we do for all $f \in L^p_u$), from the proof of Lemma 1.3.1.6, we get

$$|\{Mf > \lambda\}| \leq \frac{1}{\lambda}\int_{\{Mf>\lambda\}} |f|d\mu. \tag{2.1.5}$$

2. (2.1.4) holds for all nonnegative L^1-submartingales too, since we have only used the fact that $(|f_n|)_{n\geq 0}$ is a L^1-submartingale.

Now discuss the weak (1,1) type of square operator S.

Theorem 2.1.2 *The square operator S is of weak type* $(1,1)$.

Proof. For $f \in L^1$ and $\lambda > 0$ given, define the stopping time $\tau = \inf\{n : f_n| > \lambda\}$. Then from Lemma 1.3.1.7 we get

$$|\{Sf > \lambda\}| \leq |\{S_{\tau-1}f > \lambda\}| + |\{\tau < \infty\}|$$
$$\leq \frac{1}{\lambda^2}||S_{\tau-1}f||_2^2 + \frac{1}{\lambda}||f||_1 \leq \frac{3}{\lambda}||f||_1 .$$

This proves the theorem. □

Now Doob's maximal inequality which tells the L^p-boundedness of M, $p > 1$, is a consequence of (2.1.5).

Theorem 2.1.3 *Let $f = (f_n)_{n\geq 0} \in L^p$, $1 < p \leq \infty$. Then*

$$||f||_p \leq ||Mf||_p \leq p'||f||_p. \tag{2.1.6}$$

Proof. Integrating both sides of (2.1.5) with respect to λ, we get

$$\int_\Omega (Mf)^p d\mu = p \int_0^\infty \lambda^{p-1}|\{Mf > \lambda\}|d\lambda$$
$$\leq p \int_0^\infty \lambda^{p-2} \int_{\{Mf>\lambda\}} |f|d\mu d\lambda = p' \int_\Omega |f|(Mf)^{p-1}d\mu$$
$$\leq p'||f||_p \left(\int_\Omega (Mf)^p d\mu\right)^{\frac{1}{p'}} .$$

Since $\int_\Omega (Mf)^p d\mu < \infty$ could be assumed (for example by considering finite martingale at first and then letting $n \to \infty$), we get (2.1.6). The proof is finished. □

Remark The assertion of the theorem holds for nonnegative submartingales, too.

Now we consider the behavior of S on L^p martingales, $1 < p < \infty$. The following reasoning will give the L^p-boundedness of S and the L^p-equivalence between M and S simultaneously. The so-called stopping time argument will play a remarkable role in the book. Here we will have the first example in which the stopping time argument is very useful. In addition, we will need an auxiliary space ${}_2K_p$, actually it is not a new sapce, just a new description of L^p, for $2 \leq p < \infty$. First we give the definition of ${}_2K_p$.

Definition 2.1.4 *Let $2 \leq p \leq \infty$, $f = (f_n)_{n\geq 0}$ be a L^1-martingale. f is said to be in ${}_2K_p$, if there exists $\gamma \in L^p_+$ such that*

$$E(|f - f_{n-1}|^2|\mathcal{F}_n) \leq E(\gamma^2|\mathcal{F}_n), \quad \forall n, \tag{2.1.7}$$

here f on the left-hand side means f_∞. (We use same symbol f to denote the martingale $f = (f_n)_{n\geq 0}$ and it's end function $f_\infty = \lim_{n\to\infty} f_n$, when it exists pointwise, as usual.) We define a norm in ${}_2K_p$ by

$$\|f\|_{{}_2K_p} = \inf\{\|\gamma\|_p : \gamma \text{ runs through all possible ones}\}. \tag{2.1.8}$$

Remarks

1. ${}_2K_\infty$ is just the BMO (=Bounded Mean Oscillation), another main object of the book (see Chapter 4). The definition of ${}_2K_p$ comes from the one of BMO obviously.

2. Assume that $f = (f_n)_{n\geq 0} \in L^1$, and (2.1.7) hold. Then it is easy to see that $f \in L^2$. In fact, putting $n = 0$ in (2.1.7) we get $E(|f_\infty|^2|\mathcal{F}_0) \leq E(\gamma^2|\mathcal{F}_0)$, $E(|f_\infty|^2) \leq E(\gamma^2)$. Thus we have

$$\begin{aligned} E(|f_n|^2)^{\frac{1}{2}} &\leq E(|f_\infty - f_n|^2)^{\frac{1}{2}} + E(|f_\infty|^2)^{\frac{1}{2}} \\ &\leq E(E(|f_\infty - f_n|^2|\mathcal{F}_{n+1}))^{\frac{1}{2}} + E(|f_\infty|^2)^{\frac{1}{2}} \\ &\leq E(E(\gamma^2|\mathcal{F}_{n+1}))^{\frac{1}{2}} + E(|f_\infty|^2)^{\frac{1}{2}} \leq 2E(\gamma^2)^{\frac{1}{2}}. \end{aligned}$$

This shows $f \in L^2$. (We will show $f \in L^p$ in Chapter 3.) The index 2 does not play the role in these arguments, so we can consider the space ${}_aK_p$, $1 \leq a \leq p \leq \infty$,

$${}_aK_p = \{f \in L^a_u : \exists\gamma \in L^p_+, \text{ s.t. } E(|f - f_{n-1}|^a|\mathcal{F}_n) \leq E(\gamma^a|\mathcal{F}_n),\ \forall n\}. \tag{2.1.9}$$

Write ${}_aK_\infty$ as BMO_a.

The following lemma is the main step of our approach to the behavior of S.

Lemma 2.1.5 *Let $\{\varepsilon_k\}$ be a sequence consisting of $\{+1, -1\}$, T_ε be the operator defined by $\varepsilon = \{\varepsilon_k\}$*

$$T_\varepsilon f = g = (g_n)_{n\geq 0}, \quad g_n = \sum_0^n \varepsilon_k \Delta_k f, \quad f = (f_n)_{n\geq 0}. \tag{2.1.10}$$

Then for $2 < p < \infty$,

$$\|Mg\|_p \leq \sqrt{e}\sqrt{\frac{p^3}{p-2}}\|f\|_{{}_2K_p}. \tag{2.1.11}$$

Proof. $g = (g_n)_{n\geq 0}$ is still a martingale, and when $f \in {}_2K_p$ we have also $g \in {}_2K_p$, since

$$\begin{aligned} E(|g - g_{n-1}|^2|\mathcal{F}_n) &= E\left(\sum_{k=n}^\infty |\Delta_k g|^2 \Big| \mathcal{F}_n\right) \\ &= E\left(\sum_{k=n}^\infty |\Delta_k f|^2 \Big| \mathcal{F}_n\right) = E(|f - f_{n-1}|^2|\mathcal{F}_n). \end{aligned}$$

Let γ be any one such that $E(|f - f_{n-1}|^2|\mathcal{F}_n) \le E(\gamma^2|\mathcal{F}_n)$, for all n. For $\lambda > 0$ given and $\alpha > 0$ determined later, define stopping times

$$\tau_1 = \inf\{n : |g_n| > \alpha\lambda\}, \quad \tau_2 = \inf\{n : |g_n| > (\alpha+1)\lambda\}.$$

Then $\tau_1 \le \tau_2$, $\{\tau_1 < \infty\} \in \mathcal{F}_{\tau_1} \subset \mathcal{F}_{\tau_2}$, and

$$\{Mg > (\alpha+1)\lambda\} = \{\tau_2 < \infty\} \subset \{\tau_1 < \infty, |g_{\tau_2} - g_{\tau_1 - 1}| > \lambda\},$$

here we have used the fact $|g_{\tau_1-1}| \le \alpha\lambda$, even on $\{\tau_1 = 0\}$, since according to the convention, $g_{\tau_1-1} = 0$ on $\{\tau_1 = 0\}$. Now denote $|\{Mg > \lambda\}|$ by $\sigma(\lambda)$. We get

$$\begin{aligned}
\sigma((\alpha+1)\lambda) &\le \frac{1}{\lambda^2}\int_{\{\tau_1<\infty\}} |g_{\tau_2} - g_{\tau_1-1}|^2 d\mu \\
&= \frac{1}{\lambda^2}\int_{\{\tau_1<\infty\}} |E(g - g_{\tau_1-1}|\mathcal{F}_{\tau_2})|^2 d\mu \\
&\le \frac{1}{\lambda^2}\int_{\{\tau_1<\infty\}} E(|g - g_{\tau_1-1}|^2|\mathcal{F}_{\tau_1}) d\mu.
\end{aligned}$$

Since for any stopping time τ we have

$$\begin{aligned}
E(|g - g_{\tau-1}|^2|\mathcal{F}_\tau) &= E\left(\sum_{n=0}^{\infty} |g - g_{n-1}|^2 \chi_{\{\tau=n\}} \Big| \mathcal{F}_\tau\right) \\
&= \sum_{n=0}^{\infty} E(|g - g_{n-1}|^2|\mathcal{F}_n)\chi_{\{\tau=n\}} \\
&= \sum_{n=0}^{\infty} E(|f - f_{n-1}|^2|\mathcal{F}_n)\chi_{\{\tau=n\}} \\
&= E(|f - f_{\tau-1}|^2|\mathcal{F}_\tau) \le E(\gamma^2|\mathcal{F}_\tau),
\end{aligned}$$

we get

$$\sigma((\alpha+1)\lambda) \le \frac{1}{\lambda^2}\int_{\{\tau_1<\infty\}} \gamma^2 d\mu,$$

and hence

$$\begin{aligned}
p\int_0^\infty \lambda^{p-1}\sigma((\alpha+1)\lambda)d\lambda &\le p\int_0^\infty \lambda^{p-3}\int_{\{Mg>\alpha\lambda\}} \gamma^2 d\mu d\lambda \\
&= \frac{p}{p-2}\int_\Omega \left(\frac{Mg}{\alpha}\right)^{p-2} \gamma^2 d\mu \\
&\le \frac{p}{p-2}\frac{1}{\alpha^{p-2}} ||Mg||_p^{p-2}||\gamma||_p,
\end{aligned}$$

$$||Mg||_p^p \le \frac{p}{p-2}\frac{(\alpha+1)^p}{\alpha^{p-2}} ||Mg||_p^{p-2}||\gamma||_p^2.$$

Setting $\alpha = p$, and taking "inf" over all possible γ, we get (2.1.11). The proof is finished. □

Now define the Hardy spaces as follows, $0 < p \leq \infty$,

$$H_p^* = \{f = (f_n)_{n\geq 0} : \|f\|_{H_p^*} = \|Mf\|_p < \infty\}, \tag{2.1.12}$$

$$H_p^S = \{f : (f_n)_{n\geq 0} : \|f\|_{H_p^S} = \|S(f)\|_p < \infty\}, \tag{2.1.13}$$

and give their characterizations for $1 \leq p < \infty$.

Theorem 2.1.6 *For $2 < p < \infty$, we have $H_p^S = H_p^* = {}_2K_p$ with the equivalent norms.*

Proof. Because of

$$E(|f - f_{n-1}|^2|\mathcal{F}_n) = E((S(f))^2 - (S_{n-1}(f))^2|\mathcal{F}_n) \leq E((S(f))^2|\mathcal{F}_n), \quad \forall n,$$

$$E(|f - f_{n-1}|^2|\mathcal{F}_n) \leq E((2Mf)^2|\mathcal{F}_n), \quad \forall n,$$

we see $H_p^S \subset {}_2K_p$, $H_p^* \subset {}_2K_p$ with the embedding being continuous. Basing on Lemma 2.1.5, and considering the identity operator I, we see

$$\|Mf\|_p \leq \sqrt{\frac{ep^3}{p-2}}\|f\|_{{}_2K_p}.$$

And considering the case where $\varepsilon = (r_k(t))_{k\geq 0}$ is Rademacher's system, we get

$$\int_\Omega \Big|\sum_k r_k(t)\Delta_k f(\omega)\Big|^p d\mu \leq C_p\|f\|^p_{{}_2K_p},$$

and hence

$$\int_\Omega\int_0^1 \Big|\sum_k r_k(t)\Delta_k f(\omega)\Big|^p dtd\mu = \int_0^1\int_\Omega \Big|\sum_k r_k(t)\Delta_k f(\omega)\Big|^p d\mu dt \leq C_p\|f\|^p_{{}_2K_p}.$$

Since

$$\int_0^1 \Big|\sum_k r_k(t)\Delta_k f(\omega)\Big|^p dt \approx \Big(\sum_k |\Delta_k f|^2\Big)^{\frac{p}{2}}, \tag{2.1.14}$$

we get

$$\|S(f)\|_p \leq C_p\left(\int_\Omega\int_0^1\Big|\sum_k r_k(t)\Delta_k f(\omega)\Big|^p dtd\mu\right)^{\frac{1}{p}} \leq C_p\|f\|_{{}_2K_p}.$$

This proves ${}_2K_p \subset H_p^S$, with the embedding being continuous. The proof is finished. □

Remarks

1. Notice that T_ε is self adjoint and $T_\varepsilon^2 = I$(the identity operator). Let $1 < p \le 2$, and $f = (f_n)_{n\ge 0} \in L^p$ be a finite martingale, then for any $g = (g_n)_{n\ge 0} \in L^{p'}$, $\frac{1}{p} + \frac{1}{p'} = 1$,

$$|E(T_\varepsilon fg)| = |E(fT_\varepsilon g)| \le \|f\|_p \|T_\varepsilon g\|_{p'} \le C\|f\|_p\|g\|_{p'}.$$

This shows that

$$\|T_\varepsilon f\|_p \le C\|f\|_p, \quad 1 < p \le 2, \ \forall f \in L^p \text{ being finite ones.}$$

By taking limits, T_ε could be extended to whole L^p, $1 < p \le 2$, and kept to be L^p-bounded. Meanwhile, we have

$$\|f\|_p = \|T_\varepsilon T_\varepsilon f\|_p \le C\|T_\varepsilon f\|_p, \quad 1 < p < \infty.$$

Thus we get at last

$$C\|f\|_p \le \|T_\varepsilon f\|_p \le C\|f\|_p, \quad 1 < p < \infty. \tag{2.1.15}$$

2. From (2.1.14) and (2.1.15), we have

$$C\|f\|_p \le \|S(f)\|_p \le C\|f\|_p, \quad 1 < p < \infty. \tag{2.1.16}$$

Theorem 2.1.7 *For $1 < p < \infty$, we have $H_p^* = H_p^S = L^p$, with*

$$\|Mf\|_p \approx \|S(f)\|_p \approx \|f\|_p. \tag{2.1.17}$$

With the same proof, ${}_aK_p$, $1 \le a < p < \infty$, could be shown to be equivalent to H_p (denoting H_p^S or H_p^*, since they are the same), too.

Now consider the case $p = 1$. First, we formulate Davis' decomposition as a lemma.

Lemma 2.1.8 *Let $f = (f_n)_{n\ge 0}$ be a martingale, then f could be decomposed as $f = g + h$, such that (denoting $d_n = \Delta_n f$, $n \ge 0$)*

$$g = (g_n)_{n\ge 0}, \quad |\Delta g| \le 4M_{n-1}d, \quad n \ge 0, \tag{2.1.18}$$

$$h = (h_n)_{n\ge 0}, \quad E\Big(\sum_0^\infty |\Delta_k h|\Big) \le 4E(Md). \tag{2.1.19}$$

Proof. Let $f = (f_n)_{n\ge 0}$, $f_n = \sum_0^n d_k$ be given. Set

$$\Delta_n g = d_n\chi_{\{|d_n|\le 2M_{n-1}d\}} - E(d_n\chi_{\{|d_n|\le 2M_{n-1}d\}}|\mathcal{F}_{n-1}), \quad n \ge 1,$$

$$\Delta_0 g = g_0 = 0,$$

$$\Delta_n h = d_n\chi_{\{|d_n|> 2M_{n-1}d\}} - E(d_n\chi_{\{|d_n|> 2M_{n-1}d\}}|\mathcal{F}_{n-1}), \quad n \ge 1,$$

$$\Delta_0 h = h_0 = d_0.$$

Obviously, we have $|\Delta_n g| \leq 4M_{n-1}d$. And since

$$|d_n|\chi_{\{|d_n|>2M_{n-1}d\}} \leq (2|d_n| - 2M_{n-1}d)\chi_{\{|d_n|>2M_{n-1}d\}} \\ \leq 2(M_n d - M_{n-1}d),$$

we have

$$E\left(\sum_{n=0}^{\infty}|\Delta_n h|\right) \leq E\left(\sum_{n=0}^{\infty}(2(M_n d - M_{n-1}d) + E(2(M_n d - M_{n-1}d)|\mathcal{F}_{n-1}))\right) \\ \leq 4E\left(\sum_{n=0}^{\infty}(M_n d - M_{n-1}d)\right) = 4E(Md).$$

This completes the proof. □

Theorem 2.1.9 *For any martingale $f = (f_n)_{n\geq 0}$, we have*

$$C\|Mf\|_1 \leq \|S(f)\|_1 \leq C\|Mf\|_1. \tag{2.1.20}$$

Proof. Making Davis' decomposition for f, we get $f = g + h$, with g, h satisfying (2.1.18), (2.1.19). Notice that

$$Md \leq \min(2Mf, S(f)) = (2Mf) \wedge S(f),$$

$$\max(Mh, S(h)) = Mh \vee S(h) \leq \sum_{1}^{\infty}|\Delta_n h|,$$

$$E(Mh) \vee E(S(h)) \leq \sum_{0}^{\infty} E(|\Delta_n h|) \leq 4E(Md) \leq 4(2E(Mf) \wedge E(S(f)).$$

We will make estimates for g and h separately. But before doing it, we want to examine so-called stopped martingales. Let $g = (g_n)_{n\geq 0}$ be any martingale, and τ be any stopping time. Then $g^{(\tau)} = (g_{n\wedge\tau})_{n\geq 0}$ is a martingale still, called stopped martingale at τ, of which the martingale difference, maximal function and square function are respectively (noticing the $\mathcal{F}_{n-1}$-measurability of $\chi_{\{n\leq\tau\}}$)

$$g_{n\wedge\tau} = \sum_{k=0}^{n}\chi_{\{k\leq\tau\}}\Delta_k g, \quad \Delta_n g^{(\tau)} = \chi_{\{n\leq\tau\}}\Delta_n g, \tag{2.1.21}$$

$$Mg^{(\tau)} = \sup_n |g_{n\wedge\tau}| = \sup_{n\leq\tau}|g_n| = M_\tau g, \tag{2.1.22}$$

$$S(g^{(\tau)}) = \left(\sum_{k=0}^{\infty}\chi_{\{k\leq\tau\}}|\Delta_k g|^2\right)^{\frac{1}{2}} = \left(\sum_{k=0}^{\tau}|\Delta_k g|^2\right)^{\frac{1}{2}} = S_\tau(g). \tag{2.1.23}$$

Now make the estimates for g. Define a suitable stopping time τ by making use of g's predictability as follows. We have

$$|g_n| \leq M_{n-1}g + 4M_{n-1}d = \rho_{n-1}, \quad n \geq 0.$$

For $\lambda > 0$, define

$$\tau = \inf\{n : \rho_n > \lambda\}.$$

We have

$$\begin{aligned}|\{S(g) > \lambda\}| &\le |\{S(g) > \lambda, \tau = \infty\}| + |\{\tau < \infty\}| \\ &\le |\{S(g^{(\tau)}) > \lambda\}| + |\{\tau < \infty\}| \\ &\le \frac{1}{\lambda^2}E(S(g^{(\tau)})^2) + |\{\tau < \infty\}| \\ &= \frac{1}{\lambda^2}E(|g^{(\tau)}|^2) + |\{\tau < \infty\}|.\end{aligned}$$

Notice that $\{\tau = \infty\} = \{M\rho \le \lambda\} \subset \{Mg \le \lambda\}$, and

$$|g_\tau| \le \rho_{\tau-1} \le \lambda,$$

even on $\{\tau = 0\}$ (since $g_0 = 0$). Thus we get

$$|\{S(g) > \lambda\}| \le \frac{1}{\lambda^2}\int_{\{Mg\le\lambda\}}(Mg)^2 d\mu + 2|\{\tau < \infty\}|,$$

$$\begin{aligned}||S(g)||_1 &\le \int_0^\infty \frac{1}{\lambda^2}\int_{\{Mg\le\lambda\}}(Mg)^2 d\mu d\lambda + C||Mg||_1 + C||Md||_1 \\ &\le C||Mg||_1 + C||Mf||_1.\end{aligned}$$

And hence

$$\begin{aligned}||S(f)||_1 &\le ||S(g)||_1 + ||S(h)||_1 \le C||Mg||_1 + C||Mf||_1 \\ &\le C||Mf||_1 + C||Mh||_1 \le C||Mf||_1.\end{aligned}$$

Analogously, according to the fact

$$S_n(g) = \left(\sum_{k=0}^{n-1}|\Delta_k g|^2 + |\Delta_n g|^2\right)^{\frac{1}{2}} \le S_{n-1}(g) + 4M_{n-1}d = \rho_{n-1},$$

define $\tau = \inf\{n : \rho_n > \lambda\}$, we can get

$$||Mg||_1 \le C||S(g)||_1 + C||S(f)||_1.$$

$$\begin{aligned}||Mf||_1 &\le ||Mg||_1 + ||Mh||_1 \le C||S(g)||_1 + C||S(f)||_1 \\ &\le C||S(f)||_1 + C||S(h)||_1 \le C||S(f)||_1.\end{aligned}$$

The proof is finished. □

Remark Henceforth, we denote $H_p^* = H_p^S$ by H_p, $1 \le p < \infty$. H_1 is a Banach space. Only to prove its completeness. Let $\{f^{(k)}\}_k$ be a Cauchy sequence in H_1.

Then $f_n^{(k)} \to f_n$ in L^1, $0 \leq n \leq \infty$. Assume that $\{k_j\}$ makes $\lim_{j\to\infty} f_n^{(k_j)} = f_n$, pointwise for all n. Then for $\varepsilon > 0$, there is j_0 such that

$$\|S(f - f^{(k_{j_0})}\|_1 = \int_\Omega \left(\sum_0^\infty \lim_{j\to\infty} |\Delta_n(f^{(k_j)} - f^{(k_{j_0})})|^2\right)^{\frac{1}{2}} d\mu$$

$$\leq \lim_{j\to\infty} \int_\Omega \left(\sum_0^\infty |\Delta_n(f^{(k_j)} - f^{(k_{j_0})}|^2\right)^{\frac{1}{2}} d\mu \leq \varepsilon.$$

This means $f^{(k_j)} \to f$ in H_1, and hence $f^{(k)} \to f$ in H_1. Meanwhile, it is easy to show that the space of all finite martinges which are in L^∞ is a dense subspace of H_1.

2.2 Dual spaces of H_p, Fefferman's inequality

We give an elementary inequality first, which we will use several times.

Lemma 2.2.1 *Let A, B be positive numbers such that $B^2 \geq A^2$. Then for $0 < p < \infty$, we have*

$$(B^2 - A^2)B^{p-2} \geq B^p - A^p \geq \frac{p}{2}(B^2 - A^2)B^{p-2}, \quad 0 < p \leq 2, \tag{2.2.1}$$

$$(B^2 - A^2)B^{p-2} \leq B^p - A^p \leq \frac{p}{2}(B^2 - A^2)B^{p-2}, \quad p \geq 2. \tag{2.2.2}$$

Proof. This is the consequence of following elementary inequality by taking $t = \left(\frac{B}{A}\right)^2$, $\alpha = \frac{p}{2}$,

$$(t-1)t^{\alpha-1} \geq t^\alpha - 1 \geq \alpha(t-1)t^{\alpha-1}, \quad t \geq 1,\ 0 < \alpha \leq 1, \tag{2.2.3}$$

$$(t-1)t^{\alpha-1} \leq t^\alpha - 1 \leq \alpha(t-1)t^{\alpha-1}, \quad t \geq 1,\ \alpha \geq 1, \tag{2.2.4}$$

which follows from the comparison of the derivatives of $t^\alpha - 1$ and $\alpha(t-1)t^{\alpha-1}$. □

In what follows, we will show that $(H_p^S)' = {}_2K_{p'}$, $1 \leq p \leq 2$, $\frac{1}{p} + \frac{1}{p'} = 1$. As a result, we get Fefferman's duality theorem $H_1' = BMO_2$.

Theorem 2.2.2 *Let $f \in H_p^S$, $1 \leq p \leq 2$, $\varphi \in {}_2K_{p'}$. Then*

$$|E(f_n\bar{\varphi}_n)| \leq \sqrt{\frac{2}{p}}\|f\|_{H_p^S}\|\varphi\|_{{}_2K_{p'}}, \quad \forall n. \tag{2.2.5}$$

Proof. Since

$$|f_n| = \left|\sum_{k=0}^n \Delta_k f\right| \leq \sqrt{n+1}S_n(f),$$

$$|\Delta_n\varphi|^2 \le E(|\varphi-\varphi_{n-1}|^2|\mathcal{F}_n) \le E(\gamma^2|\mathcal{F}_n),$$

we see that $f_n \in L^p$, $\varphi_n \in L^{p'}$, and hence $E(f_n\bar{\varphi}_n)$, $E(\Delta_k f\bar{\Delta}_k\varphi)$ have meaning. Owing to the orthogonality in the sense: for all $f\in L^p$, $\varphi\in L^{p'}$, $\mu>\nu\ge n$,

$$E(\Delta_\nu f\Delta_\mu\bar{\varphi}|\mathcal{F}_n) = E(\Delta_\nu f E(\Delta_\mu\bar{\varphi}|\mathcal{F}_\nu)|\mathcal{F}_n) = 0, \tag{2.2.6}$$

we get

$$E(f_n\overline{\varphi_n}) = \sum_{k=0}^{n} E(\Delta_k f\Delta_k\bar{\varphi}), \tag{2.2.7}$$

$$E(|\varphi_m-\varphi_{n-1}|^2|\mathcal{F}_n) = E\left(\sum_{k=n}^{m}|\Delta_k\varphi|^2\middle|\mathcal{F}_n\right). \tag{2.2.8}$$

Thus, we have

$$\begin{aligned}|E(f_n\overline{\varphi_n})| &= \left|\sum_{k=0}^{n} E\left(\Delta_k f S_k(f)^{\frac{p}{2}-1}\Delta_k\bar{\varphi}S_k(f)^{1-\frac{p}{2}}\right)\right|\\ &\le \left(\sum_{k=0}^{n} E(|\Delta_k f|^2 S_k(f)^{p-2})\right)^{\frac{1}{2}}\left(\sum_{k=0}^{n} E(|\Delta_k\varphi|^2 S_k(f)^{2-p}\right)^{\frac{1}{2}}\\ &= \sqrt{AB}.\end{aligned}$$

Applying (2.2.1), we get

$$\begin{aligned}A &= \sum_{k=0}^{n} E\left([S_k(f)^2-S_{k-1}(f)^2]S_k(f)^{p-2}\right)\\ &\le \frac{2}{p}\sum_{k=0}^{n} E\left(S_k(f)^p-S_{k-1}(f)^p\right) \le \frac{2}{p}E(S_n(f)^p).\end{aligned} \tag{2.2.9}$$

In order to estimate B, denote $Q_k = S_k(f)^{(2-p)} - S_{k-1}(f)^{2-p}$. Notice that $Q_k\ge 0$. We have

$$\begin{aligned}B &= \sum_{k=0}^{n}\sum_{m=0}^{k}(E(Q_m|\Delta_k\varphi|^2) = \sum_{m=0}^{n}\sum_{k=m}^{n} E(Q_m E(|\Delta_k\varphi|^2|\mathcal{F}_m))\\ &= \sum_{m=0}^{n} E\left(Q_m E(|\varphi_n-\varphi_{m-1}|^2|\mathcal{F}_m)\right) \le \sum_{m=0}^{n} E(Q_m E(\gamma^2|\mathcal{F}_m))\\ &= E\left(\sum_{m=0}^{n} Q_m\gamma^2\right) = E\left(S_n(f)^{2-p}\gamma^2\right) \le E\left(S_n(f)^p\right)^{\frac{2-p}{p}} E(\gamma^{p'})^{\frac{2p-2}{p}}.\end{aligned}$$

Finally, we get

$$|E(f_n\bar{\varphi}_n)| \le \sqrt{\frac{2}{p}}E(S_n(f)^p)^{\frac{1}{p}}E(\gamma^{p'})^{\frac{1}{p'}},$$

$$|E(f_n\bar{\varphi}_n)| \leq \sqrt{\frac{2}{p}}\|f\|_{H_p^S}\|\bar{\varphi}\|_{{}_2K_{p'}}.$$

The proof is finished. □

Remark From this theorem, we see that each $\varphi \in {}_2K_{p'}$ could yield a bounded linear functional on H_p^S,

$$l\varphi(f) = \langle f, \varphi\rangle = \lim_{n\to\infty} E(f_n\bar{\varphi}_n), \quad \forall f \in H_p^S,\ 1 \leq p \leq 2,$$

$$|l_\varphi(f)| \leq \sqrt{\frac{2}{p}}\|\varphi\|_{{}_2K_{p'}}\|f\|_{H_p^S}.$$

Here $\lim\limits_{n\to\infty} E(f_n\overline{\varphi}_n) = \lim\limits_{n\to\infty} \sum_0^n E(\Delta_k f\Delta_k\bar{\varphi})$ exists owing to the absolute convergence of $\sum_0^\infty E(\Delta_k f\Delta_k\bar{\varphi})$.

Now we want to show that each l in $(H_p^S)'$ comes from some $\varphi \in {}_2K_{p'}$ as l_φ defined as above. The idea is embedding H_p^S into a large space of which the dual space could be easily charcterized. Let $L^p(l^2)$ denote the space of sequences $\theta = (\theta_0, \theta_1, \cdots)$ of random variables, with the norm defined by

$$\|\theta\|_{L^p(l^2)} = E\Big(\Big(\sum_0^\infty |\theta_\nu|^2\Big)^{\frac{p}{2}}\Big)^{\frac{1}{p}}. \tag{2.2.10}$$

Lemma 2.2.3 *Let $1 \leq p \leq \infty$. Then $L^p(l^2)' = L^{p'}(l^2)$.*

Proof. Let $\theta \in L^p(l^2)$, $\sigma \in L^{p'}(l^2)$. Then

$$|\langle\theta,\sigma\rangle| = \Big|E\Big(\sum_0^\infty \bar{\sigma}_\nu\theta_\nu\Big)\Big| \leq E\Big(\Big(\sum_0^\infty |\sigma_\nu|^2\Big)^{\frac{p'}{2}}\Big)^{\frac{1}{p'}} E\Big(\Big(\sum_0^\infty |\theta_\nu|^2\Big)^{\frac{p}{2}}\Big)^{\frac{1}{p}}.$$

This means that each $\sigma \in L^{p'}(l^2)$ could yield a bounded linear functional l_σ on $L^p(l^2)$, and $\|l_\sigma\| \leq \|\sigma\|_{L^{p'}(l^2)}$. On the contrary, let l be a bounded linear functional on $L^p(l^2)$. Consider $p > 1$ at first. For all $f \in L^p$, consider $\theta = (0, \cdots 0, f, 0 \cdots)$, $\theta_n = f$, $\theta_k = 0$, $k \neq n$. Set $l_n(f) = l(\theta)$. Then $|l_n(f)| \leq \|l\|\,\|f\|_p$. Thus we get a bounded linear functional l_n on L^p. So there exists $\sigma_n \in L^{p'}$, such that $l_n(f) = E(\bar{\sigma}_n f)$. From the linearity of l,we see that for any θ such that $\theta_\nu = 0, \nu > n$, we have

$$l(\theta) = \sum_{\nu=0}^n E(\bar{\sigma}_\nu\theta_\nu).$$

Now letting

$$\theta_\nu = \sigma_\nu\left(\sum_{\mu=0}^n |\sigma_\mu|^2\right)^{\frac{p'-2}{2}},$$

we get $E\Big(\big(\sum_0^n |\theta_\nu|^2\big)^{\frac{p}{2}}\Big) = E\Big(\big(\sum_0^n |\sigma_\mu|^2\big)^{\frac{p'}{2}}\Big)$, and

$$E\Big(\big(\sum_{\nu=0}^{n} |\sigma_\nu|^2\big)^{\frac{p'}{2}}\Big) \le \|l\| E\Big(\big(\sum_{\mu=0}^{n} |\sigma_\mu|^2\big)^{\frac{p'}{2}}\Big)^{\frac{1}{p}},$$

$$E\Big(\big(\sum_{0}^{n} |\sigma_\nu|^2\big)^{\frac{p'}{2}}\Big)^{\frac{1}{p'}} \le \|l\|.$$

Letting $n \to \infty$, we see $\sigma \in L^{p'}(l^2)$, and $\|\sigma\|_{L^{p'}(l^2)} \le \|l\|$.

Now consider the case $p = 1$. Notice that $L^q(l^2) \subset L^1(l^2)$, $q > 1$, with the embedding being continuous. So, each bounded linear functional l on $L^1(l^2)$ automatically yields a bounded linear functional l on $L^q(l^2)$ and hence there exists $\sigma \in L^{q'}(l^2)$ such that $l(\theta) = \sum_0^\infty E(\bar\sigma_\nu \theta_\nu)$, $\theta = (\theta_\nu) \in L^q(l^2)$, and $\|\sigma\|_{L^{q'}(l^2)} \le \|l\|$, for all $q' < \infty$. Letting $q \to 1$, we see $\|\sigma\|_{L^\infty(l^2)} \le \|l\|$. The proof is finished. □

We need another lemma which connects the space $L^q(l^2)$ and the space ${}_2K_q$, $q \ge 2$.

Lemma 2.2.4 *Let $2 \le g \le \infty$, and $\sigma = (\sigma_0, \sigma_1, \cdots) \in L^q(l^2)$ such that $\|\sigma\|_{L^q(l^2)} \le B$. Then the following martingale $\varphi = (\varphi_n)_{n\ge 0}$ with*

$$\varphi_n = \sum_{\nu=0}^{n} (E(\sigma_\nu|\mathcal{F}_\nu) - E(\sigma_\nu|\mathcal{F}_{\nu-1})), \quad \varphi_0 = E_0(\sigma_0|\mathcal{F}_0), \tag{2 2.11}$$

here $E(\sigma_0|\mathcal{F}_{-1})$ is meant as 0 as usual, satisfies

$$\|\varphi\|_{{}_2K_q} \le 2g'B. \tag{2 2.12}$$

Proof. $\varphi = (\varphi_n)_{n\ge 0}$ is a martingale, since $E(E(\sigma_\nu|\mathcal{F}_\nu) - E(\sigma_\nu|\mathcal{F}_{\nu-1})) = 0$, $\nu \ge 1$. For $\nu \ge n+1$, we have

$$\begin{aligned} E(|\Delta_\nu\varphi|^2|\mathcal{F}_n) &= E\left(E(|\Delta_\nu\varphi|^2|\mathcal{F}_{\nu-1})|\mathcal{F}_n\right) \\ &= E\left(|E(\sigma_\nu|\mathcal{F}_\nu)|^2 - |E(\sigma_\nu|\mathcal{F}_{\nu-1})|^2|\mathcal{F}_n\right) \\ &\le E\left(E(|\sigma_\nu|^2|\mathcal{F}_\nu)\mathcal{F}_n\right) = E(|\sigma_\nu|^2|\mathcal{F}_n). \end{aligned}$$

Denote

$$g = \Big(\sum_0^\infty |\sigma_\nu|^2\Big)^{\frac{1}{2}}, \quad Mg = \sup_n E(g|\mathcal{F}_n).$$

We have

$$\begin{aligned} |\Delta_n\varphi|^2 &= |E(\sigma_n|\mathcal{F}_n)|^2 + |E(\sigma_n|\mathcal{F}_{n-1})|^2 \\ &\quad - E(\sigma_n|\mathcal{F}_n)E(\bar\sigma_n|\mathcal{F}_{n-1}) - E(\bar\sigma_n|\mathcal{F}_n)E(\sigma_n|\mathcal{F}_{n-1}) \\ &\le E(|\sigma_n|^2|\mathcal{F}_n) + 3(Mg)^2, \end{aligned}$$

and hence

$$\sum_{\nu=n}^{\infty} E\left(|\Delta_\nu \varphi|^2|\mathcal{F}_n\right) \leq \sum_{\nu=n}^{\infty} E(|\sigma_\nu|^2|\mathcal{F}_n) + 3E\left((Mg)^2|\mathcal{F}_n\right)$$
$$\leq E\left((2Mg)^2|\mathcal{F}_n\right).$$

This proves $\varphi \in {}_2K_q$, and

$$\|\varphi\|_{{}_2K_q} \leq 2\|Mg\|_q \leq 2q'\|g\|_q \leq 2q'B.$$

The proof is finished. □

Now we are in the position to get the second part of Fefferman's duality theorem.

Theorem 2.2.5 *Let $l \in (H_p^S)'$, $1 \leq p \leq 2$. Then there exists $\varphi \in {}_2K_{p'}$, such that $l(f) = \lim E(f_n\bar{\varphi}_n)$, and*

$$\|\varphi\|_{{}_2K_{p'}} \leq 2p\|l\|. \tag{2.2.13}$$

Proof. Consider H_p^S as a subspace of $L^p(l^2)$ which consists of those $\theta = (\theta_0, \theta_1, \cdots)$ with

$$\theta_\nu = E(f|\mathcal{F}_\nu) - E(f|\mathcal{F}_{\nu-1}), \quad \nu \geq 0 \quad (\theta_0 = E(f|\mathcal{F}_0)), \ f \in H_p^S.$$

Since this embedding is norm-preserving, l is also a bounded linear functional on this subspace. By means of Hahn-Banach's theorem, l can be extended as a bounded linear functional on the whole $L^p(l^2)$. So, from Lemma 2.2.3, there exists $\sigma \in L^{p'}(l^2)$ such that

$$l(\theta) = \sum_0^{\infty} E(\bar{\sigma}_\nu\theta_\nu), \quad \|\sigma\|_{L^{p'}(l^2)} \leq \|l\|.$$

In particular, restricted on this subspace, even smaller one, i.e. for all $\theta = (\theta_0, \theta_1, \cdots)$, $\theta_\nu = E(f|\mathcal{F}_\nu) - E(f|\mathcal{F}_{\nu-1})$, $\nu \geq 0$ ($\theta_0 = E(f|\mathcal{F}_0)$), with $f \in H_p^S$ and stopped at n (notice that in this case $\theta_\nu = 0$, $\nu \geq n+1$, and $\theta_\nu = E(f_n|\mathcal{F}_\nu) - E(f_n|\mathcal{F}_{\nu-1})$, $0 \leq \nu \leq n$), we have

$$l(f_n) = l(\theta) = \sum_{\nu=0}^{n} E(\bar{\sigma}_\nu(E(f_n|\mathcal{F}_\nu) - E(f_n|\mathcal{F}_{\nu-1})))$$
$$= \sum_{\nu=0}^{n} E(f_n[E(\bar{\sigma}_\nu|\mathcal{F}_\nu) - E(\bar{\sigma}_\nu|\mathcal{F}_{\nu-1})]) = E(f_n\bar{\varphi}_n),$$

where $\varphi = (\varphi_n)_{n\geq 0}$, $\varphi_n = \sum_{\nu=0}^{n}(E(\sigma_\nu|\mathcal{F}_\nu) - E(\sigma_\nu|\mathcal{F}_{\nu-1}))$. From Lemma 2.2.4, $\varphi \in {}_2K_{p'}$, and $\|\varphi\|_{{}_2K_{p'}} \leq 2p\|l\|$. Since $\|f - f_n\|_{H_p^S} \to 0$, we have

$$l(f) = \lim_{n\to\infty} l(f_n) = \lim_{n\to\infty} E(f_n\bar{\varphi}_n), \quad \forall f \in H_p^S.$$

The proof of the theorem is finished. □

2.3 From Fefferman's inequality to Davis' inequality

We will see in the next chapter that Davis' inequality is a fundamental inequality from which the L^p- and even the L^Φ-inequlity between M and S follow. We will show in this section that Fefferman's inequality implies Davis'. First, we want to indicate that all inequalities we obtained in §2.1 have their conditional versions. For example, we have

Proposition 2.3.1 *For all martingales $f = (f_n)_{n\geq 0}$ with respect to $\{\mathcal{F}_n\}_{n\geq 0}$, we have*

$$|E_0(f_n\bar{\varphi}_n)| \leq \sqrt{2}E_0(S(f))||\varphi||_{BMO_2}, \quad a.e. \tag{2.3.1}$$

The proof is just the same as in Theorem 2.2.2.

We need some criterions for $f \in BMO_2$.

Lemma 2.3.2 *Let $\{\theta_\nu\}_{\nu\geq 0}$ and $\{\varepsilon_\nu\}_{\nu\geq 0}$ be two sequences of random variables satisfying*

$$\{\theta_\nu\}_{\nu\geq 0} \quad \textit{adapted and} \quad |\theta_\nu| \leq 1, \quad \forall\nu, \quad \textit{and} \quad \sum_0^\infty |\varepsilon_\nu| \leq B.$$

Then

$$\varphi = \sum_{\nu=0}^\infty \theta_\nu E(\varepsilon_\nu|\mathcal{F}_\nu)$$

is in BMO_2, and

$$||\theta||_{BMO_2} \leq \sqrt{5}B. \tag{2.3.2}$$

Proof. The series defining φ is convergent, a.e. absolutely, since

$$E(|\varphi|) \leq \sum_{\nu=0}^\infty E(E(|\varepsilon_\nu||\mathcal{F}_\nu)) = E\Big(\sum_{\nu=1}^\infty |\varepsilon_\nu|\Big) \leq B.$$

Define

$$\psi_n = \sum_{\nu=n}^\infty \theta_\nu E(\varepsilon_\nu|\mathcal{F}_\nu), \quad n \geq 0.$$

We have

$$\varphi - \varphi_{n-1} = \psi_n - E(\psi_n|\mathcal{F}_{n-1}), \quad n \geq 0,$$

(the term with index -1 is meant 0 as usual), and hence

$$\begin{aligned} E(|\varphi - \varphi_{n-1}|^2|\mathcal{F}_n) \leq{}& E(|\psi_n|^2|\mathcal{F}_n) + |E(\psi_n|\mathcal{F}_{n-1})|^2 \\ &+ 2|E(\psi_n|\mathcal{F}_n)E(\psi_n|\mathcal{F}_{n-1})|. \end{aligned}$$

Since

$$E(|\psi_n||\mathcal{F}_n) \leq E\Big(\sum_{\nu=n}^{\infty} E(|\varepsilon_\nu||\mathcal{F}_\nu)\Big|\mathcal{F}_n\Big) = E\Big(\sum_{\nu=n}^{\infty} |\varepsilon_\nu|\Big|\mathcal{F}_n\Big) \leq B,$$

we have also $E(|\psi_n||\mathcal{F}_{n-1}) \leq B$. So it remains to estimate $E(|\psi_n|^2|\mathcal{F}_n)$. We have

$$\begin{aligned} E(|\psi_n|^2|\mathcal{F}_n) &\leq \sum_{\nu=n}^{\infty}\sum_{\mu=n}^{\infty} E(E(|\varepsilon_\nu||\mathcal{F}_\nu)E(|\varepsilon_\mu||\mathcal{F}_\mu)|\mathcal{F}_n) \\ &\leq 2\sum_{\nu=n}^{\infty}\sum_{\mu=\nu}^{\infty} E(E(|\varepsilon_\nu|\,|\mathcal{F}_\nu)|\varepsilon_\mu||\mathcal{F}_n) \\ &\leq 2B\sum_{\nu=n}^{\infty} E(E(|\varepsilon_\nu||\mathcal{F}_\nu)|\mathcal{F}_n) \leq 2B^2. \end{aligned}$$

Thus, we get

$$E(|\varphi - \varphi_{n-1}|^2|\mathcal{F}_n) \leq 5B^2.$$

(2.3.2) follows. The proof is finished. □

Remark We will show in §4.2 that each $\varphi \in BMO_2$ has such an expression with $\theta_\nu \equiv 1$, for all ν.

Lemma 2.3.3 *Let $\{\theta_\nu\}$ be a sequence of random variables. Then for all $\varepsilon > 0$,*

$$\varphi = \sum_{\nu=0}^{\infty} \theta_{\nu-1}\left\{E\left(\frac{1}{M\theta+\varepsilon}\Big|\mathcal{F}_\nu\right) - E\left(\frac{1}{M\theta+\varepsilon}\Big|\mathcal{F}_{\nu-1}\right)\right\} \tag{2.3.3}$$

is in BMO_2, and $\|\varphi\|_{BMO_2} \leq \sqrt{2}$.

Proof. Denote

$$g_\nu = E\left(\frac{1}{M\theta+\varepsilon}\Big|\mathcal{F}_\nu\right), \quad \Delta_\nu g = g_\nu - g_{\nu-1},\ \nu \geq 0,$$

$$\varphi_n = \sum_{\nu=0}^{n} \theta_{\nu-1}\Delta_\nu g, \quad n \geq 0 \quad (\varphi_0 = 0).$$

Then $\varphi = (\varphi_n)_{n\geq 0}$ is a martingale. For $N \geq n$, we have

$$\begin{aligned} E(|\varphi_N - \varphi_{n-1}|^2|\mathcal{F}_n) &= \sum_{\nu=n}^{N} E(|\theta_{\nu-1}|^2|\Delta_\nu g|^2|\mathcal{F}_n) \\ &\leq \sum_{\nu=n}^{N} E((M_{\nu-1}\theta)^2|\Delta_\nu g|^2|\mathcal{F}_n). \end{aligned}$$

When $\nu = n$, we have

$$E((M_{n-1}\theta)^2|\Delta_n g|^2|\mathcal{F}_n)$$
$$\leq E\Big((M_{n-1}\theta)^2 \max\Big(E\Big(\frac{1}{M\theta+\varepsilon}\Big|\mathcal{F}_n\Big)^2, E\Big(\frac{1}{M\theta+\varepsilon}\Big|\mathcal{F}_{n-1}\Big)^2\Big)\Big|\mathcal{F}_n\Big) \leq 1.$$

When $\nu \geq n+1$, we have

$$\begin{aligned} E((M_{\nu-1}\theta)^2|\Delta_\nu g|^2|\mathcal{F}_n) &= E((M_{\nu-1}\theta)^2 E(|\Delta_\nu g|^2|\mathcal{F}_{\nu-1})|\mathcal{F}_n) \\ &= E((M_{\nu-1}\theta)^2\{E(|g_\nu|^2|\mathcal{F}_{\nu-1}) - |g_{\nu-1}|^2\}|\mathcal{F}_n) \\ &= E((M_{\nu-1}\theta)^2\{|g_\nu|^2 - |g_{\nu-1}|^2\}|\mathcal{F}_n) \\ &\leq E((M_\nu\theta)^2|g_\nu|^2 - (M_{\nu-1}\theta)^2|g_{\nu-1}|^2|\mathcal{F}_n). \end{aligned}$$

Thus, we get

$$\sum_{\nu=n}^{N} E((M_{\nu-1}\theta)^2|\theta_\nu g|^2|\mathcal{F}_n)$$
$$\leq 1 + \sum_{\nu=n+1}^{N}\{E((M_\nu\theta)^2|g_\nu|^2|\mathcal{F}_n) - E((M_{\nu-1}\theta)^2|g_{\nu-1}|^2|\mathcal{F}_n)\}$$
$$\leq 1 + E((M_N\theta)^2|g_N{}^2|\mathcal{F}_n).$$

Since

$$M_N\theta g_N = E\Big(\frac{M_N\theta}{M\theta+\varepsilon}\Big|\mathcal{F}_n\Big) \leq 1,$$

we get $E(|\varphi_N - \varphi_{n-1}|^2|\mathcal{F}_n) \leq 2$. The proof of the lemma is finished. □

Now we can get Davis' inequality in conditional form by making use of Fefferman's inequality.

Theorem 2.3.4 *For all martingales $f = (f_n)_{n\geq 0}$, we have*

$$CE_0(Mf) \leq E_0(S(f)) \leq CE_0(Mf). \tag{2.3.4}$$

Proof. Without loss of generality, we can assume that f is a finite martingale. Let $f = (f_m)_{m\geq 0}$ be a martingale stopped at n. Set

$$F_\nu = \{M_{\nu-1}f < |f_\nu| = M_n f\},$$

$$\theta_\nu = \operatorname{sgn} f_\nu = \bar{f}_\nu/|f_\nu|.$$

Then we have

$$\begin{aligned} E_0(M_n f) &= \sum_{\nu=0}^{n} E_0(f_\nu\theta_\nu E(\chi_{F_\nu}|\mathcal{F}_\nu)) \\ &= \sum_{\nu=0}^{n} E_0(f_n\theta_\nu E(\chi_{F_\nu}|\mathcal{F}_\nu)) = E_0(f_n\varphi_n), \end{aligned}$$

where

$$\varphi_n = \sum_{\nu=0}^{n} \theta_\nu E(\chi_{F_\nu} | \mathcal{F}_\nu),$$

which is not a martingale. But for each n, φ_n could yield a BMO_2 martingale as shown in Lemma 2.3.2, of which the norm is

$$\|\varphi\|_{BMO_2} \leq \sqrt{5} \left\| \sum_{\nu=0}^{n} \chi_{F_\nu} \right\|_\infty \leq \sqrt{5}.$$

From Fefferman's inequality formulated in Proposition 2.3.1, we see

$$|E_0(f_n \varphi_n)| \leq \sqrt{2}\sqrt{5} E_0(S(f_n)) \leq \sqrt{10} E_0(S(f)).$$

This proves the left-hand side inequality of (2.3.4).

Now turn to the right-hand side inequality of (2.3.4). We have

$$\begin{aligned} S_n(f)^2 &= \sum_0^n (f_k - f_{k-1})(\bar{f}_k - \bar{f}_{k-1}) \\ &= |f_n|^2 + \sum_0^n f_{k-1}(\bar{f}_{k-1} - \bar{f}_k) + \sum_0^n \bar{f}_{k-1}(f_{k-1} - f_k), \end{aligned}$$

and hence

$$E_0\left(\frac{(S_n f)^2}{Mf + \varepsilon}\right) \leq E_0\left(\frac{|f_n|^2}{Mf + \varepsilon}\right) + 2\left| E_0\left(\sum_0^n \bar{f}_{k-1}(f_{k-1} - f_k)/(Mf + \varepsilon)\right)\right|.$$

But we have

$$\begin{aligned} \left|\sum_0^n E_0\left(\frac{\bar{f}_{k-1}\Delta_k f}{Mf + \varepsilon}\right)\right| &= \left|\sum_0^n E_0\left(\bar{f}_{k-1}\Delta_k f \left\{ E\left(\frac{1}{Mf + \varepsilon}\Big|\mathcal{F}_k\right)\right.\right.\right. \\ &\qquad \left.\left.\left. - E\left(\frac{1}{Mf + \varepsilon}\Big|\mathcal{F}_{k-1}\right)\right\}\right)\right| = E_0(f_n \varphi_n), \end{aligned}$$

where

$$\varphi_n = \sum_{k=0}^{n} \bar{f}_{k-1}\left\{ E\left(\frac{1}{Mf + \varepsilon}\Big|\mathcal{F}_k\right) - E\left(\frac{1}{Mf + \varepsilon}\Big|\mathcal{F}_{k-1}\right)\right\}, \quad n \geq 0,$$

is just a BMO_2 martingale with its norm $\|\varphi\|_{BMO_2} \leq \sqrt{2}$, from Lemma 2.3.3. Making use of Fefferman's inequality again, we get

$$E_0\left(\frac{S_n(f)^2}{Mf + \varepsilon}\right) \leq E_0(Mf) + 4E_0(S_n(f)),$$

$$\begin{aligned} E_0(S_n(f)) &\leq E_0(Mf + \varepsilon)^{\frac{1}{2}} E_0\left(\frac{S_n(f)^2}{Mf + \varepsilon}\right)^{\frac{1}{2}} \\ &\leq E_0(Mf + \varepsilon)^{\frac{1}{2}}(E_0(Mf) + 4E_0(S_n(f))^{\frac{1}{2}}, \end{aligned}$$

$$E_0(S(f)) \leq (2+\sqrt{5})E_0(Mf).$$

The proof of the theorem is finished. □

2.4 Davis' decomposition for H_p, the Spaces $\mathcal{P}_p$ and $\mathcal{A}_p$

Definition 2.4.1 *The adapted process $(\lambda_n)_{n\geq 0}$ is said to be L^p-predictable, $0 < p \leq \infty$, if there is some nonnegative, adapted and increasing process $(\gamma_n)_{n\geq 0}$ such that*

$$|\lambda_n| \leq \gamma_{n-1}, \quad n \geq 0, \quad E(\gamma_\infty^p) < \infty. \tag{2.4.1}$$

And define

$$\mathcal{P}_p = \{\textit{martingale } f = (f_n)_{n\geq 0} : f \textit{ is } L^p\textit{-predictable} \},$$

$$\|f\|_{\mathcal{P}_p} = \inf_\gamma\{\|\gamma_\infty\|_p\}, \quad 0 < p \leq \infty. \tag{2.4.2}$$

Remark "inf" in the definition of $\|\cdot\|_{\mathcal{P}_p}$ is attainable. In fact, let $\{(\gamma_n^{(k)})_{n\geq 0}\}_k$ be a family of admissible controls of f such that $\lim_k \|\gamma_\infty^{(k)}\|_p = \|f\|_{\mathcal{P}_p}$. Then $(\gamma_n^{(f)})_{n\geq 0}$ with $\gamma_n^{(f)} = \inf_k \gamma_n^{(k)}$, is an admissible control of f satistying $\|f\|_{\mathcal{P}_p} = \|\gamma_\infty^{(f)}\|_p$, which will be called the optimal control of f. When $1 \leq p \leq \infty$, $\mathcal{P}_p$ is a Banach space. Only the completeness has to be proved. Let $\{f^{(k)}\}$ be a Cauchy sequence in $\mathcal{P}_p$. Choose a subsequence $\{k_j\}$ and a family $\{(\gamma_n^{(j)})_{n\geq 0}\}$ such that

$$|f_n^{(k_{j+1})} - f_n^{(k_j)}| \leq \gamma_{n-1}^{(j)}, \quad \forall n,\ \forall j,$$

$$\|\gamma_\infty^{(j)}\|_p \leq 2^{-j}.$$

Then $\sum_j (f_\infty^{(k_{j+1})} - f_\infty^{(k_j)})$ converges to $f_\infty - f_\infty^{(k_1)}$, and $f_n = \lim_j f_n^{(k_j)}$ in L^p. But by taking a subsequence again, we could have

$$f_n = \lim_j f_n^{(k_j)}, \quad \text{a.e.}$$

So, we get

$$|f_n - f_n^{(k_l)}| \leq \sum_{j\geq l} \gamma_{n-1}^{(j)}, \quad \forall n,\ \forall l.$$

Since $\left\| \sum_{j\geq l} \gamma_\infty^{(j)} \right\|_p \leq 2^{-l+1}$, $\left(\sum_{j\geq l} \gamma_n^{(j)}\right)_{n\geq 0}$ is an admissible control of $f - f^{(k_l)}$, and $\|f - f^{(k_l)}\|_{\mathcal{P}_p} \leq 2^{-l+1} \to 0$. Since $\{f^{(k)}\}$ is a Cauchy sequence, we have $\|f - f^{(k)}\|_{\mathcal{P}_p} \to 0$. This proves the assertion.

Definition 2.4.2 *An adapted process* $(\lambda_n)_{n\geq 0}$ *is said to be* L^p*-variation integrable, if* $\|\sum|\Delta_n\lambda|\|_p < \infty$. *Define*

$$\mathcal{A}_p = \Big\{\text{martingale } f = (f_n)_{n\geq 0}: \|f\|_{\mathcal{A}_p} = \Big\|\sum|\Delta_n f|\Big\|_p < \infty\Big\}, \quad 1 \leq p \leq \infty. \tag{2.4.3}$$

By making use of the idea of Davis' decomposition, H_p, $1 \leq p < \infty$, could be expressed as a sum of $\mathcal{P}_p$ and $\mathcal{A}_p$. We need a L^p-estimate for $\sum_\nu E(\varepsilon_\nu|\mathcal{F}_\nu)$ formulated as a lemma.

Lemma 2.4.3 *Let* $(\varepsilon_\nu)_{\nu\geq 0}$ *be a nonnegative process, not necessarily being adapted. Then*

$$\Big\|\sum_\nu E(\varepsilon_\nu|\mathcal{F}_\nu)\Big\|_p \leq p\Big\|\sum_\nu \varepsilon_\nu\Big\|_p, \quad 1 \leq p < \infty. \tag{2.4.4}$$

Proof. We have

$$\begin{aligned}\Big\|\sum_\nu E(\varepsilon_\nu|\mathcal{F}_\nu)\Big\|_p &= \sup_{\gamma:\|\gamma\|_{p'}\leq 1}\Big|E\Big(\sum_\nu E(\varepsilon_\nu|\mathcal{F}_\nu)\gamma\Big)\Big| \\ &= \sup_\gamma\Big|\sum_\nu E(\varepsilon_\nu E(\gamma|\mathcal{F}_\nu))\Big| \leq \sup_\gamma E\Big(\sum_\nu \varepsilon_\nu M\gamma\Big) \\ &\leq \sup_\gamma \|M\gamma\|_{p'}\Big\|\sum_\nu \varepsilon_\nu\Big\|_p \leq p\Big\|\sum_\nu \varepsilon_\nu\Big\|_p.\end{aligned}$$

The proof is complete. □

Remark For $\sum_\nu E(\varepsilon_\nu|\mathcal{F}_\nu)$, more general Φ-estimates could be done. See Ch. 3.

Theorem 2.4.4 *Let* $1 \leq p < \infty$. *Then each* $f \in H_p$ *could be decomposed as* $f = g + h$, *with* $g \in \mathcal{P}_p$, $h \in \mathcal{A}_p$, *and*

$$\|g\|_{\mathcal{P}_p} \leq (13 + 4p)\|f\|_{H_p}, \tag{2.4.5}$$

$$\|h\|_{\mathcal{A}_p} \leq (4 + 4p)\|f\|_{H_p}. \tag{2.4.6}$$

Proof. The proof of Davis' decomposition will work well for this task. Let $(\lambda_n)_{n\geq 0}$ be any increasing control process of $f = (f_n)_{n\geq 0}$, that is $|f_n| \leq \lambda_n$, for all n. Define

$$\Delta_n h = \Delta_n f\chi_{\{\lambda_n > 2\lambda_{n-1}\}} - E(\Delta_n f\chi_{\{\lambda_n > 2\lambda_{n-1}\}}|\mathcal{F}_{n-1}), \quad n \geq 0,$$

$$\Delta_n g = \Delta_n f\chi_{\{\lambda_n \leq 2\lambda_{n-1}\}} - E(\Delta_n f\chi_{\{\lambda_n \leq 2\lambda_{n-1}\}}|\mathcal{F}_{n-1}), \quad n \geq 0.$$

Then the martingales $h = (h_n)_{n\geq 0}$, $h_n = \sum_0^n \Delta_k h$, and $g = (g_n)_{n\geq 0}$, $g_n = \sum_k^n \Delta_k g$ are just what we want. In fact, when $\lambda_n > 2\lambda_{n-1}$, we have

$$\lambda_n = 2\lambda_n - \lambda_n \leq 2\lambda_n - 2\lambda_{n-1},$$

so

$$|\Delta_n h| \le 2\lambda_n \chi_{\{\lambda_n > 2\lambda_{n-1}\}} + 2E(\lambda_n \chi_{\{\lambda > 2\lambda_{n-1}\}} | \mathcal{F}_{n-1})$$
$$\le 4(\lambda_n - \lambda_{n-1}) + 4E(\lambda_n - \lambda_{n-1} | \mathcal{F}_{n-1}), \quad n \ge 0. \tag{2.4.7}$$

Meanwhile,

$$|\Delta_n g| \le 8\lambda_{n-1},$$

$$|g_n| \le |f_{n-1}| + |h_{n-1}| + |\Delta_n g|$$
$$\le 9\lambda_{n-1} + 4\lambda_{n-1} + 4\sum_0^{n-1} E(\lambda_k - \lambda_{k-1} | \mathcal{F}_{k-1})$$
$$\le 13\lambda_{n-1} + 4\sum_0^{n-1} E(\lambda_k - \lambda_{k-1} | \mathcal{F}_{k-1}). \tag{2.4.8}$$

Denote

$$\gamma_n = \sum_0^n E(\lambda_k - \lambda_{k-1} | \mathcal{F}_{k-1}), \quad n \ge 0.$$

From Lemma 2.4.3, we see

$$||\gamma_\infty||_p \le p \Big\| \sum_k (\lambda_k - \lambda_{k-1}) \Big\|_p = p||\lambda_\infty||_p.$$

From (2.4.7) and (2.4.8), we get

$$||h||_{\mathcal{A}_p} \le 4||\lambda_\infty||_p + 4p||\lambda_\infty||_p = (4 + 4p)||\lambda_\infty||_p,$$

$$||g||_{\mathcal{P}_p} \le 13||\lambda_\infty||_p + 4p||\lambda_\infty||_p = (13 + 4p)||\lambda_\infty||_p.$$

Since $f \in H_p$, $(\lambda_n)_{n \ge 0}$ could be taken as $(M_n f)_{n \ge 0}$. This completes the proof of the theorem. □

2.5 Another proof of Fefferman's theorem (via Davis' decomposition and atomic decomposition)

Davis' decomposition of H_1 gives us another proof of Fefferman's H_1-BMO duality theorem, once we get the duals of $\mathcal{P}_1$ and $\mathcal{A}_1$. For the former, we need the concept of atoms and the atomic decomposition of $\mathcal{P}_1$. This topic will be studied a litter more in §7.2.1.

Definition 2.5.1 *A martingale $f = (f_n)_{n \ge 0}$ is said to be of jump bounded, if $\sup_n ||\Delta_n f||_\infty < \infty$. Define*

$$BD = \{\textit{martingale } f = (f_n)_{n \ge 0} : ||f||_{BD} = \sup_n ||\Delta_n f||_\infty < \infty\}. \tag{2.5.1}$$

Theorem 2.5.2 *The dual of $\mathcal{A}_1$ is BD. That is to say, each $\varphi \in BD$ yields a bounded linear functional l_φ satisfying $\|l_\varphi\| \leq \|\varphi\|_{BD}$; on the contrary, each bounded linear Functional l on $\mathcal{A}_1$ just comes from some $\varphi \in BD$ and $\|\varphi\|_{BD} \leq 2\|l\|$. The action of l_φ on $f \in \mathcal{A}_1$ is as follows*

$$l_\varphi(f) = E\Big(\sum_0^\infty \Delta_n f \Delta_n \varphi\Big). \tag{2.5.2}$$

Proof. Let $\varphi \in BD$, $f \in \mathcal{A}_1$. Then $E(\sum_n |\Delta_n f \Delta_n \varphi|) < \infty$, and hence (2.5.2) defines a bounded linear functional l_φ on $\mathcal{A}_1$, and $\|l_\varphi\| \leq \|\varphi\|_{BD}$. Now we prove the inverse. Consider a larger space

$$L^1(l^1) = \Big\{\text{process} \quad \xi = (\xi_n)_{n\geq 0} : \ \|\xi\|_{L^1(l^1)} = E\Big(\sum_0^\infty |\xi_n|\Big) < \infty\Big\}. \tag{2.5.3}$$

It is obviously a Banach space, and its dual space is

$$L^\infty(l^\infty) = \{\text{process} \quad \eta = (\eta_n)_{n\geq 0} : \|\eta\|_{L^\infty(l^\infty)} = \sup_n \|\eta_n\|_\infty < \infty\}. \tag{2.5.4}$$

This can be seen as follows, say to prove $L^1(l^1)' \subset L^\infty(l^\infty)$. Let $l \in L^1(l^1)'$, $f \in L^1$. Define $l_n(f) = l(\xi)$, where $\xi = (0, \cdots f, 0 \cdots)$(all the components except the n-th, are 0). We have

$$|l_n(f)| = |l(\xi)| \leq \|l\| \, \|\xi\|_{L^1(l^1)} = \|l\| \, \|f\|_1.$$

So, there exists a $\eta_n \in L^\infty$, such that

$$l_n(f) = E(f\eta_n), \quad \|\eta_n\|_\infty \leq \|l\|.$$

Now for any $\xi \in L^1(l^1)$, consider $\xi^{(n)} = (\xi_0, \xi_1, \cdots, \xi_n, 0 \cdots) \in L^1(l^1)$. We have

$$l(\xi^{(n)}) = \sum_{k=0}^n E(\xi_k \eta_k), \quad \forall n.$$

Since $\|\xi - \xi^{(n)}\|_{L^1(l^1)} \to 0$, we have

$$l(\xi) = \lim_{n\in\infty} \sum_0^n E(\xi_k\eta_k) = \sum_0^\infty E(\xi_k\eta_k).$$

This shows that $\eta = (\eta_k)_{k\geq 0}$, which is in $L^\infty(l^\infty)$, makes $l(\xi) = \sum_0^\infty E(\xi_k\eta_k)$, and $\|\eta\|_{L^\infty(l^\infty)} \leq \|l\|$.

Now return to the proof of the assertion $\mathcal{A}_1' \subset BD$. $\mathcal{A}_1$ is embedded isometrically into $L^1(l^1)$ with the image

$$\{\text{process} \quad \xi = (\xi_n)_{n\geq 0} \in L^1(l^1) : \xi_n = \Delta_n f, \ \text{for some } f = (f_n)_{n\geq 0} \in \mathcal{A}_1\}.$$

So each $l \in \mathcal{A}_1'$ could be extended to whole $L^1(l^1)$ preserving the same bound, in symbols l still. From the preceding arguments, we see that there exists $\eta = (\eta_n)_{n\geq 0} \in L^\infty(l^\infty)$ such that for all $f = (f_n)_{n\geq 0} \in \mathcal{A}_1$,

$$l(f) = \sum_0^\infty E(\Delta_n f \eta_n) = \sum_0^\infty E(\Delta_n f\{E(\eta_n|\mathcal{F}_n) - E(\eta_n|\mathcal{F}_{n-1})\})$$

$$= \sum_{0}^{\infty} E(\Delta_n f \Delta_n g),$$

where

$$g = (g_n)_{n \geq 0}, \quad \Delta_n g = E(\eta_n | \mathcal{F}_n) - E(\eta_n | \mathcal{F}_{n-1}), \quad n \geq 0,$$

satisfies $\|g\|_{BD} \leq 2\|\eta\|_{L^\infty(l^\infty)} \leq 2\|l\|$. The proof is finished. □

We now discuss the dual space of $\mathcal{P}_1$ by making use of the atomic decomposition of $\mathcal{P}_1$. The classical atomic decomposition says that H_1 is linearly generated by so-called atoms. In martingale setting the corresponding fact holds for $\mathcal{P}_1$ but not for H_1 in general. A so-called classical atom is a simple function on $\mathbf{R}^n$ which is supported in a cube Q, has a vanishing moment, and has its size like $|Q|^{-1}$ in some suitable sense. What is the martingale version of atoms? When we introduced the concept of stopping time in §1.2, we mentioned that it is such a concept which is intimately related to the classical Calderón-Zygmund decomposition. The latter tells us that each $f \in L^1$ could be decomposed as a sum of a good function g and a bad function h, the latter of which is supported on a disjoint union $\bigcup_k Q_k$ of cubes, and $h_{Q_k} = 0$ (h_{Q_k} denotes the average of h on Q_k), for all k. In martingale setting, each $f \in L^1_u$ could be written as $f = f^{(\tau)} + (f - f^{(\tau)}) = g + h$, where τ is some stopping time, and g is good and h is bad, and the latter is supported on the set $\{\tau < \infty\}$, and has vanishing random average $h_\tau = (f - f^{(\tau)})_\tau = 0$. From this fact, we see that a disjont union of cubes has its counterpart $\{\tau < \infty\}$ in martingale setting. Suggested by this we have following definition.

Definition 2.5.3 *A bounded measurable function a is called a p-atom, $0 < p \leq 1$, if there is a stopping time τ such that*

$$a_n \chi_{\{n \leq \tau\}} = 0, \quad \forall n, \tag{2.5.5}$$

$$\|a\|_\infty \leq |\{\tau < \infty\}|^{-\frac{1}{p}}. \tag{2.5.6}$$

Remark The condition (2.5.5) is equivalent to a simpler one, that is

$$a_\tau = 0. \tag{2.5.7}$$

(2.5.5) obviously implies (2.5.7). On the contrary, we have

$$a_\tau \chi_{\{\tau < n\}} + a_n \chi_{\{\tau \geq n\}} = a_{\tau \wedge n} = E(a_\tau | \mathcal{F}_n), \quad \forall n.$$

So, (2.5.7) implies $a_\tau \chi_{\{\tau < n\}} = 0 = E(a_\tau | \mathcal{F}_n)$, and hence $a_n \chi_{\{n \leq \tau\}} = 0$ for all n. The condition (2.5.7) means that

$$\operatorname{supp} a \subset \{\tau < \infty\},$$

and the random average $a_\tau = 0$. In addition, all atoms a fulfil $a_0 = 0$, a.e., which follows from (2.5.5).

Lemma 2.5.4 *All* 1*-atoms are in the unit ball of* $\mathcal{P}_1$.

Proof. Let a be an atom and τ be its associated stopping time. Consider the process $(\gamma_n)_{n\geq 0}$ with $\gamma_n = \|a\|_\infty \chi_{\{\tau<n+1\}}$. Then it is an adapted increasing and nonnegative process and

$$E(\gamma_\infty) = E(\|a\|_\infty \chi_{\{\tau<\infty\}}) \leq 1,$$

$$|a_n| = |a_n|\chi_{\{n\leq\tau\}} + |a_n|\chi_{\{\tau<n\}} = |a_n|\chi_{\{\tau<n\}} \leq \gamma_{n-1}, \quad \forall n \geq 0.$$

This proves $a \in \mathcal{P}_1$ and $\|a\|_{\mathcal{P}_1} \leq 1$. The proof is finished. □

The dual space of $\mathcal{P}_1$ is a version of BMO. Now we define it.

Definition 2.5.5 *Denote*

$$bmo_1 = \{ \textit{ martingale } f = (f_n)_{n\geq 0} \in L_u^1 : \\ \|f\|_{bmo_1} = \sup_n \|E(|f - f_n||\mathcal{F}_n)\|_\infty < \infty\}.$$

Remark bmo_1 is a linear space of the equivalence classes of martingales (modulo $\mathcal{F}_0$-martingales).

Lemma 2.5.6 *We have*

$$\|f\|_{bmo_1} = \sup_n \|f - f_\tau\|_1 |\{\tau < \infty\}|^{-1}, \tag{2.5.8}$$

where τ *runs over all stopping times.*

Proof. Let $f \in bmo_1$, τ be any stopping time. Then

$$\|E(|f - f_\tau||\mathcal{F}_\tau)\|_\infty \leq \|f\|_{bmo_1},$$

and

$$\|f - f_\tau\|_1 = \int_{\{\tau<\infty\}} |f - f_\tau| d\mu \\ = \int_{\{\tau<\infty\}} E(|f - f_\tau||\mathcal{F}_\tau) d\mu \leq \|f\|_{bmo_1} |\{\tau < \infty\}|.$$

This proves

$$\sup_\tau \|f - f_\tau\|_1 |\{\tau < \infty\}|^{-1} \leq \|f\|_{bmo_1}.$$

On the contrary, denote $\beta = \sup_\tau \|f - f_\tau\|_1 |\{\tau < \infty\}|^{-1} < \infty$. Let τ be any stopping time and $F \in \mathcal{F}_\tau$. Define the stopping time

$$\tau_F = \begin{cases} \tau, & x \in F, \\ \infty, & x \notin F. \end{cases}$$

We have

$$\frac{1}{|F|}\int_F |f-f_\tau|d\mu \le \|f-f_{\tau_F}\|_1|\{\tau_F<\infty\}|^{-1}\le \beta,$$
$$\|E(|f-f_\tau||\mathcal{F}_\tau)\|_\infty \le \beta.$$

This proves the remaining half of the assertion. The proof is finished. □

The action of 1-atoms on bmo_1, or reciprocally the action of bmo_1 on 1-atoms, is very natural and simple, as shown by following lemma. Before giving the lemma, we want to show that for all 1-atmos a and $\varphi \in bmo_1$, $E(a\varphi)$ has meaning which is independent of the choice of φ in it's equivalence class. In fact for all φ_0 being $\mathcal{F}_0$-measurable, $E(a\varphi_0) = E(a_0\varphi_0) = 0$, and hence $E(a\varphi) = E(a(\varphi+\varphi_0))$.

Lemma 2.5.7 *Let a be 1-atom, and $\varphi \in L_u^1$. Then*

$$\frac{1}{2}\|\varphi\|_{bmo_1} \le \sup_{1\text{-atom } a} |E(a\varphi)| \le \|\varphi\|_{bmo_1}. \tag{2.5.9}$$

Proof. Let a be a 1-atom, τ be its associated stopping time. Then we have

$$\begin{aligned} E(a\varphi) &= |E((a-a_\tau)\varphi)| = |E(a\varphi)-E(a_\tau\varphi_\tau)| = |E(a(\varphi-\varphi_\tau))| \\ &\le \|\varphi-\varphi_\tau\|_1|\{\tau<\infty\}|^{-1} \le \|\varphi\|_{bmo_1}. \end{aligned}$$

On the contrary, let τ be any stopping time and $\varphi \in L_u^1$. Set $f = \operatorname{sgn}(\varphi-\varphi_\tau)$, $a = \frac{1}{2|\{\tau<\infty\}|}(f-f_\tau)$, then a is a 1-atom, and

$$E(|\varphi-\varphi_\tau|) = E(f(\varphi-\varphi_\tau)) = E((f-f_\tau)\varphi) = 2|\{\tau<\infty\}|E(a\varphi).$$

This means

$$\|\varphi\|_{bmo_1} = \sup_\tau \|\varphi-\varphi_\tau\|_1|\{\tau<\infty\}|^{-1} \le 2\sup_a |E(a\varphi)|.$$

The proof is finished. □

Now we show that all 1-atoms generate $\mathcal{P}_1$.

Theorem 2.5.8 *Denote*

$$H_1^{(a)} = \Big\{f\in L^1 : f=\sum_1^\infty \lambda_j a_j, \{a_j\}_{j\ge1} \ \textit{sequence of 1-atoms, and}$$
$$\{\lambda_j\}_{j\ge1} \ \textit{sequence of positive numbers}, \ \sum_1^\infty \lambda_j<\infty\Big\}. \tag{2.5.10}$$

Then $\mathcal{P}_1 = H_1^{(a)}$, and for $f\in\mathcal{P}_1$, there exists such decomposition satisfying

$$\frac{1}{8}\sum_1^\infty \lambda_j \le \|f\|_{\mathcal{P}_1} \le \sum_1^\infty \lambda_j. \tag{2.5.11}$$

Proof. Lemma 2.5.4 tells us that each series $\sum_1^\infty \lambda_j a_j$ as above converges to a $f \in \mathcal{P}_1$ and $\|f\|_{\mathcal{P}_1} \leq \sum_1^\infty \lambda_j$. Now prove that for all $f \in \mathcal{P}_1$, there exists a suitable decomposition $f = \sum_1^\infty \lambda_j a_j$ such that $\sum_1^\infty \lambda_j \leq 8\|f\|_{\mathcal{P}_1}$. Let $f \in \mathcal{P}_1$ and $(r_n)_{n\geq o}$ be any nonnegative increasing and adapted process such that $|f_n| \leq r_{n-1}$, $n \geq 0$, and $\|r_\infty\|_1 < \infty$. For all $j \in \mathbf{Z}$, define the stopping time $T_j = \inf\{n : r_n > 2^j\}$. Then $\{T_j\}$ is an increasing sequence of stopping times such that $T_j \to \infty$, a.e. otherwise we would have $T_j \leq M$ on a set E of positive measure, for all j, this means $r_M \geq 2^j$, for all j on E, this would contradict to $\|r_\infty\|_1 < \infty$. And hence $\lim_{j\to\infty} f_{T_j} = f$, a.e. Meanwhile we have

$$\lim_{j\to-\infty} |f_{T_j}| \leq \lim_{j\to-\infty} r_{T_j-1} \leq \lim_{j\to-\infty} 2^j = 0.$$

Thus we get f's decomposition

$$f = \sum_{j=-\infty}^{\infty} (f_{T_j} - f_{T_{j-1}})\chi_{\{T_{j-1}<\infty\}} = \sum_{j=-\infty}^{\infty} \lambda_j a_j = \sum_{k=1}^{\infty} \mu_k b_k,$$

where

$$\lambda_j = 2^{j+1}|\{T_{j-1} < \infty\}|,$$
$$a_j = 2^{-j-1}|\{T_{j-1} < \infty\}|^{-1}\chi_{\{T_{j-1}<\infty\}}(f_{T_j} - f_{T_{j-1}}), \quad \forall j.$$

Noting that each a_j is a 1-atom with T_{j-1} as its associated stopping time, and (denoting $\sigma(\lambda) = |\{r_\infty > \lambda\}|$)

$$\sum_{-\infty}^{\infty} \lambda_j = \sum_{-\infty}^{\infty} 2^{j+1}|\{T_{j-1} < \infty\}| = \sum_{-\infty}^{\infty} 2^{j+1}\sigma(2^{j-1})$$
$$\leq 8\sum_{-\infty}^{\infty} \int_{2^{j-1}}^{2^j} \sigma(\lambda)d\lambda = 8\|r_\infty\|_1.$$

In particular, taking $(r_n)_{n\geq 0}$ such that $\|f\|_{\mathcal{P}_1} = \|r_\infty\|_1$, we get $\sum\limits_{-\infty}^{\infty} \lambda_j \leq 8\|f\|_{\mathcal{P}_1}$. The proof is finished. □

Corollary 2.5.9 *Denote $L_0^\infty = \{f \in L^\infty : f_0 = 0\}$. Then $L_0^\infty \subset \mathcal{P}_1$ with the embedding being continuous and dense.*

Proof. Let $f = (f_n)_{n\geq 0} \in L_0^\infty$. Set $r_n = \|f\|_\infty$, for all n. Then $|f_n| \leq r_{n-1}$, for all n. This means that f is L^∞-predictable. And $\|f\|_{\mathcal{P}_1} \leq \|r_\infty\|_1 = \|f\|_\infty$. This proves that L_0^∞ is embedded into $\mathcal{P}_1$ continuously. Now let $f \in \mathcal{P}_1$ be given, and $f = \sum_1^\infty \lambda_j a_j$ be any decomposition. Denote $f^{(N)} = \sum_1^N \lambda_j a_j$, then $f^{(N)} \in L_0^\infty$ and

$$\|f - f^{(N)}\|_{\mathcal{P}_1} \leq \sum_{N+1}^{\infty} \lambda_j \to 0.$$

This proves the density of L_0^∞ in $\mathcal{P}_1$. The proof is finished. □

Lemma 2.5.10 *For all $f \in L_0^\infty \subset \mathcal{P}_1$, and for all $\varphi \in bmo_1$, we have*

$$|E(f\varphi)| \leq 8\|f\|_{\mathcal{P}_1}\|\varphi\|_{bmo_1}. \tag{2.5.12}$$

Proof. Let $f \in L_0^\infty$. Make f's decomposition $f = \sum_{-\infty}^{\infty} \lambda_j a_j$, made as in Lemma 2.5.8. Then rearrange $\sum_{-\infty}^{\infty} \lambda_j a_j$ as $\sum_0^\infty \mu_k b_k$, for example $\lambda_0 = \mu_1$, $\lambda_1 = \mu_2$, $\lambda_{-1} = \mu_3, \lambda_2 = \mu_4, \cdots$, then we have

$$\lim_{k\to\infty} \sum_1^\infty \mu_k b_k = f, \quad \text{a.e., and boundedly (say by } C\|f\|_\infty).$$

Since $\varphi \in L^1$, we have

$$E(f\varphi) = \lim_{n\to\infty} \sum_{k=1}^n \mu_k E(b_k\varphi) = \sum_1^\infty \mu_k E(b_k\varphi),$$

$$|E(f\varphi)| \leq \sum_1^\infty \mu_k |E(b_k\varphi)| \leq \sum_1^\infty \mu_k \|\varphi\|_{bmo_1} \leq 8\|f\|_{\mathcal{P}_1}\|\varphi\|_{bmo_1}.$$

The proof of the lemma is finished. □

Remark The lemma shows that each $\varphi \in bmo_1$ yields a bounded linear functional l_φ on $\mathcal{P}_1$ with $\|l_\varphi\| \leq 8\|\varphi\|_{bmo_1}$. In addition we see that if $f \in \mathcal{P}_p$, $1 < p < \infty$, $\varphi \in bmo_1 \bigcap L^{p'}$. Then

$$l_\varphi(f) = E(f\varphi). \tag{2.5.13}$$

In fact for $f \in \mathcal{P}_p \subset \mathcal{P}_1$, let $f = \sum_1^\infty \lambda_j a_j$ be its atomic decomposition made as in Lemma 2.5.8, and $f^{(N)} = \sum_1^N \lambda_j a_j$, we have $f^{(N)} \to f$ in L_p. Thus (since $f^{(N)} \in L_0^\infty$)

$$l_\varphi(f) = \lim_{N\to\infty} l_\varphi(f^{(N)}) = \lim_{N\to\infty} E(f^{(N)}\varphi) = E(f\varphi).$$

The assertion is proved.

Now we are in the position to prove Fefferman's duality theorem.

Theorem 2.5.11 *We have $H_1' = BMO_1$. That is to say, we have*

$$|E(f\varphi)| \leq C\|f\|_{H_1}\|\varphi\|_{BMO_1}, \quad \forall f \in L^p,\ p > 1,\ \varphi \in BMO_1, \tag{2.5.14}$$

and hence each $\varphi \in BMO_1$ yields a bounded linear functional l_φ on H_1 such that

$$\|l_\varphi\| \leq C\|\varphi\|_{BMO_1};$$

On the contrary, each $l \in H_1'$ must arise from some $\varphi \in BMO_1$ in the preceding way, and

$$\|\varphi\|_{BMO_1} \leq 4\|l\|.$$

Proof. Notice that $\varphi \in BMO_1$ if and only if $\varphi \in bmo_1 \bigcap BD$, and

$$\|\varphi\|_{BD} \vee \|\varphi\|_{bmo_1} \leq \|\varphi\|_{BMO_1} \leq \|\varphi\|_{BD} + \|\varphi\|_{bmo_1}. \tag{2.5.15}$$

Now let $\varphi \in BMO_1$, $f \in L^p \subset H_1$. Make f's Davis decomposition $f = g + h$,

$$\|g\|_{\mathcal{P}_1} \leq 17\|f\|_{H_1}, \quad \|h\|_{\mathcal{A}_1} \leq 8\|f\|_{H_1}.$$

We will see that $\varphi \in L^{p'}$ (see § 4.1). Thus

$$E(f\varphi) = E(g\varphi) + E(h\varphi),$$

and by Lemma 2.5.10, Theorem 2.5.2,

$$\begin{aligned}|E(f\varphi)| &\leq |E(g\varphi)| + |E(h\varphi)| \leq 8\|g\|_{\mathcal{P}_1}\|\varphi\|_{bmo_1} + \|h\|_{\mathcal{A}_1}\|\varphi\|_{BD} \\ &\leq C\|f\|_{H_1}\|\varphi\|_{BMO_1}.\end{aligned}$$

On the contrary, let $l \in H_1'$ be arbitrary. Since $H_2 \subset H_1$, there exists $\varphi \in L^2$ such that

$$l(f) = E(f\varphi) = \sum_0^\infty E(\Delta_n f \Delta_n \varphi), \quad \forall f \in H_2 \subset H_1.$$

In particular, for all 1-atoms a, we have

$$|E(\varphi a)| = |l(a)| \leq \|l\| \, \|a\|_{H_1} \leq \|l\|.$$

From Lemma 2.5.7, we see $\|\varphi\|_{bmo_1} \leq 2\|l\|$. Meanwhile, for all $f \in H_2 \bigcap \mathcal{A}_1$, we have also

$$\Big|\sum_0^\infty E(\Delta_n f \Delta_n \varphi)\Big| \leq \|l\| \, \|f\|_{H_1} \leq \|l\| \, \|f\|_{\mathcal{A}_1}.$$

But $H_2 \bigcap \mathcal{A}_1$ is dense in $\mathcal{A}_1$. In fact, the set of all finite martingales is obviously dense in $\mathcal{A}_1$. And for f a martingale stopped at n, let $g \in L^\infty(\mathcal{F}_n)$ be such that $\|f_n - g\|_1 \leq \frac{\varepsilon}{2n}$, then

$$\begin{aligned}E\Big(\sum_{k=0}^n |\Delta_k(f_n - g)|\Big) &\leq \sum_{k=0}^n E(|(f_n - g)_k| + \sum_{k=0}^n E(|(f_n - g)_{k-1}|) \\ &\leq 2n\|f_n - g\|_1 \leq \varepsilon.\end{aligned}$$

Thus for any $f \in \mathcal{A}_1$ and $\varepsilon > 0$, assume n being such that

$$\|f - f_n\|_{\mathcal{A}_1} = E\Big(\sum_{n+1}^\infty |\Delta_k f|\Big) \leq \varepsilon,$$

and $g \in L^\infty(\mathcal{F}_n)$ being such that $\|f_n - g\|_{\mathcal{A}_1} \leq \varepsilon$, then

$$\|f - g\|_{\mathcal{A}_1} \leq \|f - f_n\|_{\mathcal{A}_1} + \|f_n - g\|_{\mathcal{A}_1} \leq 2\varepsilon.$$

This proves the assertion. Now l which was first defined on $H_2 \bigcap \mathcal{A}_1$ could be extended to whole $\mathcal{A}_1$, and hence $\|\varphi\|_{BD} \leq 2\|l\|$. At last, we get

$$\|\varphi\|_{BMO_1} \leq \|\varphi\|_{bmo_1} + \|\varphi\|_{BD} \leq 4\|l\|.$$

The proof of the theorem is complete. □

2.6 Weak compactness and convergence in H_1

In this section, we want to establish the H_1 version of several familiar facts about the weak compactness and convergence in L^1. The main ones are characterizing weak relative compact subsets of H_1 and proving the weak sequence completeness of H_1. We need BMO's following characterization formulated in §4.2, i.e.: $\varphi \in BMO = BMO_1 = BMO_2$, if and only if there exist a random varialbe η and a sequence $\{\xi_n\}_0^\infty$ of random variables such that

$$\varphi = \eta + \sum_0^\infty E(\xi_n|\mathcal{F}_n), \tag{2.6.1}$$

$$\| |\eta| + \sum_0^\infty |\xi_n| \|_\infty \approx \|\varphi\|_{BMO} = \|\varphi\|_*. \tag{2.6.2}$$

Basing on it, the duality between H_1 and BMO could be written as

$$\langle f, \varphi\rangle = E(f\eta) + \sum_0^\infty E(f_n\xi_n), \quad \forall f = (f_n)_{n\geq 0} \in H_1, \ \varphi \in BMO, \tag{2.6.3}$$

where the series is absolutely convergent (see Corollary 4.2.10). Denote the weak topology in H_1 by $\sigma(H_1, BMO)$. (Meanwhile, the weak *topology in BMO, will be denoted by $\sigma(BMO, H_1)$.)

At first, we examine the relation between the uniform integrability in L^1 and in H_1.

Lemma 2.6.1 *Let K be a set of measurable processes $X = (X_n)_{n\geq 0}$ such that $\sup_{X\in K} E(MX) \leq C$. Suppose that for all measurable functions S from Ω to $\mathbf{Z}^+$, $K_S = \{X_S : X \in K\}$, with*

$$X_S = \sum_{n=0}^\infty X_n \chi_{\{S=n\}},$$

is uniformly integrable in L^1. Then $MK = \{MX : X \in K\}$ is uniformly integrable in L^1, too.

Proof. Suppose that MK were not uniformly integrable. Then there exists a disjoint set sequence $\{B_n\}$ and $\{X^{(n)}\} \subset K$ such that $\int_{B_n} MX^{(n)} d\mu > 2\varepsilon$ for some $\varepsilon > 0$, by means of Lemma 1.3.2.6. So there exists a measurable function S_n from B_n to $\mathbf{Z}^+$ such that $\int_{B_n} |X_{S_n}^{(n)}| d\mu \geq \varepsilon$. This can be seen as follows. For any measurable process X satisfying $MX < \infty$, a.e., and any $\varepsilon > 0$, define $S = \inf\{m : |X_m| > MX - \varepsilon\}$, which is a measurable function from Ω to $\mathbf{Z}^+$. For such S, we have

$$\int_\Omega |X_S| d\mu \geq \int_\Omega MX d\mu - \varepsilon.$$

This proves the assertion. Now let S be any measurable function from Ω to $\mathbf{Z}^+$ such that $S|_{B_n} = S_n$. Then

$$\int_{B_n} |X_S^{(n)}| d\mu = \int_{B_n} |X_{S_n}^{(n)}| d\mu \geq \varepsilon.$$

This would contradict to the uniform integrability of K_S. The proof is finished. □

Now we can get one of our main theorems.

Theorem 2.6.2 *Let $K \subset H_1$. Then K is weak relative compact if and only if $MK = \{Mf : f \in K\}$ is uniformly integrable.*

Proof. Assume MK is uniformly integrable. We want to show that for all $\varphi \in BMO$, for all $\varepsilon > 0$, there exists N_0, such that when $N \geq N_0$ we have

$$\Big|E\Big(\sum_{n=N+1}^{\infty} f_n \xi_n\Big)\Big| \leq \varepsilon, \quad \forall f \in K, \tag{2.6.4}$$

where $\{\xi_n\}$ is the associated sequence of φ which occurs in (2.6.1–2.6.2). In fact, $\sum_0^N |\xi_n| \to \sum_0^\infty |\xi_n|$, a.e., and hence in measure. So for all $\delta > 0$, there exists N_0 such that when $N \geq N_0$, we have

$$|F_N| = \Big|\Big\{\omega : \sum_{N+1}^{\infty} |\xi_n| > \frac{\varepsilon}{2C}\Big\}\Big| \leq \delta,$$

where C is chosen such that $C \geq \max(\sup_{f \in K} E(Mf), \| |\eta| + \sum_0^\infty |\xi_n| \|_\infty)$. When δ is chosen such that

$$\int_F Mf d\mu \leq \frac{\varepsilon}{2C}, \quad \forall f \in K, \quad \text{provided} \quad |F| \leq \delta,$$

then for all $f \in K$, we have

$$\begin{aligned}\Big|E\Big(\sum_{N+1}^{\infty} f_n \xi_n\Big)\Big| &\leq \Big(\int_{F_N} + \int_{F_N^c}\Big) \sum_{N+1}^{\infty} Mf |\xi_n| d\mu \\ &\leq C \int_{F_N} Mf d\mu + \frac{\varepsilon}{2C} \int_\Omega Mf d\mu \leq \varepsilon.\end{aligned}$$

This proves (2.6.4). Now we can conclude the weak relative compactness of K from the uniform integrability of MK. The uniform integrability of MK implies that $K \subset L^1$ is weak relative compact in the topology $\sigma(L^1, L^\infty)$. (See for example Dunford-Schwartz [1] p.294, Corollary IV.8.11.) That means for any infinite subset of K, there exists a subsequence $\{f^{(i)}\} \subset K$ and $f \in L^1$ such that for all $g \in L^\infty$

$\lim_i E(f^{(i)}g) = E(fg)$. From the uniformity given by (2.6.4), for this subsequence $\{f^{(i)}\}$, we have

$$\begin{aligned}\lim_{i\to\infty}\langle f^{(i)},\varphi\rangle &= \lim_{i\to\infty} E(f^{(i)}\eta) + \lim_{i\to\infty}\sum_0^\infty E(f_n^{(i)}\xi_n)\\ &= E(f\eta) + \sum_0^\infty \lim_{i\to\infty} E(f_n^{(i)}\xi_n)\\ &= E(f\eta) + \sum_0^\infty \lim_{i\to\infty} E(E(f^{(i)}\xi_n|\mathcal{F}_n))\\ &= E(f\eta) + \sum_0^\infty \lim_{i\to\infty} E(f^{(i)}\xi_n) = E(f\eta) + \sum_0^\infty E(f\xi_n)\\ &= E(f\eta) + \sum_0^\infty E(f_n\xi_n) = \langle f,\varphi\rangle. \end{aligned} \tag{2.6.5}$$

We claim $f \in H_1$. In fact, for any measurable S from Ω to $\mathbf{Z}^+$ taking

$$\eta = 0, \quad \xi_n = \operatorname{sgn} f_n \chi_{\{S=n\}},$$

and substituting it into (2.6.5), we get

$$\begin{aligned}E(|f_S|) &= \sum_0^\infty E(|f_n|\chi_{\{S=n\}}) = \sum_0^\infty E(f_n\xi_n)\\ &\le \sup_i \Big|\sum_0^\infty E(f_n^{(i)}\xi_n)\Big| \le \sup_i \sum_0^\infty E(Mf^{(i)}|\xi_n|) \le C.\end{aligned}$$

Taking "sup" over S, we see $f \in H_1$. This proves one half of the theorem.

Now assume that K is weak relative compact in $\sigma(H_1, BMO)$. Let S be any measurable function from Ω to $\mathbf{Z}^+$. Then we can show that K_S is weak relative compact in $\sigma(L^1, L^\infty)$. In fact assume $\{f^{(i)}\}$ being a convergent sequence of H_1 in the topology $\sigma(H_1, BMO)$ with the limit $f \in H_1$. Then for all S, and $g \in L^\infty$,

$$E(f_S^{(i)}g) = E\Big(\sum_0^\infty f_n^{(i)} g\chi_{\{S=n\}}\Big) = E\Big(\sum_0^\infty f_n^{(i)}\xi_n\Big)$$

is nothing, but $\langle f^{(i)},\varphi\rangle$ with $\varphi = \sum E(\xi_n|\mathcal{F}_n) \in BMO$, where $\xi_n = g\chi_{\{S=n\}}$ satisfying $\sum_0^\infty |\xi_n| \le \|g\|_\infty$, and hence

$$E(f_S^{(i)}g) = \langle f^{(i)},\varphi\rangle \to \langle f,\varphi\rangle.$$

From the weak relative compactness characterization in L_1 (as cited above), we see that K_S is uniformly integrable for all S. Noticing that K is bounded in H_1 owing to its weak relative compactness, from Lemma 2.6.1, we see that MK is uniformly integrable. The proof of the theorem is finished. □

A similar characterization of weak relative compactness in H_1 can be formulated in terms of square functions.

Theorem 2.6.2′ *Let $K \subset H_1$. Then K is weak relative compact, if and only if $SK = \{S(f) : f \in K\}$ is uniformly integrable.*

Proof. First we want to show that for any $K \subset H_1$, MK and SK are or are not uniformly integrable simultaneously. This follows from Lemma 1.3.2.3 and Theorem 3.2.5. In fact, assume MK is uniformly integrable. Then there exists a convex function $\Phi(u)$ of moderate growth, satisfying $\lim_{u\to\infty} \frac{\Phi(u)}{u} = \infty$, such that $\sup_{f\in K} E(\Phi(Mf)) \leq C$ (see Lemma 1.3.2.3), and hence

$$\sup_{f\in K} E(\Phi(Sf)) \leq C \sup_{f\in K} E(\Phi(Mf)) \leq C \quad \text{(see Theorem 3.2.5)}.$$

This proves that SK is uniformly integrable. Now the theorem follows from Theorem 2.6.2 immediately. □

As a result, we can get the weak sequence completeness of H_1.

Theorem 2.6.3 *Let $\{f^{(n)}\}$ be a sequence in H_1 such that $\langle f^{(n)}, g\rangle$ has a limit, for all $g \in BMO$. Then there exists $f \in H_1$ such that $\lim_{n\to\infty} f^{(n)} = f$ in $\sigma(H_1, BMO)$.*

Proof. $\{f^{(n)}\}$ defines a sequence of bounded linear functionals on BMO, which satisfies $\sup_n |\langle f^{(n)}, g\rangle| < \infty$, for all $g \in BMO$, so $\{f^{(n)}\}$ is a bounded set in H_1. As we have shown in the proof of theorem 2.6.2 that for all measurable S from Ω to $\mathbf{Z}^+$, $\{f_S^{(n)}\}$ is convergent in $\sigma(L^1, L^\infty)$. So $\{f_S^{(n)}\}$ and hence $\{Mf^{(n)}\}$ is uniformly integrable. From Theorem 2.6.2, $\{f^{(n)}\}$ is weak relative compact in H_1. Take a subsequence $\{f^{(n_k)}\}$ which converges to some $f \in H_1$ in $\sigma(H_1, BMO)$. Then f must be the limit of $\{f^{(n)}\}$ itself. The proof is finished. □

At last, as for the strong convergence in H_1, we have

Theorem 2.6.4 *Let $\{f^{(n)}\} \subset H_1$ be such that $\lim_{n\to\infty} f^{(n)} = f$ both with respect to $\sigma(H_1, BMO)$ and L^1−norm. Then the convergence holds in H_1-norm, too.*

Proof. Theorem 2.6.2 tells us that $\{Mf^{(n)}\}$ is uniformly integrable since it is convergent in $\sigma(H_1, BMO)$. Theorem 2.1.1 tells us that $\{Mf^{(n)}\}$ is convergent to Mf in measure, since $\lim_{n\to\infty} f^{(n)} = f$ in L^1 and

$$|\{|Mf^{(n)} - Mf| > \delta\}| \leq |\{M(f^{(n)} - f) > \delta\}| \leq \frac{c}{\delta}\|f^{(n)} - f\|_1.$$

Now $\lim_{n\to\infty} f^{(n)} = f$ in H_1 follows from the uniform integrability and the convergence in measure. The proof is finished. □

2.7 The spaces $h_p(0 < p \leq \infty)$

Besides $\mathcal{P}_p$ (defined in §2.4), there is another version of H_p, i.e. h_p, which is concerned with the conditional square function operator, and has some "predictability", too. In this section we will introduce its definition and characterize it. In the next section, we will give its dual space.

Definition 2.7.1 *For any martingale* $f = (f_n)_{n\geq 0}$,

$$f \to \sigma(f) = \Big(\sum_1^\infty E(|\Delta_n f|^2|\mathcal{F}_{n-1})\Big)^{\frac{1}{2}}, \tag{2.7.1}$$

is called the conditional square function operator. Define

$$h_p = \{f = (f_n)_{n\geq 0}: \ \|f\|_{h_p} = \|\sigma(f)\|_p < \infty\}, \quad 0 < p \leq \infty. \tag{2.7.2}$$

Remark For $1 \leq p < \infty$, $\|\cdot\|_{h_p}$ is a norm, of which the trigonometric inequality follows from Minkowski's inequality concerning the conditional expectation. The completeness of h_p can be seen by an argument proving the completeness of H_1. So, h_p is a Banach space.

We want to give various characterizations for h_p in order to get more tools to study it. A natural candidate which might characterize h_p is ${}_2k_p$, $2 \leq p \leq \infty$,

$$\begin{aligned}{}_2k_p = \{&f = (f_n)_{n\geq 0} \in L^2 : \exists \gamma \in L^p_+, \text{ s.t.} \\ &E(|f - f_n|^2 \mid \mathcal{F}_n) \leq E(\gamma^2|\mathcal{F}_n), \forall\, n\}.\end{aligned} \tag{2.7.3}$$

But another candidate is not so natural, which is

$${}_2l_p = \{f = (f_n)_{n\geq 0} : \|f\|_{{}_2l_p} < \infty\}, \quad -\infty \leq p \leq \infty,\ p \neq 0, \tag{2.7.4}$$

where

$$\|f\|_{{}_2l_p} = \inf_{r\in R}\Big\{E\Big(\sum_{n=1}^\infty r_{n-1}^{1-\frac{2}{p}}|\Delta_n f|^2\Big)^{\frac{1}{2}}\Big\}, \quad 0 < p \leq 2, \tag{2.7.5}$$

$$\|f\|_{{}_2l_p} = \sup_{r\in R}\Big\{E\Big(\sum_{n=1}^\infty r_{n-1}^{1-\frac{2}{p}}|\Delta_n f|^2\Big)^{\frac{1}{2}}\Big\}, \quad 2 \leq p \leq \infty,\ -\infty \leq p < 0, \tag{2.7.6}$$

with

$$\begin{aligned}R = \{&\text{nonnegative increasing adapted process;} \\ &r = (r_n)_{n\geq 0}: \ E(r_\infty) \leq 1\}.\end{aligned} \tag{2.7.7}$$

The definition of ${}_2l_p$ is a little strange, but after we have given an equivalent expression for $\|\cdot\|_{{}_2l_p}$, it looks clear that ${}_2l_p$ is equivalent to ${}_2k_p$, and hence to h_p. For $2 \leq p \leq \infty$, or $-\infty \leq p < 0$, denote α such that

$$\frac{1}{\alpha} = \frac{1}{2} - \frac{1}{p} \geq 0, \tag{2.7.8}$$

and define

$$G_2^p = \Big\{\text{adapted process } (g_n)_{n\geq 0} : g_n \in L^\alpha, E\Big(\Big(\sum_{n=0}^{\infty} |g_n|^2\Big)^{\frac{\alpha}{2}}\Big) \leq 1\Big\}. \tag{2.7.9}$$

Then it is easy to show that R and G_2^p are one to one corresponding to each other. In fact

$$(g_n)_{n\geq 0} \in G_2^p \Longrightarrow r = (r_n)_{n\geq 0} \in R, \text{ with } r_n = \Big(\sum_{k=0}^{n} |g_k|^2\Big)^{\frac{\alpha}{2}},$$

$$(r_n)_{n\geq 0} \in R \Longrightarrow g = (g_n)_{n\geq 0} \in G_2^p, \text{ with } |g_n|^2 = r_n^{1-\frac{2}{p}} - r_{n-1}^{1-\frac{2}{p}}.$$

We claim that for $2 \leq p \leq \infty$ or $-\infty \leq p < 0$,

$$\|f\|_{{}_2l_p} = \sup_{r\in R} E\left(\sum_{n=1}^{\infty} r_{n-1}^{1-\frac{2}{p}} |\Delta_n f|^2\right)^{\frac{1}{2}} = \sup_{g\in G_2^p} E\left(\sum_{n=0}^{\infty} |g_n|^2 |f^{(n)}|^2\right)^{\frac{1}{2}}, \tag{2.7.10}$$

where $f^{(n)}$ denotes $f - f_n = f_\infty - f_n$, a function not a process. It follows from

$$\begin{aligned} E\left(\sum_{n=0}^{\infty} |g_n f^{(n)}|^2\right) &= E\left(\sum_{n=0}^{\infty} |g_n|^2 E\left(\sum_{k=n+1}^{\infty} |\Delta_k f|^2 | \mathcal{F}_n\right)\right) \\ &= E\left(\sum_{k=1}^{\infty} |\Delta_k f|^2 \sum_{n=0}^{k-1} |g_n|^2\right) = E\left(\sum_{n=1}^{\infty} r_{n-1}^{1-\frac{2}{p}} |\Delta_n f|^2\right). \end{aligned}$$

Remark Notice that all of these spaces are the spaces modulo $\mathcal{F}_0$-finite martingales, so we can consider only those martingales with $f_0 = 0$.

Now we give the equivalence of h_p to the other two for some indices.

Theorem 2.7.2 *For $0 < p < \infty$, we have $h_p = {}_2l_p$. More precisely, for any martingale $f = (f_n)_{n\geq 0}$ with $f_0 = 0$, we have, when $0 < p \leq 2$,*

$$\left(\frac{p}{2}\right)^{\frac{1}{2}} \|f\|_{{}_2l_p} \leq \|\sigma(f)\|_p \leq \|f\|_{{}_2l_p}; \tag{2.7.11}$$

when $2 \leq p < \infty$, we have

$$\left(\frac{2}{p}\right)^{\frac{1}{2}} \|\sigma(f)\|_p \leq \|f\|_{{}_2l_p} \leq \|\sigma(f)\|_p. \tag{2.7.12}$$

Proof. Assume $0 < p \le 2$, $\|\sigma(f)\|_p = 1$, Then $(\sigma_{n+1}(f)^p)_{n\ge 0} \in R$. From (2.2.1), we have

$$\|f\|_{{}_2l_p} \le E\Big(\sum_{n=1}^{\infty} \sigma_n(f)^{p-2}(\sigma_n(f)^2 - \sigma_{n-1}(f)^2)\Big)^{\frac{1}{2}}$$
$$\le \Big(\frac{2}{p}\Big)^{\frac{1}{2}} E\Big(\sum_{n=1}^{\infty}(\sigma_n(f)^p - \sigma_{n-1}(f)^p)\Big)^{\frac{1}{2}} \le \Big(\frac{2}{p}\Big)^{\frac{1}{2}}.$$

Assume $f \in {}_2l_p$. For any $r = (r_n)_{n\ge 0} \in R$, denote $r = r_\infty$. We have (since $1-\frac{2}{p} \le 0$)

$$\sum_{n=1}^{\infty} r_{n-1}^{1-\frac{2}{p}}(\sigma_n(f)^2 - \sigma_{n-1}(f)^2) \ge r^{1-\frac{2}{p}}\sigma(f)^2,$$

from which we get

$$\|\sigma(f)\|_p \le \inf_{r\in R} E\Big(\sum_{n=1}^{\infty} r_{n-1}^{1-\frac{2}{p}}|\Delta_n f|^2\Big)^{\frac{1}{2}} = \|f\|_{{}_2l_p}.$$

Now consider the case $2 \le p < \infty$. Assume $f \in {}_2l_p$. At first, we want to show $\sigma_n(f) \in L^p$, for all n. Since

$$\sup_{r_{n-1}} E\left(r_{n-1}^{1-\frac{2}{p}} E(|\Delta_n f|^2)|\mathcal{F}_{n-1})\right)^{\frac{1}{2}} \le \|f\|_{{}_2l_p}, \quad \forall n,$$

where sup is taken over all $r_{n-1} \in L^1_+(\mathcal{F}_{n-1})$ and $E(r_{n-1}) \le 1$, we get

$$\|E(|\Delta_n f|^2|\mathcal{F}_{n-1})^{\frac{1}{2}}\|_p^2 = \|E(|\Delta_n f|^2|\mathcal{F}_{n-1})\|_{\frac{p}{2}} \le \|f\|_{{}_2l_p}^2.$$

This proves the assertion. So, if f is a finite martingale, then $\sigma(f) \in L^p$. Assume $\|\sigma(f)\|_p = 1$. Then $(\sigma_{n+1}(f)^p)_{n\ge 0} \in R$. From (2.2.2), we get

$$\|f\|_{{}_2l_p} \ge E\left(\sum_{n=1}^{\infty} \sigma_n(f)^{p-2}(\sigma_n(f)^2 - \sigma_{n-1}(f)^2)\right)^{\frac{1}{2}} \ge \Big(\frac{2}{p}\Big)^{\frac{1}{2}}.$$

Assume $f \in h_p, r = (r_n)_{n\ge 0} \in R$, then

$$E\left(\sum_{n=1}^{\infty} r_{n-1}^{1-\frac{2}{p}}(\sigma_n(f)^2 - \sigma_{n-1}(f)^2)\right)^{\frac{1}{2}} \le E(r^{1-\frac{2}{p}}\sigma(f)^2)^{\frac{1}{2}} \le \|\sigma(f)\|_p.$$

This proves the theorem. □

As for ${}_2k_p$, we have

Theorem 2.7.3 *Let $2 \le p < \infty$. Then $h_p = {}_2k_p = {}_2l_p$. More precisely, for any martingale $f = (f_n)_{n\ge 0}$ with $f_0 = 0$, we have*

$$\|f\|_{{}_2l_p} \le \|f\|_{{}_2k_p} \le \|\sigma(f)\|_p \le \Big(\frac{p}{2}\Big)^{\frac{1}{2}}\|f\|_{{}_2l_p}. \tag{2.7.13}$$

Proof. For any $p, 2 \leq p \leq \infty$, we have $h_p \subset {}_2k_p$. In fact

$$E(|f^{(n)}|^2|\mathcal{F}_n) = E\left(\sum_{n+1}^{\infty} E(|\Delta_k f|^2|\mathcal{F}_{k-1})\Big|\mathcal{F}_n\right) \leq E(\sigma(f)^2|\mathcal{F}_n),\ \forall n,$$

$$\|f\|_{{}_2k_p} \leq \|\sigma(f)\|_p.$$

Now assume $f \in {}_2k_p$, $r \in L^p_+$ be any one that occurs in the definition of ${}_2k_p$. Then

$$\|f\|_{{}_2l_p} = \sup_{g \in G^p_2} E\Big(\sum_n |g_n|^2 |f^{(n)}|^2\Big)^{\frac{1}{2}} = \sup_g E\Big(\sum_n |g_n|^2 E(|f^{(n)}|^2|\mathcal{F}_n)\Big)^{\frac{1}{2}}$$
$$\leq \sup_g E\Big(\sum_n |g_n|^2 r^2\Big)^{\frac{1}{2}} \leq \sup_g E\Big(\Big(\sum_n |g_n|^2\Big)^{\frac{\alpha}{2}}\Big)^{\frac{1}{\alpha}} E(r^p)^{\frac{1}{p}} \leq \|r\|_p.$$

Taking "inf" over all posssible r, we get $\|f\|_{{}_2l_p} \leq \|f\|_{{}_2k_p}$. The last inequality in (2.7.13) has been obtained in (2.7.12). The proof is finished. □

Remarks

1. How about ${}_2l_p$, for $p < 0$? We will show in next section that ${}_2l_p$ is equivalent to the so-called Lipschitz space.

2. ${}_2k_p$, ${}_2l_p$ have the following extensions. Let $1 \leq a \leq p \leq \infty$, or $-\infty \leq p < 0$, $a \neq \infty$, and α be such that

$$\frac{1}{\alpha} = \frac{1}{a} - \frac{1}{p}. \tag{2.7.14}$$

Set

$$G^p_a = \Big\{\text{adapted process } g = (g_n)_{n\geq 0} : E\Big(\Big(\sum_0^{\infty} |g_n|^a\Big)^{\frac{\alpha}{a}}\Big) \leq 1\Big\}. \tag{2.7.15}$$

Define

$${}_al_p = \Big\{f = (f_n)_{n\geq 0} \in L^a_u : \|f\|_{{}_al_p} = \sup_{g\in G^p_a} E\Big(\sum_n |g_n|^a |f^{(n)}|^a\Big)^{\frac{1}{a}} < \infty\Big\}, \tag{2.7.16}$$

$$\begin{aligned} {}_ak_p = \{f = (f_n)_{n\geq 0} \in L^a_u : \exists r \in L^p_+, \text{ s.t.} \\ E(|f - f_n|^a|\mathcal{F}_n) \leq E(r^a|\mathcal{F}_n), \forall\, n\}, \end{aligned} \tag{2.7.17}$$

$$\|f\|_{{}_ak_p} = \inf\{\|r\|_p : r \text{ runs over all possible ones}\}. \tag{2.7.18}$$

Then we have still

$${}_al_p = {}_ak_p, \quad 1 \leq a \leq p \leq \infty,\ a \neq \infty. \tag{2.7.19}$$

We will establish the equivalence between ${}_al_\Phi$ and ${}_ak_\Phi$ in §3.3.

2.8 Dual spaces of h_p and $\mathcal{P}_p$

First we define the Lipschitz space ${}_2\lambda_\beta$, and characterize it. Let $\mathcal{I}^{(n)}$ be the set of all $\mathcal{F}_n$-atoms, $n \geq 0$. (It could be empty, of course.) Denote

$$\omega_n = \sum |I^{(n)}|\chi_{I^{(n)}}, \tag{2.8.1}$$

where the sum is taken over all $I^{(n)} \in \mathcal{I}^{(n)}$. Then the Lipschitz space ${}_2\lambda_\beta$, $\beta \geq 0$, can be defined as

$$\begin{aligned} {}_2\lambda_\beta = \{f = (f_n)_{n\geq 0} \in L^2 : \\ \|f\|_{{}_2\lambda_\beta} = \sup_n \|\omega_n^{-\beta} E(|f^{(n)}|^2|\mathcal{F}_n)^{\frac{1}{2}}\|_\infty < \infty\}, \end{aligned} \tag{2.8.2}$$

where $f^{(n)} = f - f_n = f_\infty - f_n$ In the dyadic case, ${}_2\lambda_\beta$ is just the classical one: f is said to be in dyadic Lipschitz β class, if

$$\left(\frac{1}{|I|}\int_I |f - f_I|^2 d\mu\right)^{\frac{1}{2}} \leq C2^{-n\beta},\ \forall n,\ \forall I \quad \text{dyadic interval, } |I| = 2^{-n}.$$

Now we characterize it by the space ${}_2l_p$ for $p < 0$.

Lemma 2.8.1 *For $-\infty \leq p < 0$, or $p = \infty$, set $\beta = -\frac{1}{p}$. We have ${}_2\lambda_\beta = {}_2l_p$. More precisely,*

$$\left(1 - \frac{2}{p}\right)^{-\frac{1}{2}} \|f\|_{{}_2l_p} \leq \|f\|_{{}_2\lambda_\beta} \leq \|f\|_{{}_2l_p}. \tag{2.8.3}$$

Proof. Let $f \in {}_2l_p$, m and $F \in \mathcal{F}_m$ be fixed. Set $g = (g_n)$ with

$$g_m = |F|^{-\frac{1}{\alpha}}\chi_F,\quad g_n = 0, n \neq m,\quad \frac{1}{\alpha} = \frac{1}{2} - \frac{1}{p}.$$

Then $(g_n)_{n\geq 0} \in G_2^p$. Since $f \in {}_2l_p$, we have

$$\left(|F|^{-\frac{2}{\alpha}} \int_F |f^{(m)}|^2 d\mu\right)^{\frac{1}{2}} \leq \|f\|_{{}_2l_p},$$

$$\left(|F|^{-1} \int_F E(|f^{(m)}|^2|\mathcal{F}_m) d\mu\right)^{\frac{1}{2}} \leq \|f\|_{{}_2l_p}|F|^{-\frac{1}{p}} = \|f\|_{{}_2l_p}|F|^\beta. \tag{2.8.4}$$

Since $\beta \geq 0$, for $x \notin \bigcup I^{(m)}$, we get $E(|f^{(m)}|^2|\mathcal{F}_m) = 0$ at x, by taking $F \ni x$, $F \in \mathcal{F}_m$, $F \bigcap I^{(m)} = \emptyset$, for all $\mathcal{F}_m$-atoms $I^{(m)}$, and $|F| \to 0$. When F is a $\mathcal{F}_m$-atom, $E(|f^{(m)}|^2|\mathcal{F}_m)^{\frac{1}{2}}$ is constant on F and dominated by ω_m^β. In a word, we get

$$E(|f^{(m)}|^2|\mathcal{F}_m)^{\frac{1}{2}} \leq \|f\|_{{}_2l_p}\omega_m^\beta,$$

$$\|f\|_{{}_2\lambda_\beta} \leq \|f\|_{{}_2l_p}.$$

Now let $f \in {}_2\lambda_\beta$. We have

$$\|f\|_{{}_2l_p} = \sup_{g \in G_2^p} E\left(\sum_0^\infty |g_n|^2 |f^{(n)}|^2\right)^{\frac{1}{2}} \leq \|f\|_{{}_2\lambda_\beta} E\left(\sum_0^\infty |g_n|^2 \omega_n^{2\beta}\right)^{\frac{1}{2}}.$$

Denote $r_n = \left(\sum_{k \leq n} |g_k|^2\right)^{\frac{\alpha}{2}}$, then we have

$$\int_\Omega r_n d\mu \leq 1, \quad r_n \omega_n \leq 1, \quad \omega_n \leq r_n^{-1}, \quad \forall n.$$

Thus we get

$$\sum_0^\infty |g_n|^2 \omega_n^{2\beta} \leq \sum_0^\infty \left(r_n^{1-\frac{2}{p}} - r_{n-1}^{1-\frac{2}{p}}\right) r_n^{\frac{2}{p}} = \sum_0^\infty \left(r_n^{\frac{p-2}{p}} - r_{n-1}^{\frac{p-2}{p}}\right) r_n^{\frac{p-2}{2p}(q-2)},$$

where q satisfies

$$\frac{p-2}{2p}(q-2) = \frac{2}{p}, \quad q = \frac{2p}{p-2} \leq 2.$$

So from (2.2.1), we get

$$\begin{aligned} \|f\|_{{}_2l_p} &\leq \|f\|_{{}_2\lambda_\beta} \sup_{g \in G_2^p} E\left(\sum_0^\infty |g_n|^2 \omega_n^{2\beta}\right)^{\frac{1}{2}} \\ &\leq \|f\|_{{}_2\lambda_\beta} \left(\frac{2}{q}\right)^{\frac{1}{2}} E(r_\infty)^{\frac{1}{2}} = \left(1 - \frac{2}{p}\right)^{\frac{1}{2}} \|f\|_{{}_2l_p}. \end{aligned}$$

The proof is finished. □

Now we can give the dual space of h_p, $0 < p \leq 2$.

Theorem 2.8.2 *For $0 < p \leq 2$, we have $h_p' = {}_2l_{p'}$. That is to say, when $1 < p \leq 2$, $h_p' = h_{p'}$; $h_1' = bmo_2$; when $0 < p < 1$, $h_p' = {}_2\lambda_\beta$, with $\beta = -\frac{1}{p'} = \frac{1}{p} - 1$. Furthermore, the mapping $\varphi \to l_\varphi$ from ${}_2l_{p'}$ to h_p' satisfies*

$$\|\varphi\|_{{}_2l_{p'}} \leq \|l_\varphi\| \leq \left(\frac{2}{p}\right)^{\frac{1}{2}} \|\varphi\|_{{}_2l_{p'}}. \tag{2.8.5}$$

Proof. We have seen that $h_p = {}_2l_p$, $0 < p < \infty$. Now, for all $f \in {}_2l_p$, for all $\varphi \in {}_2l_{p'}$, $r = (r_n)_{n \geq 0} \in R$, we have (say $f_0 = \varphi_0 = 0$)

$$\begin{aligned} \left|E\Big(\sum_1^\infty \Delta_n f \Delta_n \varphi\Big)\right| &\leq E\Big(\sum_1^\infty |\Delta_n f||\Delta_n \varphi|\Big) \\ &\leq E\Big(\sum_{n=1}^\infty r_{n-1}^{1-\frac{2}{p}} |\Delta_n f|^2\Big)^{\frac{1}{2}} E\Big(\sum_{n=1}^\infty r_{n-1}^{1-\frac{2}{p'}} |\Delta_n \varphi|^2\Big)^{\frac{1}{2}}. \end{aligned} \tag{2.8.6}$$

For $\varepsilon > 0$, choose $(r_n) \in R$ such that

$$E\Big(\sum_1^\infty r_{n-1}^{1-\frac{2}{p}}|\Delta_n f|^2\Big)^{\frac{1}{2}} \le \|f\|_{{}_2l_p} + \varepsilon.$$

Making $\varepsilon \to 0$, we get

$$\Big|E\Big(\sum_1^\infty \Delta_n f \Delta_n \varphi\Big)\Big| \le \|f\|_{{}_2l_p}\|\varphi\|_{{}_2l_{p'}} \le \Big(\frac{2}{p}\Big)^{\frac{1}{2}}\|\sigma(f)\|_p\|\varphi\|_{{}_2l_{p'}}. \tag{2.8.7}$$

This means that ${}_2l_{p'}$ can be embedded into $h_p'(0 < p \le 2)$ continuously.

On the contrary, since $L_0^2 = h_2 \subset h_p$, $0 < p \le 2$, each $l \in h_p'$ yields a bounded linear functional on L_0^2. So there exists a $\varphi = (\varphi_n) \in L^2$ (modulo $\mathcal{F}_0$-finite martingales, say $\varphi_0 = 0$ without loss of generality) such that

$$l(f) = E\left(\sum_0^\infty \Delta_n f \Delta_n \varphi\right), \quad \forall f \in L_0^2.$$

Let N be fixed, and $(r_n) \in R$ be such that $r_n \in L^\infty$, for all n. We have

$$\begin{aligned} E\left(\sum_1^N r_{n-1}^{1-\frac{2}{p'}}|\Delta_n\varphi|^2\right) &= E\left(\sum_1^N r_{n-1}^{1-\frac{2}{p'}}\Delta_n\overline{\varphi}\Delta_n\varphi\right) \\ &= E\left(\sum_1^N \Delta_n f \Delta_n\varphi\right), \end{aligned} \tag{2.8.8}$$

where

$$f = \sum_1^N r_{n-1}^{1-\frac{2}{p'}}\Delta_n\overline{\varphi} \in L_0^2,$$

since $r_n \in L^\infty$, and $\Delta_n\varphi \in L^2$, and

$$\|f\|_{{}_2l_p} \le E\left(\sum_1^N r_{n-1}^{1-\frac{2}{p}}\left|r_{n-1}^{1-\frac{2}{p'}}\Delta_n\varphi\right|^2\right)^{\frac{1}{2}} = E\left(\sum_1^N r_{n-1}^{1-\frac{2}{p'}}|\Delta_n\varphi|^2\right)^{\frac{1}{2}} < \infty.$$

Thus we get

$$\begin{aligned} E\left(\sum_1^N r_{n-1}^{1-\frac{2}{p'}}|\Delta_n\varphi|^2\right) &= l(f) \le \|l\|\ \|\sigma(f)\|_p \\ &\le \|l\|\ \|f\|_{{}_2l_p} \le \|l\|E\left(\sum_1^N r_{n-1}^{1-\frac{2}{p'}}|\Delta_n\varphi|^2\right)^{\frac{1}{2}}, \end{aligned}$$

$$\begin{aligned} \|\varphi\|_{{}_2l_{p'}} &\le \lim_{N\to\infty}\|\varphi_N\|_{{}_2l_{p'}} \\ &= \lim_{N\to\infty}\sup_{(r_n)\in R, r_n\in L^\infty} E\left(\sum_1^N r_{n-1}^{1-\frac{2}{p'}}|\Delta_n\varphi|^2\right)^{\frac{1}{2}} \le \|l\|. \end{aligned}$$

This completes the proof. □

Now consider the case $2 \leq p < \infty$.

Theorem 2.8.3 *For $2 \leq p < \infty$, we have $h'_p = h_{p'}$. Furthermore the mapping $\varphi \to l_\varphi$ from $h_{p'}$ to h'_p satisfies*

$$\left(\frac{1}{p-1}\right)^{\frac{1}{2}} \|\sigma(\varphi)\|_{p'} \leq \|l\| \leq \left(\frac{2(p-1)}{p}\right)^{\frac{1}{2}} \|\sigma(\varphi)\|_{p'}. \tag{2.8.9}$$

Proof. From (2.8.7) we have

$$\left|E\Big(\sum_1^\infty \Delta_n f \Delta_n \varphi\Big)\right| \leq \|f\|_{2l_p}\|\varphi\|_{2l_{p'}} \leq \left(\frac{2}{p'}\right)^{\frac{1}{2}} \|\sigma(\varphi)\|_{p'}\|\sigma(f)\|_p.$$

This means $h_{p'} \subset h'_p$. On the contrary, since $L_0^p \subset h_p$ (this is an elementary fact which may be proved by a couple of ways, see for example §3.2 or §5.1), each $l \in h'_p$ yields a bounded linear functional on L_0^p. So there exists $\varphi = (\varphi_n)_{n \geq 0} \in L_0^{p'}$ such that

$$l(f) = E\left(\sum_1^\infty \Delta_n f \Delta_n \varphi\right), \quad f \in L_0^p.$$

For any N fixed, consider φ_N. We claim $\varphi_N \in h_{p'}$. In fact

$$\rho_n = E(|\Delta_n\varphi|^2|\mathcal{F}_{n-1})^{\frac{1}{2}} \in L^{p'}.$$

This can be seen as follows. Denote

$$\rho_n^{(M)} = \rho_n \chi_{\{\rho_n \leq M\}} = \rho_n \chi_F,$$

then it is $\mathcal{F}_{n-1}$-measurable, and

$$\begin{aligned} E(\rho_n^{(M)p'}) &= E(\rho_n^{p'-2}\rho_n^2\chi_{\{\rho_n \leq M\}}) = E(\rho_n^{p'-2}\chi_F E(|\Delta_n\varphi|^2|\mathcal{F}_{n-1})) \\ &= E(\rho_n^{p'-2}\chi_F \Delta_n\overline{\varphi}\Delta_n\varphi). \end{aligned} \tag{2.8.10}$$

Set

$$f_m = 0,\ m < n; \quad f_m = \rho_n^{p'-2}\chi_F\Delta_n\overline{\varphi}, \quad m \geq n.$$

Then $f = (f_m)_{m \geq 0}$ is a martingale, and

$$\sigma(f)^2 = \rho_n^{2(p'-2)}\chi_F E(|\Delta_n\varphi|^2|\mathcal{F}_{n-1}) = \rho_n^{(M)2(p'-1)},$$

$$\|\sigma(f)\|_p^p = \|\rho_n^{(M)}\|_{p'}^{p'}.$$

Thus we get

$$\|\rho_n^{(M)}\|_{p'}^{p'} = l(f) \leq \|l\|\ \|\sigma(f)\|_p = \|l\|\ \|\rho_n^{(M)}\|_{p'}^{\frac{p'}{p}},$$

$$\|\rho_n^{(M)}\|_{p'} \le \|l\|, \quad \|\rho_n\|_{p'} \le \|l\|.$$

This proves $\varphi_N \in h_{p'}$. Noticing that

$$\sigma(\varphi_N) = \sigma_N(\varphi), \; \sigma_n(\varphi_N) = \sigma_n(\varphi), \; n \le N,$$

$$\left(\|\sigma_N(\varphi)\|_{p'}^{-p'} \sigma_{n+1}(\varphi)^{p'}\right)_{n=0}^{N-1} \in R,$$

we get

$$\|\sigma_N(\varphi)\|_{p'}^2 \le \|\varphi_N\|_{2l_{p'}}^2 \le E\left(\sum_{n=1}^N \|\sigma_N(\varphi)\|_{p'}^{2-p'} \sigma_n(\varphi)^{p'-2} |\Delta_n\varphi|^2\right),$$

$$\|\sigma_N(\varphi)\|_{p'}^{p'} \le E\left(\sum_{n=1}^N \sigma_n^{p'-2}(\varphi)\Delta_n\overline{\varphi}\Delta_n\varphi\right) = E\left(\sum_{n=1}^N \Delta_n f \Delta_n\varphi\right), \tag{2.8.11}$$

where

$$f = \sum_1^N \sigma_n(\varphi)^{p'-2}\Delta_n\overline{\varphi} \in h_p,$$

$$\|\sigma(f)\|_p \le (p-1)^{\frac{1}{2}} \|\sigma_N(\varphi)\|_{p'}^{\frac{p'}{p}},$$

which can be seen as follows. We have

$$\begin{aligned}
\sigma(f)^2 &= \sum_1^N \sigma_n(\varphi)^{2(p'-2)} E(|\Delta_n\varphi|^2|\mathcal{F}_{n-1}) \\
&= \sum_1^N \sigma_n(\varphi)^{2(p'-2)}(\sigma_n(\varphi)^2 - \sigma_{n-1}(\varphi)^2) \\
&\le \frac{2}{q}\sum_1^N(\sigma_n(\varphi)^q - \sigma_{n-1}(\varphi)^q) = \frac{2}{q}\sigma_N(\varphi)^q,
\end{aligned}$$

with q satisfying

$$q - 2 = 2(p'-2), \quad q = 2p' - 2 \le 2, \quad \frac{pq}{2} = p',$$

and hence

$$\|\sigma(f)\|_p \le \left(\frac{2}{q}\right)^{\frac{1}{2}} E(\sigma_N(\varphi)^{p'})^{\frac{1}{p}} = (p-1)^{\frac{1}{2}}\|\sigma_N(\varphi)\|_{p'}^{\frac{p'}{p}}.$$

Now (2.8.11) gives

$$\|\sigma_N(\varphi)\|_{p'}^{p'} \le l(f) \le (p-1)^{\frac{1}{2}}\|l\| \, \|\sigma_N(\varphi)\|_{p'}^{\frac{p'}{p}},$$

$$\|\sigma(\varphi)\|_{p'} \leq (p-1)^{\frac{1}{2}}\|l\|.$$

This completes the proof of the theorem. □

Remark The same idea can be used to study the dual space of $H_p^S, 0 < p < \infty$. Of course, only the case $0 < p \leq 1$ is interesting. We have $(H_p^S)' = {}_2\Lambda_\beta$, with $\beta = \frac{1}{p} - 1$, $0 < p \leq 1$, where

$$\begin{aligned} {}_2\Lambda_\beta = \{f = (f_n) \in L^2 : \\ \|f\|_{{}_2\Lambda_\beta} = \sup_n \|\omega_n^{-\beta} E(|f^{(n-1)}|^2|\mathcal{F}_n)^{\frac{1}{2}}\|_\infty < \infty\}, \end{aligned} \tag{2.8.12}$$

which can be characterized by ${}_2L_p, -\infty \leq p < 0$. The spaces ${}_2L_p$ are defined by

$${}_2L_p = \{f = (f_n) \in L^2 : \|f\|_{{}_2L_p} < \infty\}, \tag{2.8.13}$$

$$\|f\|_{{}_2L_p} = \inf_{(r_n)\in R} E\left(\sum_0^\infty r_n^{1-\frac{2}{p}}|\Delta_n f|\right)^{\frac{1}{2}}, \quad 0 < p \leq 2, \tag{2.8.14}$$

$$\|f\|_{{}_2L_p} = \sup_{(r_n)\in R} E\left(\sum_0^\infty r_n^{1-\frac{2}{p}}|\Delta_n f|\right)^{\frac{1}{2}}, \quad -\infty \leq p < 0,\ 2 \leq p \leq \infty. \tag{2.8.15}$$

Meanwhile, the index 2 in the spaces ${}_2L_p$, ${}_2K_p$, and ${}_2\Lambda_\beta$ can be replaced by a, $1 \leq a \leq p$. We omit the detail for them, since we will study ${}_aL_\Phi$, ${}_aK_\Phi$ in §3.3.

Now we discuss the dual space of $\mathcal{P}_p$, $0 < p \leq \infty$. We need some preparations. First we want to make a decomposition for $\mathcal{P}_p$. We have done the atomic decomposition for $\mathcal{P}_1$ in §2.5, which works well for $\mathcal{P}_p$, $0 < p \leq 1$, obviously. Thus we have

$$\begin{aligned} \mathcal{P}_p = \Big\{f = (f_n)_{n\geq 0},\ f_0 = 0 : \ f = \sum_1^\infty \lambda_k a_k \\ \text{with } a_k\text{'s being } p\text{-atoms, and } \sum |\lambda_k|^p < \infty\Big\}, \end{aligned} \tag{2.8.16}$$

$$\|f\|_{\mathcal{P}_p} \approx \inf\left\{\left(\sum_1^\infty |\lambda_k|^p\right)^{\frac{1}{p}} : \ \{\lambda_k\} \text{ runs through all possible ones}\right\}. \tag{2.8.17}$$

As we will show in §7.2.2 that $\mathcal{P}_p' = {}_1\lambda_\beta$, $\beta = \frac{1}{p} - 1$, $0 < p \leq 1$. Now we make a decomposition for $\mathcal{P}_p$, $1 \leq p \leq \infty$.

Theorem 2.8.4 *Let $1 \leq p \leq \infty$, $f = (f_n)_{n\geq 0} \in \mathcal{P}_p$. Then there exist processes $(g_m)_{m\geq 0}$, $(\psi_m)_{m\geq 0}$ such that*

$$f = \sum_{m=0}^\infty g_m \psi_m, \tag{2.8.18}$$

$$g_m \geq 0, \quad g_m \in L^\infty(\mathcal{F}_{k_m}), \quad E\Big(\Big(\sum_0^\infty g_m\Big)^p\Big)^{\frac{1}{p}} \leq 4\|f\|_{\mathcal{P}_p}, \tag{2.3.19}$$

$$E(\psi_m|\mathcal{F}_{k_m}) = 0, \quad \sup\|\psi_m\|_\infty \leq \frac{3}{2}. \tag{2.3.20}$$

Inversely, each expression (2.8.18) *with* (g_m), (ψ_m) *satisfying* (2.8.19), (2.8.20) *respectively, must be in* $\mathcal{P}_p$, *and*

$$\|f\|_{\mathcal{P}_p} \leq \frac{3}{2}\left\|\sum_0^\infty g_m\right\|_p. \tag{2.8.21}$$

Proof. Let $(r_n)_{n\geq 0}$ be any nonnegative increasing adapted process such that

$$|f_n| \leq r_{n-1}, \ \|f\|_{\mathcal{P}_p} = \|r_\infty\|_p.$$

For $a > 1$ determined later, for all $i \in \mathbf{Z}$, define stopping times

$$T_i = \inf\{n : r_n > a^i\}.$$

Then $\{T_i\}$ is an increasing sequence of stopping times. Denote

$$B^{(i)} = \{T_i < T_{i+1}\}, B_k^{(i)} = \{r_{k-1} \leq a^i < r_k \leq a^{i+1}\}. \tag{2.8.22}$$

Then each $B_k^{(i)} \in \mathcal{F}_k$, $\{B_k^{(i)}\}$ is disjoint with respect to each index for another index fixed, and $B^{(i)} = \bigcup_k B_k^{(i)}$. We have

$$\begin{aligned} f &= \sum_{-\infty}^\infty (f_{T_{i+1}} - f_{T_i})\chi_{B^{(i)}} = \sum_{-\infty}^\infty \sum_{k=0}^\infty (f_{T_{i+1}} - f_{T_i})\chi_{B_k^{(i)}} \\ &= \sum_{-\infty}^\infty \sum_{k=0}^\infty a^{i+1}\chi_{B_k^{(i)}} a^{-i-1}\chi_{B_k^{(i)}}(f_{T_{i+1}} - f_{T_i}) = \sum g_m\psi_m, \end{aligned}$$

where each g_m and ψ_m has following form respectively

$$g_m = a^{i+1}\chi_{B_k^{(i)}}, \quad \text{for some} \quad k, i,$$

$$\psi_m = a^{-i-1}\chi_{B_k^{(i)}}(f_{T_{i+1}} - f_{T_i}), \quad \text{for some} \quad k, i.$$

Now we prove that (g_m), (ψ_m) are what we want to find. In fact, let (k, i) be corresponding with m, then for (ψ_m), we have

$$|f_{T_{i+1}}| \leq r_{T_{i+1}-1} \leq a^{i+1}, \quad |f_{T_i}| \leq a^i,$$

$$\|\psi_m\|_\infty \leq 1 + \frac{1}{a},$$

$$\begin{aligned} E(\psi_m|\mathcal{F}_k) &= a^{-i-1}E(f_{T_{i+1}} - f_{T_i}|\mathcal{F}_k)\chi_{B_k^{(i)}} \\ &= a^{-i-1}(f_{T_{i+1}\wedge k} - f_{T_i\wedge k})\chi_{B_k^{(i)}} = 0, \end{aligned}$$

and for (g_m), we have

$$g_m \geq 0, \quad g_m \in L^\infty(\mathcal{F}_k), \quad \sum g_m \leq a^2(a-1)^{-1} r_\infty, \tag{2.8.23}$$

the last estimate in (2.8.23) follows from

$$\sum g_m = \sum_{i,k} a^{i+1} \chi_{B_k^{(i)}} \leq \sum_{i \leq j(x)} a^{i+1} \leq a^{j(x)+2}(a-1)^{-1} \leq a^2(a-1)^{-1} r_\infty,$$

where

$$j(x) = \sup\{i : a^i < r_\infty(x)\}.$$

And hence

$$E\Big(\Big(\sum_m g_m\Big)^p\Big)^{\frac{1}{p}} \leq a^2(a-1)^{-1}\|r_\infty\|_p = a^2(a-1)^{-1}\|f\|_{\mathcal{P}_p}.$$

For the sake of simplicity, take $a = 2$. This completes the proof of the first assertion of the theorem.

On the contrary, let (g_m), (ψ_m) satisfy (2.8.18)-(2.8.20). Noticing

$$E(g_m \psi_m | \mathcal{F}_n) = E(g_m E(\psi_m | \mathcal{F}_{k_m}) | \mathcal{F}_n) = 0, \quad k_m \geq n,$$

$$E(g_m \psi_m | \mathcal{F}_n) = g_m E(\psi_m | \mathcal{F}_n), \quad k_m < n,$$

we get

$$|f_n| = \left| \sum_{m: k_m < n} g_m E(\psi_m | \mathcal{F}_n) \right| \leq \frac{3}{2} \sum_{m: k_m < n} g_m = \rho_{n-1},$$

$$\|f\|_{\mathcal{P}_p} \leq \|\rho_\infty\|_p \leq \frac{3}{2} \left\| \sum_m g_m \right\|_p.$$

The proof of the theorem is finished. □

Remark From (2.8.23) we see $\sum |g_m \psi_m| \leq \sup \|\psi_m\|_\infty \sum |g_n| \leq C r_\infty$. This shows that for all $f \in L_0^\infty \subset \mathcal{P}_p$ the series in (2.8.18) converges boundedly pointwise.

We will show that there exists some duality between $\mathcal{P}_p$ and ${}_1l_{p'}$, so we want to do some further discussion on ${}_1l_p$. Obviously, the set G_1^p that occured in the definition of ${}_1l_p$ could be smaller, say like following : for $1 \leq p \leq \infty$,

$$\widetilde{G}_1^p = \Big\{\text{finite adapted process } g = (g_n) : \\ g_n \in L^\infty(\mathcal{F}_n), \ E\Big(\Big(\sum |g_n|\Big)^{p'}\Big) \leq 1\Big\}. \tag{2.8.24}$$

(Compare (2.8.24) with (2.7.15).) Now denote

$$F^p = \Big\{\text{finite sum } f = \sum g_m \psi_m : \ g_m \in L^\infty(\mathcal{F}_{k_m}),$$
$$E\Big(\Big(\sum |g_m|\Big)^{p'}\Big) \le 1, \quad E(\psi_m|\mathcal{F}_{k_m}) = 0, \ \sup \|\psi_m\|_\infty \le 1\Big\}, \qquad (2.8.25)$$

and give a new description of $\|\cdot\|_{{}_1l_p}$.

Lemma 2.8.5 *For $\varphi = (\varphi_n)_{n\ge 0} \in L^1_u$, we have*

$$\frac{1}{2}\|\varphi\|_{{}_1l_p} \le \sup_{f\in F^p} |E(f\varphi)| \le \|\varphi\|_{{}_1l_p}. \qquad (2.8.26)$$

Proof. Let $\varphi \in L^1_u$, $f \in F^p$. We have

$$E(f\varphi) = E\left(\sum_m g_m\psi_m\varphi\right) = E\left(\sum_n \sum_{m:k_m=n} g_m\psi_m(\varphi - \varphi_n)\right),$$

here neither convergence nor integrability trouble occurs. So

$$|E(f\varphi)| \le E\left(\sum_n \sum_{m:k_m=n} |g_m\varphi^{(n)}|\right) \sup_m \|\psi_m\|_\infty$$
$$\le \sup_{(h_n)\in \tilde{G}^p_1} E\left(\sum_n |h_n\varphi^{(n)}|\right) \le \|\varphi\|_{{}_1l_p},$$

where

$$h = (h_n)_{n\ge 0} \in \tilde{G}^p_1, \quad \text{with} \quad h_n = \sum_{m:k_m=n} |g_m|.$$

Now for $\varphi \in L^1_u$, let $g = (g_n)_{n\ge 0} \in \tilde{G}^p_1$ be arbitrary. Then (denote $\xi_n = \operatorname{sgn}(g_n\varphi^{(n)})$)

$$E\left(\sum_0^\infty |g_n||\varphi^{(n)}|\right) = E\left(\sum_0^\infty g_n\varphi^{(n)}\xi_n\right) = E\left(\sum_0^\infty g_n\varphi^{(n)}(\xi_n - E(\xi_n|\mathcal{F}_n)\right)$$
$$= E\left(\sum_0^\infty g_n\varphi^{(n)}\psi_n\right) = E\left(\sum_0^\infty g_n\psi_n\varphi\right) = 2E(f\varphi).$$

We claim that $f = \frac{1}{2}\sum g_n\psi_n \in F^p$. In fact

$$g_n \in L^\infty(\mathcal{F}_n), \quad E\left(\Big(\sum |g_m|\Big)^{p'}\right) \le 1,$$

$$E(\psi_n|\mathcal{F}_n) = 0, \quad \sup_n \|\psi_n\|_\infty \le 2.$$

This proves $\|\varphi\|_{{}_1l_p} \le 2\sup_{f\in F^p} |E(f\varphi)|$. The proof of the lemma is finished. □

Now we can get the characterization of $\mathcal{P}'_p \bigcap L^1_u$.

Theorem 2.8.6 *Let* $1 \leq p \leq \infty$. *Then* $\mathcal{P}_p' \bigcap L_u^1 = {}_1l_{p'}$, *More precisely, we have*

$$\frac{1}{2}\|\varphi\|_{{}_1l_{p'}} \leq \|l_\varphi\| \leq 6\|\varphi\|_{{}_1l_{p'}}. \tag{2.8.27}$$

Proof. Assume that $l \in \mathcal{P}_p'$ is generated from $\varphi = (\varphi_n)_{n\geq 0} \in L_u^1$ by $l(f) = l_\varphi(f) = E(f\varphi)$. As shown by the second part of Theorem 2.8.4, $F^{p'}$ is contained in the unit ball of $\mathcal{P}_p$. So we get

$$\|\varphi\|_{{}_1l_{p'}} \leq 2 \sup_{f\in F^{p'}} |l_\varphi(f)| \leq 2 \sup_{f\in\mathcal{P}_p, \|f\|_{\mathcal{P}_p}\leq 1} \|l_\varphi\|\,\|f\|_{\mathcal{P}_p} \leq 2\|l_\varphi\|.$$

On the contrary, we want to show that each $\varphi \in {}_1l_{p'}$ can define a bounded linear functional on $\mathcal{P}_p$. When $p = \infty$, it is obvious, since $\|f\|_{\mathcal{P}} = \|f\|_\infty$, and $\|\varphi\|_1 \leq \|\varphi\|_{{}_1l_1}$. Now consider the case $1 \leq p < \infty$. By the homogeneity, from (2.8.26) we get: For $f = \sum g_m\psi_m$, $g_m \in L^\infty(\mathcal{F}_{k_m})$, $E(\psi_m|\mathcal{F}_{k_m}) = 0$,

$$|E(f\varphi)| \leq \sup_m \|\psi_m\|_\infty E\left(\left(\sum |g_m|\right)^p\right)^{\frac{1}{p}} \|\varphi\|_{{}_1l_{p'}}. \tag{2.8.28}$$

Now let $f \in L_0^\infty \subset \mathcal{P}_p$, $\varphi \in {}_1l_{p'}$. Then f can be decomposed as $f = \sum g_m\psi_m$ with the series converging boundedly. Denoting $f^{(N)} = \sum_1^N g_n\psi_n$, and noticing ${}_1l_{p'} \subset L_u^1$, we see

$$E(f\varphi) = \lim_{N\to\infty} E(f^{(N)}\varphi). \tag{2.8.29}$$

Since $f^{(N)} \in F^{p'}$ modulo a constant, we have

$$|E(f^{(N)}\varphi)| \leq \sup_m \|\psi_m\|_\infty E\left(\left(\sum |g_m|\right)^p\right)^{\frac{1}{p}} \|\varphi\|_{{}_1l_{p'}}.$$

From (2.8.29), (2.8.19), (2.8.20), we get

$$|E(f\varphi)| \leq 6\|f\|_{\mathcal{P}_p}\|\varphi\|_{{}_1l_{p'}}. \tag{2.8.30}$$

(2.8.30) says that each $\varphi \in {}_1l_{p'}$ yields a bounded linear functional on a dense subspace of $\mathcal{P}_p$. This proves ${}_1l_{p'} \subset \mathcal{P}_p'$. Meanwhile ${}_1l_{p'} \subset L_u^1$. This completes the proof of the theorem. □

Remark In general, we could not expect $\mathcal{P}_p' = {}_1l_{p'}$. For example, assume $\mathcal{F}_0$=trivial one, $\mathcal{F}_1 = \mathcal{F}_2 = \ldots = \mathcal{F}$ on any $(\Omega, \mathcal{F}, \mu)$. Then ${}_1l_{p'} = L^1$ (modulo constants), $\mathcal{P}_p = L_0^\infty$. In general $(L_0^\infty)' \neq L^1$.

Notes to Chapter 2

§2.1. The weak type (1,1) inequality is often called Kolmogorof's inequality. The type (p, p) of M is due to J. Doob. The weak type (1,1), and type (p, p) of the square function operator S are due to Burkholder [1]. It may be said that the deep studies about S began from D. G. Austin's work which said that $S(f)$ is L^2-integrable on the set $\{Mf \leq \lambda\}$ for all $f \in L^1$, for all $\lambda > 0$. The famous inequality $C\|Mf\|_p \leq \|Sf\|_p \leq C\|Mf\|_p$ is due to Burkholder-Gundy [1], the proof of which given here is due to the author following an idea of Stroock [1]. The spaces ${}_2K_p$ and ${}_1K_p$ were introduced by Garsia [1], and he established the equivalence between ${}_1K_p$ and L^p. Davis' decomposition and Davis' inequality are due to Davis [1].

§2.2. Lemma 2.2.1 is taken from Garsia [1]. Theorem 2.2.2 is due to C. Fefferman but the proof of which is due to Herz [2] and Garsia [1]. Lemma 2.2.4 and the proof of Theorem 2.2.5 are due to Garsia [1].

§2.3. The idea from Fefferman's inequality to Davis' inequality is due to Garsia [1]. All results collected here are due to Garsia [1].

§2.4. The introduction of the spaces $\mathcal{P}_p$ and $\mathcal{A}_p$, and all results here are due to Garsia [1].

§2.5. The idea to prove Fefferman's inequality by atomic decomposition is due to Herz [1], and in classical case, is due to C. Fefferman, and Coifman [1], too. In martingale setting Bernard-Maisonneuve [1] did such practice, they defined the concept of atoms, and gave the atomic decomposition of H_1 in various cases, dyadic case and continuous martingale case etc.

§2.6. All results collected here are due to Dellacherie-Meyer-Yor [1].

§2.7-2.8. The h_p, ${}_2l_p$, and ${}_2\lambda_\beta$ were introduced and studied first by Herz [2], meanwhile ${}_2K_p$ was defined by Garsia [1]. All the results obtained in these two sections are due to Herz [1] essentially, but many of the proofs here are different from his. It should be indicated that it was C. Herz who discovered the atomic decomposition of spaces first in martingale setting as shown in Theorem 2.5.8 and Theorem 2.8.4.

3 Φ-inequalities on Martingales

The L^p-inequalities introduced in the preceding chapter, concerning various operators on martingales (such as maximal operator, square operator and conditioned square operator, etc.), could be extended to so-called Φ-inequalities. In this chapter, we will study them according to Φ's different situations: Φ is convex, or convex in a strict sense, or concave, or merely of moderate growth (we will call such Φ "general Φ" for the simplicity). §3.1 will be devoted to some elementary properties of convex functions, especially to the moderate growth property, and two related indices. §3.2–3.5 will be devoted to the general theory of Φ -inequalities in four cases, which may be useful in some domains other than martingale, then to Φ-inequalities on martingales. The last section §3.6 will be devoted to the rearrangement technique in martingale setting. This technique is almost equivalent to "distribution" one with slight differences between them. Roughly speaking, by means of the latter, we could get general Φ-inequalities, and by means of the former we lose some generality of Φ, but as a compensation, we get better constants. In this section, we keep the assumptions imposed on $(\Omega, \mathcal{F}, \mu, \{\mathcal{F}_n\}_{n\geq 0})$ as before.

3.1 Convex functions with moderate growth

In this chapter by convex function we mean an increasing convex function $\Phi(u)$ on $\mathbf{R}^+$ with $\Phi(0) = 0$. It is well known that any convex function like this has the form

$$\Phi(u) = \int_0^u \varphi(t)dt, \quad \forall u > 0, \tag{3.1.1}$$

where $\varphi(t)$ is a positive, increasing, finite valued, (say) left continuous function on $\mathbf{R}^+$. We want to study some elementary properties of such convex functions by means of so-called Young's complementary function. But, Young's complementary function (the definition will be introduced soon) need not be finite valued. So for the sake of convenience, we will adopt a slight generalization of a convex function, that is, $\varphi(t)$ in (3.1.1) could be permitted to be infinite valued on some interval (a, ∞) with $0 < a \leq \infty$. In (and only in) this section the corresponding generalized convex functions will be used to help in the studies on convex functions. For any $\varphi(t)$ of the generalized kind (that is, they are, maybe, infinite valued), we can define its inverse

$$\psi(s) = \inf\{t : \varphi(t) \geq s\}, \tag{3.1.2}$$

with the same convention: $\inf\{\emptyset\} = \infty$. It is easy to see that $\psi(s)$ is still of the same kind. It's integral $\Psi(v) = \int_0^v \psi(s)ds$, is a generalized convex function, will be called Young's complementary function of Φ. In the category of generalized convex functions, the complementarity is reflexive. An increasing continuous function φ on $\mathbf{R}^+$ is called to be of moderate growth, if for all $\alpha > 1$, there exists a constant C_α, such that

$$\Phi(\alpha u) \leq C_\alpha \Phi(u), \quad \forall u > 0^{\dagger}.$$

We will see soon that for a convex function, this property is intimately connected with the following index

$$p_\Phi = \sup_{0<u<\infty} \frac{u\varphi(u)}{\Phi(u)}. \tag{3.1.3}$$

We will also see that another index of a generalized convex function Φ will play the role of characterizing the moderate growth property of Φ's complementary Ψ, it is

$$q_\Phi = \inf_{0<u<a} \frac{u\varphi(u)}{\Phi(u)} \quad (\Phi \equiv \infty \text{ on } (a, \infty)). \tag{3.1.4}$$

The following theorem collects some properties of generalized convex functions.

Theorem 3.1.1 *Let $\Phi(u)$ be a generalized convex function, and $\Psi(v)$ be its Young's complementary function. Then we have*

(a) *$uv \leq \Phi(u) + \Psi(v)$, $\forall u, v$, and the equality holds if and only if $v = \varphi(u)$ or $u = \psi(v)$.*

(b) $\Phi(\lambda u) \leq \lambda \Phi(u)$, $\forall \lambda \leq 1$, $\forall u > 0$.

(c) *The following two assertions are equivalent:*

(1) *Φ is of moderate growth,*

(2) $p_\Phi < \infty$.

(d) *If $p = p_\Phi < \infty$, then for all $\lambda > 1$, we have $\Phi(\lambda u) \leq \lambda^p \Phi(u)$.*

(e) *If $p = p_\Phi < \infty$, then $\Phi(u)/u^p$ is decreasing; if $q = q_\Phi < \infty$ then $\Phi(u)/u^q$ is increasing.*

(f) *In any case. we have $p_\Phi = q'_\Psi$ and $p_\Psi = q'_\Phi$, with $'$ denoting conjugate index. So, Ψ is of moderate growth, if $q_\Phi > 1$.*

---

† It is easy to see that we have an equivalent definition: For some $\alpha > 1$ (usually $\alpha = 2$) $\exists C$, s.t. $\Phi(\alpha u) \leq C\Phi(u)$, $\forall u > 0$.

Proof. (a) This is just the famous Young's inequality. Thinking $\Phi(u)$ as the area under the curve represented by $v = \varphi(u)$, and analogously for $\Psi(v)$, then it is geometrically obvious that the area of the rectangle $\{0, (u, 0), (u, v), (0, v)\}$ is always less than or equals to the sum of $\Phi(u)$ and $\Psi(v)$, and that the equality holds if the point (u, v) is on the curve. The argument is valid in spite of $\Phi(u)$ or $\Psi(v)$ were not finite valued. The assertion is proved.

(b) When $\Phi(u) < \infty$, the inequality follows from the convexity. When $\Phi(u) = \infty$, there is nothing to prove.

(c) Suppose that $\Phi(u)$ were of moderate growth. Then for all u,

$$u\varphi(u) \leq \int_u^{2u} \varphi(t)dt = \Phi(2u) - \Phi(u) \leq (C-1)\Phi(u),$$

so $p_\Phi \leq C - 1 < \infty$. Now, from $p = p_\Phi < \infty$, we have

$$p\Phi(u) \geq u\varphi(u) \geq (p+1)\int_{\frac{p}{p+1}u}^{u} \varphi(t)dt = (p+1)\left(\Phi(u) - \Phi\left(\frac{p}{p+1}u\right)\right),$$

$$\Phi(u) \leq (p+1)\Phi\left(\frac{p}{p+1}u\right),$$

$$\Phi\left(\frac{p+1}{p}u\right) \leq (p+1)\Phi(u).$$

This proves that Φ is of moderate growth.

(d) From $\dfrac{u\varphi(u)}{\Phi(u)} \leq p$ we have

$$\int_u^{\lambda u} \frac{d\Phi(t)}{\Phi(t)} \leq p\int_u^{\lambda u} \frac{dt}{t},$$

$$\log\frac{\Phi(\lambda u)}{\Phi(u)} \leq \log \lambda^p,$$

$$\Phi(\lambda u) \leq \lambda^p \Phi(u).$$

(e) From $\Phi(\lambda u) \leq \lambda^p\Phi(u)$, we have

$$\frac{\Phi(\lambda u)}{(\lambda u)^p} \leq \frac{\Phi(u)}{u^p}, \quad \forall \lambda \geq 1, \quad \forall u > 0.$$

This means that $\frac{\Phi(u)}{u^p}$ is decreasing. Analogously, on $[0, a)$, we have

$$\frac{u\varphi(u)}{\Phi(u)} \geq q \Rightarrow \Phi(\lambda u) \geq \lambda^q \Phi(u)$$

$$\Rightarrow \frac{\Phi(\lambda u)}{(\lambda u)^q} \geq \frac{\Phi(u)}{u^q}, \quad \forall \lambda \geq 1, \quad \forall u \in [0, a).$$

And on (a,∞), nothing has to be proved.

(f) Since when Φ takes infinite value, $p_\Phi=\infty$, $q_\Psi=1$, $p_\Phi=q'_\Psi$ holds naturally, so it is enough to consider the case $\Phi(u)<\infty$ for all u. We want to show

$$\inf_u \frac{u\varphi(u)}{\Psi(\varphi(u))}=\inf_{v:\psi(v)<\infty}\frac{v\psi(v)}{\Psi(v)}=q_\Psi. \tag{3.1.5}$$

Once it is proved, then we would have

$$1=\frac{\Phi(u)+\Psi(\varphi(u))}{u\varphi(u)}=\inf_u\frac{\Phi(u)}{u\varphi(u)}+\sup_u\frac{\Psi(\varphi(u))}{u\varphi(u)}=\frac{1}{p_\Phi}+\frac{1}{q_\Psi},$$

and hence the assertion.

Now prove (3.1.5). Noticing that $\psi(\varphi(u))\le u$, for all u, we have

$$\inf_u\frac{u\varphi(u)}{\Psi(\varphi(u))}\ge\inf_u\frac{\psi(\varphi(u))\varphi(u)}{\Psi(\varphi(u))}=\inf_{v\in\{\varphi(u):u\in R^+\}}\frac{v\psi(v)}{\Psi(v)}\ge q_\Psi.$$

On the contrary, for all $\alpha>q_\Psi$, there exists v_0 such that $\frac{v_0\psi(v_0)}{\Psi(v_0)}<\alpha$. Without loss of generality, v_0 could be assumed to be a discoutinuity point of $\psi(v)$ (otherwise, the situation is similar, and simpler). Notice that $\varphi(u)$ takes a constant value v_0 on the interval $(\psi(v_0),\psi(v_0+)]$, so for all $u\in(\psi(v_0),\psi(v_0+)]$, we have

$$\frac{u\varphi(u)}{\Psi(\varphi(u))}=\frac{uv_0}{\Psi(v_0)}\to\frac{v_0\psi(v_0)}{\Psi(v_0)}<\alpha,\quad\text{when } u\to\psi(v_0),$$

and hence

$$\inf_u\frac{u\varphi(u)}{\Psi(\varphi(u))}<\alpha,\quad \inf\frac{u\varphi(u)}{\Psi(\varphi(u))}\le q_\Psi.$$

This completes the proof of the assertion. The theorem is proved. □

Remark When $1<q_\Phi\le p_\Phi<\infty$, then $1<q_\Psi\le p_\Psi<\infty$. In this case, Φ and Ψ are both usual convex functions.

Now consider Orlicz space L^Φ defined by a generalized convex function. Let $(\Omega,\mu,\mathcal{F})$ be any σ-finite(nonnegative) measure space, Φ be any generalized convex function with its complementary function Ψ . Then

$$L^\Phi=\left\{f\text{ measurable: } N_\Phi(f)=\sup_{g:E(\Psi(|g|))\le1}\left|\int_\Omega fgd\mu\right|<\infty\right\}. \tag{3.1.6}$$

It is a Banach space, on which there is another equivalent norm, that is

$$\|f\|_\Phi=\inf\left\{\lambda>0:\ E\left(\Phi\left(\frac{|f|}{\lambda}\right)\right)\le1\right\}. \tag{3.1.7}$$

That this is really a norm follows from the fact

$$\Phi\left(\frac{|f+g|}{a+b}\right)\le\frac{a}{a+b}\Phi\left(\frac{|f|}{a}\right)+\frac{b}{a+b}\Phi\left(\frac{|g|}{b}\right),$$

where f, g are measurable functions, and $a, b \in \mathbf{R}^+$. That the two norms are equivalent are indicated by

$$\| f \|_\Phi \leq N_\Phi(f) \leq 2 \| f \|_\Phi . \tag{3.1.8}$$

Some other facts we will need are listed as follows:

$$E\left(\Phi\left(\frac{|f|}{\| f \|_\Phi}\right)\right) \leq 1, \tag{3.1.9}$$

$$E\left(\Phi\left(\frac{|f|}{N_\Phi(f)}\right)\right) \leq 1, \tag{3.1.10}$$

$$|E(fg)| \leq \max(E(\Psi(|g|)), 1) N_\Phi(f). \tag{3.1.11}$$

The proof may be seen, for example, in Zygmund [1] or Zaanen [1]. Here we only want to point that when $p_\Phi < \infty$, then $L^\Phi = \{f : E(\Phi(| f |)) < \infty\}$; and that in the finite measure space case, when $\lim_{u \to \infty} \varphi(u) = l < \infty$, then $L^\Phi$is essentially L^1, and when $\varphi(u) \equiv \infty$ on (a, ∞) then L^Φ is essentially L^∞.

3.2 Convex Φ-inequalities

We begin with a lemma which establishes the equivalence between convex Φ-inequalities and a kind of simple integral inequalities involving two measurable functions.

Lemma 3.2.1 *Let W, Y be two nonnegative measurable functions, Φ be any convex function with its derivative φ (say $\varphi(0) = \varphi(0+)$). Then following inequalities are equivalent:*

$$\int_{\{W > \lambda\}} (W - \lambda) d\mu \leq \int_{\{W > \lambda\}} Y d\mu, \quad \forall \lambda > 0, \tag{3.2.1}$$

$$\int_\Omega \Phi(W) d\mu \leq \int_\Omega \varphi(W) Y d\mu, \quad \forall \text{ convex } \Phi. \tag{3.2.2}$$

Proof. Let $\varphi(u) = \chi_{\{u > \lambda\}}$, then $\Phi(u) = (u - \lambda)^+$, and so from (3.2.2) we get (3.2.1). Now suppose (3.2.1) holds. Integrating two sides of (3.2.1) on $[0, \infty)$ with respect to the measure $d\varphi(\lambda)$, we get

$$\int_0^\infty \int_{\{W > \lambda\}} (W - \lambda) d\mu d\varphi(\lambda) \leq \int_0^\infty \int_{\{W > \lambda\}} Y d\mu d\varphi(\lambda).$$

The term on the left-hand side is

$$\begin{aligned}\int_\Omega \int_0^{W-} (W - \lambda) d\varphi(\lambda) d\mu &= \int_\Omega \left\{ (W - \lambda)\varphi(\lambda) \,|_0^{W-} + \int_0^{W-} \varphi(\lambda) d\lambda \right\} d\mu \\ &= \int_\Omega \{\Phi(W) - \varphi(0) W\} d\mu.\end{aligned}$$

And the term on the right-hand side is

$$\int_\Omega Y \int_0^{W-} d\varphi(\lambda) d\mu = \int_\Omega \{Y\varphi(W) - \varphi(0)Y\} d\mu.$$

Since (3.2.1) holds for $\lambda = 0$, too, we have

$$\begin{aligned}\int_\Omega \Phi(W) d\mu &\le \int_\Omega Y\varphi(W) d\mu + \varphi(0)\int_\Omega W d\mu - \varphi(0)\int_\Omega Y d\mu \\ &\le \int_\Omega Y\varphi(W) d\mu.\end{aligned}$$

This proves (3.2.1) $\Rightarrow$ (3.2.2). The proof of the lemma is finished. □

Remark We have following conditioned version of the assertion in the lemma: following (3.2.3) and (3.2.4) are equivalent

$$E_0(\chi_{\{W>\lambda\}}(W-\lambda)) \le E_0(\chi_{\{W>\lambda\}}Y), \quad \forall \lambda > 0, \tag{3.2.3}$$

$$E_0(\Phi(W)) \le E_0(Y\varphi(W)), \quad \forall \text{ convex } \Phi, \tag{3.2.4}$$

where $E_0 = E(\cdot \mid \mathcal{F}_0)$.

Lemma 3.2.2 *Let Y be any nonnegative measurable function, $A = (A_n)_{n\ge 0}$ be any nonnegative, increasing, adapted process satisfying*

$$E(A_\infty - A_{T-1} \mid \mathcal{F}_T) \le E(Y \mid \mathcal{F}_T), \quad \textit{for all stopping times } T; \tag{3.2.5}$$

or $A = (A_n)_{n\ge 0}$ be any nonnegative, increasing, predictable process with $A_0 \equiv 0$, and such that

$$E(A_\infty - A_T \mid \mathcal{F}_T) \le E(Y \mid \mathcal{F}_T), \quad \textit{for all stopping times } T. \tag{3.2.6}$$

Then

$$E_0(\chi_{\{A_\infty>\lambda\}}(A_\infty - \lambda)) \le E_0(\chi_{\{A_\infty>\lambda\}}Y), \quad \forall \lambda > 0. \tag{3.2.7}$$

Proof. In the first case, for λ given , we define a stopping time $T = \inf\{n : A_n > \lambda\}$. For some λ and some $\omega \in \Omega$, we could have $T(\omega) = 0$ of course. In this case, we mean $A_{T(\omega)-1}(\omega) = 0$. So in any case, we have $A_{T-1} \le \lambda$, and

$$A_\infty - \lambda \le A_\infty - A_{T-1}, \quad \text{a.e.}$$

Then we get

$$\begin{aligned}E_0(\chi_{\{A_\infty>\lambda\}}(A_\infty - \lambda)) &\le E_0(\chi_{\{T<\infty\}}(A_\infty - A_{T-1})) \\ &= E_0(\chi_{\{T<\infty\}}E(A_\infty - A_{T-1} \mid \mathcal{F}_T)) \\ &\le E_0(\chi_{\{T<\infty\}}E(Y \mid \mathcal{F}_T)) = E_0(\chi_{\{T<\infty\}}Y).\end{aligned}$$

In the second case, the proof remains unchanged provided the stopping time T is defined as $T = \inf\{n : A_{n+1} > \lambda\}$, owing to the fact $A_T \le \lambda$ (even $T = 0$, since

$A_0 = 0$ is assumed). The proof is thus finished. □

Remarks

1. Lenglart-Lepingle-Pratelli [1] reformulated the conditions (3.2.5) and (3.2.6) as following

$$E(A_\infty - A_{T-1}) \le E(Y\chi_{\{T<\infty\}}), \quad \text{for all stopping times } T, \tag{3.2.8}$$

$$E(A_\infty - A_T) \le E(Y\chi_{\{T<\infty\}}), \text{ for predictable } A \text{ with } A_0 \equiv 0, \ \forall T. \tag{3.2.9}$$

These are not new conditions, but equivalent ones.

2. It is sufficient to verify (3.2.5) or (3.2.6) for stopping times taking constant values n.

3. The condition (3.2.6) comes from Potential Theory. Let $(Q_n)_{n\ge 0}$ be a potential generated by an increasing predictable process $A = (A_n)_{n\ge 0}$ with $A_0 \equiv 0$, i.e. $Q_n = E(A_\infty - A_n \mid \mathcal{F}_n)$. Then the condition (3.2.6) means that Q_n has a domination Y. Two lemmas say that a domination of potential itself implies some domination of its associated predictable process.

4. Lemmas 1 and 2 are called Garsia-Neveu Lemmas.

Now we deduce the convex Φ-inequality from (3.2.2) after Dellacherie [1]. This approach is better than the previous one, it could give the best constant in the obtained inequality.

Lemma 3.2.3 *Let Φ be any convex function, U, V be two nonnegative measurable functions satisfying*

$$E(U\varphi(U)) < \infty, \quad E(U\varphi(U)) \le E(V\varphi(U)) + C_V, \tag{3.2.10}$$

where C_V is a constant depending on V. Then with the same constant, we have

$$E(\Phi(U)) \le E(\Phi(V)) + C_V. \tag{3.2.11}$$

Proof. Substituting u, v by $U(\omega)$, $V(\omega)$ respectively in the following inequalities, then integrating terms of both-hand sides

$$u\varphi(u) = \Phi(u) + \Psi(\varphi(u)), \quad \forall u,$$

$$v\varphi(u) \le \Phi(v) + \Psi(\varphi(u)), \quad \forall u, v,$$

we get

$$E(U\varphi(U)) = E(\Phi(U)) + E(\Psi(\varphi(U))),$$

$$E(V\varphi(U)) \le E(\Phi(V)) + E(\Psi(\varphi(U))).$$

Since $E(\Psi(\varphi(U))) \le E(U\varphi(U)) < \infty$, we get (3.2.11). The proof is finished. □

Remarks

1. The conditioned version of the lemma holds, too. That is to say E could be replaced by E_0 in (3.2.10) and (3.2.11), and C_V is a $\mathcal{F}_0$ measurable function.

2. Φ in the lemma could be a generalized one. And when $C_V = 0$, we have $\|U\|_\Phi \le \|V\|_\Phi$, as shown in following theorem (there only the $p_\Phi < \infty$ case is discussed.)

Now we deduce convex Φ-inequalities for processes from the preceding lemmas.

Theorem 3.2.4 *Let $A = (A_n)_{n\ge 0}$ be a nonnegative increasing, adapted or predictable process, and $A_0 \equiv 0$ in the second case, satisfying (3.2.5) or (3.2.6) respectively, Φ be any convex function, with $p = p_\Phi < \infty$. Then the following assertions hold*

(a) $E(\Phi(A_\infty)) \le E(Y\varphi(A_\infty)$,
(b) $E((A_\infty\varphi(A_\infty)) \le E(pY\varphi(A_\infty))$,
(c) $E(\Phi(A_\infty)) \le E(\Phi(pY))$,
(d) $\|A_\infty\|_\Phi \le p\|Y\|_\Phi$.

Proof. (a) It follows from Lemmas 3.2.1, 3.2.2.

(b) It follows from (a) because of $A_\infty\varphi(A_\infty) \le p\Phi(A_\infty)$.

(d) (c) implies (d) by making use of the homogeneity: Consider a new process $(\frac{A_n}{p\|Y\|_\Phi})_{n\ge 0}$ which satisfies (3.2.5) or (3.2.6) likewise with a new dominant $\frac{Y}{p\|Y\|_\Phi}$. So from (c) we have

$$E\left(\Phi\left(\frac{A_\infty}{p\|Y\|_\Phi}\right)\right) \le E\left(\Phi\left(pY\frac{1}{p\|Y\|_\Phi}\right)\right) \le 1.$$

From this we get $\|A_\infty\| \le p\|Y\|_\Phi$.

(c) It remains to prove (b) $\Rightarrow$ (c). When $E(A_\infty\varphi(A_\infty)) < \infty$, it follows from Lemma 3.2.3 immediately . Otherwise, we consider the process $A\wedge N = (A_n\wedge N)_{n\ge 0}$. It has the same properties and the same dominant as A. The last assertion can be seen as follows. Noting $u \wedge N$ is a concave function, so

$$E(A_\infty \wedge N \mid \mathcal{F}_n) \le E(A_\infty \mid \mathcal{F}_n) \wedge N, \quad \forall n,$$

$$E(A_\infty \wedge N \mid \mathcal{F}_n) - A_{n-1} \wedge N \le E(A_\infty \mid \mathcal{F}_n) \wedge N - A_{n-1} \wedge N. \tag{3.2.12}$$

It is easy to verify

$$E(A_\infty \wedge N \mid \mathcal{F}_n) - A_{n-1} \wedge N \le E(Y \mid \mathcal{F}_n), \quad \forall n,$$

from (3.2.12). Hence (c) holds for $A_\infty \wedge N$. Making $N \to \infty$, we complete the proof of (b) $\Rightarrow$ (c).

The proof of the theorem is finished. □

Remarks

1. Let $A = (A_n)_{n\geq 0}$ be as above with the dominant $Y = \gamma$, a constant. Then from the Lemma 3.2.1, we have

$$E(A_\infty^n) \leq n!\gamma^n, \qquad \forall n,$$
$$E(\varepsilon^{\lambda A_\infty}) \leq \frac{1}{1-\lambda\gamma}, \quad 0 < \lambda < \frac{1}{\gamma},$$

by taking $\Phi(u) = u^n$ and $\Phi(u) = \varepsilon^{\lambda u} - 1$ respectively.

2. (a), (b) and (c) of the theorem can be formulated in terms of E_0 replacing E.

Theorem 3.2.5 *Let Φ be any convex function. Then there exists a constant C such that*

$$C\|Mf\|_\Phi \leq \|S(f)\|_\Phi \leq C\|Mf\|_\Phi, \quad \textit{for all martingales } f = (f_n)_{n\geq 0}.$$

Proof. For $f = (f_n)_{n\geq 0}$ given, consider two increasing adapted processes $Mf = (M_n f)_{n\geq 0}$ and $S(f) = (S_n(f))_{n\geq 0}$ with

$$M_n f = \sup_{0\leq k\leq n} |f_k|, \quad S_n(f) = \left(|f_0|^2 + \sum_1^n |\Delta_k f|^2\right)^{\frac{1}{2}}, \quad n \geq 0.$$

In order to prove the assertion, it is enough to verify (3.2.5), for $A = S(f)$, $Y = CMf$, or $A = Mf$, $Y = CS(f)$, and for $T = n$, for all n. Consider a new increasing family of sub-σ-fields $\{\mathcal{F}'_m\}_{m\geq 0}$ with $\mathcal{F}'_m = \mathcal{F}_{n+m}, n \in \mathbf{Z}^+$ fixed. Let $f = (f_m)_{m\geq 0}$ be any martingale with respect to the $(\Omega, \mathcal{F}, \mu, \{\mathcal{F}_m\}_{m\geq 0})$. Let

$$g'_m = f_{m+n} - f_{n-1}, \quad m \geq 0.$$

Then $(g'_m)_{m\geq 0}$ is a martingale with respect to $(\Omega, \mathcal{F}, \mu, \{\mathcal{F}'_m\}_{m\geq 0})$. In fact, we have

$$E(g'_{m+1} \mid \mathcal{F}'_m) = E(f_{n+m+1} - f_{n-1} \mid \mathcal{F}_{n+m}) = f_{n+m} - f_{n-1} = g'_m, \quad \forall m \geq 0.$$

From Davis' inequality formulated in §2.3 (see Theorem 2.3.4), we have

$$\begin{aligned} E(Mf - M_{n-1}f \mid \mathcal{F}_n) &\leq E(\sup_{m\geq 0} |f_{n+m} - f_{n-1}| |\mathcal{F}_n) \\ &= E(Mg' \mid \mathcal{F}'_0) \leq CE(S(g') \mid \mathcal{F}'_0) \\ &= CE(\sqrt{S(f)^2 - S_{n-1}(f)^2} \mid \mathcal{F}_n) \\ &\leq CE(S(f) \mid \mathcal{F}_n), \quad \forall n. \end{aligned}$$

Analogously, we have also

$$\begin{aligned} E(S(f) - S_{n-1}(f) \mid \mathcal{F}_n) &\leq E(\sqrt{S(f)^2 - S_{n-1}(f)^2} \mid \mathcal{F}_n) \\ &= E(S(g') \mid \mathcal{F}'_0) \\ &\leq CE(Mg' \mid \mathcal{F}'_0) \\ &\leq CE(Mf \mid \mathcal{F}_n), \quad \forall n. \end{aligned}$$

This verifies (3.2.5) for $A = Mf$, $Y = CS(f)$ or $A = S(f)$, $Y = CMf$. The proof of the theorem is thus finished. □

Theorem 3.2.6 *Let Φ be any convex function with $p = p_\Phi < \infty$. Then*

$$\| \, |f_0|^2 + \sigma^2(f)\|_\Phi \leq p\|S^2(f)\|_\Phi, \quad \forall f = (f_n)_{n\geq 0}. \tag{3.2.13}$$

Proof. For $f = (f_n)_{n\geq 0}$ given, consider the process $A = (A_n) \geq 0$,

$$A_n = | f_0 |^2 + \sum_1^n E(| \Delta_k f |^2 | \mathcal{F}_{k-1}), \quad n \geq 1, \ A_0 \equiv 0,$$

and $Y = (Y_n)_{n\geq 0}$,

$$Y_n = | f_0 |^2 + \sum_1^n | \Delta_k f |^2, \quad n \geq 1, \ Y_0 \equiv 0.$$

Then, for $n \geq 1$, we have

$$\begin{aligned} E(A_\infty - A_n \mid \mathcal{F}_n) &= E\left(\sum_{n+1}^\infty E(| \Delta_k f |^2 | \mathcal{F}_{k-1}) \,\Big|\, \mathcal{F}_n\right) = E\left(\sum_{n+1}^\infty | \Delta_k f |^2 \Big| \mathcal{F}_n\right) \\ &= E(Y_\infty - Y_n \mid \mathcal{F}_n) \leq E(Y_\infty \mid \mathcal{F}_n), \end{aligned}$$

and also

$$\begin{aligned} E(A_\infty \mid \mathcal{F}_0) &= E\left(| f_0 |^2 + \sum_1^\infty E(| \Delta_k f |^2 | \mathcal{F}_{k-1}) \,\Big|\, \mathcal{F}_0\right) \\ &= E\left(| f_0 |^2 + \sum_1^\infty | \Delta_k f |^2 \Big| \mathcal{F}_0\right) = E(Y \mid \mathcal{F}_0). \end{aligned}$$

Noting that A is predictable, Theorem 3.2.4 gives

$$\|A_\infty\|_\Phi \leq p\|Y_\infty\|_\Phi.$$

The proof of the theorem is finished. □

Remark Applying the theorem to the case $\Phi(u) = u^{\frac{p}{2}}$, $p \geq 2$, we get

$$\| \, |f_0|^2 + \sigma(f)\|_p \leq \sqrt{\frac{p}{2}}\|S(f)\|_p, \quad p \geq 2.$$

Theorem 3.2.6′ *Let Φ be any convex function with $p = p_\Phi < \infty$. Then*

$$\| \, |f_0|^2 + \sigma^2(f)\|_\Phi \leq p\| \, | f |^2 \, \|_\Phi, \quad \forall f = (f_n)_{n\geq 0}. \tag{3.2.14}$$

Proof. As in Theorem 3.2.6, let $A = (A_n)_{n\geq 0}$. But set $Y = | f |^2$ in this case. Then we have

$$E(A_\infty \mid \mathcal{F}_0) = E(|f_0|^2 + \sigma^2(f) \mid \mathcal{F}_0) = E(| f |^2 | \mathcal{F}_0),$$

$$E(A_\infty - A_n \mid \mathcal{F}_n) = E\left(\sum_{n+1}^{\infty} \mid \Delta_k f \mid^2 \middle| \mathcal{F}_n\right) = E(\mid f - f_n \mid^2 \mid \mathcal{F}_n)$$
$$= E(\mid f \mid^2 - \mid f_n \mid^2 \mid \mathcal{F}_n) \le E(\mid f \mid^2 \mid \mathcal{F}_n), \quad \forall n \ge 1.$$

This proves the theorem. □

Remark Applying the theorem to the case $\Phi(u) = u^{\frac{p}{2}}$, $p \ge 2$, we get

$$\| \, |f_0|^2 + \sigma(f) \|_p \le \sqrt{\frac{p}{2}} \|f\|_p, \quad p \ge 2.$$

Now we want to consider a stronger convex Φ-inequality on martingales . For any martingale $f = (f_n)_{n \ge 0}$, define

$$Mf \vee Sf = \max(Mf, S(f)), \quad Mf \wedge Sf = \min(Mf, S(f)). \tag{3.2.15}$$

Could we still have a convex Φ-inequality between $M \wedge S$ and $M \vee S$? The answer is affirmative. Before answering it, we give the L^Φ-estimate of the measurable function having the form $\sum_0^\infty E(h_n | \mathcal{F}_n)$ with $(h_n)_{n \ge 0}$ any nonnegative process in the following lemma, which is the extension of Lemma 2.4.3.

Lemma 3.2.7 *Let Φ be a convex function with* $p = p_\Phi < \infty$, $h \ge 0$, $h \in L^\Phi$, *and* $(\varepsilon_n)_{n \ge 0}$ *be any process satisfying* $\sum \mid \varepsilon_n \mid \le 1$. *Then* $g = \sum E(h\varepsilon_n \mid \mathcal{F}_n)$ *satisfies*

$$\|g\|_\Phi \le p_\Phi \|h\|_\Phi. \tag{3.2.16}$$

Proof. Set

$$W_n = \sum_{k=0}^{n} E(h \mid \varepsilon_k \| \mathcal{F}_k), \quad W = W_\infty,$$

$$Y_n = \sum_0^n h \mid \varepsilon_k \mid, \quad Y = Y_\infty.$$

Define the stopping time

$$T = \inf\{n : W_n > \lambda\}, \quad \forall \lambda > 0.$$

We have

$$\int_{\{W>\lambda\}} (W - \lambda) d\mu \le \int_{\{T<\infty\}} E(W - W_{T-1} \mid \mathcal{F}_T) d\mu$$
$$= \int_{\{T<\infty)\}} E(Y - Y_{T-1} \mid \mathcal{F}_T) d\mu \le \int_{\{W>\lambda\}} Y d\mu.$$

Then the Theorem 3.2.4 gives the desired result,

$$\|g\|_\Phi \le \|W\|_\Phi \le p\|Y\|_\Phi \le p\|h\|_\Phi.$$

The proof is complete. □

Theorem 3.2.8 *Let Φ be any convex function with $p = p_\Phi < \infty$. Then*

$$E(\Phi(Mf \vee Sf)) \leq CE(\Phi(Mf \wedge Sf)), \quad \forall f = (f_n)_{n\geq 0}.$$

Proof. First, we want to establish the following inequality

$$E((Mf \vee Sf))^2 \mid \mathcal{F}_0) \leq CE((Mf \wedge Sf)^2 \mid \mathcal{F}_0), \quad \forall f = (f_n)_{n\geq 0}. \tag{3.2.17}$$

In fact, remember, we have the following Davis' inequalities

$$E(Mf - M_{n-1}f \mid \mathcal{F}_n) \leq CE(S(f) \mid \mathcal{F}_n), \quad \forall n,$$

$$E(S(f) - S_{n-1}(f) \mid \mathcal{F}_n) \leq CE(Mf \mid \mathcal{F}_n), \quad \forall n.$$

By making use of Theorem 3.2.4, we get (for $\Phi(u) = u^2$)

$$E_0((Mf)^2) \leq CE_0(MfS(f)), E_0(S(f)^2) \leq CE_0(MfS(f)),$$

and hence

$$\begin{aligned} E_0(\max((Mf)^2, S(f)^2)) &\leq E_0((Mf)^2 + S(f)^2) \leq CE_0(MfS(f)) \\ &= CE_0(Mf \vee Sf \cdot Mf \wedge Sf) \\ &\leq CE_0(Mf \vee Sf)^2)^{\frac{1}{2}} E_0((Mf \wedge Sf)^2)^{\frac{1}{2}}. \end{aligned}$$

When $E_0((Mf \vee Sf)^2) < \infty$, a.e., (3.2.17) follows immediately. In the general case, by passage to limits, (3.2.17) holds. Now we want to prove that for any martingale $g = (g_n)_{n\geq 0}$ such that there exists a nonnegative, increasing and adapted process $D = (D_n)_{n\geq 0}$ satisfying $\mid \Delta_n g \mid\leq D_{n-1}$, for all n, we have

$$E_0(Mg \vee Sg) \leq CE_0(Mg \wedge Sg + D_\infty). \tag{3.2.18}$$

For the simplicity, denote $M \vee S$ and $M \wedge S$ by K and k respectively. Let $g = (g_n)_{n\geq 0}$ be given like this, and $\lambda > 0$ be given. Set $\rho_n = k_{n-1}(g) + D_{n-1}$, and define a stopping time by

$$\tau = \inf\{n : \rho_{n+1} > \lambda\}.$$

Then

$$\{\rho_\infty > \lambda\} = \{\tau < \infty\},$$

$$k_\tau(g) \leq k_{\tau-1}(g) + D_{\tau-1} = \rho_\tau \leq \lambda,$$

and

$$\{\tau = \infty\} = \{\rho_\infty \leq \lambda\} \subset \{k(g) \leq \lambda\}.$$

We have

$$\begin{aligned} E_0(\chi_{\{K(g)>\lambda\}}) &\leq E_0(\chi_{\{K(g)>\lambda\}\cap\{\tau=\infty\}}) + E_0(\chi_{\{\tau<\infty\}}) \\ &\leq E_0(\chi_{\{K(g^{(\tau)})>\lambda\}}) + E_0(\chi_{\{\rho_\infty>\lambda\}}) \end{aligned}$$

$$\begin{aligned}
&\leq \frac{1}{\lambda^2} E_0(K(g^{(\tau)})^2) + E_0(\chi_{\{\rho_\infty > \lambda\}}) \\
&\leq \frac{c}{\lambda^2} E_0(k(g^{(\tau)})^2) + E_0(\chi_{\{\rho_\infty > \lambda\}}) \\
&= \frac{c}{\lambda^2} E_0(k(g^{(\tau)})^2 \chi_{\{\tau = \infty\}}) \\
&\quad + \frac{c}{\lambda^2} E_0(k(g^{(\tau)})^2 \chi_{\{\tau < \infty\}}) + E_0(\chi_{\{\rho_\infty > \lambda\}}) \\
&\leq \frac{c}{\lambda^2} E_0(k(g)^2 \chi_{\{k(g) \leq \lambda\}}) + c E_0(\chi_{\{\rho_\infty > \lambda\}}). \qquad (3.2.19)
\end{aligned}$$

So, for all $F \in \mathcal{F}_0$,

$$\int_F \chi_{\{K(g) > \lambda\}} d\mu \leq \frac{c}{\lambda^2} \int_F k(g)^2 \chi_{\{k(g) \leq \lambda\}} d\mu + c \int_F \chi_{\{\rho_\infty > \lambda\}} d\mu.$$

Integrating both sides with respect to $\lambda \in (0, \infty)$, we get

$$\int_F K(g) d\mu \leq c \int_F k(g) d\mu + \int_F \rho_\infty d\mu = c \int_F (k(g) + D_\infty) d\mu.$$

This gives (3.2.18).

That (3.2.18) implies for all convex functions Φ, for all $g = (g_n)_{n \geq 0}$ as above,

$$E_0(\Phi(K(g))) \leq C E_0(\Phi(k(g) + D_\infty)), \qquad (3.2.20)$$

is just a routine. For n fixed, consider $\{\mathcal{F}'_m\}_{m \geq 0} = \{\mathcal{F}_{n+m}\}_{m \geq 0}$ and a martingale $g' = (g'_m)$, $g'_m = g_{m+n} - g_{n-1}$, $m \geq 0$. Then $D' = (D'_m)$, $D'_m = D_{m+n}$ plays the same role. Applying (3.2.18) to this case, and noticing

$$\begin{aligned}
K(g) - K_{n-1}(g) &\leq \max(Mg - M_{n-1}g, S(g) - S_{n-1}(g)) \\
&\leq \max(Mg', S(g')) = K(g'),
\end{aligned}$$

$$k(g') \leq \min(2Mg, S(g)) \leq 2k(g),$$

we get

$$\begin{aligned}
E(K(g) - K_{n-1}(g) \mid \mathcal{F}_n) &\leq E(K(g') \mid \mathcal{F}'_0) \\
&\leq C E(k(g') + D'_\infty \mid \mathcal{F}'_0) \\
&\leq C E(k(g) + D_\infty \mid \mathcal{F}_n), \quad \forall n.
\end{aligned}$$

From this, the Theorem (3.2.4) gives (3.2.20).

Now we are in the position to prove (3.2.17). Let $f = (f_n)_{n \geq 0}$ be any martingale. Making Davis' decomposition $f = g + h$, with

$$| \Delta_n g | \leq C M_{n-1} d \quad (d_n = \Delta_n f),$$

$$\sum_0^\infty | \Delta_n h | \leq C \sum_0^\infty \{M_n d - M_{n-1} d + E(M_n d - M_{n-1} d \mid \mathcal{F}_{n-1})\},$$

and noting that

$$Md \leq \min(2Mf, S(f)),$$

$$\max(Mh, S(h)) \le \sum_0^\infty | \Delta_n h |,$$

$$K(f) \le K(g) + K(h),$$

$$\begin{aligned} k(g) &\le \min(Mf + Mh, S(f) + S(h)) \\ &\le \min(Mf, S(f)) + \sum_0^\infty | \Delta_n h |, \end{aligned}$$

$$\begin{aligned} E\Big(\Phi\Big(\sum_0^\infty | \Delta_n h |\Big)\Big) &\le CE(\Phi(Md)) + CE\Big(\Phi\Big(\sum_0^\infty E(M_n d - M_{n-1} d | \mathcal{F}_{n-1})\Big)\Big) \\ &\le CE(\Phi(Md)) \le CE(\Phi(k(f))), \end{aligned}$$

we finally get

$$\begin{aligned} E(\Phi(K(f)) &\le CE(\Phi(K(g))) + CE(\Phi((K(h))) \\ &\le CE(\Phi(k(g))) + CE(\Phi(Md))) + CE(\Phi(k(f))) \\ &\le CE(\Phi(k(f))). \end{aligned}$$

Here we have used the fact $E\Big(\Phi\Big(\sum_0^\infty E(M_n d - M_{n-1} d \mid \mathcal{F}_{n-1})\Big)\Big) \le CE(\Phi(Md))$, which follows from Lemma 3.2.7. This proves the theorem. □

3.3 Convex Φ-inequalities with $q_\Phi > 1$

Doob's inequality on maximal function operators holds only for $p > 1$, it could not be extended to any convex Φ-inequalities of course. The same phenomenon happened in inequalities among ${}_1K_p$, ${}_1L_p$ and L^p in Chapter 2. But, these L^p inequalities could be extended to Φ-inequalities for some convex functions Φ, that are those Φ with $q_\Phi > 1$. In this section, the Φ-inequalities and the Φ-extensions of ${}_aK_p$ and ${}_aL_p$ for such Φ will be investigated.

Lemma 3.3.1 *Let Φ be any convex function such that $1 < q_\Phi \le p_\Phi < \infty$. (f, g) is a pair of nonnegative measurable functions satisfying (with $\alpha \ge \beta > 0$ constants)*

$$\lambda \mid \{f > \alpha\lambda\} \mid \le \int_{\{f > \beta\lambda\}} g \, d\mu, \quad \forall \lambda > 0. \tag{3.3.1}$$

Then

$$E\left(\frac{f}{\alpha}\varphi\Big(\frac{f}{\alpha}\Big)\right) \le q'_\Phi C_{\alpha,\beta,\Phi} E\left(g\,\varphi\Big(\frac{f}{\alpha}\Big)\right). \tag{3.3.2}$$

And when $\alpha = \beta$, the condition $p_\Phi < \infty$ is not necessary, and $C_{\alpha,\beta,\Phi} = 1$.

Proof. Integrating (3.3.1) with respect to the measure $d\varphi(\lambda)$ on $(0,\infty)$, we get

$$\int_0^\infty \int_{\{f>\alpha\lambda\}} \lambda d\mu d\varphi(\lambda) \le \int_0^\infty \int_{\{f>\beta\lambda\}} g d\mu d\varphi(\lambda),$$

$$\int_\Omega \int_0^{\frac{f}{\alpha}} \lambda d\varphi(\lambda) d\mu \le \int_\Omega \int_0^{\frac{f}{\beta}} d\varphi(\lambda) g d\mu,$$

$$\int_\Omega \frac{f}{\alpha}\varphi\Big(\frac{f}{\alpha}\Big)d\mu - \int_\Omega \Phi\Big(\frac{f}{\alpha}\Big)d\mu \le \int_\Omega g\varphi\Big(\frac{f}{\beta}\Big)d\mu.$$

Noticing that

$$\Phi\Big(\frac{f}{\alpha}\Big) \le \frac{1}{q_\Phi}\frac{f}{\alpha}\varphi\Big(\frac{f}{\alpha}\Big) = \frac{1}{q}\frac{f}{\alpha}\varphi\Big(\frac{f}{\alpha}\Big),$$

$$\varphi(\gamma u) \le p_\Phi \frac{\Phi(\gamma u)}{\gamma u} \le p(\gamma u)^{p-1}\frac{\Phi(u)}{u^p} \le pq^{-1}\gamma^{p-1}\varphi(u), \quad \forall u > 0, \gamma \ge 1,$$

we get (3.3.2) immediately. When $\alpha = \beta$, the assertions are obvious. The proof is finished. □

Remark We have a conditioned version of the lemma as follows: (3.3.3) implies (3.3.4)

$$\lambda E_0(\chi_{\{f>\alpha\lambda\}}) \le E_0(g\chi_{\{f>\beta\lambda\}}), \quad \forall \lambda > 0, \tag{3.3.3}$$

$$E_0\left(\frac{f}{\alpha}\varphi\Big(\frac{f}{\alpha}\Big)\right) \le q' C_{\alpha,\beta,\Phi} E_0\left(g\varphi\Big(\frac{f}{\alpha}\Big)\right). \tag{3.3.4}$$

In fact, it comes from

$$\int_F \left(\frac{f}{\alpha}\varphi\Big(\frac{f}{\alpha}\Big) - \Phi\Big(\frac{f}{\alpha}\Big)\right) d\mu \le \int_F g\varphi\Big(\frac{f}{\beta}\Big)d\mu, \quad \forall F \in \mathcal{F}_0.$$

Lemma 3.3.2 *Let Φ and (f,g) be as in Lemma* 3.3.1. *Then*

$$||f||_\Phi \le q' C_{\alpha,\beta,\Phi} ||g||_\Phi. \tag{3.3.5}$$

Proof. Let (f,g) be any pair such that (3.3.1) holds. Then $(f \wedge N, g)$ satisfies (3.3.1) with same constants, for all $N > 0$. In fact, when $N > \alpha\lambda$, we have

$$\{f \wedge N > \alpha\lambda\} = \{f > \lambda\alpha\},$$

$$\{f \wedge N > \beta\lambda\} = \{f > \beta\lambda\},$$

when $N \le \alpha\lambda$, we have $|\{f \wedge N > \alpha\lambda\}| = 0$. Since $E\left(\frac{f\wedge N}{\alpha}\varphi\Big(\frac{f\wedge N}{\alpha}\Big)\right) < \infty$ from (3.3.2) and Lemma 3.2.3, we get

$$E\left(\Phi\left(\frac{f \wedge N}{\alpha}\right)\right) \le E(\Phi(q' C_{\alpha,\beta,\Phi} g)). \tag{3.3.6}$$

Since the condition is homogeneous (that is to say, (f,g) satisfies (3.3.1) implies so does (cf, cg)), by applying (3.3.6) to (cf, cg) with $c = (q' C_{\alpha,\beta,\Phi} \|g\|_\Phi)^{-1}$, we get

$$E\left(\Phi\left(\frac{cf}{\alpha}\right)\right) \leq 1,$$

$$\|f\|_\Phi \leq \frac{\alpha}{c} = q' C_{\alpha,\beta,\Phi} \|g\|_\Phi.$$

The proof of the lemma is finished. □

Now Doob's inequality has its Φ-extension.

Theorem 3.3.3 *Let $f = (f_n)_{0 \leq n \leq \infty}$ be a nonnegatve submartingale, Φ be any convex function with $q_\Phi > 1$. Then*

$$\|Mf\|_\Phi \leq q'_\Phi \|f\|_\Phi. \tag{3.3.7}$$

Proof. Since $f = (f_n)_{0 \leq n \leq \infty}$ is a nonnegative submartingale, we have

$$\lambda \mid \{Mf > \lambda\} \mid \leq \int_{\{Mf > \lambda\}} f_\infty d\mu.$$

By making use of Lemma 3.3.2, (3.3.7) follows from the preceding inequality. □

Remarks

1. The theorem may be applied to any martingale in L^1.
2. The argument is valid for generalized Φ. But it gives nothing new, since when $\Phi(u) \equiv \infty$ on some (a, ∞), L^Φ is nothing but L^∞.

Now we consider the Φ-extensions of spaces ${}_aK_p$ and ${}_aL_p$, where $0 < a < \infty$, and Φ are convex functions. First we give several definitions.

$$ {}^aL^\Phi = \{ \text{ measurable } f : \|f\|_{{}^aL^\Phi} = \| \mid f \mid^a \|_\Phi^{\frac{1}{a}} < \infty\}, \tag{3.3.8}$$

$$\begin{aligned} {}_aK_\Phi = \{ & \text{ measurable } f : \exists \text{ at least one adapted process } \theta = (\theta_n(f))_{n \geq 0}, \\ & \text{s.t. } \exists \gamma \geq 0, \gamma \in {}^aL^\Phi, \text{ satisfying} \\ & E(\mid f - \theta_{n-1} \mid^a \mid \mathcal{F}_n) \leq E(\gamma^a \mid \mathcal{F}_n),\ n = 0, 1, \cdots\}. \end{aligned} \tag{3.3.9}$$

Noticing the convention $\theta_{-1} \equiv 0$ and the condition (3.3.9) with $n = -1$, we see ${}_aK_\Phi \subset L^a$. Obviously, ${}_aK_\Phi$ is a linear space of measurable functions on which a quasi-norm† could be defined.

$$\|f\|_{{}_aK_\Phi} = \inf_\theta\{\|f\|^{(\theta)}_{{}_aK_\Phi}\} = \inf_\theta\{\inf_\gamma\{\|\gamma^a\|_\Phi^{\frac{1}{a}}\}\}. \tag{3.3.10}$$

† "quasi-norm" means positive homogeneity and generalized trigonometric inequality.

It is worthy to be noticed that when $1 \le a < \infty$ and $q_\Phi > 1$, an equivalent quasi-norm on ${}_aK_\Phi$ could be defined

$$\|f\|^*_{{}_aK_\Phi} = \inf\{ \|\gamma^a\|_\Phi^{\frac{1}{a}} : \gamma \text{ runs through } {}^aL^\Phi, \quad \text{s.t. } E(| f - f_{n-1} |^a | \mathcal{F}_n) \le E(\gamma^a \mid \mathcal{F}_n), \quad \forall n \ge 0\}, \tag{3.3.11}$$

here $f_n = E(f|\mathcal{F}_n)$. The equivalence can be seen as follows. We have

$$\begin{aligned} E(| f - f_{n-1} |^a | \mathcal{F}_n) &\le 2^{a-1}E(| f - \theta_{n-1} |^a | \mathcal{F}_n) + 2^{a-1} | f_{n-1} - \theta_{n-1} | \\ &\le 2^{a-1}E(\gamma^a \mid \mathcal{F}_n) + 2^{a-1}E(\gamma^a \mid \mathcal{F}_{n-1}) \\ &\le 2^a E(M(\gamma^a) \mid \mathcal{F}_n) \end{aligned}$$

and

$$\|M(\gamma^a)\|_\Phi \le q'_\Phi \|\gamma^a\|_\Phi,$$

so,

$$\|f\|_{{}_aK_\Phi} \le \|f\|^*_{{}_aK_\Phi} \le 2(q'_\Phi)^{\frac{1}{a}} \|f\|_{{}_aK_\Phi}.$$

From now on, when $1 \le a < \infty$ and $q_\Phi > 1$, we always take $\theta = (f_n)_{n \ge 0}$ and adopt the quasi-norm $\|\cdot\|^*_{{}_aK_\Phi}$ with the script $*$ taken off.

In order to define the Φ-extension of ${}_aL_p$, define a set of processes at first:

$$G_a^\Phi = \left\{\text{adapted processes} : E\left(\Psi\left(\sum | g_n |^a \right)\right) \le 1\right\}, \tag{3.3.12}$$

where Ψ is the complementary function of Φ, $0 < a < \infty$. Define

$$\begin{aligned} {}_aL_\Phi = \Big\{ & \text{measurable function } f : \exists \text{ adapted process } \theta = (\theta_n)_{n \ge 0}, \\ & \text{s.t. } \|f\|^{(\theta)}_{{}_aL_\Phi} = \sup_{(g_n) \in G_a^\Phi} E\left(\sum | g_n |^a | f - \theta_{n-1} |^a \right)^{\frac{1}{a}} < \infty\Big\}, \end{aligned} \tag{3.3.13}$$

$$\|f\|_{{}_aL_\Phi} = \inf_\theta \{\|f\|^{(\theta)}_{{}_aL_\Phi}\}. \tag{3.3.14}$$

Analogously, by means of the convention $\theta_{-1} \equiv 0$ and a suitable choice of $g = (g_n)_{n \ge 0}$ ($g_0 \equiv 1$, $g_n \equiv 0$, $n \ge 1$), we see ${}_aL_\Phi \subset L^a$.

When $\Phi(u) = u^{\frac{p}{a}}, 0 < a \le p$, we return to ${}_aK_p$ and ${}_aL_p$.

We want to characterize ${}_aK_\Phi$ and ${}_aL_\Phi$, at first ${}_aK_\Phi$.

Theorem 3.3.4 *Let* $1 \le a < \infty$, *and* Φ *be such that* $1 < q_\Phi \le p_\Phi < \infty$. *Then* ${}_aK_\Phi = {}^aL^\Phi$. *More precisely, we have*

$$\frac{1}{2(q')^{\frac{1}{a}}} \|f\|_{{}_aK_\Phi} \le \|f\|_{{}^aL^\Phi} \le \|Mf\|_{{}^aL^\Phi} \le C\|f\|_{{}_aK_\Phi}. \tag{3.3.15}$$

Proof. We have

$$(Mf)^a \le \sup_n E(| f |^a | \mathcal{F}_n) = M(| f |^a),$$

$$E(|f - f_{n-1}|^a | \mathcal{F}_n) \le 2^{a-1}(E(|f|^a | \mathcal{F}_n) + E((Mf)^a | \mathcal{F}_n))$$
$$\le 2^a E(M(|f|^a) | \mathcal{F}_n),$$

$$\|f\|_{{}_aK_\Phi} \le \|2^a M(|f|^a)\|_{\Phi}^{\frac{1}{a}} \le 2(q'_{\Phi})^{\frac{1}{a}} \| |f|^a \|_{\Phi}^{\frac{1}{a}} = 2(q')^{\frac{1}{a}} \|f\|_{{}^aL^\Phi}.$$

This proves the first inequality in (3.3.15). Now prove the third one . Let $\alpha > 0$ be determined later. For all $\lambda > 0$, define stopping times

$$T_1 = \inf\left\{n : |f_n|^a > \frac{1}{2^{a-1}}\alpha\lambda\right\},$$

$$T_2 = \inf\{n : |f_n|^a > (\alpha+1)\lambda\}.$$

Then $T_1 \le T_2$, and

$$\{T_2 < \infty\} \subset \left\{T_1 < \infty, |f_{T_2} - f_{T_1 - 1}|^a \ge \frac{\lambda}{2^{a-1}}\right\}.$$

So, we have

$$\begin{aligned}
\lambda |\{(Mf)^a > (\alpha+1)\lambda\}| &\le 2^{a-1} \int_{\{T_1<\infty\}} |f_{T_2} - f_{T_1-1}|^a \, d\mu \\
&= 2^{a-1} \int_{\{T_1<\infty\}} |E(f - f_{T_1-1} | \mathcal{F}_{T_2})|^a \, d\mu \\
&\le 2^{a-1} \int_{\{T_1<\infty\}} E(|f - f_{T_1-1}|^a | \mathcal{F}_{T_1}) d\mu \\
&\le 2^{a-1} \int_{T_1<\infty} E(\gamma^a | \mathcal{F}_{T_1}) d\mu \\
&= 2^{a-1} \int_{\{Mf > 2^{1-a}\alpha\lambda\}} \gamma^a d\mu.
\end{aligned}$$

Applying Lemma 3.3.1 and letting $\alpha = p$, we get

$$E\left(\frac{(Mf)^a}{(p+1)}\varphi\left(\frac{(Mf)^a}{(p+1)}\right)\right) \le q'CE\left(\gamma^a \varphi\left(\frac{(Mf)^a}{(p+1)}\right)\right).$$

Applying Lemma 3.3.2, we get

$$\|(Mf)^a\|_\Phi \le q'C\|\gamma^a\|_\Phi.$$

Taking "inf" in the right-hand side, we get the third inequality in (3.3.15). The proof of the theorem is complete. □

Remark The theorem could not be improved to include the case $p_\Phi = \infty$ or $q_\Phi = 1$. For example, we have known that ${}_1K_\infty (= BMO) \not\subset L^\infty$ in general, this means that the third inequality in (3.3.15) fails to be true for $p_\Phi = \infty$. Garsia [1] showed that ${}_1K_1 \not\subset H_1$, and we will show that $L^1 \not\subset {}_1K_1$, these mean that the first and the third both are not true any longer for $q_\Phi = 1$. The example to show $L^1 \not\subset {}_1K_1$ may be taken as follows.

Consider $(\Omega, \mathcal{F}, \mu, \{\mathcal{F}_n\}_{n\geq 1})$ with $\Omega = (0,1]$, $\mu = dx$, $\mathcal{F}_1 =$ trivial, $\mathcal{F}_n$ generated by n atoms

$$\left(0, \frac{1}{2^{n-1}}\right], \left(\frac{1}{2^{n-1}}, \frac{1}{2^{n-2}}\right], \cdots, \left(\frac{1}{2}, 1\right].$$

Denote $I_k = \left(\frac{1}{2^k}, \frac{1}{2^{k-1}}\right]$. Let

$$f = \sum_1^\infty c_k \chi_{I_k} \in L^1, \quad c_k \text{ will be determined later.}$$

Then

$$f_n = \sum_1^{n-1} c_k \chi_{I_k} + c_n' \chi_{(0, \frac{1}{2^{n-1}}]}, \quad c_n' = 2^{n-1} \sum_{k\geq n} \frac{c_k}{2^k}.$$

We have

$$| f - f_{n-1} | = | c_{n-1} - c_{n-1}' | \chi_{I_{n-1}} + \sum_{k\geq n} | c_k - c_{n-1}' | \chi_{I_k},$$

$$E(| f - f_{n-1} | \mid \mathcal{F}_n)\chi_{I_{n-1}} = | c_{n-1} - c_{n-1}' | \chi_{I_{n-1}}, \quad \forall n.$$

Let $\gamma = \sum_1^\infty d_k \chi_{I_k}$ be any nonnegative function, such that

$$E(| f - f_{n-1} | \mid \mathcal{F}_n) \leq E(\gamma \mid \mathcal{F}_n), \quad \forall n,$$

then $d_n \geq | c_n - c_n' |$, for all n. Now let $c_n = \frac{2^n}{n^2}$, then

$$c_n' = 2^{n-1} \sum_{k\geq n} \frac{1}{k^2} \approx n c_n,$$

$$\sum \frac{d_n}{2^n} \geq \sum \frac{c_n' - c_n}{2^n} \approx \sum \frac{1}{n} = \infty.$$

This means $\gamma \notin L_1$, so $f \notin {}_1K_1$. The proof of the assertion is complete. □

Theorem 3.3.5 *Let $0 < a < 1$, and Φ be as in Theorem 3.3.4. Then ${}_aK_\Phi = {}^aL^\Phi$. More precisely, we have*

$$\|f\|_{{}_aK_\Phi} \leq \|f\|_{{}^aL^\Phi} \leq \|M(| f |^a)\|_\Phi^{\frac{1}{a}} \leq C\|f\|_{{}_aK_\Phi}. \tag{3.3.16}$$

Proof. Let $f \in {}^aL^\Phi$. Taking $\theta = (\theta_n)$, $\theta_n \equiv 0$, we have

$$\|f\|_{{}_aK_\Phi} \leq \|f\|_{{}_aK_\Phi}^{(\theta)} \leq \| \, | f |^a \, \|_{\frac{\Phi}{a}}^{\frac{1}{a}} = \|f\|_{{}^aL^\Phi} \leq \|M(| f |^a)\|_{\frac{\Phi}{a}}^{\frac{1}{a}}.$$

This proves the first and the second inequality in (3.3.16).

Now let $f \in {}_aK_\Phi$, $\theta = (\theta_n)_{n\geq 0}$ and $\gamma \in {}^aL^\Phi$ be given arbitrarily in (3.3.9). Since $0 < a < 1$, we have

$$||f|^a - |\theta_{n-1}|^a| \leq |f - \theta_{n-1}|^a, \tag{3.3.17}$$

and hence

$$E(||f|^a - |\theta_{n-1}|^a| \mid \mathcal{F}_n) \leq E(\gamma^a \mid \mathcal{F}_n), \quad \forall n,$$

$$\begin{aligned} &E(||f|^a - (|f|^a)_{n-1}| \,|\mathcal{F}_n) \\ &\quad \leq E(||f|^a - |\theta_{n-1}|^a| \,|\mathcal{F}_n) + E(||f|^a - |\theta_{n-1}|^a| \,|\mathcal{F}_{n-1}) \\ &\quad \leq E(\gamma^a \mid \mathcal{F}_n) + E(\gamma^a \mid \mathcal{F}_{n-1}) \leq 2E(M(\gamma^a) \mid \mathcal{F}_n)). \end{aligned}$$

That means ${}_aK_\Phi \subset \{f : |f|^a \in {}_1K_\Phi\}$. And by means of the inequalities proved for $a = 1$, we get

$$\|M(|f|^a)\|_\Phi \leq C\|M(\gamma^a)\|_\Phi,$$

$$\|M(|f|^a)\|_\Phi^{\frac{1}{a}} \leq C\|M(\gamma^a)\|_\Phi^{\frac{1}{a}}.$$

Taking "inf" over all possible γ, we get the third inequality in (3.3.16). The proof is finished. □

Now characterize ${}_aL_\Phi$.

Theorem 3.3.6 *Let $0 < a < \infty$, Φ be any convex function with $q_\Phi > 1$. Then ${}_aL_\Phi = {}_aK_\Phi$. More precisely, we have*

$$2^{-\frac{1}{a}}\|f\|_{{}_aL_\Phi} \leq \|f\|_{{}_aK_\Phi} \leq (q')^{\frac{1}{a}}\|f\|_{{}_aL_\Phi}. \tag{3.3.18}$$

Proof. Let $f \in {}_aK_\Phi$, $\theta = (\theta_n)_{n\geq 0}$ and $\gamma \in {}^aL^\Phi$ be given arbitrarily in (3.3.9). Denote $f^{(n-1)} = f - \theta_{n-1}$. Then

$$E(|f^{(n-1)}|^a \mid \mathcal{F}_n) \leq E(\gamma^a \mid \mathcal{F}_n), \quad \forall n.$$

For the same θ, we have

$$\begin{aligned} \|f\|_{{}_aL_\Phi}^{(\theta)} &= \sup_{(g_n)\in G_a^\Phi} E\Big(\sum |g_n|^a |f^{(n-1)}|^a\Big)^{\frac{1}{a}} \\ &= \sup E\Big(\sum |g_n|^a E(|f^{(n-1)}|^a \mid \mathcal{F}_n)\Big)^{\frac{1}{a}} \leq \sup E\Big(\sum |g_n|^a \gamma^a\Big)^{\frac{1}{a}} \\ &\leq \sup \max E\Big(\Psi\Big(\sum |g_n|^a\Big), 1\Big)^{\frac{1}{a}} N_\Phi(\gamma^a)^{\frac{1}{a}} \leq 2^{\frac{1}{a}}\|\gamma^a\|_\Phi^{\frac{1}{a}}. \end{aligned}$$

For fixed θ taking "inf" over γ, and then taking "inf" over θ, we get the first inequality in (3.3.18).

Now let $f \in {}_aL_\Phi$, and $\theta = (\theta_n)_{n\geq 0}$ be any one such that

$$\|f\|^{(\theta)}_{{}_aL_\Phi} = \sup_{(g_n)\in G^\Phi_a} E\Big(\sum |g_n|^a| f^{(n-1)}|^a\Big)^{\frac{1}{a}} < \infty.$$

For N fixed, consider the process

$$(\gamma^a_n)^N_{n=0}, \quad \gamma^a_n = E(| f^{(n-1)}|^a| \mathcal{F}_n).$$

Now we want to estimate $E((M_N\gamma)^a h)$ for any $h \geq 0$, $E(\Psi(h)) \leq 1$. In order to do that, make a partition of Ω

$$\Omega = \bigcup_{n=0}^N F_n, \quad F_n = \{\omega \in \Omega : M_{n-1}\gamma < \gamma_n = M_N\gamma\}.$$

Then we have

$$\begin{aligned} E((M_N\gamma)^a h) &= \sum_{n=0}^N E(\gamma^a_n h\chi_{F_n}) = \sum_{n=0}^N E(\gamma^a_n E(h\chi_{F_n} \mid \mathcal{F}_n) \\ &= E\Big(\sum_{n=0}^N | f^{(n-1)}|^a E(h\chi_{F_n} \mid \mathcal{F}_n)\Big). \end{aligned}$$

Denote $g^a_n = E(h\chi_{F_n} \mid \mathcal{F}_n)$. Then $(g_n)^N_0$ is an adapted process, and by means of Lemma 3.2.7 we have

$$\Big\|\sum g^a_n\Big\|_\Psi \leq p_\Psi\|h\|_\Psi \leq p_\Psi = q'_\Phi.$$

So

$$E\left(\Psi\left(\frac{\sum g^a_n}{q'_\Phi}\right)\right) \leq E\left(\Psi\left(\frac{\sum g^a_n}{\|\sum g^a_n\|_\Psi}\right)\right) \leq 1.$$

This means

$$\left(\frac{E(h\chi_{F_n} \mid \mathcal{F}_n)^{\frac{1}{a}}}{(q')^{\frac{1}{a}}}\right)^N_{n=0} \in G^\Phi_a.$$

Now we have

$$\begin{aligned} \|(M_N\gamma)^a\|_\Phi &\leq \sup_{E(\Psi(h))\leq 1} E((M_N\gamma)^a h) \\ &\leq q'_\Phi \sup_h E\Big(\sum_0^N | f^{(n-1)}|^a (q'_\Phi)^{-1}E(h\chi_{F_n} \mid \mathcal{F}_n)\Big) \\ &\leq q' \sup_{(g_n)\in G^\Phi_a} E\Big(\sum_0^N | f^{n-1)}|^a| g_n|^a\Big) \leq q'(\|f\|^{(\theta)}_{{}_aL_\Phi})^a. \end{aligned}$$

Letting $N \to \infty$, and noticing that $(M\gamma)^a = \sup_n E(| f^{(n-1)}|^a| \mathcal{F}_n)$ satisfies

$$E(| f^{(n-1)}|^a| \mathcal{F}_n) \leq E((M\gamma)^a \mid \mathcal{F}_n), \quad \forall n,$$

we get

$$\|f\|^{(\theta)}_{{}_aK_\Phi} \le \|(M\gamma)^a\|_\Phi^{\frac{1}{a}} \le q'^{\frac{1}{a}}\|f\|^{(\theta)}_{{}_aL_\Phi}.$$

Taking "inf" over θ, we get the second inequality in (3.3.18). The proof of the theorem is finished. □

Remark When $1 \le a < \infty$, $q_\Phi > 1$, we usually take $\theta = (f_n)_{n\ge 0}$ in the definitions of ${}_aK_\Phi$ and ${}_aL_\Phi$. In this case (3.3.18) become

$$2^{-\frac{1}{a}}\|f\|_{{}_aL_\Phi} \le \|f\|_{{}_aK_\Phi} \le \|(f_a^\#)^a\|_\Phi^{\frac{1}{a}} \le (q'_\Phi)^{\frac{1}{a}}\|f\|_{{}_aL_\Phi}, \tag{3.3.19}$$

where $f_a^\#$ is the sharp function in martingale setting

$$f_a^\# = \sup_n E(|\, f - f_{n-1}\,|^a |\, \mathcal{F}_n)^{\frac{1}{a}}. \tag{3.3.20}$$

For the Φ-extensions ${}_ak_\Phi$ of ${}_ak_p$ and ${}_al_\Phi$ of ${}_al_p$ which are defined by (3.3.9) and (3.3.13) respectively, but with a modification, i.e. θ_{n-1} is replaced by θ_n, we have following

Theorem 3.3.7′ *Let $0 < a < \infty$, Φ be such that $q_\Phi > 1$. Then ${}_al_\Phi = {}_ak_\Phi$, and*

$$2^{-\frac{1}{a}}\|f\|_{{}_al_\Phi} \le \|f\|_{{}_ak_\Phi} \le (q')^{\frac{1}{a}}\|f\|_{{}_al_\Phi}. \tag{3.3.21}$$

When $1 \le a < \infty$, $q_\Phi > 1$, we have

$$2^{-\frac{1}{a}}\|f\|_{{}_al_\Phi} \le \|f\|_{{}_ak_\Phi} \le \|(\tilde f_a^\#)^a\|_\Phi^{\frac{1}{a}} \le (q')^{\frac{1}{a}}\|f\|_{{}_al_\Phi}, \tag{3.3.22}$$

where

$$\tilde f_a^\# = \sup E(|\, f - f_n\,|^a |\, \mathcal{F}_n)^{\frac{1}{a}}. \tag{3.3.23}$$

The proof remains unchanged.
Finally we will give a kind of result without the restriction $q_\Phi > 1$.

Theorem 3.3.7 *Let Φ be a convex function with $p_\Phi < \infty$. Then*

$${}_2K_\Phi = {}_2L_\Phi = {}_2H_\Phi = \{f : \|S(f)^2\|_\Phi < \infty\}.$$

More precisely,

$$\frac{1}{2p_\Phi}\|S(f)^2\|_\Phi \le \frac{1}{2}\|f\|^2_{{}_2L_\Phi} \le \|f\|^2_{{}_2K_\Phi} \le \|S(f)^2\|_\Phi. \tag{3.3.24}$$

Proof. Let $f \in {}_2H_\Phi$. Then from

$$E(|\, f - f_{n-1}\,|^2 |\, \mathcal{F}_n) = E(S(f)^2 - S_{n-1}(f)^2 \mid \mathcal{F}_n) \le E(S(f)^2 \mid \mathcal{F}_n),$$

we get the third inequality in (3.3.24).

Let $f \in {}_2K_\Phi$, and γ be any dominant of f in the definition. Then

$$\|f\|^2_{{}_2L_\Phi} = \sup_{(g_n)\in G_2^\Phi} E\Big(\sum | g_n |^2 | f^{(n-1)} |^2\Big) \le \sup_g E\Big(\sum | g_n |^2 \gamma^2\Big)$$
$$\le \sup_g \max\Big(E\Big(\Psi\Big(\sum | g_n |^2\Big)\Big), 1\Big) N_\Phi(\gamma^2) \le 2\|\gamma^2\|_\Phi.$$

This gives the second inequality in (3.3.24).

In order to get the first inequality in (3.3.24), we want to reformulate the definition of ${}_2L_\Phi$ first. Denote

$$R^\Phi = \{\text{nonnegative increasing adapted } (r_n)_{n\ge 0} : \quad E(\Psi(r_\infty) \le 1\}. \tag{3.3.25}$$

Then we have the equivalent definition

$$\|f\|_{{}_2L_\Phi} = \sup_{(r_n)\in R^\Phi} E\Big(\sum r_n | \Delta_n f |^2\Big)^{\frac{1}{2}}. \tag{3.3.26}$$

The equivalence can be seen as follows: Making a correspondence between R^Φ and G_2^Φ by $r_n = \sum_{k\le n} | g_k |^2$, we get

$$E\Big(\sum_{k=0}^\infty | g_k |^2 | f^{(k-1)} |^2\Big) = E\Big(\sum_{k=0}^\infty | g_k |^2 \sum_{n=k}^\infty | \Delta_n f |^2\Big)$$
$$= E\Big(\sum_{n=0}^\infty \sum_{k=0}^n | g_k |^2 | \Delta_n f |^2\Big) = E\Big(\sum_{n=0}^\infty r_n | \Delta_n f |^2\Big),$$

that gives the equivalence. Now we consider all martingales $\rho = (\rho_n)_{n\ge 0}$ with $\rho \ge 0$, $E(\Psi(\rho)) \le 1$. We have

$$\|M\rho\|_\Psi \le q'_\Psi \|\rho\|_\Psi \le p_\Phi,$$

provided $q'_\Psi > 1$ (i.e. $p_\Phi < \infty$). This means $(p_\Phi^{-1} M_n\rho)_{n\ge 0} \in R^\Phi$, so

$$\|S(f)^2\|_\Phi \le \sup_{\rho: E(\Psi(\rho)\le 1)} E\Big(\rho \sum_0^\infty | \Delta_n f |^2\Big)$$
$$= \sup_\rho E\Big(\sum_0^\infty E(\rho \mid \mathcal{F}_n) | \Delta_n f |^2\Big)$$
$$\le p_\Phi \sup_\rho E\Big(\sum_0^\infty \frac{M_n \rho}{p_\Phi} | \Delta_n f |^2\Big)$$
$$\le p_\Phi \|f\|^2_{{}_2L_\Phi}.$$

This completes the proof. □

Theorem 3.3.8′ *Let Φ be as in Theorem 3.3.8. Then*

$${}_2k_\Phi = {}_al_\Phi = {}_2h_\Phi = \{f : \|\sigma(f)^2\|_\Phi < \infty\}.$$

More precisely, we have

$$\frac{1}{2p_\Phi}\|\sigma(f)^2\|_\Phi \le \frac{1}{2}\|f\|^2_{2l_\Phi} \le \|f\|^2_{2k_\Phi} \le \|\sigma(f)^2\|_\Phi. \tag{3.3.27}$$

Proof. The proof is essentially the same as in the preceding theorem. The slight difference occurs in the first inequality. Now we use the following equivalence of $\|\cdot\|_{2l_\Phi}$:

$$\begin{aligned}\|f\|_{2l_\Phi} &= \sup_{(g_n)\in G_2^\Phi} E\Big(\sum_0^\infty | g_n |^2 | f^{(n)} |^2\Big)^{\frac{1}{2}}\\ &= \sup_{(r_n)\in R^\Phi} E\Big(\sum_0^\infty r_{n-1} | \Delta_n f |^2\Big)^{\frac{1}{2}}.\end{aligned} \tag{3.3.28}$$

From it we get

$$\begin{aligned}\|\sigma(f)^2\|_\Phi &\le \sup_{\rho:E(\Psi(\rho))\le 1} E\Big(\rho\sum_0^\infty E(| \Delta_n f |^2 | \mathcal{F}_{n-1})\Big)\\ &\le \sup_\rho E\Big(\sum_0^\infty M_{n-1}\rho E(| \Delta_n f |^2 | \mathcal{F}_{n-1})\Big) \le p_\Phi\|f\|^2_{2l_\Phi}.\end{aligned}$$

This proves the theorem. □

3.4 Concave Φ-inequalities

Let $\Phi(u)$ be an increasing concave function from $\mathbf{R}^+$ to $\mathbf{R}^+$, with $\Phi(0) = 0$, $\varphi(u)$ be its derivative normalized to be left continuous with $\varphi(0) = \varphi(0+)$. Notice that $\varphi(\infty) = \lim_{u\to\infty} \varphi(u)$ exists, and

$$\Phi(2u) \le 2\Phi(u), \quad p = \sup \frac{u\varphi(u)}{\Phi(u)} \le 1,$$

since $\varphi(u)$ is decreasing. Furthermore, we have

$$\begin{aligned}\Phi(u) &= \int_0^u \varphi(t)dt = \lim_{a\to 0, b\to\infty} \int_a^b \varphi(t)d(u\wedge t)\\ &= \lim_{a\to 0, b\to\infty}\Big\{\varphi(t)(u\wedge t)\,|_a^b - \int_a^b (u\wedge t)d\varphi(t)\Big\}\\ &= u\varphi(\infty) - \int_0^\infty (u\wedge t)d\varphi(t),\end{aligned} \tag{3.4.1}$$

where $-d\varphi$ is a nonnegative Lebesgue-Stieltjes measure.

Lemma 3.4.1 *Let $\Phi(u)$ be any concave function as above, W, Y be two nonnegative measurable functions. Then*

$$E(\Phi(W)) \le CE(\Phi(Y)), \quad \forall\Phi, \tag{3.4.2}$$

if and only if (with the same constant C)

$$E(W \wedge \lambda) \le CE(Y \wedge \lambda), \quad \forall\lambda > 0. \tag{3.4.3}$$

Proof. Obviously (3.4.2) implies (3.4.3), since $u \wedge \lambda$ is concave. Setting $u = W$ and Y in (3.4.1), and integrating with respect to the measure $d\mu$, we get from (3.4.3),

$$\begin{aligned} E(\Phi(W)) &= \varphi(\infty)E(W) - \int_0^\infty E(W \wedge \lambda)d\varphi(\lambda) \\ &\le C\varphi(\infty)E(Y) - C\int_0^\infty E(Y \wedge \lambda)d\varphi(\lambda) + \varphi(\infty)(E(W) - CE(Y)) \\ &\le CE(\Phi(Y)). \end{aligned}$$

This proves the lemma. □

Remark Replacing E by E_0 in both (3.4.2) and (3.4.3), the lemma remains to be true.

For some concave function in strict sense, we could have (3.4.2) under some weaker condition.

Definition 3.4.2 *Let $\Phi(u)$be a concave function as above. It is called a strict concave function, if*

$$p = p_\Phi = \sup \frac{u\varphi(u)}{\Phi(u)} < 1. \tag{3.4.4}$$

Lemma 3.4.3 *Let Φ be a strict concave function, W, Y be two nonnegative measurable functions satisfying*

$$|\{W > \lambda\}| \le \frac{C}{\lambda}E(Y \wedge \lambda), \quad \forall\lambda > 0. \tag{3.4.5}$$

Then

$$E(\Phi(W)) \le \frac{C}{(1-p)}E(\Phi(Y)). \tag{3.4.6}$$

Proof. Notice that $\frac{\Phi(u)}{u^p}$ is decreasing ,the proof of which has occurred in the Theorem 3.1.1. Integrating both sides of (3.4.5) with respect to the measure $d\Phi(\lambda)$, we get

$$\begin{aligned} E(\Phi(W)) &\le C\int_0^\infty \frac{1}{\lambda}\Big\{\int_{\{Y\le\lambda\}} Y d\mu + \int_{\{Y>\lambda\}} \lambda d\mu\Big\} d\Phi(\lambda) \\ &\le C\int_\Omega Y\int_Y^\infty \frac{d\Phi(\lambda)}{\lambda} d\mu + CE(\Phi(Y)). \end{aligned}$$

Now estimate $\int_Y^\infty \frac{d\Phi(\lambda)}{\lambda}$. We have

$$\begin{aligned}\int_Y^\infty \frac{d\Phi(\lambda)}{\lambda} &= \int_Y^\infty \int_\lambda^\infty \frac{dt}{t^2} d\Phi(\lambda) = \int_Y^\infty \int_Y^t d\Phi(\lambda)\frac{dt}{t^2}\\ &= \int_Y^\infty \frac{\Phi(t)}{t^p}\frac{dt}{t^{2-p}} - \frac{\Phi(Y)}{Y}\\ &\le \frac{\Phi(Y)}{Y^p}\int_Y^\infty \frac{dt}{t^{2-p}} - \frac{\Phi(Y)}{Y} = \frac{p}{1-p}\frac{\Phi(Y)}{Y}.\end{aligned}$$

Finally we get

$$E(\Phi(W)) \le \left(\frac{p}{1-p}+1\right) CE(\Phi(Y)) = \frac{C}{1-p}E(\Phi(Y)).$$

The lemma is thus proved. □

Remark From the preceding argument we see that any conditions which imply

$$\int_u^\infty \frac{\Phi(t)}{t^2}dt \le C\frac{\Phi(u)}{u}, \quad \forall u > 0, \tag{3.4.7}$$

are sufficient for the truth of the lemma, for example, the following condition on Φ will be in the case

$$\Phi(\alpha u) \le \beta\Phi(u), \quad \text{for all } u, \text{ for some } \alpha, \beta, \quad 1 < \beta < \alpha. \tag{3.4.8}$$

In fact, (3.4.8) implies

$$\begin{aligned}\int_u^\infty \frac{\Phi(t)}{t^2}dt &= \sum_{k=0}^\infty \int_{\alpha^k u}^{\alpha^{k+1}u} \frac{\Phi(t)}{t^2}dt = \sum_{k=0}^\infty \int_u^{\alpha u} \frac{\Phi(\alpha^k s)}{(\alpha^k s)^2}\alpha^k ds\\ &\le \sum_{k=0}^\infty \left(\frac{\beta}{\alpha}\right)^k \int_u^{\alpha u} \frac{\Phi(s)}{s^2}ds \le C\frac{\Phi(u)}{u}.\end{aligned}$$

But the condition

$$\Phi(\alpha u) < \alpha\Phi(u), \quad \text{for all } u, \text{ for some } \alpha > 1, \tag{3.4.9}$$

would not be the case, as shown by $\Phi(u) = \frac{u}{\log(1+u)}$. Obviously, (3.4.8) is implied by (3.4.4). We do not know if the converse assertion is also true.

Lemma 3.4.4 *Let $(X_n)_{n\ge0}$ be a nonnegative adapted and convergent process, and $B = (B_n)_{n\ge0}$ be a nonnegative increasing predictable process without any condition on B_0. Suppose that for all stopping times T, we have*

$$E(X_T\chi_{\{T>0\}}) \le CE(B_T\chi_{\{T>0\}}). \tag{3.4.10}$$

Then we have

$$E(X_\infty \wedge \lambda) \le (C+1)E(B_\infty \wedge \lambda). \quad \forall\lambda > 0. \tag{3.4.11}$$

Proof. For $\lambda > 0$ given, define the stopping time $T = \inf\{n : B_{n+1} > \lambda\}$. Then $B_T\chi_{\{T>0\}} \leq \lambda$. Noticing that

$$X_\infty \wedge \lambda \leq X_T\chi_{\{T=\infty\}} + \lambda\chi_{\{T<\infty\}},$$

and

$$B_T\chi_{\{T>0\}} \leq \min(B_\infty, \lambda) = B_\infty \wedge \lambda,$$

$$\lambda \mid \{T < \infty\} \mid \leq \int_{\{T<\infty\}} \lambda d\mu + \int_{\{T=\infty\}} B_\infty d\mu = E(B_\infty \wedge \lambda),$$

we get

$$\begin{aligned} E(X_\infty \wedge \lambda) &\leq E(X_T\chi_{\{T>0\}}) + E(\lambda\chi_{\{T<\infty\}}) \\ &\leq CE(B_T\chi_{\{T>0\}}) + E(B_\infty \wedge \lambda) \leq (C+1)E(B_\infty \wedge \lambda). \end{aligned}$$

This proves the lemma. □

Lemma 3.4.5 *Let $(X_n)_{n\geq 0}$ be a nonnegative adapted process, and $B = (B_n)_{n\geq 0}$ be a nonnegative increasing predictable process satisfying*

$$E(X_\tau \mid \mathcal{F}_0) \mid\leq E(B_\tau \mid \mathcal{F}_0), \quad \textit{for all stopping times } \tau. \tag{3.4.12}$$

Then we have

$$\mid \{MX > c\} \mid \leq \frac{1}{c}E(B_\infty \wedge d) + \mid \{B_\infty > d\} \mid, \quad \forall c, d > 0. \tag{3.4.13}$$

Proof. Define two stopping times

$$T = \inf\{n : B_{n+1} > d\}, \quad S = \inf\{n : X_n > c\}.$$

Without loss of generality, we can assume that $(X_n)_{n\geq 0}$ is convergent, otherwise we consider a finite process X. We have

$$|\{MX > c\}| \leq |\{MX > c, T = \infty\}| + \mid \{T < \infty\} \mid,$$

$$\begin{aligned} \mid \{MX > c, T = \infty\} \mid &\leq \mid \{M_T X > c, T > 0\} \mid \\ &\leq \mid \{X_S > c, S \leq T, T > 0\} \mid \leq \mid \{X_{S\wedge T} > c, T > 0\} \mid \\ &\leq \frac{1}{c}E(X_{S\wedge T}\chi_{\{T>0\}}) = \frac{1}{c}E(E(X_{S\wedge T} \mid \mathcal{F}_0)\chi_{\{T>0\}}) \\ &\leq \frac{1}{c}E(E(B_{S\wedge T} \mid \mathcal{F}_0)\chi_{\{T>0\}}) \leq \frac{1}{c}E(B_T\chi_{\{T>0\}}) \\ &\leq \frac{1}{c}E(B_\infty \wedge d), \end{aligned}$$

and

$$\mid \{T < \infty\} \mid = \mid \{B_\infty > d\} \mid.$$

This proves the lemma. □

Remark The condition (3.4.12) is stronger than (3.4.10), the conclusion of the former should be stronger than that of the latter. But (3.4.13) is faced to MX not as (3.4.11) faced to X itself, so they are not comparable.

Now we apply preceding lemmas to martingales.

Theorem 3.4.6 *Let $\Phi(u)$ be a concave function as above. Then for all martingales $f = (f_n)_{n\geq 0}$ with $f_0 = 0$, we have*

$$E(\Phi(S(f)^2) \leq 2E(\Phi(\sigma(f)^2)). \tag{3.4.14}$$

Proof. Denote

$$W = (W_n)_{n\geq 0}, \quad W_n = \sum_1^n |\Delta_k f|^2, \quad W_0 = 0,$$

$$Y = (Y_n)_{n\geq 0}, \quad Y_n = \sum_1^n E(|\Delta_k f|^2 | \mathcal{F}_{k-1}), \quad Y_0 = 0.$$

Then Y is predictable. In addition, we have, for all stopping times T

$$E(W_T) = E(S_T(f)^2) = E(\sigma_T(f)^2) = E(Y_T).$$

That is to say Lemmas 3.4.4 and 3.4.1 may be applied , and we get (3.4.14). The proof is finished. □

Remark Since we have $E_0(W_T) = E_0(Y_T)$, we have (3.4.14) with E replaced by E_0, too.

Theorem 3.4.7 *The condition are as in Theorem 3.4.6. Then*

$$E(\Phi((Mf)^2)) \leq 5E(\Phi(\sigma(f)^2)). \tag{3.4.15}$$

Proof. Denote

$$W = (M_n f)^2_{n\geq 0}, \quad Y = (\sigma_n(f)^2)_{n\geq 0}.$$

Since for all stopping times T,

$$W_T = (M_T f)^2 = M(f^{(T)})^2, \quad Y_T = \sigma(f^{(T)})^2,$$

we have

$$E(W_T) = E(M(f^{(T)})^2) \leq 4E(| f^{(T)} |^2) = 4E(\sigma(f^{(T)})^2) = 4E(Y_T).$$

Then, Lemmas 3.4.4 and 3.4.1 give the desired result. □

Remark We have also

$$E_0(\Phi((Mf)^2)) \leq 5E_0(\Phi(\sigma(f)^2)).$$

Theorem 3.4.8 *Let $Z = (Z_n)_{n\geq 0}$ be a nonnegative submartingale, $Z = M + A$ be it's Doob Decomposition, where M is a martingale in L^1 and A is a nonnegative increasing and predictable process. Then for any strict concave Φ, we have*

$$E(\Phi(MZ)) \leq CE(\Phi(Z_0 + A_\infty)). \tag{3.4.16}$$

Proof. Denote $X = Z$ and $B = Z_0 + A$, then B is nonnegative, increasing and predictable. For all stopping times T, we have

$$\begin{aligned} E(Z_T \mid \mathcal{F}_0) &= E(M_T \mid \mathcal{F}_0) + E(A_T \mid \mathcal{F}_0) \\ &= E(M_0 + A_T \mid \mathcal{F}_0) = E(Z_0 + A_T \mid \mathcal{F}_0). \end{aligned}$$

Applying Lemma 3.4.5 and 3.4.3, we get (3.4.16). The proof is finished. □

Remark From $E(Z_T \mid \mathcal{F}_0) = E(Z_0 + A_T \mid \mathcal{F}_0)$, by making use of Lemmas 3.4.4 and 3.4.1, we get for all concave Φ

$$E(\Phi(Z_\infty)) \leq CE(\Phi(z_0 + A_\infty)). \tag{3.4.17}$$

3.5 General Φ-inequalities

In this section, we will consider those Φ, which are continuous and increasing functions from $\mathbf{R}^+$ to $\mathbf{R}^+$ with $\Phi(0) = 0$, and satisfy the following moderate growth condition

$$\Phi(\alpha f) \leq C_\alpha \Phi(\lambda), \quad \forall \lambda > 0, \text{ some } \alpha > 1. \tag{3.5.1}$$

We call these Φ "general Φ" as we mentioned before.

In this section, we will give some typical examples of general Φ-inequalities on martingales by means of distribution function inequalities. Essentially, they could be obtained by another technique (called rearrangement technique) along almost the same way, see next section. In order to avoid repetitions, the details of some inequalities will be put in this section, others in next section.

We begin with the so-called good λ-inequality.

Definition 3.5.1 *Let (f, g) be a pair of nonnegative measurable functions on $(\Omega, \mathcal{F}, \mu)$. We say that it satisfies the good λ-inequality, if there is $\alpha > 1$, and for all $\beta > 0$ small enough, there exist constants ε_β satisfying $\lim_{\beta \to 0} \varepsilon_\beta = 0$, such that*

$$\mid \{f > \alpha\lambda, g \leq \beta\lambda\} \mid \leq \varepsilon_\beta \mid \{f > \lambda\} \mid, \quad \forall \lambda > 0, \tag{3.5.2}$$

or, essentially the same thing,

$$\mid \{f > \alpha\lambda\} \mid \leq \varepsilon_\beta \mid \{f > \lambda\} \mid + \delta_\beta \mid \{g > \beta\lambda\} \mid . \tag{3.5.3}$$

Lemma 3.5.2 *Let Φ be general one , and (f,g) be a pair satisfying good λ-inequality. Then for $\beta > 0$ small enough, we have*

$$E(\Phi(f)) \le C_\alpha C_{\beta^{-1}} \delta_\beta (1 - C_\alpha \varepsilon_\beta)^{-1} E(\Phi(g)). \tag{3.5.4}$$

Proof. First, we show that for any Φ as above and any h being nonnegative measurable, we have

$$E(\Phi(h)) = \int_0^\infty \sigma(\lambda) d\Phi(\lambda) = \int_0^\infty \mid \{h > \lambda\} \mid d\Phi(\lambda). \tag{3.5.5}$$

In fact it follows from Fubini's theorem immediately

$$\begin{aligned} E(\Phi(h)) &= \int_\Omega \int_0^h d\Phi(\lambda) d\mu \\ &= \int_0^\infty \int_\Omega \chi_{\{h>\lambda\}} d\mu d\Phi(\lambda) = \int_0^\infty \sigma(\lambda) d\Phi(\lambda). \end{aligned}$$

Now integrating both sides of (3.5.3) with respect to the measure $d\Phi(\lambda)$ on $(0, \infty)$, we get

$$E\left(\Phi\left(\frac{f}{\alpha}\right)\right) \le \varepsilon_\beta E(\Phi(f)) + \delta_\beta E\left(\Phi\left(\frac{g}{\beta}\right)\right),$$

$$E(\Phi(f)) \le C_\alpha \varepsilon_\beta E(\Phi(f)) + C_\alpha C_{\beta^{-1}} \delta_\beta E(\Phi(g)).$$

For $\beta > 0$ small enough, we would have $C_\alpha \varepsilon_\beta < 1$. If $E(\Phi(f)) < \infty$, it would give the desired result. In general case, we consider the pair $(f \wedge N, g)$ which satisfies still (3.5.2) or (3.5.3). In fact , when $N > \alpha\lambda$, we have

$$\{f \wedge N > \alpha\lambda\} = \{f > \alpha\lambda\}, \quad \{f \wedge N > \lambda\} = \{f > \lambda\};$$

and when $N \le \alpha\lambda$, we have

$$\{f \wedge N > \alpha\lambda\} = \emptyset.$$

Applying what has just been proved to $(f \wedge N, g)$, then making $N \to \infty$, we get (3.5.4). This proves the lemma. □

Remark With the condition

$$E(\chi(\{f > \alpha\lambda\}) \mid \mathcal{F}_0) \le \varepsilon_\beta E(\chi(\{f > \lambda\}) \mid \mathcal{F}_0) + \delta_\beta E(\chi(\{g > \beta\lambda\}) \mid \mathcal{F}_0), \tag{3.5.6}$$

we have (3.5.4) with E replaced by $E_0 = E(\cdot \mid \mathcal{F}_0)$. In fact, for all $F \in \mathcal{F}_0$, we could prove

$$\int_F \Phi(f) d\mu \le C_\alpha C_{\beta^{-1}} \delta_\beta (1 - C_\alpha \varepsilon_\beta)^{-1} \int_F \Phi(g) d\mu, \tag{3.5.7}$$

which gives the desired modified (3.5.4). For the proof of (3.5.7), we notice that by denoting

$$\begin{aligned}\sigma_F(h,\lambda) &= |\{x \in F : |h| > \lambda\}| \\ &= |\{x : |h|\chi_F > \lambda\}|, \quad \forall\lambda > 0, F \in \mathcal{F}_0,\end{aligned} \tag{3.5.8}$$

then we have also (let h be nonnegative measurable)

$$\int_F \Phi(h)d\mu = \int_\Omega \Phi(h\chi_F)d\mu = \int_0^\infty \sigma_F(h,\lambda)d\Phi(\lambda).$$

Now we give some sufficient conditions which could imply good λ-inequalities associated with processes.

Lemma 3.5.3 *Let A, B be two nonnegative increasing adapted processes; or A, B be two nonnegative increasing predictable processes, with $A_0 = B_0 = 0$; or A be adapted one and B be predictable one as above. Suppose that for the stopping times defined as follows*

$$T = \inf\{n : B_n > \beta\lambda\}, \quad \forall\lambda > 0,$$

$$S = \inf\{n : A_n > \lambda\}, \quad \forall\lambda > 0,$$

(When we deal with predictable processes, the definition of the corresponding stopping times will be modified slightly as usual.) there would exist positive constants a, q, such that

$$E((A_{T-1} - A_{T\wedge(S-1)})^q) \le aE(B_{T-1}^q\chi_{\{S<\infty\}}), \tag{3.5.9}$$

or when A, B are predictable, such that

$$E((A_T - A_{T\wedge S})^q) \le aE(B_T^q\chi_{\{S<\infty\}}), \tag{3.5.10}$$

or only when B is predictable, such that

$$E((A_T - A_{T\wedge(S-1)})^q) \le aE(B_T^q\chi_{\{S<\infty\}}). \tag{3.5.11}$$

Then for all $\alpha > 1$, and $\beta > 0$ small enough, the pair (A_∞, B_∞) satisfies the good λ-inequality.

Proof. At first consider the case when A, B are both merely adapted. We have

$$\begin{aligned}|\{A_\infty > \alpha\lambda, B_\infty \le \beta\lambda\}| &= |\{A_{T-1} > \alpha\lambda, T = \infty\}| \le |\{A_{T-1} > \alpha\lambda\}| \\ &\le |\{A_{T-1} - A_{T\wedge(S-1)} > (\alpha-1)\lambda\}| \\ &\le \frac{1}{(\alpha-1)^q\lambda^q}E((A_{T-1} - A_{T\wedge(S-1)})^q) \\ &\le \frac{a}{(\alpha-1)^q\lambda^q}E(B_{T-1}^q\chi_{\{S<\infty\}}) \\ &\le \frac{a\beta^q}{(\alpha-1)^q}|\{A_\infty > \lambda\}|.\end{aligned}$$

When A and B, or only B, are predictable, the proof remains unchanged, only noticing that

$$A_{T\wedge S} \leq A_S \leq \lambda, \text{ or } A_{T\wedge(S-1)} \leq A_{S-1} \leq \lambda,$$

$$B_T \leq \beta\lambda.$$

The proof is finished. □

Remark When replacing E by $E_0 = E(\cdot \mid \mathcal{F}_0)$ in (3.5.9), (3.5.10) and (3.5.11), we get the good λ-inequality formulated in (3.5.6).

The following two general Φ-inequalities on martingales will be discussed in detail.

Theorem 3.5.4 *Let $f = (f_n)_{n\geq 0}$ be a martingale with it's difference process $(\Delta_n f)_{n\geq 0}$ satisfying*

$$\mid \Delta_n f \mid \leq D_{n-1}, \quad \forall n, \tag{3.5.12}$$

where $D = (D_n)_{n\geq 0}$ is a nonnegative increasing adapted process. Then both pairs $(Mf, S(f) + D_\infty)$ and $(S(f), Mf + D_\infty)$ satisfy the good λ-inequality, and hence for any general Φ, there exists a constant C depending on Φ but independing of martingales such that

$$E(\Phi(Mf)) \leq CE(\Phi(S(f))) + CE(\Phi(D_\infty)), \tag{3.5.13}$$

$$E(\Phi(S(f))) \leq CE(\Phi(Mf)) + CE(\Phi(D_\infty)). \tag{3.5.14}$$

Proof. We have

$$S_n(f) = \Big(\sum_0^{n-1} \mid \Delta_k f \mid^2 + \mid \Delta_n f \mid^2\Big)^{\frac{1}{2}} \leq S_{n-1}(f) + D_{n-1} = \rho_{n-1}.$$

For $\beta > 0$ and $\lambda > 0$, define the stopping time

$$\tau = \inf\{n : \rho_n > \beta\lambda\}.$$

Consider the stopping martingale $f^{(\tau)} = (f_n^{(\tau)})_{n\geq 0} = (f_{n\wedge\tau})_{n\geq 0}$. Define another stopping time

$$T = \inf\{n : \mid f_n^{(\tau)} \mid > \lambda\}.$$

For $\alpha > 1$, we have

$$\begin{aligned} \mid \{Mf > \alpha\lambda\} \mid &\leq \mid \{Mf > \alpha\lambda, \tau = \infty\} \mid + \mid \{\tau < \infty\} \mid \\ &\leq \mid \{Mf^{(\tau)} > \alpha\lambda\} \mid + \mid \{\tau < \infty\} \mid \\ &\leq \mid \{Mf^{(\tau)} - M_{T-1}(f^{(\tau)}) > (\alpha - 1)\lambda\} \mid + \mid \{\tau < \infty\} \mid . \end{aligned}$$

Now consider a new family of σ-fields $\{\mathcal{F}'_n\}_{n\geq 0}$ with$\mathcal{F}'_n = \mathcal{F}_{n+T}$, and a new process

$$g = f^{(\tau)} - f^{(\tau)}_{T-1} = (g'_n), \quad g'_n = f^{(\tau)}_{n+T} - f^{(\tau)}_{T-1}.$$

Then it is a martingale with respect to$(\Omega, \mathcal{F}, \mu, \{\mathcal{F}'_n\}_{n\geq 0})$. Noticing that

$$Mf^{(\tau)} - M_{T-1}(f^{(\tau)}) \leq \sup_{m\geq T} \mid f^{(\tau)}_m - f^{(\tau)}_{T-1} \mid = Mg',$$

(Since when $Mf^{(\tau)} = \mid f^{(\tau)}_m \mid$ for some $m \leq T-1$, then the left-hand side equals to zero.) and

$$\begin{aligned} S(g') &= S(f^{(\tau)} - f^{(\tau)}_{T-1}) \leq S(f^{(\tau)})\chi_{\{T<\infty\}} \\ &= S_\tau(f)\chi_{\{T<\infty\}} \leq \rho_{\tau-1}\chi_{\{T<\infty\}} \leq \beta\lambda\chi_{\{T<\infty\}}, \end{aligned}$$

then applying Davis' inequality we get

$$\begin{aligned} \mid \{Mf^{(\tau)} > \alpha\lambda\} \mid &\leq \mid \{Mg' > (\alpha-1)\lambda\} \mid \leq \frac{1}{(\alpha-1)\lambda} E(E(Mg' \mid \mathcal{F}_T)) \\ &\leq \frac{c}{(\alpha-1)\lambda} E(E(S(g') \mid \mathcal{F}_T)) \leq \frac{c\beta}{\alpha-1} \mid \{T<\infty\} \mid \\ &= \frac{c\beta}{\alpha-1} \mid \{Mf^{(\tau)} > \lambda\} \mid \leq \frac{c\beta}{\alpha-1} \mid \{Mf > \lambda\} \mid . \end{aligned}$$

Finally we get

$$\mid \{Mf > \alpha\lambda\} \mid \leq \frac{c\beta}{\alpha-1} \mid \{Mf > \lambda\} \mid + \mid \{S(f) + D_\infty > \beta\lambda\} \mid,$$

which is just the good λ-inquality for $(Mf, S(f) + D_\infty)$.

Analogously, basing on

$$\mid f_n \mid \leq M_{n-1}f + D_{n-1} = \rho_{n-1},$$

and defining first for $(\rho)_{n\geq 0}$, and then for $(S_n(f^{(\tau)}))_{n\geq 0}$ two stopping times τ and T, and noticing that

$$S(f^{(\tau)}) - S_{T-1}(f^{(\tau)}) \leq \sqrt{S(f^{(\tau)})^2 - S_{T-1}(f^{(\tau)})^2} = S(f^{(\tau)} - f^{(\tau)}_{T-1}),$$

$$\sup_{n\geq T} \mid f^{(\tau)}_n - f^{(\tau)}_{T-1} \mid \leq 2Mf^{(\tau)}\chi_{\{T<\infty\}} \leq 2\beta\lambda\chi_{\{T<\infty\}},$$

we get the good λ-inequality for $(S(f), Mf + D_\infty)$.

Then, applying Lemma 3.5.2, we get (3.5.13) and (3.5.14). This proves the theorem. □

Remarks

1. We could have (3.5.13), (3.5.14) with E replaced by E_0.

2. The lemma 3.5.3 could be used to deduce Theorem 3.5.4 with $A = Mf$ or $S(f)$ and $B = (\rho_n)_{n\geq 0}$. Here we do not apply Lemma 3.5.3 apparently, instead we give a whole proof with the same idea.

Now we want to strengthen the Theorem 3.5.4 to get a general Φ-inequality between $Mf \vee Sf$ and $Mf \wedge Sf$ occurred in Theorem 3.2.8.

Theorem 3.5.5 *Let $f = (f_n)_{n\geq 0}$ be a martingale with it's difference $(\Delta_n f)_{n\geq 0}$ having a predictable dominant, that is meant as in Theorem 3.5.4. Then for $Mf \vee Sf$ and $Mf \wedge Sf$, and general Φ, we have*

$$E(\Phi(Mf \vee Sf)) \leq CE(\Phi(Mf \wedge Sf + D_\infty)). \tag{3.5.15}$$

Proof. We have obtained in (3.2.17)

$$E((Mf \vee SF)^2 \mid \mathcal{F}_0) \leq CE((Mf \wedge Sf)^2 \mid \mathcal{F}_0), \quad \forall f = (f_n)_{n\geq 0}, \tag{3.5.16}$$

with the constant being independent of families of σ-fields, and of $f = (f_n)_{n\geq 0}$. Now let $f = (f_n)_{n\geq 0}$ be any martingale as stated in the theorem. Consider two associated processes, one is

$$A(f) = K(f) = K_n(f), \quad K_n(f) = M_n f \vee S_n f, \quad n \geq 0,$$

another is

$$B(f) = (B_n(f)) = (k_{n-1}(f) + D_{n-1}(f)), \quad k_n(f) = M_n f \wedge S_n f, \quad n \geq 0,$$

which is predictable with $B_0 = (k_{-1}(f) + D_{-1}(f) = 0$ by convention. Now we have

$$\begin{aligned} k_n(f) &= \min(M_n f, S_n(f)) \leq \min(M_{n-1} f + D_{n-1}, S_{n-1} + D_{n-1}) \\ &= k_{n-1}(f) + D_{n-1} = B_n. \end{aligned}$$

For $\lambda > 0$, define stopping times

$$T = \inf\{n : B_n > \beta\lambda\}, \qquad S = \inf\{n : A_n > \lambda\}.$$

Consider the stopped martingale $f^{(T)}$, and a new family of σ-fields $\{\mathcal{F}'_m\}_{m\geq 0} = \{\mathcal{F}_{m+S}\}_{m\geq 0}$. Then

$$g' = f^{(T)} - f^{(T)}_{S-1} = (g'_m)_{m\geq 0}, \quad g'_m = f^{(T)}_{m+S} - f^{(T)}_{S-1},$$

is a martingale with respect to $(\Omega, \mathcal{F}, \mu, \{\mathcal{F}'_m\}_{m\geq 0})$ and we have

$$\begin{aligned} K_T(f) - K_{T\wedge(S-1)}(f) &\leq \max(M_T f - M_{T\wedge(S-1)} f, S_T(f) - S_{T\wedge(S-1)}(f)) \\ &\leq \max(\sup_{m\geq 0} \mid f^{(T)}_{m+S} - f^{(T)}_{S-1} \mid, S(f^{(T)} - f^{(T)}_{S-1})) = K(g'), \end{aligned}$$

$$\begin{aligned} k(g') &= \min(\sup_{m\geq 0} \mid f^{(T)}_{m+S} - f^{(T)}_{S-1} \mid, S(f^{(T)} - f^{(T)}_{S-1})) \\ &\leq \min(2M_T f, S_T(f))\chi_{\{S<\infty\}} \\ &\leq Ck_T(f)\chi_{\{S<\infty\}}. \end{aligned}$$

Now applying (3.5.16) to $g' = (g'_m)_{m\geq 0}$ on $(\Omega, \mathcal{F}, \mu, \{\mathcal{F}'_m\}_{m\geq 0})$, we get

$$E((A_T - A_{T\wedge(S-1)})^2) \leq E(K(g')^2) \leq CE(k(g')^2)$$
$$\leq CE(k_T(f)^2\chi_{\{S<\infty\}}) \leq CE(B_T^2\chi_{\{S<\infty\}}).$$

Then by making use of Lemma 3.5.3 and 3.5.2, we get the desired inequality (3.5.15). The proof is finished. □

Remark We could have (3.5.15) with E replaced by E_0.

There are some other general Φ-inequalities on martingales. For examples for nonnegative martingales $f = (f_n)$, we have always $E(\Phi(Sf)) \leq CE(\Phi(Mf))$ without added conditions. In addition we have also

$$E(\Phi(M_a(f))) \leq CE(\Phi(f_a^\#)), \quad \forall f = (f_n)_{n\geq 0},$$

where $1 \leq a < \infty$, and

$$M_a(f) = \sup_n E(|f|^a|\mathcal{F}_n)^{\frac{1}{a}}, \tag{3.5.17}$$

$$f_a^\# = \sup_n \rho_n = \sup_n E(|f - f_{n-1}|^a|\mathcal{F}_n)^{\frac{1}{a}}. \tag{3.5.18}$$

We will formulate them as theorems, but omit the details, since we will obtain them by the rearrangement technique in the next section.

Theorem 3.5.6 *Let $f = (f_n)_{n\geq 0}$ be a nonnegative martingale. Then $(S(f), Mf)$ satisfies the good λ-inequality with $\alpha > 1$, $0 < \beta < \sqrt{\alpha^2 - 1}$, and hence the general Φ-inequality*

$$E(\Phi(Sf)) \leq CE(\Phi(Mf)), \quad \forall f = (f_n)_{n\geq 0} \geq 0. \tag{3.5.19}$$

Theorem 3.5.7 *For general Φ, we have*

$$E(\Phi(M_a(f))) \leq C_{a,\Phi}E(\Phi(f_a^\#)), \quad \forall f. \tag{3.5.20}$$

Remarks

1. In particular, when $a = 1$ and $\Phi(u) = u$, we have

$$E(M(f)) \leq CE(f^\#).$$

2. Since $f_a^\# \leq CM_a(f)$, the opposite one of (3.5.20) holds naturally.
3. It seems to be a little difficult to deal with the following definition of #-function

$$\tilde{\tilde{f}}{}_a^\# = \sup_n \inf_\theta E(|\, f - \theta_{n-1}\,|^a|\, \mathcal{F}_n)^{\frac{1}{a}}, \tag{3.5.21}$$

where "$\inf_\theta$" is taken over all adapted processes. The difficulty lies in the fact, that this time we could only obtain (for $a \geq 1$)

$$E(|\, f - f_{n-1}\,|^a|\, \mathcal{F}_n)^{\frac{1}{a}} \leq E(|\, f - \theta_{n-1}\,|^a|\, \mathcal{F}_n)^{\frac{1}{a}} + E(|\, f - \theta_{n-1}\,|^a|\, \mathcal{F}_{n-1})^{\frac{1}{a}}.$$

which could not give the desired equivalence of $\tilde{\tilde{f}}^{\#}_a$ and $f^{\#}_a$. But we still have the following result.

Theorem 3.5.7′ *There exists a constant $C = C_{a,\Phi}$ only depending on the indices shown, where $1 \le a < \infty$ and Φ is general, such that for all $f = (f_n)_{n\ge 0} \in L^a_u$ satisfying*

$$\int_\Omega \Phi(f^{\#}_{a,\theta})d\mu < \infty, \quad f^{\#}_{a,\theta} = \sup_n E(| f - \theta_{n-1} |^a | \mathcal{F}_n)^{\frac{1}{a}},$$

where $\theta = (\theta_n)_{n\ge}$ is an adapted process depending on f, we have

$$\int_\Omega \Phi(M_a(f))d\mu \le C_{a,\Phi} \int_\Omega \Phi(f^{\#}_{a,\theta})d\mu. \tag{3.5.22}$$

Proof. Define three stopping times as follows

$$T = \inf\{n :| f_n |> (\alpha+1)\lambda\},$$

$$S = \inf\{n :| \theta_{n-1} |> \alpha\lambda\},$$

$$R = \inf\{n : \rho_n > \beta\lambda\},$$

$$\rho_n = E(| f - \theta_{n-1} |^a | \mathcal{F}_n)^{\frac{1}{a}}.$$

Now we have

$$\begin{aligned}\{T<\infty\} \subset \{T<\infty, S\le T, S<R\} &\bigcup \{T<\infty, T<S, T<R\}\\ &\bigcup \{T<\infty, S\le T, R\le S\} \bigcup \{T<\infty, T<S, R\le T\},\end{aligned}$$

$$\{T<\infty, S\le T, S<R\} \subset \{S<R, S\le T, | f_T - \theta_{S-1} |> \lambda\},$$

$$\{T<\infty, T<S, T<R\} \subset \{T<R, | f_T - \theta_{T-1} |> \lambda\},$$

$$\{T<\infty, S\le T, R\le S\} \bigcup \{T<\infty, T<S, R\le T\} \subset \{R<\infty\}.$$

So, we get

$$\begin{aligned}| \{T<\infty\} | &\le \frac{1}{\lambda}\left\{\int_{\{S<R\}\cap\{S\le T\}} | f_T - \theta_{S-1} | d\mu \right.\\ &\qquad \left. + \int_{\{T<R\}} | f_T - \theta_{T-1} | d\mu\right\} + | \{R<\infty\} |\\ &\le \frac{1}{\lambda}\left\{\int_{\{S<R\}\cap\{S\le T\}} E(E(| f - \theta_{S-1} || \mathcal{F}_T) | \mathcal{F}_S)d\mu \right.\\ &\qquad \left. + \int_{\{T<R\}} E(| f - \theta_{T-1} || \mathcal{F}_T)d\mu\right\} + | \{R<\infty\} |\\ &\le \beta | \{S<\infty\} | + \beta | \{T<\infty\} | + | \{R<\infty\} | .\end{aligned} \tag{3.5.23}$$

That is

$$|\{Mf > (\alpha+1)\lambda\}| \le \varepsilon_\beta |\{M\theta > \alpha\lambda\}| + \delta_\beta |\{f^\#_{a,\theta} > \beta\lambda\}|. \tag{3.5.24}$$

From it we get

$$\int_\Omega \Phi(Mf)d\mu \le \varepsilon_\beta \int_\Omega \Phi(M\theta)d\mu + \delta_\beta \int_\Omega \Phi(f^\#_{a,\theta})d\mu.$$

But we have

$$|\theta_{n-1}| \le E(|f-\theta_{n-1}|^a|\mathcal{F}_n)^{\frac{1}{a}} + E(|f|^a|\mathcal{F}_n)^{\frac{1}{a}}, \tag{3 5.25}$$

$$\begin{aligned} E(|f|^a|\mathcal{F}_n)^{\frac{1}{a}} &\le E(|f-f_n|^a|\mathcal{F}_n)^{\frac{1}{a}} + |f_n| \\ &\le 2E(|f-\theta_{n-1}|^a|\mathcal{F}_n)^{\frac{1}{a}} + |f_n|. \end{aligned} \tag{3.5.26}$$

That is

$$M\theta \le M_a(f) + f^\#_{a,\theta},$$

$$M_a(f) \le 2f^\#_{a,\theta} + Mf,$$

$$\int_\Omega \Phi(M_a(f))d\mu \le c\varepsilon_\beta \int_\Omega \Phi(M_a(f))d\mu + c\delta_\beta \int_\Omega \Phi(f^\#_{a,\theta})d\mu.$$

Since $\lim_{\beta\to 0} \varepsilon_\beta = 0$. we get (3.5.22). The proof is finished. □

3.6 Rearrangement inequalities

Let f be any measurable function on any measure space $(\Omega, \mathcal{F}, \mu)$. As well known, the so-called distribution function and decreasing rearrangement function of f are as follows respectively

$$\sigma_f(\lambda) = |\{\omega \in \Omega : |f(\omega)| > \lambda\}|, \quad \lambda > 0. \tag{3.6.1}$$

$$f^*(t) = \inf\{\lambda : \sigma_f(\lambda) \le t\}, \quad t > 0. \tag{3.6.2}$$

Habitually, people prefer to use distribution functions to estimate the functions itself. How about the rearrangement function technique? In this section, we will show that essentially these two techniques are equivalent. Of course, there are some differences between them. For example , so-called good λ-inequality, a typical example of distribution function inequalities,

$$|\{|F| > \alpha\lambda\}| \le \varepsilon_\beta |\{|F| > \lambda\}| + |\{|G| > \beta\lambda\}|, \quad \forall\lambda > 0, \tag{3.6.3}$$

where (F, G) is a measurable function pair on $(\Omega, \mathcal{F}, \mu)$, $\lim_{\beta\to 0}\varepsilon_\beta = 0$, and $\alpha > 1$, has following rearranged version

$$F^*(t) \le CG^*\left(\frac{t}{2}\right) + F^*(2t), \quad \forall t > 0. \tag{3.6.4}$$

It can not be said that one of them implies the other. But both of them follow from almost the same way. As for their applications, roughly speaking, (3.6.3) implies the general Φ-inequalities, and (3.6.4) implies only convex Φ-inequalities, but as a compensation, the implied Φ could be of exponential growth, and the implied constants are sharper in the case $p_\Phi < \infty$.

First we want to devote some words to the Hardy's average operators T and T^*, and to the maximal rearrangement operator $**$. Let $(\Omega, \mathcal{F}, \mu)$ be any measure space. Then the maximal rearrangement operator is defined by

$$h^{**}(t) = \sup_{F:|F|\ge t} \frac{1}{|F|}\int_F |h| \, d\mu, \quad t > 0. \tag{3.6.5}$$

With this definition, we have

$$h^*(t) \le h^{**}(t) \le \frac{1}{t}\int_0^t h^*(\tau)d\tau, \quad t > 0, \tag{3.6.6}$$

and when $(\Omega, \mathcal{F}, \mu)$ is atom-free, we have the identity

$$h^{**}(t) = \frac{1}{t}\int_0^t h^*(\tau)d\tau, \quad t > 0. \tag{3.6.7}$$

One may adopt (3.6.7) as the definition of $**$ as we do in this section, since we do not need the sublinearity of $**$ which is the advantage of the definition (3.6.5).

Hardy's average operators are defined as

$$Tg(t) = \frac{1}{t}\int_0^t g(\tau)d\tau, \quad t > 0, \tag{3.6.8}$$

$$T^*g(t) = \int_t^\infty \frac{g(\tau)}{\tau}d\tau, \quad t > 0. \tag{3.6.9}$$

We want to investigate the convex Φ-boundedness of T and T^*.

Lemma 3.6.1 *Let $\Phi(u)$ be any convex function. Then*

$$\|Tf\|_\Phi \le q'_\Phi \|f\|_\Phi, \quad \forall f \in L^1_{\mathrm{loc}}(0,\infty), \tag{3.6.10}$$

$$\|T^*g\|_\Phi \le p_\Phi \|g\|_\Phi, \quad \forall g \in L^1\left(\varepsilon, \infty, \frac{dt}{t}\right), \quad \forall \varepsilon > 0. \tag{3.6.11}$$

Proof. First consider T. We want to get

$$|\{|Tf| > \lambda\}| \le \frac{1}{\lambda}\int_{\{|Tf|>\lambda\}} |f| \, dx, \quad \forall \lambda > 0. \tag{3.6.12}$$

Without loss of generality, we can assume that f is nonnegative and decreasing on $(0,\infty)$, otherwise we consider f^* replacing f, since

$$| Tf(x) | \leq Tf^*(t), \text{ and } \|f^*\|_\Phi = \|f\|_\Phi.$$

Now for $\lambda > 0$ given, let x_0 be the solution of $Tf(x) = \lambda$. Then

$$\{Tf(x) > \lambda\} = (0, x_0), \quad x_0 = \frac{1}{\lambda}\int_0^{x_0} f(t)dt.$$

It is nothing but

$$| \{Tf > \lambda\} | = \frac{1}{\lambda}\int_{\{Tf>\lambda\}} f(t)dt.$$

Now arguing as in Lemmas 3.3.1 and 3.3.2, we get (denote $\varphi = \Phi'$)

$$E(Tf\varphi(Tf)) \leq q'_\Phi E(f\varphi(Tf)),$$

$$E(\Phi(Tf)) \leq E(\Phi(q'_\Phi f)).$$

and hence (3.6.10) by homogeneity.

Now consider T^*. We want to get

$$\int_{\{|T^*g|>\lambda\}} (| T^*g | - \lambda)dx \leq \int_{\{|T^*g|>\lambda\}} | g | dx, \quad \forall\lambda > 0. \tag{3.6.13}$$

Obviously, g could be assumed nonnegative. For $\lambda > 0$ given, let x_0 be as small as possible such that $\lambda = \int_{x_0}^\infty \frac{g(t)}{t}dt$. Then $\{T^*g > \lambda\} = (0, x_0)$, and

$$\begin{aligned}\int_{\{T^*g>\lambda\}} (T^*g - \lambda)dx &= \int_0^{x_0}\left(\int_x^{x_0}\frac{g(t)}{t}dt + \int_{x_0}^\infty \frac{g(t)}{t}dt\right)dx - \lambda x_0\\ &= \int_0^{x_0} g(t)dt = \int_{\{T^*>\lambda\}} g(x)dx.\end{aligned}$$

Arguing as in Lemmas 3.2.1 and 3.2.3, we get

$$E(T^*g\varphi(T^*g)) \leq E(p_\Phi g\varphi(T^*g)),$$

$$E(\Phi(T^*g)) \leq E(\Phi(p_\Phi g)),$$

and hence (3.6.11). The proof is finished. □

Remark T^* restricted on the set $\{f :$ nonnegative and decreasing $\}$ looks like being Φ-bounded for general Φ, since it can be shown to be Φ-bounded for those Φ satisfying $\Phi(u+v) \leq \Phi(u) + \Phi(v)$ (for example, any increasing (to infinity) and concave Φ is of the case). In fact, for all f in the set, we have

$$\begin{aligned}\int_0^\infty \Phi\left(\int_x^\infty \frac{f(t)}{t}dt\right)dx &\leq \int_0^\infty \sum_{k=0}^\infty \Phi\left(\int_{2^kx}^{2^{k+1}x}\frac{f(t)}{t}dt\right)dx\\ &\leq \sum_{k=0}^\infty \int_0^\infty \Phi(f(2^kx))dx = 2\int_0^\infty \Phi(f(x))dx.\end{aligned}$$

We do not know if it holds without any added conditions imposed on Φ.

What happens when $q_\Phi = 1$ with the operator T, or $p_\Phi = \infty$ with T^*? We only consider a simple kind of Young's pair of convex functions, that is $(\Phi_\alpha, \Psi_\alpha)$ where

$$\Phi_\alpha(u) = \int_0^u \varphi_\alpha(t)dt \text{ with } \varphi_\alpha(t) = 1 + \log^{+\alpha} t, \quad 0 \le \alpha < \infty. \tag{3.6.14}$$

Obviously, Φ_α satisfies following estimate

$$\frac{u}{2}\left(1 + \log^{+\alpha}\frac{u}{2}\right) \le \Phi_\alpha(u) \le u(1 + \log^{+\alpha} u), \tag{3.6.15}$$

and Ψ_α is like $\left(e^{(u-1)^{\frac{1}{\alpha}}} - 1\right)\chi_{\{u\ge1\}}$.

Lemma 3.6.2 *Let $A > 0, \alpha \ge 1$, and $(\Phi_\alpha, \Psi_\alpha)$ be as above. Then there exists $C_{A,\alpha}$ such that*

$$||Tf||_{\Phi_{\alpha-1}} \le C_{A,\alpha}||f||_{\Phi_\alpha}, \quad \forall f, \ \operatorname{supp} f \subset [0, A]. \tag{3.6.16}$$

$$||T^*g||_{\Psi_\alpha} \le C_{A,\alpha}||g||_{\Psi_{\alpha-1}}, \quad \forall g, \ \operatorname{supp} g \subset [0, A]. \tag{3.6.17}$$

Proof. Without loss of generality A could be assumed to be 1. First consider T. It is enough to prove

$$\int_0^1 | Tf | (1 + \log^{+(\alpha-1)} | Tf |)dx$$
$$\le B\int_0^1 | f | (1 + \log^{+\alpha} | f |)dx + D, \quad \forall f, \operatorname{supp} f \subset [0, 1]. \tag{3.6.18}$$

Once it is proved, applying it to $\frac{f}{||f||_{\Phi_\alpha}}$, we get

$$\int_0^1 \Phi_{\alpha-1}\left(\frac{C\,| Tf |}{||f||_{\Phi_\alpha}}\right)dx \le 1, \quad ||Tf||_{\Phi_{\alpha-1}} \le C||f||_{\Phi_\alpha}.$$

Now turn to (3.6.18). Without loss of generality, assume $||f||_1 \ge 1$. We have

$$\begin{aligned}
\int_0^1 (Tf)^*(t)\log^{+(\alpha-1)}(Tf)^*(t)dt &\le \int_0^1 \frac{1}{t}\int_0^t f^*(\tau)d\tau \log^{\alpha-1}\frac{||f||_1}{t}dt \\
&= \int_0^1\int_\tau^1 \frac{1}{t}\log^{\alpha-1}\frac{||f||_1}{t}dt f^*(\tau)d\tau \\
&\le C_\alpha\int_0^1 f^*(t)\log^\alpha\frac{||f||_1}{t}dt \\
&= C_\alpha\left(\int_{\{f^*(t)\le(\frac{||f||_1}{t})^{\frac{1}{2}}\}} + \int_{\{f^*(t)>(\frac{||f||_1}{t})^{\frac{1}{2}}\}}\right) \\
&\le C_\alpha||f||_1 + C_\alpha\int_0^1 f^*(t)\log^{+\alpha} f^*(t)dt \\
&= C_\alpha\int_0^1 f^*(1 + \log^{+\alpha} f^*(t))dt.
\end{aligned}$$

Thus (3.6.18) and hence (3.6.16) is proved.

Now prove (3.6.17) by duality. Since Φ_α is of moderate growth,

$$S = \{\text{all simple functions on } [0,1]\}$$

is dense in $L^{\Phi_\alpha}(0,1)$. Notice as well that the set $\{f \in L^{\Phi_\alpha}(0,1) : \|f\|_{\Phi_\alpha} \le 1\}$ is the same as $\{f \in L^{\Phi_\alpha}(0,1) : \int_0^1 \Phi_\alpha(|f|)dx \le 1\}$. So we get

$$\begin{aligned}\|T^*g\|_{\Psi_\alpha} &\le C\sup\left\{\left|\int_0^1 T^*gfdx\right| : f \in S, \int_0^1 \Phi_\alpha(|f|)dx \le 1\right\}\\ &= C\sup\left\{\left|\int_0^1 gTfdx\right| : f \in S, \|f\|_{\Phi_\alpha} \le 1\right\}\\ &\le C\|g\|_{\Psi_{\alpha-1}}\sup_f \|Tf\|_{\Phi_{\alpha-1}} \le C\|g\|_{\Psi_{\alpha-1}}.\end{aligned}$$

The proof of the lemma is finished. □

Remark It is well known that $h \in L^\Psi(X)$, if and only if there is $\theta_h > 0$ such that $\int_X \Psi(\theta_h |h|)dx < \infty$, for any measure space X and any convex function Ψ. In the case of the lemma,

$$\Psi_\alpha(u) = \int_1^u e^{(t-1)^{\frac{1}{\alpha}}}dt\chi_{\{u\ge1\}} \ge Ce^{c(u-1)^{\frac{1}{\alpha}}}\chi_{\{u\ge2\}}.$$

So (3.6.17) is an exponential type inequality on martingales.

Now study some rearranged inequalities on martingales. First, see Doob's maximal inequality.

Theorem 3.6.3 *Let $f = (f_n)_{n\ge0} \in L^1$. Then*

$$(Mf)^*(t) \le f^{**}(t), \quad \forall t > 0. \tag{3.6.19}$$

Proof. By taking the limit, it is enough to assume $f \in L^1_u$. Let $t > 0$, and $\lambda_0 > f^{**}(t)$. We have

$$\begin{aligned}\sigma_{Mf}(\lambda_0) = |\{Mf > \lambda_0\}| &\le \frac{1}{\lambda_0}\int_{\{Mf>\lambda_0\}} |f| d\mu\\ &= \frac{1}{\lambda_0}\int_0^{\sigma_{Mf}(\lambda_0)} f^*(s)ds,\end{aligned}$$

and hence (noticing $\lambda_0 > f^{**}(t)$)

$$(\lambda_0 t)^{-1}\int_0^t f^*(s)ds < 1 \le (\lambda_0\sigma_{Mf}(\lambda_0))^{-1}\int_0^{\sigma_{Mf}(\lambda_0)} f^*(s)ds.$$

This implies $\sigma_{Mf}(\lambda_0) \le t$ since f^* is decreasing. So

$$(Mf)^*(t) = \inf\{\lambda : \sigma_{Mf}(\lambda) \le t\} \le \lambda_0.$$

Letting $\lambda_0 \to f^{**}(t)$, we get (3.6.19). The proof is finished. □

Remark Unlike the classical case, we could not expect $f^{**}(t) \le C(Mf)^*(t)$ in general, as shown by following simple example. Assume $\mathcal{F}_0$ is trivial but $\mathcal{F}_1 = \cdots = \mathcal{F}$. Let $f \in L^1$ with $f_0 = 0$. Then $Mf = \mid f \mid$. We could not get $f^{**}(t) \le Cf^*(t)$ of course.

Corollary 3.6.4 *Let $\Phi(u)$ be convex with $q_\Phi > 1$, and Φ_α be as in Lemma 3.6.2. Then*

$$||Mf||_\Phi \le q'||f||_\Phi, \quad \forall f = (f_n)_{n\ge 0},$$

$$||Mf||_{\Phi_{\alpha-1}} \le C||f||_{\Phi_\alpha}, \quad \forall f = (f_n)_{n\ge 0}. \tag{3.6.20}$$

In what follows, we want to establish the rearrangement versions of good λ-inequalities obtained in §3.5.

Lemma 3.6.5 *Let (A, B) be a pair of nonnegative, increasing processes. In addition, A is adapted and B is predictable with $B_0 = 0$. Assume a, $q > 0$ such that for all stopping times T, τ.*

$$E((A_T - A_{T\wedge(\tau-1)})^q) \le a^q E(B_T^q \chi_{\{\tau<\infty\}}). \tag{3.6.21}$$

Then

$$A_\infty^*(t) \le 4^{\frac{1}{q}} a B_\infty^*\left(\frac{t}{2}\right) + A_\infty^*(2t), \quad \forall t > 0. \tag{3.6.22}$$

Proof. Let $t > 0$. It is enough to consider the case $0 < A_\infty^*(2t)B_\infty^*(\frac{t}{2}) < \infty$. The infinite case is obvious. The zero case could be avoided by taking the limits. Define stopping times

$$T = \inf\left\{n : B_{n+1} > B_\infty^*\left(\frac{t}{2}\right)\right\}, \quad \tau = \inf\{n : A_n > A_\infty^*(2t)\}.$$

Then we have, with $C = 4^{\frac{1}{q}} a$,

$$\left\{A_\infty > CB_\infty^*\left(\frac{t}{2}\right) + A_\infty^*(2t)\right\} \subset \{T < \infty\} \bigcup \left\{A_T > CB_\infty^*\left(\frac{t}{2}\right) + A_\infty^*(2t)\right\}.$$

Notice that $A_{T\wedge(\tau-1)} \le A_\infty^*(2t)$ (even when $\tau = 0$, since $A_{T\wedge(\tau-1)}$ is meant as 0 on the set $\{\tau = 0\}$) so

$$\left\{A_T > CB_\infty^*\left(\frac{t}{2}\right) + A_\infty^*(2t)\right\} \subset \left\{A_T - A_{T\wedge(\tau-1)} > CB_\infty^*\left(\frac{t}{2}\right)\right\}.$$

Thus we get

$$\begin{aligned}
&\left|\left\{A_\infty > CB_\infty^*\left(\frac{t}{2}\right) + A_\infty^*(2t)\right\}\right| \\
&\le \mid \{T < \infty\} \mid + \left(CB_\infty^*\left(\frac{t}{2}\right)\right)^{-q} E((A_T - A_{T\wedge(\tau-1)})^q) \\
&\le \frac{t}{2} + a^q \left(CB_\infty^*\left(\frac{t}{2}\right)\right)^{-q} E(B_T^q \chi_{\{\tau<\infty\}}) \le t.
\end{aligned}$$

This proves (3.6.22). The proof is finished. □

Now applying this lemma to various pairs $(A(f)), B(f))$ associated with a martingale $f = (f_n)_{n\geq 0}$, we get

Theorem 3.6.6 *Let $f = (f_n)_{n\geq 0}$ be a martingale having a predictable control $D = (D_n)_{n\geq 0}$ in the sense as before (see §3.2). Then for $(A(f)), B(f))$, with $A(f)) = (A_n(f))_{n\geq 0}$, $B(f) = (B_n(f))_{n\geq 0}$, being respectively*

$$A_n(f) = M_n f, \quad B_n(f) = S_{n-1}(f) + D_{n-1}, \quad n \geq 0,$$

$$A_n(f) = S_n f, \quad B_n(f) = M_{n-1}(f) + D_{n-1}, \quad n \geq 0,$$

$$A_n(f) = M_n f \vee S_n f, \quad B_n(f) = M_{n-1}(f) \wedge S_{n-1}(f) + D_{n-1}, \quad n \geq 0,$$

we have

$$A^*_\infty(t) \leq C B^*_\infty\left(\frac{t}{2}\right) + A^*_\infty(2t), \quad \forall t > 0.$$

Proof. Only verify (3.6.21). We have, say $A_\infty = Mf, \quad B_\infty = Sf + D_\infty$

$$\begin{aligned} E((A_T - A_{T\wedge(\tau-1)})^2) &= E((M_T - M_{T\wedge(\tau-1)})^2) \\ &\leq E(M(f^{(T)} - f^{(T)}_{\tau-1})^2) \leq CE(S(f^{(T)} - f^{(T)}_{\tau-1})^2) \\ &= CE(S_T(f)^2 - S_{T\wedge(\tau-1)}(f)^2) \leq CE(B_T(f)^2 \chi_{\{\tau<\infty\}}). \end{aligned}$$

The same argument works well in the case $A_\infty = Mf \vee Sf$, $B_\infty = Mf \wedge Sf + D_\infty$. The proof is finished. □

For nonnegative martingales $f = (f_n)_{n\geq 0}$, we have rearranged inequality of the pair (Sf, Mf) without the "predictability".

Theorem 3.6.7 *Let $f = (f_n)_{n\geq 0}$ be a nonnegative martingale. Then*

$$(Sf)^*(t) \leq 3(Mf)^*\left(\frac{t}{2}\right) + (Sf)^*(2t), \quad \forall t > 0. \tag{3.6.23}$$

Proof. For $t > 0$, define stopping times

$$\tau = \inf\left\{n : | f_n | > (Mf)^*\left(\frac{t}{2}\right)\right\}, \quad T = \inf\{n : S_n f > (Sf)^*(2t)\}.$$

It is enough to prove

$$(Sf)^{*2}(t) \leq 9(Mf)^{*2}\left(\frac{t}{2}\right) + (Sf)^{*2}(2t), \quad \forall t > 0. \tag{3.6.24}$$

We have

$$\left\{S(f)^2 > 9(Mf)^{*2}\left(\frac{t}{2}\right) + (Sf)^{*2}(2t)\right\}$$
$$\subset \{\tau < \infty\} \bigcup \left\{S_{\tau-1}(f)^2 > 9(Mf)^{*2}\left(\frac{t}{2}\right) + (Sf)^{*2}(2t)\right\},$$

$$\left\{S_{\tau-1}(f)^2 > 9(Mf)^{*2}\left(\frac{t}{2}\right) + (Sf)^{*2}(2t)\right\}$$
$$\subset \left\{T < \tau, S_{\tau-1}(f)^2 - S_T(f)^2 > 8(Mf)^{*2}\left(\frac{t}{2}\right)\right\},$$

here we have used the fact that on the set $\{T < \tau\}$, we have

$$S_T(f)^2 \le S_{T-1}(f)^2 + (f_T - f_{T-1})^2 \le (Mf)^{*2}\left(\frac{t}{2}\right) + (Sf)^{*2}(2t).$$

Noticing an identity (see (1.3.1.7) in §1.3.1)

$$\sum_{k=T+1}^{\tau-1} (\Delta_k f)^2 = 2\sum_{k=T+1}^{\tau} f_{k-1}(f_{k-1} - f_k) - f_T^2 - f_{\tau-1}^2 + 2f_{\tau-1}f_\tau, \qquad (3.6.25)$$

and the fact

$$E\Big(\sum_{k=T+1}^{\tau} f_{k-1}\Delta_k f \mid \mathcal{F}_T\Big) = E\Big(\sum_1^{\infty} E(\chi_{\{T+1\le k\le \tau\}} f_{k-1}\Delta_k f \mid \mathcal{F}_{k-1}) \mid \mathcal{F}_T\Big) = 0,$$

we get

$$\left|\left\{T < \tau, S_{\tau-1}(f)^2 - S_T(f)^2 > 8(Mf)^{*2}\left(\frac{t}{2}\right)\right\}\right|$$
$$\le \left(8(Mf)^{*2}\left(\frac{t}{2}\right)\right)^{-1} E((S_{\tau-1}(f)^2 - S_T(f)^2)\chi_{\{T<\tau\}})$$
$$\le \left(8(Mf)^{*2}\left(\frac{t}{2}\right)\right)^{-1} E(E(2f_{\tau-1}f_\tau \mid \mathcal{F}_T)\chi_{\{T<\tau\}})$$
$$\le \frac{1}{4} \mid \{T < \infty\} \mid \le \frac{t}{2}.$$

This completes the proof of the theorem. □

Another important example is the inequality between $M_a f$ and $f_a^{\#}$. Let $1 \le a < \infty$, the definitions of $M_a(f)$ and $f_a^{\#}$ are

$$M_a(f) = M(\mid f\mid^a)^{\frac{1}{a}}, \ f_a^{\#} = \sup \rho_n = \sup E(\mid f - f_{n-1}\mid^a\mid \mathcal{F}_n)^{\frac{1}{a}}. \qquad (3.6.26)$$

Theorem 3.6.8 *Let $1 \le a < \infty$. Then for all martingales $f = (f_n)_{n\ge 0}$, we have*

$$(Mf)^*(t) \le 4f_a^{\#*}\left(\frac{t}{2}\right) + (Mf)^*(2t), \quad \forall t > 0, \qquad (3.6.27)$$

$$(M_a(f)^*(t) \le 6f_a^{\#*}\left(\frac{t}{4}\right) + (M_a(f))^*\left(\frac{5t}{4}\right), \quad \forall t > 0. \qquad (3.6.28)$$

Proof. For $t > 0$, define stopping times

$$S = \inf\{n : | f_n | > (Mf)^*(2t)\},$$

$$T = \inf\left\{n : | f_n | > 4f_a^{\#*}\left(\frac{t}{2}\right) + (Mf)^*(2t)\right\},$$

$$R = \inf\left\{n : \rho_n > f_a^{\#*}\left(\frac{t}{2}\right)\right\}.$$

Notice that $S \leq T$, and

$$\{T < \infty\} = \left\{Mf > 4f_a^{\#*}\left(\frac{t}{2}\right) + (Mf)^*(2t)\right\},$$

$$\{S < \infty\} = \{Mf > (Mf)^*(2t)\},$$

$$\{R < \infty\} = \left\{f_a^{\#} > f_a^{\#*}\left(\frac{t}{2}\right)\right\},$$

$$| \{S < \infty\} | \leq 2t, \quad | \{R < \infty\} | \leq \frac{t}{2}.$$

We have

$$\begin{aligned} \{T < \infty\} &= \{T < \infty, S < R\} \bigcup \{T < \infty, R \leq S\} \\ &\subset \{R < \infty\} \bigcup \{T < \infty, S < R\}, \end{aligned}$$

$$\{T < \infty, S < R\} \subset \left\{S < R,\ | f_T - f_{S-1} | > 4f_a^{\#*}\left(\frac{t}{2}\right)\right\}.$$

And so

$$\begin{aligned} | \{T < \infty, S < R\} | &\leq \left(4f_a^{\#*}\left(\frac{t}{2}\right)\right)^{-1} \int_{\{S<R\}} | E(f - f_{S-1} \mid \mathcal{F}_T) | \, d\mu \\ &\leq \left(4f_a^{\#*}\left(\frac{t}{2}\right)\right)^{-1} \int_{\{S<R\}} E(|f - f_{S-1}|^a|\mathcal{F}_S)^{\frac{1}{a}} d\mu \\ &\leq \left(4f_a^{\#*}\left(\frac{t}{2}\right)\right)^{-1} \|\rho_S \chi_{\{S<R\}}\|_\infty \, | \{S < \infty\} | \leq \frac{t}{2}. \end{aligned}$$

This proves $| \{T < \infty\} | \leq t$, and hence (3.6.27).

From

$$E(| f |^a | \mathcal{F}_n)^{\frac{1}{a}} \leq E(| f - f_{n-1} |^a | \mathcal{F}_n)^{\frac{1}{a}} + | f_{n-1} |,$$

$$M_a(f) \leq f_a^{\#} + Mf,$$

and similarly

$$Mf \leq f_a^{\#} + M_a(f),$$

we get

$$\begin{aligned}(M_a(f))^*(t) &\leq f_a^{\#*}\left(\frac{t}{4}\right) + (Mf)^*\left(\frac{3}{4}t\right)\\ &\leq f_a^{\#*}\left(\frac{t}{4}\right) + 4f_a^{\#*}\left(\frac{3}{8}t\right) + (Mf)^*\left(\frac{6}{4}t\right)\\ &\leq 6f_a^{\#*}\left(\frac{t}{4}\right) + (M_a(f))^*\left(\frac{5}{4}t\right).\end{aligned}$$

The proof is finished. □

What Φ-inequalities could (3.6.4) imply? Now we discuss it.

Theorem 3.6.9 *Let $(\Omega, \mathcal{F}, \mu)$ be a σ-finite measure space, (F, G) be a measurable pair on it such that (3.6.4) holds . Then we have*

$$F_*(t) \leq 2CG^*\left(\frac{t}{2}\right) + C\int_t^\infty \frac{G^*(s)}{s}ds + F^*(\infty), \quad \forall t > 0, \tag{3.6.29}$$

$$F^{**}(t) \leq 8CG^{**}(t) + F^*(t), \quad \forall t > 0, \tag{3.6.30}$$

with the same C as in (3.6.4).

Proof. By forward iterating (3.6.4), we get

$$\begin{aligned}F^*(t) &\leq C\sum_{k=0}^{\infty} G^*(2^{k-1}t) + \lim_{s\to\infty} F^*(s)\\ &\leq 2CG^*\left(\frac{t}{2}\right) + C\sum_{k=2}^{\infty} G^*(2^{k-1}t) + F^*(\infty)\\ &\leq 2CG^*\left(\frac{t}{2}\right) + C\int_t^\infty \frac{G^*(s)}{s}ds + F^*(\infty).\end{aligned}$$

By backward iterating (3.6.4), we get

$$F^*(2^{-n}t) \leq C\sum_{k=1}^{n} G^*(2^{-k-1}t) + F^*(t),$$

$$\begin{aligned}F^{**}(t) &= \sum_{n=1}^{\infty}\frac{1}{t}\int_{2^{-n}t}^{2^{-n+1}t} F^*(s)ds \leq \sum_{n=1}^{\infty} 2^{-n}F^*(2^{-n}t)\\ &\leq 8C\sum_{k=1}^{\infty} 2^{-k-2}G^*(2^{-k-1}t) + F^*(t) \leq 8CG^{**}(t) + F^*(t).\end{aligned}$$

The proof is finished. □

Now appealing to the Φ-boundedness of T^* (see Lemmas 3.6.1,3.6.2), we get convex Φ-inequalities of all preceding process pairs $(A(f), B(f))$ associated with a martingale $f = (f_n)_{n\geq 0}$, or with better constants, or with exponential type Φ. We take the inequalities between Mf and Sf as examples.

Theorem 3.6.10 *Let Φ be a convex function with $p_\Phi < \infty$. Then we have*

$$\|Mf\|_\Phi \approx \|Sf\|_\Phi, \quad \|Mf \vee Sf\|_\Phi \approx \|Mf \wedge Sf\|_\Phi, \quad \forall f = (f_n)_{n\geq 0}, \qquad (3.6.31)$$

with the equivalent constants less than Cp_Φ^2.

Proof. For $f = (f_n)_{n\geq 0}$ given, make Davis' decomposition to get $f = g + h$, with (denoting $\Delta_k f$ by d_k)

$$| \Delta_n g | \leq M_{n-1}d, \quad n \geq 0,$$

$$\sum_0^\infty | \Delta_n h | \leq \sum_0^\infty (2(M_n d - M_{n-1}d) + 2E(M_n d - M_{n-1}d \mid \mathcal{F}_{n-1})). \qquad (3.6.32)$$

Notice that

$$Md \leq 2Mf \wedge Sf, \quad Mh \vee Sh \leq \sum_0^\infty | \Delta_n h |,$$

$$Mf \vee Sf \leq Mg \vee Sg + Mh \vee Sh,$$

$$Mg \wedge Sg \leq Mf \wedge Sf + \sum_0^\infty | \Delta_n h | .$$

Since g has a predictable control $D = (M_n d)_{n\geq 0}$, we get

$$\begin{aligned} \|Mf \vee Sf\|_\Phi &\leq \|Mg \vee Sg\|_\Phi + \Big\| \sum_0^\infty |\Delta_n h| \Big\|_\Phi \\ &\leq Cp_\Phi(\|Mg \wedge Sg\|_\Phi + \|Md\|_\Phi) + \Big\| \sum_0^\infty |\Delta_n h| \Big\|_\Phi \\ &\leq Cp_\Phi\Big(\|Mf \wedge Sf\|_\Phi + \Big\| \sum_0^\infty |\Delta_n h| \Big\|_\Phi\Big). \end{aligned}$$

But the convexity lemma gives

$$\Big\| \sum_0^\infty E(M_n d - M_{n-1}d \mid \mathcal{F}_{n-1})\Big\|_\Phi \leq p_\Phi \|Md\|_\Phi.$$

This proves $\|Mf \vee Sf\|_\Phi \leq Cp_\Phi^2 \|Mf \wedge Sf\|_\Phi$. The proof is finished. □

Remark For nonnegative martingale f, we have $\|Sf\|_\Phi \leq Cp_\Phi \|Mf\|_\Phi$. We do not know if the constant Cp_Φ^2 could be replaced by Cp_Φ in other cases.

Notes to Chapter 3

§3.1. The index q_Φ was introduced in Dellacherie [1], there "$p_\Phi = q'_\Phi$ for any convex Φ" was proposed as a problem. Long [5] gave an affirmative answer to it. Here the nice simple proof of the assertion "$p_\Phi = q'_\Phi$" was communicated to the author by P.D. Chen. The remaining assertions that occurred in Theorem 3.1.1. were formulated in C.S. Chou [1].

§3.2. Lemmas 3.2.1, 3.2.2, are usually called Garsia-Neveu Lemmas. Lemma 3.2.3 and Theorem 3.2.4 were first formulated in Dellacherie [1]. Theorem 3.2.5 is due to Burkholder-Davis-Gundy [1]. Theorem 3.2.7 was formulated in Lenglart-Lepingle-Pratelli [1], but the case $\Phi(u) = u^p$, $1 \leq p < \infty$, was due to L. Chevalier.

§3.3. Lemmas 3.3.1, 3.3.2, stimulated by Dellacherie [1], have not been formulated explicitly before. Theorem 3.3.3 is due to Dellacherie [1]. The spaces ${}_aK_\Phi$, ${}_aL_\Phi$ and the results about them are due to Long [6]. Theorems 3.3.8, 3.3.8′, have not been formulated before.

§3.4. Lemma 3.4.1 is due to Burkholder [2]. Lemma 3.4.3 occurred in Burkholder [2] implicitly, but was formulated explicitly in L-L-P [1]. Theorems 3.4.6, 3.4.7 are due to Burkholder-Gundy [1].

§3.5. The good λ-inequality, which should be considered as a method rather than an inequality, was discovered by D. L. Burkholser, R.F. Gundy etc., see for example, Burkholder-Gundy [1], Burkholder [2], [4]. Lemma 3.5.2 is due to Burkholder [2]. As for the sufficient conditions to get good λ-inequality, the idea is well known. Here the Lemma 3.5.3, modulo some slight supplements, was formulated in L-L-P [1]. Theorems 3.5.4, 3.5.6 are due to Burkholder [2]. Theorems 3.5.7, 3.5.7′ are due to Long [6].

§3.6. The concept of the rearrangement function was introduced by Hardy-Littlewood, and recognized by C. Herz, Bennett-Sharpley [1] etc., for it's applications in operator's boundedness estimate. Lemmas 3.6.1, 3.6.2 have not been seen before in publications. The proof of Theorem 3.6.3 is borrowed from Bagby [1]. Lemma 3.6.5 is a rearrangement version of Lemma 3.5.3. Theorems 3.6.6, 3.6.7, 3.6.8 are new, but as we have mentioned, the proofs are parallel to those in good λ-inequality case. Theorem 3.6.9 is due to Bagby-Kurtz [1]. In Theorem 3.6.10 the implied constant estimate is new.

4 *BMO* Martingales

Let $(\Omega, \mathcal{F}, \mu, \{\mathcal{F}_n\}_{n\geq 0})$ be as defined in §2.1. The Banach spaces BMO_a, $1 \leq a < \infty$, are defined as follows†

$$BMO_a = \{f = (f_n)_{n\geq 0} \in L_u^a : \\ \|f\|_{BMO_a} = \sup_n \|E(|f - f_{n-1}|^a|\mathcal{F}_n)^{\frac{1}{a}}\|_\infty < \infty\},$$

here the f in $|f - f_{n-1}|^a$ means f_∞. (We often use the same symbol f to denote a martingale $f = (f_n)_{n\geq 0}$ or f_∞ when it has meaning.) Sometimes, for the sake of simplicity, we assume $f_0 = 0$ for $f \in BMO_a$. Obviously, this is only a non-essential convention. With $f - f_0$ replacing f, we get a new BMO martingale $f - f_0$ from the original $f \in BMO$ without any influence to all problems we want to consider owing to the fact

$$\|f - f_0\|_{BMO_a} + \|f_0\|_\infty \approx \|f\|_{BMO_a}.$$

This fact can be seen as follows. Only the case $n = 0$ in the definition is concerned:

$$E(|f|^a|\mathcal{F}_0)^{\frac{1}{a}} \leq E(|f - f_0|^a|\mathcal{F}_0)^{\frac{1}{a}} + E(|f_0|^a|\mathcal{F}_0)^{\frac{1}{a}} \leq 3E(|f|^a|\mathcal{F}_0)^{\frac{1}{a}}.$$

We have shown that both of BMO_1 and BMO_2 are the Banach dual space of H_1 in §2.5, §2.2. Basing mainly on this duality and something else, the space BMO played aremarkable role in classical analysis. We shall show the similar significance of BMO in martingale setting in this chapter and next ones. For example, we will show that BMO is a good space in operator actions (see this chapter and Chapter 5), and that BMO can be used in weight theory (see Chapter 6). But in this chapter, we mainly study the BMO itself. What we want to introduce are following: John-Nirenberg's theorem (one of the most important theorems in BMO theory) which implied $BMO_a = BMO_1$, for all $a \geq 1$; some other characterizations of BMO; the BLO decomposition of BMO; Carleson measure theory and commutator theory related to BMO; the distance of $f \in BMO$ to L^∞, etc.

---

† BMO, BLO are abbreviations of "Bounded Mean Oscillation" and "Bounded Lower Oscillation" respectively.

4.1 John-Nirenberg's theorem

The inventors of the classical BMO discovered that the definition condition of BMO can be improved by itself, that is to say,the boundedness of mean oscillation of 1-degree implies the higher degree boundedness. The following lemma is the main step to approach the problem.

Lemma 4.1.1 *Let* $f \in BMO_1$, $\|f\|_{BMO_1} \leq 1$. *Then for all* α, $0 < \alpha < \frac{1}{e}$, *we have*

$$E\left(\exp\left(\alpha \sup_n |f_n - f_0|\right) \Big| \mathcal{F}_0\right) \leq K_\alpha < \infty, \quad \text{a.e.} \tag{4.1.1}$$

Proof. Let $g = (g_n)_{n\geq 0}$, $g_n = f_n - f_0$, $Mg = \sup_{n\geq 0} |g_n|$. For $\lambda, \mu > 0$, define stopping times T and τ

$$T = \inf\{n : |g_n| > \lambda\}, \quad \tau = \inf\{n : |g_n| > \lambda + \mu\}.$$

Obviously, $T > 0$, a.e., since $g_0 \equiv 0$, and $T \leq \tau$. For any $F \in \mathcal{F}_0$, define

$$T_F = \begin{cases} T, & \omega \in F, \\ \infty, & \omega \notin F, \end{cases} \qquad \tau_F = \begin{cases} \tau, & \omega \in F, \\ \infty, & \omega \notin F. \end{cases}$$

Since $T_F \leq \tau_F$, and $|g_{T_F - 1}\chi_F| \leq \lambda$, we have

$$\begin{aligned} \sigma_F(\lambda + \mu) &= |\{Mg > \lambda + \mu\} \cap F| = |\{\tau_F < \infty\}| \\ &\leq |\{T_F < \infty, |g_{\tau_F} - g_{T_F - 1}| = |f_{\tau_F} - f_{T_F - 1}| \geq \mu\}|. \end{aligned}$$

Noticing two well known facts, one is that for any stopping time τ, $\tau \neq 0$, a.e. and any adapted process $g = (g_n)_{n\geq 0}$, $g_{\tau - 1}$ is $\mathcal{F}_\tau$-measurable, which follows from (let Δ be any Borel set in $\mathbf{C}$)

$$\begin{aligned} g_{\tau-1}^{-1}(\Delta) \cap \{\tau \leq k\} &= \left(\sum_1^\infty g_{n-1}\chi_{\{\tau = n\}} + g\chi_{\{\tau = \infty\}}\right)^{-1}(\Delta) \cap \{\tau \leq k\} \\ &= \bigcup_1^k g_{n-1}^{-1}(\Delta) \cap \{\tau = n\} \in \mathcal{F}_k, \quad \forall k; \end{aligned}$$

and another is that for two stopping times T and τ with $T \leq \tau$, we have (since g and g_{T-1} are in L^1)

$$E(|g - g_{T-1}| | \mathcal{F}_T)\chi_{(T<\infty)} = E(E(|g - g_{T-1}| | \mathcal{F}_\tau) | \mathcal{F}_T)\chi_{(T<\infty)},$$

we get

$$\begin{aligned} \sigma_F(\lambda + \mu) &\leq \frac{1}{\mu} \int_{\{T_F < \infty\}} |f_{\tau_F} - f_{T_F - 1}| d\mu \\ &= \frac{1}{\mu} \int_{\{T_F < \infty\}} |E(f - f_{T_F - 1} | \mathcal{F}_{\tau_F})| d\mu \\ &\leq \frac{1}{\mu} \int_{\{T_F < \infty\}} E(|f - f_{T_F - 1}| | \mathcal{F}_{T_F})\, d\mu \\ &\leq \frac{1}{\mu} |\{T_F < \infty\}| = \frac{1}{\mu} \sigma_F(\lambda). \end{aligned}$$

Taking $\mu = e$, $\lambda = ke$, we get

$$\sigma_F((k+1)e) \leq \frac{1}{e}\sigma_F(ke) \leq |F|e^{-k}, \quad k = 1, 2, \cdots.$$

Since $\sigma_F(\lambda)$ is decreasing, for $\lambda \in [e, \infty)$, say $ke \leq \lambda < (k+1)e$, for some $k = 1, 2, \cdots$, we have

$$\sigma_F(\lambda) \leq \sigma_F(ke) \leq |F|e^2 e^{-\frac{\lambda}{e}}.$$

This inequality holds for $0 < \lambda < e$, too, since

$$|F|e^2 e^{-\frac{\lambda}{e}} \geq |F|e \geq \sigma_F(\lambda).$$

So, for all α, $0 < \alpha < \frac{1}{e}$, we have

$$\begin{aligned} \frac{1}{|F|}\int_F e^{\alpha M g} d\mu &\leq 1 + \frac{1}{|F|}\int_0^\infty \sigma_F(\lambda) d e^{\alpha\lambda} \\ &\leq 1 + \alpha e^2 \int_0^\infty e^{(\alpha - \frac{1}{e})\lambda} d\lambda = K_\alpha < \infty. \end{aligned}$$

This proves the lemma. □

Theorem 4.1.2 *Let $f \in BMO_1$. Then for all α, $0 < \alpha < (e\|f\|_{BMO_1})^{-1}$, we have*

$$\sup_n \left\| E\Big(\sup_{m \geq n} \exp(\alpha|f_m - f_n|)|\mathcal{F}_n\Big)\right\|_\infty \leq K_\alpha < \infty, \quad \text{a.e.} \tag{4.1.2}$$

Proof. By considering $(\|f\|_{BMO_1})^{-1}f$, we could assume $\|f\|_{BMO_1} = 1$. For n fixed, consider a new martingale g with respect to a new family $\{\mathcal{F}'_m\}_{m\geq 0}$ with $\mathcal{F}'_m = \mathcal{F}_{n+m}$, and

$$g = f - f_{n-1}, \quad g_m = f_{n+m} - f_{n-1}, \quad m \geq 0.$$

We have

$$E(|g - g_{m-1}|\ |\mathcal{F}'_m) = E(|f - f_{n+m-1}|\ |\mathcal{F}_{n+m}) \leq 1, \quad m \geq 0.$$

That is to say $\|g\|_{BMO_1} \leq 1$ with respect to the new family. Applying Lemma 4.1.1, we get

$$E\Big(\exp\Big(\alpha \sup_{m\geq 0}|g_m - g_0|\Big)\Big|\mathcal{F}'_0\Big) \leq K_\alpha < \infty,$$

and hence

$$\sup_n \left\| E\Big(\exp\Big(\alpha \sup_{m\geq 0}|f_{n+m} - f_n|\Big)\Big|\mathcal{F}_n\Big)\right\|_\infty \leq K_\alpha < \infty. \tag{4.1.3}$$

The theorem is thus proved. □

Remarks

1. $f_m - f_n$ in (4.1.2) could be replaced by $f_m - f_{n-1}$, since $|\Delta f| \le \|f\|_{BMO_1}$.

2. The theorem is called John-Nirenberg's theorem, another form of which is stated in terms of local distribution function of Mf. We have already got this form in Lemma 4.1.1: For n and $F \in \mathcal{F}_n$, denoting $\sigma_F(\lambda) = |\{(M(f - f_n) > \lambda\} \bigcap F|$,

$$\sigma_F(\lambda) \le e^2 e^{-(e\|f\|_{BMO_1})^{-1}\lambda}|F|, \quad \forall f \in BMO_1, \quad \forall \lambda > 0. \tag{4.1.4}$$

3. The theorem implies in particular

$$\sup_n \|E(|f - f_{n-1}|^a|\mathcal{F}_n)^{\frac{1}{a}}\|_\infty \le K_a\|f\|_{BMO_1}, \quad \forall a \ge 1. \tag{4.1.5}$$

This follows from

$$\begin{aligned}\frac{1}{|F|}\int_F |f - f_n|^a d\mu &\le \frac{1}{|F|}\int_0^\infty \sigma_F(\lambda)d\lambda^a \\ &\le K_a \int_0^\infty \lambda^{a-1}e^{-\frac{\lambda}{e\|f\|_{BMO_1}}}d\lambda \le K_a\|f\|^a_{BMO_1},\end{aligned}$$

$$\sup_n \|E(|f - f_n|^a|\mathcal{F}_n)^{\frac{1}{a}}\|_\infty \le K_a\|f\|_{BMO_1}.$$

Henceforth, we write BMO for any BMO_a, $a \ge 1$, and write $\|f\|_*$ for f's BMO norm in all equivalent forms, unless otherwise stated.

Obviously, we could not expect to raise the power of Mf in (4.1.1), as shown by the typical example $f(x) = \log \frac{1}{|x|}$ in the classical case.But for the square function $S(f)$ of $f \in BMO$, we could do better as shown in next theorem.

Theorem 4.1.3 *Let $f \in BMO_2$. Then for all α, $0 < \alpha < \|f\|^{-2}_{BMO_2}$, we have*

$$E(\exp(\alpha S(f)^2)|\mathcal{F}_0) \le (1 - \alpha\|f\|^2_{BMO_2})^{-1}. \tag{4.1.6}$$

Proof. Noticing that$(S_n(f)^2)_{n\ge 0}$ is a nonnegative increasing adapted process, and satisfies

$$E(S(f)^2 - S_{n-1}(f)^2|\mathcal{F}_n) = E(|f - f_{n-1}|^2|\mathcal{F}_n) \le \|f\|^2_{BMO_2},$$

by making use of convexity lemmas (see §3.2, Lemma 3.2.1 and 3.2.2) to $\Phi(u) = e^{\alpha u} - 1$, with α determined later, we get

$$E(\exp(\alpha S(f)^2) - 1|\mathcal{F}_0) \le \alpha\|f\|^2_{BMO_2}E(\exp \alpha S(f)^2|\mathcal{F}_0).$$

When $E(\exp \alpha S(f)^2|\mathcal{F}_0) < \infty$, we get (4.1.6) immediately. In general, we consider $(S_n(f)^2 \wedge N)_{n\ge 0}$. Since

$$S(f)^2 \wedge N - S_{n-1}(f)^2 \wedge N \le S(f)^2 - S_{n-1}(f)^2,$$

$$E(S(f)^2 \wedge N - S_{n-1}(f)^2 \wedge N|\mathcal{F}_n) \le \|f\|^2_{BMO_2},$$

we get the inequality for $S(f)^2 \wedge N$ instead of $S(f)^2$. The desired one follows by letting $N \to \infty$. The proof is finished. □

Remark The exponential integrability of $S(f)^2$ is not surprising, since $S(f)^2 \in BMO_1$ for any $f \in BMO_2$. This assertion is a consequence of the following theorem.

Theorem 4.1.4 *Let $(Q_n)_{n\geq 0}$ be a bounded potential with bound B. Then its associated increasing process $A = (A_n)_{n\geq 0}$ satisfies*

$$||A_\infty||_{BMO_1} \leq 2B. \tag{4.1.7}$$

Proof. We have

$$0 \leq Q_n = E(A_\infty - A_{n-1}|\mathcal{F}_n) \leq B.$$

By making use of a simple characterization of BMO (see Theorem 4.2.1 in next section), we get (4.1.7). □

Remark For all $f \in BMO_2$, $(S_n(f)^2)_{n\geq 0}$ is the associated increasing process of a bounded potential (i.e., (Q_n), $Q_n = E(S(f)^2 - S_{n-1}(f)^2|\mathcal{F}_n)$), the bound of which is $||f||^2_{BMO_2}$. So $||S(f)^2||_{BMO_1} \leq 2||f||^2_{BMO_2}$. As a result, we know that the square function operator is a bounded operator on BMO. That is, we have

$$E(|S(f) - S_{n-1}(f)|^2|\mathcal{F}_n)^{\frac{1}{2}} \leq E(S(f)^2 - S_{n-1}(f)^2|\mathcal{F}_n)^{\frac{1}{2}} \leq ||f||_{BMO_2},$$

and so

$$||S(f)||_{BMO_2} \leq 2||f||_{BMO_2}.$$

Analogously, the maximal operator is also a bounded one on BMO, as shown in next theorem.

Theorem 4.1.5 *We have*

$$||Mf||_{BMO_2} \leq C||f||_{BMO_2}, \quad \forall f \in BMO. \tag{4.1.8}$$

Proof. Let $f \in BMO$. For n fixed, let $g = (g_m)_{m\geq 0}$, $g_m = f_{n+m} - f_{n-1}$. We have

$$Mf - M_{n-1}f \leq Mg = \sup_{m\geq 0} |f_{n+m} - f_{n-1}|.$$

So, by making use of the equivalence between $S(f)$ and Mf in conditional form, we get

$$\begin{aligned} E(|Mf - M_{n-1}f|^2|\mathcal{F}_n) &\leq CE(S(f - f_{n-1})^2|\mathcal{F}_n) \\ &= CE(S(f)^2 - S_{n-1}(f)^2|\mathcal{F}_n) \\ &= CE(|f - f_{n-1}|^2|\mathcal{F}_n) \leq C||f||^2_{BMO_2}, \end{aligned}$$

and hence, by applying Theorem 4.2.1 once again,

$$E(|Mf - (Mf)_{n-1}|^2|\mathcal{F}_n)^{\frac{1}{2}} \leq C||f||_{BMO_2}.$$

The proof of the theorem is finished. □

4.2 Other characterizations of *BMO*

First we want to give an equivalent difinition of BMO norm which is more convenient than original one when we want to verify that f belongs to BMO for some given f. Meanwhile,the original definition of BMO is convenient to negate f's BMO property.

Theorem 4.2.1 *Let $1 \leq a < \infty$, $f \in L^a_u$. Then $f \in BMO_a$ if and only if there exists adapted process $\theta = (\theta_n)_{n\geq 0}$ such that*

$$C_\theta = \sup_n \left\| E(|f - \theta_{n-1}|^a|\mathcal{F}_n)^{\frac{1}{a}} \right\|_\infty < \infty.$$

And in any case, we have

$$|||f|||_{BMO_a} = \inf_\theta C_\theta \leq ||f||_{BMO_a} \leq 2|||f|||_{BMO_a}. \tag{4.2.1}$$

Proof. Assume $f \in BMO_a$. Then obviously

$$|||f|||_{BMO_a} \leq ||f||_{BMO_a}.$$

Now suppose $|||f|||_{BMO_a} < \infty$, and $\theta = (\theta_n)_{n\geq 0}$be any one such that $C_\theta < \infty$. Then we have

$$\begin{aligned} E(|f - f_{n-1}|^a|\mathcal{F}_n)^{\frac{1}{a}} &\leq E(|f - \theta_{n-1}|^a|\mathcal{F}_n)^{\frac{1}{a}} + |\theta_{n-1} - f_{n-1}| \\ &\leq C_\theta + E(|f - \theta_{n-1}|\ |\mathcal{F}_{n-1}) \\ &\leq C_\theta + E(|f - \theta_{n-1}|^a|\mathcal{F}_{n-1})^{\frac{1}{a}} \\ &= C_\theta + E(E(|f - \theta_{n-1}|^a|\mathcal{F}_n)|\mathcal{F}_{n-1})^{\frac{1}{a}} \leq 2C_\theta. \end{aligned}$$

Taking "inf" over all possible θ, we get the desired inequality. □

Corollary 4.2.2 *Let $f \in BMO_a$. Then $|f| \in BMO_a$, and*

$$||\,|f|\,||_{BMO_a} \leq 2||f||_{BMO_a}. \tag{4.2.2}$$

Proof. This comes from

$$E(|\,|f| - |f_{n-1}|\,|^a|\mathcal{F}_n)^{\frac{1}{a}} \leq E(|f - f_{n-1}|^a|\mathcal{F}_n)^{\frac{1}{a}} \leq ||f||_{BMO_a}.$$

□

Corollary 4.2.3 Re BMO_a *is a lattice. That means when $f, g \in$* Re BMO_a, *then $f \vee g$, $f \wedge g \in BMO_a$, and*

$$\max(||f \vee g||_{BMO_a}, ||f \wedge g||_{BMO_a}) \leq 2(||f||_{BMO_a} + ||g||_{BMO_a}). \tag{4.2.3}$$

Proof. Notice that for any $x, y \in \mathbf{R}$, we have

$$x \vee y = \frac{x+y+|x-y|}{2}, \quad x \wedge y = \frac{x+y-|x-y|}{2}.$$

Thus, we have

$$\begin{aligned} &E(|f \vee g - f_{n-1} \vee g_{n-1}|^a|\mathcal{F}_n)^{\frac{1}{a}} \\ &= E\Big(\Big|\frac{f+g+|f-g|}{2} - \frac{f_{n-1}+g_{n-1}+|f_{n-1}-g_{n-1}|}{2}\Big|^a\Big|\mathcal{F}_n\Big)^{\frac{1}{a}} \\ &\le E(|\,|f-f_{n-1}|+|g-g_{n-1}|\,|^a|\mathcal{F}_n)^{\frac{1}{a}} \\ &\le \|f\|_{BMO_a} + \|g\|_{BMO_a}. \end{aligned}$$

This proves the assertion concerning $f \vee g$. The same argument works well for $f \wedge g$. The proof is finished. □

Corollary 4.2.4 *Let $f \in \mathrm{Re}\, BMO_a$. Then for any $N > 0$, $f^{(N)} = (f \wedge N) \vee (-N) \in BMO_a$, and*

$$\|f^{(N)}\|_{BMO_a} \le 4\|f\|_{BMO_a}. \tag{4.2.4}$$

Proof. It is immediate owing to the preceding corollary. □

Remark In order to approximate $f \in BMO$ by $g \in L^\infty$, the corollary shows a very simple and natural way. It is the favorite way for the analysts, but in martingale setting, we prefer $f^{(\tau_n)}$with $\{\tau_n\}_{n\ge1}$ a suitable sequence of stopping times. We will use $f^{(\tau_n)}$ to approximate f in weak topology in §4.6.

Another simple characterization of BMO_a is as follows.

Theorem 4.2.5 *Let $f = (f_n)_{n\ge0}$ be a martingale in L^a_u, $1 \le a < \infty$. Then*

$$\|f\|_{BMO_a} = \sup_\tau |\{\tau < \infty\}|^{-\frac{1}{a}}\|f - f_{\tau-1}\|_a, \tag{4.2.5}$$

where "sup" is taken over all stopping times τ.

Proof. Assume $\|f\|_{BMO_a} < \infty$, τ is any stopping time. Then

$$\begin{aligned} \|f - f_{\tau-1}\|_a^a &= \int_{\{\tau<\infty\}} |f - f_{\tau-1}|^a d\mu \\ &= \int_{\{\tau<\infty\}} E(|f - f_{\tau-1}|^a|\mathcal{F}_\tau)d\mu \le \|f\|^a_{BMO_a}|\{\tau < \infty\}| \end{aligned}$$

This proves one half of the assertion. Conversely, assume that $\beta = \sup_\tau |\{\tau < \infty\}|^{-\frac{1}{a}}\|f - f_{\tau-1}\|_a < \infty$, and τ is any stopping time, $F \in \mathcal{F}_\tau$, $F \subset \{\tau < \infty\}$. By defining $\tau_F = \begin{cases} \tau, & \omega \in F, \\ \infty, & \omega \notin F, \end{cases}$ we get

$$\frac{1}{|F|}\int_F |f - f_{\tau-1}|^a d\mu = |\{\tau_F < \infty\}|^{-1}\|f - f_{\tau_F-1}\|_a^a \le \beta^a.$$

This proves $\|f\|_{BMO_a} \leq \beta$. The proof is finished. □

A characterization of *BMO* has occurred in §2.2 (Lemma 2.2.4 and Theorem 2.2.5) in an implicit form. It is the part of $q = \infty$ of the following theorem.

Theorem 4.2.6 *Let* $2 \leq q \leq \infty$, $\varepsilon = (\varepsilon_\nu)_{\nu \geq 1}$ *be a sequence of random variables satisfying*

$$E\Big(\Big(\sum_{\nu=1}^{\infty} |\varepsilon_\nu|^2\Big)^{\frac{q}{2}}\Big)^{\frac{1}{q}} \leq B.$$

Then $\varphi = (\varphi_n)_{n \geq 0}$ *with*

$$\varphi_n = \sum_{\nu=1}^{n} \{E(\varepsilon_\nu|\mathcal{F}_\nu) - E(\varepsilon_\nu|\mathcal{F}_{\nu-1})\}, \quad n \geq 1, \quad \varphi_0 = 0, \tag{4.2.6}$$

is in ${}_2K_q$, *and*

$$\|\varphi\|_{{}_2K_q} \leq 2q'B. \tag{4.2.7}$$

Conversely, each $\varphi \in {}_2K_q$ *with* $\varphi_0 = 0$ *has such expression with* $\varepsilon = (\varepsilon_\nu)_{\nu \geq 1}$ *satisfying*

$$E\left(\left(\sum_{\nu=1}^{\infty} |\varepsilon_\nu|^2\right)^{\frac{q}{2}}\right)^{\frac{1}{q}} \leq \sqrt{\frac{2}{q'}}\|\varphi\|_{{}_2K_q}. \tag{4.2.8}$$

Proof. The first part is just Lemma 2.2.4. The second part occurred in Theorem 2.2.5 implicitly. We sketch it here. Let $\varphi \in_{{}_2K_q}$, $\varphi_0 = 0$. Define a linear functional l_φ on $H_{q'}$ by

$$\langle f, \varphi \rangle = \lim_{n \to \infty} E(\varphi_n f_n) = \lim_{n \to \infty} E\Big(\sum_{1}^{n} \Delta_\nu f \Delta_\nu \varphi\Big), \quad \forall f \in H_{q'}.$$

Then $\|l_\varphi\| \leq \sqrt{\frac{2}{q'}}\|\varphi\|_{{}_2K_q}$, as shown by Theorem 2.2.2. By means of Lemma 2.2.3 and Theorem 2.2.5, there exists $\varepsilon = (\varepsilon_\nu)_{\nu \geq 1}$ satisfying

$$E\left(\left(\sum_{\nu=1}^{\infty} |\varepsilon_\nu|^2\right)^{\frac{q}{2}}\right)^{\frac{1}{q}} \leq \|l_\varphi\|,$$

$$l_\varphi(f_n) = E\Big(\sum_{\nu=1}^{n} f_n[E(\varepsilon_\nu|\mathcal{F}_\nu) - E(\varepsilon_\nu|\mathcal{F}_{\nu-1})]\Big) = E(f_n\varphi_n), \quad \forall f \in H_{q'}.$$

Since $f_n \in L^{q'}(\mathcal{F}_n)$ is arbitrary, we get

$$\varphi_n = \sum_{\nu=1}^{n} [E(\varepsilon_\nu|\mathcal{F}_\nu) - E(\varepsilon_\nu|\mathcal{F}_{\nu-1})], \quad \varphi_0 = 0.$$

And we have already got

$$E\Big(\Big(\sum_{\nu=1}^{\infty}|\varepsilon_\nu|^2\Big)^{\frac{q}{2}}\Big)^{\frac{1}{q}} \le \|l_\varphi\| \le \sqrt{\frac{2}{q'}}\|\varphi\|_{2K_q}.$$

This proves the theorem. □

Remark We have seen the application of this characterization of BMO (see §2.2) in the establishment of H_1-BMO duality.

A similar argument gives another characterization of BMO which we have to pay more attention to because of its applications. First we need a lemma. Let $\mathcal{H}$ be the following normed linear space consisting of sequences of random variables

$$\mathcal{H} = \Big\{\theta = (\theta_0, \theta_1, \cdots) : \|\theta\|_{\mathcal{H}} = E(M\theta),\ M\theta = \sup_{n\ge 0}|\theta_n|\Big\}. \tag{4.2.9}$$

We want to give some information of bounded linear functionals on $\mathcal{H}$. We formulate a lemma to give it.

Lemma 4.2.7 *Let $l = l(\theta)$ be a bounded linear functional on $\mathcal{H}$. Then there exist random variables ξ and $\{\varepsilon_\nu\}_{\nu\ge 0}$ such that*

$$\Big\||\xi| + \sum_{\nu=0}^{\infty}|\varepsilon_\nu|\Big\|_\infty \le \|l\|; \tag{4.2.10}$$

and for any $\theta = (\theta_\nu)_{\nu\ge 0}$ in $\mathcal{H}$ with a constant tail, say

$$\theta = (\theta_0, \theta_1, \cdots, \theta_N, \gamma, \gamma, \cdots),$$

$$l(\theta) = \sum_{\nu=0}^{N} E(\varepsilon_\nu\theta_\nu) + E\Big(\gamma\Big(\xi + \sum_{\nu=N+1}^{\infty}\varepsilon_\nu\Big)\Big). \tag{4.2.11}$$

Proof. For $N \in \mathbf{Z}^+$ fixed, L^1 could be embedded continuously in $\mathcal{H}$ in following ways:

$$L^1 \ni \theta_n \to (0, \cdots, \theta_n, 0, \cdots) \in \mathcal{H}, \quad n = 0, 1, \cdots, N,$$

$$L^1 \ni \gamma \to (0, \cdots, 0, \gamma, \gamma, \cdots) \in \mathcal{H} \quad \text{(with first } N+1 \text{ zero)}.$$

Restricting l on these component spaces, we get $N+2$ bounded linear functionals $l_0, l_1, \cdots, l_N, l_{N+1}$ on L^1, and hence $N+2$ elements of L^∞, say $\varepsilon_0, \varepsilon_1, \cdots, \varepsilon_N, \xi_N$, such that

$$l_\nu(\theta_\nu) = E(\varepsilon_\nu\theta_\nu),\ \forall\theta_\nu \in L^1,\ \nu = 0, 1, \cdots, N,\ l_{N+1}(\gamma) = E(\xi_N\gamma), \quad \forall\gamma \in L^1.$$

Now let $\theta = (\theta_0, \theta_1, \cdots, \theta_N, \gamma, \cdots)$, be any element of $\mathcal{H}$. By the linearity of l, we have

$$l(\theta) = \sum_{\nu=0}^{N} E(\varepsilon_\nu\theta_\nu) + E(\xi_N\gamma). \tag{4.2.12}$$

But writing θ as $\theta = (\theta_0, \theta_1, \cdots, \theta_N, \theta_{N+1}, \gamma, \cdots)$, with $\theta_{N+1} = \gamma$, the same argument shows

$$l(\theta) = \sum_{\nu=0}^{N+1} E(\varepsilon_\nu \theta_\nu) + E(\xi_{N+1}\gamma) = \sum_{\nu=0}^{N} E(\varepsilon_\nu \theta_\nu) + E(\varepsilon_{N+1}\gamma) + E(\xi_{N+1}\gamma).$$

Since γ is arbitrary, we get

$$\xi_N = \varepsilon_{N+1} + \xi_{N+1} = \sum_{\nu=N+1}^{M} \varepsilon_\nu + \xi_M, \quad \forall M > N. \tag{4.2.13}$$

In particuar, set $\theta_\nu = \delta \operatorname{sgn} \varepsilon_\nu$, $\nu = 0, \cdots, N$ and $\gamma = \delta \operatorname{sgn} \xi_N$, with δ any element of L^1_+, then from (4.2.12) we have

$$l(\theta) = \sum_{v=0}^{N} E(|\varepsilon_\nu|\delta + E(|\xi_N|\delta) \leq ||l|| E(\delta),$$

which gives

$$\left\| |\xi_N| + \sum_{v=0}^{N} |\varepsilon_\nu| \right\|_\infty \leq ||l||, \quad \forall N. \tag{4.2.14}$$

This shows particularly that $\sum_{v=0}^{\infty} \varepsilon_\nu$ converges absolutely almost everywhere. Taking $M \to \infty$ in (4.2.13), we see that $\xi = \lim_{M\to\infty} \xi_M$ exists, and (4.2.14) holds when $N \to \infty$. This is just (4.2.10). Furthermore, substituting $\xi_N = \xi + \sum_{N+1}^{\infty} \varepsilon_\nu$ into (4.2.12), we get (4.2.11). The proof is finished. □

Now we can establish the last characterization of BMO in this section.

Theorem 4.2.8 *$\varphi \in BMO$, if and only if there exists ξ and $\{\varepsilon_\nu\}_{\nu \geq 0}$ such that*

$$\varphi = \xi + \sum_{\nu=0}^{\infty} E(\varepsilon_\nu | \mathcal{F}_\nu), \tag{4.2.15}$$

$$|| \, |\xi| + \sum_{0}^{\infty} |\varepsilon_\nu| \, ||_\infty \approx ||\varphi||_*. \tag{4.2.16}$$

Proof. We have seen that (4.2.15) implied $\varphi \in BMO$ in §2.3 (Lemma 2.3.2) and $||\varphi||_* \leq C|| \, |\xi| + \sum_0^\infty |\varepsilon_\nu| \, ||_\infty$. Now we make the required decomposition of $\varphi \in BMO$. From Fefferman's theorem and Davis' inequality, we have for all $f \in H_1$,

$$|E(f_N\varphi)| \leq \sqrt{2}||\varphi||_{BMO_2}||S(f)||_1 \leq \sqrt{2}(2+\sqrt{5})||\varphi||_{BMO_2}||Mf||_1. \tag{4.2.17}$$

Now consider the subspace of H_1 consisting of all martingales generated by $f_N \in L^\infty(\mathcal{F}_N)$ for N fixed. It could be embedded continuously and naturally into $\mathcal{H}$. (4.2.17) shows that φ defines a bounded linear functional on this subspace. By

Hahn-Banach's theorem, φ could defines a bounded linear functional l_φ on whole $\mathcal{H}$ with $\|l_\varphi\|$ at most $\sqrt{2}(2+\sqrt{5})\|\varphi\|_{BMO_2}$. By means of Lemma 4.2.7, there exist ξ and $\{\varepsilon_\nu\}_{\nu\geq 0}$ such that (4.2.10), (4.2.11) hold. Furthermore for $\theta = (\theta_\nu)_{\nu\geq 0}$ with $\theta_\nu = E(f_N|\mathcal{F}_\nu)$, $\nu \geq 0$, noticing that $\theta_\nu = f_\nu$, $\nu \leq N$, and $= f_N$, $\nu \geq N$, and $f_N \in L^\infty \subset H_1$, we have

$$\begin{aligned}E(f_N\varphi) &= \sum_{\nu=0}^{N} E(\varepsilon_\nu f_\nu) + E\Big(f_N\Big(\xi + \sum_{\nu=N+1}^{\infty} \varepsilon_\nu\Big)\Big)\\ &= \sum_{\nu=0}^{N} E(f_N E(\varepsilon_\nu|\mathcal{F}_\nu)) + E\Big(f_N\Big(\xi + \sum_{\nu=N+1}^{\infty} E(\varepsilon_\nu|\mathcal{F}_\nu)\Big)\Big)\\ &= E\Big(f_N\Big(\xi + \sum_{\nu=0}^{\infty} E(\varepsilon_\nu|\mathcal{F}_\nu)\Big)\Big).\end{aligned}$$

Since f_N is arbitrary, we get

$$\varphi_N = E\Big(\xi + \sum_{\nu=0}^{\infty} E(\varepsilon_\nu|\mathcal{F}_\nu)|\mathcal{F}_N\Big), \quad \forall N. \tag{4.2.18}$$

$\xi+\sum_{\nu=0}^{\infty} E(\varepsilon_\nu|\mathcal{F}_\nu)$ generates at least a uniformly integrable martingale, taking limit as $N \to \infty$, we get

$$\varphi = \xi + \sum_{\nu=0}^{\infty} E(\varepsilon_\nu|\mathcal{F}_\nu), \quad \text{a.e.}$$

And we have got already

$$\|\,|\xi| + \sum |\varepsilon_\nu|\,\|_\infty \leq \|l_\varphi\| \leq C\|\varphi\|_*.$$

The proof is finished. □

Remark We have seen an application of this characterization of BMO in §2.6. There have been others. For example, we could get BLO decomposition of BMO, and study the approximation of BMO by L^∞, by making use of this characterization in next sections.

The preceding theorem is a consequence of Fefferman's theorem and Davis' inequality $\|S(f)\|_1 \leq C\|Mf\|_1$. Conversely, such decomposition of BMO implies Davis' inequality.

Theorem 4.2.9 *Suppose that each $\varphi \in BMO$ could be decomposed as $\varphi = \xi + \sum_{\nu=0}^{\infty} E(\varepsilon_\nu|\mathcal{F}_\nu)$ with $\|\,|\xi| + \sum_{\nu=0}^{\infty} |\varepsilon_\nu|\,\|_\infty \leq C\|\varphi\|_*$, then $E(Sf) \leq 2CE(Mf)$, for all f.*

Proof. Letting $f = (f_n)_{n\geq 0}$ be a finite martingale, we have

$$E(S(f)) = E\Big(\sum_0^\infty \frac{|\Delta_\nu f|^2}{S(f)}\Big) = E\Big(\sum_0^\infty \Delta_\nu f \frac{\Delta_\nu \bar{f}}{S(f)}\Big)$$

$$= E\Big(\sum_0^\infty \Delta_\nu f\Big(E\Big(\frac{\Delta_\nu \bar{f}}{S(f)}\Big|\mathcal{F}_\nu\Big) - E\Big(\frac{\Delta_\nu \bar{f}}{S(f)}\Big|\mathcal{F}_{\nu-1}\Big)\Big)\Big)$$

$$= E\Big(\sum_0^\infty \Delta_\nu f \Delta_\nu \overline{\varphi}\Big),$$

where

$$\varphi = \sum_0^\infty \left(E\left(\frac{\Delta_\nu f}{S(f)}\Big|\mathcal{F}_\nu\right) - E\left(\frac{\Delta_\nu f}{S(f)}\Big|\mathcal{F}_{\nu-1}\right)\right).$$

Since $\sum_0^\infty |\frac{\Delta_\nu f}{S(f)}|^2 = 1$, we see from Theorem 4.2.6 that $\varphi \in BMO$, and $\|\varphi\|_{BMO_2} \leq 2$. According to the assumption, there are ξ and $\{\varepsilon_\nu\}_{\nu\geq 0}$ such that

$$\varphi = \xi + \sum_{\nu=0}^\infty E(\varepsilon_\nu|\mathcal{F}_\nu), \quad \text{and} \quad \Big\||\xi| + \sum_{\nu=0}^\infty |\varepsilon_\nu|\Big\|_\infty \leq 2C.$$

Assuming that f is stopped at N, we have

$$\begin{aligned} E(S(f)) &= E\Big(\sum_{\nu=0}^\infty \Delta_\nu f \Delta_\nu \overline{\varphi}\Big) = E\Big(fE\Big(\xi + \sum_{\nu=0}^\infty E(\varepsilon_\nu|\mathcal{F}_\nu)|\mathcal{F}_N\Big)\Big) \\ &= E(f\xi) + E\Big(\sum_{\nu=0}^N fE(\varepsilon_\nu|\mathcal{F}_\nu)\Big) + E\Big(f\sum_{N+1}^\infty E(\varepsilon_\nu|\mathcal{F}_N)\Big) \\ &= E(f\xi) + \sum_{\nu=0}^\infty E(f_\nu\varepsilon_\nu), \end{aligned}$$

and hence

$$E(S(f)) \leq E\Big(Mf\Big(|\xi| + \sum_{\nu=0}^\infty |\varepsilon_\nu|\Big)\Big) \leq 2CE(Mf), \quad \text{for all } f \text{ stopped at } N,$$

which holds for all $f = (f_n)_{n\geq 0}$ of course. The proof is finished. □

As a consequence of Theorem 4.2.8 we get a new way to describe the duality H_1 and BMO. Denote this duality by $\langle\, ,\rangle$ as we do usually.

Corollary 4.2.10 *Let $\varphi \in BMO$, then there exist ξ and $\{\varepsilon_\nu\}_{\nu>0}$ with $\|\,|\xi| + \sum_0^\infty |\varepsilon_\nu|\,\|_\infty \leq C\|\varphi\|_*$ such that*

$$\langle f, \varphi\rangle = E(f\xi) + \sum_{\nu=0}^\infty E(f_\nu\varepsilon_\nu), \quad \forall f \in H_1. \tag{4.2.19}$$

Proof. For $\varphi \in BMO$, let ξ and $\{\varepsilon_\nu\}_{\nu\geq 0}$ be ones occurred in Theorem 4.2.8. Then for all $f \in L^\infty$, since $\sum_0^\infty E(\varepsilon_\nu|\mathcal{F}_\nu)$ converges in L^1 at least, we have

$$E(f\varphi) = E(f\xi) + E\left(\sum_0^\infty fE(\varepsilon_\nu|\mathcal{F}_\nu)\right) = E(f\xi) + \sum_0^\infty E(f_\nu\varepsilon_\nu).$$

For any $f \in H_1$, choose $\{f^{(n)}\} \subset L^\infty$ such that $f^{(n)} \to f$ in H_1, then

$$\langle f, \varphi \rangle = \lim_{n\to\infty} \Big(E(f^{(n)}\xi) + \sum_0^\infty E(f_\nu^{(n)}\varepsilon_\nu) \Big).$$

Since $|f_\nu^{(n)} - f_\nu| \le M(f^{(n)} - f)$ and $\|M(f^{(n)} - f)\|_1 \to 0$, we have

$$\Big| \sum_0^\infty E(f_\nu^{(n)}\varepsilon_\nu) - \sum_0^\infty E(f_\nu\varepsilon_\nu) \Big| \le E\Big(\sum_0^\infty M(f^{(n)} - f)|\varepsilon_\nu| \Big) \to 0.$$

This proves (4.2.19). The proof of the corollary is finished. □

4.3 BMO and BLO

As shown in the classical case, BMO functions can be decomposed as a difference of two functions, each of which has one-sided pointwise bounded oscillations. Such functions are called BLO functions. The situation is similar in martingale setting. In the next sections or chapters, we will benefit very much from such decomposition of BMO martingales. In this section, we introduce BLO and establish the decomposition.

Definition 4.3.1 *Let $f = (f_n)_{n\ge 0}$ be a real martingale in L_u^1. It is called in BLO, if*

$$|f_n - f_{n-1}| \le C, \quad f_n \le f + C, \quad \text{a.e}, \quad \forall n. \tag{4.3.1}$$

In this case, we define

$$\|f\|_{BLO} = \inf\{C : C \quad \textit{occurs in } (4.3.1)\}.$$

Theorem 4.3.2 *We have* $\operatorname{Re} L^\infty \subset BLO \subset BMO$, *and*

$$\frac{1}{3}\|f\|_{BMO} \le \|f\|_{BLO} \le 2\|f\|_\infty. \tag{4.3.2}$$

Proof. The second inequality is obvious. Now let $f \in BLO$. Then

$$\begin{aligned} E(|f - f_{n-1}|\,|\mathcal{F}_n) &\le E(|f - f_n|\,|\mathcal{F}_n) + |f_n - f_{n-1}| \\ &\le 2E((f_n - f)^+|\mathcal{F}_n) + \|f\|_{BLO} \\ &\le 2\|f\|_{BLO} + \|f\|_{BLO} = 3\|f\|_{BLO}. \end{aligned}$$

Here we have used the fact

$$E((f_n - f)^+|\mathcal{F}_n) = E((f_n - f)^-|\mathcal{F}_n), \quad \text{for all martingales } f \in L_u^1.$$

□

In what follows we will show that BLO generates BMO by making use of the concepts of γ-graded sequences of stopping times. First we give another description of exponential integrability of BMO martingales. We introduce a new notation, perhaps not a standard one, which reflects the connection between BMO martingales and A_p weight martingales.

Definition 4.3.3 *Let $f = (f_n)_{n\geq 0}$ be a real martingale, $\alpha > 0$, $\beta > 0$. f is said to be in $\log A_{\alpha,\beta}$, if*

$$\sup_n \|E(\exp \alpha(f - f_n)|\mathcal{F}_n)\|_\infty^{\frac{1}{\alpha}} \leq K_\alpha < \infty, \tag{4.3.3}$$

$$\sup_n \|E(\exp(-\beta(f - f_n))|\mathcal{F}_n)\|_\infty^{\frac{1}{\beta}} \leq K_\beta < \infty. \tag{4.3.4}$$

Remark (4.3.3), (4.3.4) are equivalent to

$$\sup_n \|E(\exp \alpha(f - f_n)^+|\mathcal{F}_n)\|_\infty^{\frac{1}{\alpha}} \leq K_\alpha < \infty, \tag{4.3.5}$$

$$\sup_n \|E(\exp \beta(f - f_n)^-|\mathcal{F}_n)\|_\infty^{\frac{1}{\beta}} \leq K_\beta < \infty, \tag{4.3.6}$$

respectively. In fact, this follows from

$$-1 \leq \exp(\alpha(f - f_n)) - \exp(\alpha(f - f_n)^+) \leq 0,$$

$$-1 \leq \exp(-\beta(f - f_n)) - \exp(\beta(f - f_n)^-) \leq 0.$$

Furthermore, as same as other similar case, replaceing n by T in the preceding conditions, we get equivalent ones. This can be easily checked. Finally, from (4.3.5) and (4.3.6), we see that when $\alpha' \leq \alpha$, $\beta' \leq \beta$, $\log A_{\alpha,\beta} \subset \log A_{\alpha',\beta'}$.

Lemma 4.3.4 *Let $f \in \log A_{\alpha,\beta}$. Then for $\gamma = \min(\alpha, \beta)$, we have*

$$E(\exp(\gamma|f - f_n|)|\mathcal{F}_n) \leq K_\gamma < \infty, \quad \forall n.$$

Proof. From (4.3.5), (4.3.6), we have

$$\begin{aligned}
&E(\exp(\gamma|f - f_n|)|\mathcal{F}_n) \\
&\quad = E(\exp(\gamma(f - f_n)^+)\chi_{\{f>f_n\}}|\mathcal{F}_n) + E(\exp(\gamma(f - f_n)^-)\chi_{\{f\leq f_n\}}|\mathcal{F}_n) \\
&\quad \leq 2 + E(\exp \alpha(f - f_n|)|\mathcal{F}_n) + E(\exp(-\beta(f - f_n))|\mathcal{F}_n) \\
&\quad \leq 2 + K_\alpha^\alpha + K_\beta^\beta < \infty.
\end{aligned}$$

This proves the lemma. □

Remark From Theorem 4.1.2 and this lemma, we see

$$\operatorname{Re} BMO = \bigcup_{\alpha,\beta>0} \left(\log A_{\alpha,\beta} \bigcap BD\right). \tag{4.3.7}$$

Lemma 4.3.5 *$f \in \log A_{\alpha,\beta}$ if and only if*

$$\sup_n \|E(e^{\alpha f}|\mathcal{F}_n)^{\frac{1}{\alpha}} E(e^{-\beta f}|\mathcal{F}_n)^{\frac{1}{\beta}}\|_\infty \le K_{\alpha,\beta} < \infty. \tag{4.3.8}$$

Proof. Let $f \in \log A_{\alpha,\beta}$. From (4.3.3), (4.3.4). we get

$$\sup_n \|E(e^{\alpha f}|\mathcal{F}_n)^{\frac{1}{\alpha}} E(e^{-\beta f}|\mathcal{F}_n)^{\frac{1}{\beta}}\|_\infty \le \sup \|K_\alpha e^{f_n} K_\beta e^{-f_n}\|_\infty = K_\alpha K_\beta < \infty.$$

Conversely, suppose that (4.3.8) holds. Applying Jensen's inequality for the convex functions $\varphi(u) = e^{\alpha u}$ and $e^{-\beta u}$, we get

$$e^{f_n} E(e^{-\beta f}|\mathcal{F}_n)^{\frac{1}{\beta}} \le E(e^{\alpha f}|\mathcal{F}_n)^{\frac{1}{\alpha}} E(e^{-\beta f}|\mathcal{F}_n)^{\frac{1}{\beta}} \le K_{\alpha,\beta},$$

$$\sup_n \|E(\exp(-\beta(f - f_n))|\mathcal{F}_n)\|_\infty^{\frac{1}{\beta}} \le K_{\alpha,\beta} < \infty.$$

Analogously, we also have

$$\sup_n \|E(\exp \alpha(f - f_n)|\mathcal{F}_n)\|_\infty^{\frac{1}{\alpha}} \le K_{\alpha,\beta} < \infty.$$

The proof is finished. □

Now we want to investigate how the $\log A_{\alpha,\beta}$ property changes under the action $f \to f - f_\tau$, where τ is any stopping time.

Lemma 4.3.6 *Let $f \in \log A_{\alpha,\beta}$, τ be any stopping time. Then with the same parameters, $\varphi = f - f_\tau$ satisfies* (4.3.3), (4.3.4), *and*

$$\|\varphi\|_{BMO_1} \le \|f\|_{BMO_1}. \tag{4.3.9}$$

Proof. Let $f \in \log A_{\alpha,\beta}$ and τ be given. Set $F = \{n \le \tau\}$, $F^c = \{n > \tau\}$. Then

$$\begin{aligned} E(\exp \alpha(\varphi - \varphi_n)|\mathcal{F}_n) &= E(\exp \alpha(\varphi - \varphi_n)(\chi_F + \chi_{F^c})|\mathcal{F}_n) \\ &= E(\exp(\alpha(f - f_\tau))\chi_F|\mathcal{F}_n) + E(\exp(\alpha(f - f_n))\chi_{F^c}|\mathcal{F}_n) \\ &= E(E(\exp \alpha(f - f_\tau)|\mathcal{F}_\tau)|\mathcal{F}_n)\chi_F + E(\exp \alpha(f - f_n)|\mathcal{F}_n)\chi_{F^c} \\ &\le K_{\alpha,\beta} < \infty. \end{aligned}$$

Here we have used the fact that in (4.3.3) n can be replaced by any stopping time τ. The other assertions can be proved in the same way. □

Remark Likewise, for $f \in BLO$, and any τ, denoting $\varphi = f - f_\tau$, we have also $\|\varphi\|_{BLO} \le \|f\|_{BLO}$. In fact we have

$$|\varphi_n - \varphi_{n-1}| = |f_n - f_{n-1}|\chi_{\{n>\tau\}} \le \|f\|_{BLO},$$

$$\varphi_n - \varphi = (f_\tau - f)\chi_{\{n\le\tau\}} + (f_n - f)\chi_{\{n>\tau\}} \le \|f\|_{BLO}.$$

Now turn to the γ-graded sequence of stopping times.

Definition 4.3.7 *Let $\{T_k\}_{k\geq 1}$ be an increasing sequence of stopping times, and $0<\gamma<1$. $\{T_k\}_{k\geq 1}$ is said to be a γ-graded sequence, if*

$$E(\chi_{\{T_{k+1}<\infty\}}|\mathcal{F}_{T_k}) \leq \gamma, \quad \forall k. \tag{4.3.10}$$

For any γ-graded sequence $\{T_k\}_1^\infty$, let $\{b_k(\omega)\}_{k\geq 1}$ be such that b_k is measurable with respect to $\mathcal{F}_{T_k}$, and $0 \leq b_k(\omega) \leq B$, a.e. Then

$$\varphi = \sum_{k=1}^{\infty} b_k(\omega)\chi_{\{T_k<\infty\}} \tag{4.3.11}$$

is called a γ-graded function.

Lemma 4.3.8 *Let $\{T_k\}_{k\geq 1}$be a γ-graded sequence. $0<\gamma<1$, φ be as in* (4.3.11). *Then $\varphi \in BLO$, and*

$$\|\varphi\|_{BLO} \leq \frac{2B}{1-\gamma}, \quad \|\varphi\|_{BMO_1} \leq \frac{2B}{1-\gamma}. \tag{4.3.12}$$

Furthermore, for all $\alpha>0$, such that $e^{\alpha B} \leq \frac{1}{\gamma}$, we have

$$\sup_n \|E(e^{\alpha(\varphi-\varphi_n)}|\mathcal{F}_n)\|_\infty < \infty. \tag{4.3.13}$$

Proof. For any γ-graded sequence $\{T_k\}_{k\geq 1}$, we have

$$\begin{aligned}|\{T_k<\infty\}| &= E(E(\chi_{\{T_k<\infty\}}|\mathcal{F}_{T_k-1})) = E(E(\chi_{\{T_k<\infty\}}\chi_{\{T_{k-1}<\infty\}}|\mathcal{F}_{T_k-1}))\\ &\leq \gamma E(\chi_{\{T_{k-1}<\infty\}}) \leq \gamma^{k-1}E(\chi_{\{T_1<\infty\}}) \leq \gamma^{k-1}.\end{aligned}$$

This means $|\{T_k<\infty\}| \to 0$, $T_k \to \infty$, a.e., and $\sum_1^\infty \chi_{\{T_k<\infty\}} \in L^1$.

Now for n fixed, consider the sets

$$X_n^{(m)} = \{T_1<n, T_2<n, \cdots, T_m<n, T_{m+1}\geq n\},$$

$$X_n^{(0)} = \{T_1 \geq n\}, \quad X_n^{(\infty)} = \{T_m<n, \forall m\}.$$

Obviously, $|X_n^{(\infty)}|=0$, and$\{X_n^{(m)}\}_{m\geq 0}$ forms a partition of Ω. Now we estimate

$$(\varphi_n-\varphi)\chi_{X_n^{(m)}}, \qquad m=0,1,2,\cdots.$$

Notice that $X_n^{(m)} \in \mathcal{F}_{n-1}\bigcap \mathcal{F}_{T_{m+1}}$, and that $b_k(\omega)\chi_{\{T_k<\infty\}}\chi_{X_n^{(m)}}$ are measurable with respect to $\mathcal{F}_{n-1}$ when $1\leq k\leq m$, so we get (noticing $b_k(\omega)\geq 0$),

$$\varphi_n\chi_{X_n^{(m)}} = \sum_{k=1}^{m} b_k(w)\chi_{\{T_k<\infty\}}\chi_{X_n^{(m)}} + \sum_{k=m+1}^{\infty} E(b_k\chi_{\{T_k<\infty\}}\chi_{X_n^{(m)}}|\mathcal{F}_n),$$

$$\begin{aligned}(\varphi_n-\varphi)\chi_{X_n^{(m)}} &= \sum_{k=m+1}^{\infty} E(b_k\chi_{\{T_k<\infty\}}\chi_n^{(m)}|\mathcal{F}_n) - \sum_{k=m+1}^{\infty} b_k\chi_{\{T_k<\infty\}}\chi_n^{(m)}\\ &\leq \sum_{k=m+1}^{\infty} E(E(b_k\chi_{\{T_k<\infty\}}\chi_n^{(m)}|\mathcal{F}_{T_{m+1}}|\mathcal{F}_n)\\ &\leq B\sum_{k=m+1}^{\infty} \gamma^{k-m-1} = \frac{B}{1-\gamma}.\end{aligned}$$

Analogously, we have also

$$\begin{aligned}
E(|\varphi-\varphi_{n-1}|\,|\mathcal{F}_n)\chi_{X_n^{(m)}} &= E(|\varphi-\varphi_{n-1}|\chi_{X_n^{(m)}}|\mathcal{F}_n)\\
&= E\Big(\Big|\sum_{k=m+1}^{\infty} b_k\chi_{\{T_k<\infty\}}\chi_{X_n^{(m)}} - \sum_{k=m+1}^{\infty} E(b_k\chi_{\{T_k<\infty\}}\chi_{X_n^{(m)}}|\mathcal{F}_{n-1})\Big|\,\Big|\mathcal{F}_n\Big)\\
&\le B\sum_{k=m+1}^{\infty} E(\chi_{\{T_k<\infty\}}\chi_{X_n^{(m)}}|\mathcal{F}_n) + B\sum_{k=m+1}^{\infty} E(\chi_{\{T_k<\infty\}}\chi_{X_n^{(m)}}|\mathcal{F}_{n-1})\\
&\le 2B\sum_{k=m+1}^{\infty}\|E(\chi_{\{T_k<\infty\}}|\mathcal{F}_{T_{m+1}})\|_\infty \le \frac{2B}{1-\gamma}.
\end{aligned}$$

In particular,

$$\|\varphi_n-\varphi_{n-1}\|_\infty \le \|\varphi\|_{BMO_1} \le \frac{2B}{1-\gamma}.$$

This proves (4.3.12). It remains to prove (4.3.13). Let $\alpha>0$ be determined later. We have

$$\begin{aligned}
\sum_{m=0}^{\infty} E(\exp(\alpha(\varphi-\varphi_n))|\mathcal{F}_n)\chi_{X_n^{(m)}} &= E\Big(\sum_{m=0}^{\infty}\exp\alpha(\varphi-\varphi_n)\chi_{X_n^{(m)}}\Big|\mathcal{F}_n\Big)\\
&= E\Big(\sum_{m=0}^{\infty}\exp\{\alpha(\varphi-\varphi_n)\chi_{X_n^{(m)}}\}\chi_{X_n^{(m)}}\Big|\mathcal{F}_n\Big)\\
&= E\Big(\sum_{m=0}^{\infty}\exp\Big\{\alpha\Big(\sum_{k=m+1}^{\infty} b_k\chi_{\{T_k<\infty\}}\chi_{X_n^{(m)}}\\
&\qquad - \sum_{k=m+1}^{\infty} E(b_k\chi_{\{T_k<\infty\}}\chi_{X_n^{(m)}}|\mathcal{F}_n)\Big)\Big\}\chi_{X_n^{(m)}}\Big|\mathcal{F}_n\Big)\\
&\le E\Big(\sum_{m=0}^{\infty}\exp\Big\{\alpha B\sum_{k=m+1}^{\infty}\chi_{\{T_k<\infty\}}\Big\}\chi_{X_n^{(m)}}\Big|\mathcal{F}_n\Big)\\
&= \sum_{m=0}^{\infty} E\Big(E\Big(\exp\Big\{\alpha B\sum_{k=m+1}^{\infty}\chi_{\{T_k<\infty\}}\Big\}\Big|\mathcal{F}_{T_{m+1}}\Big)\Big|\mathcal{F}_n\Big)\chi_{X_n^{(m)}}. \qquad (4.3.14)
\end{aligned}$$

It needs a uniform estimate of

$$\Big\|E\Big(\exp\Big\{\alpha B\sum_{k=m+1}^{\infty}\chi_{\{T_k<\infty\}}\Big\}\Big|\mathcal{F}_{T_{m+1}}\Big)\Big\|_\infty$$

with respect to m. This estimate follows from

$$\begin{aligned}
\exp\Big\{\alpha B\sum_{k=m+1}^{\infty}\chi_{\{T_k<\infty\}}\Big\} &= \sum_{l=0}^{\infty}\frac{(\alpha B)^l}{l!}\Big(\sum_{k=m+1}^{\infty}\chi_{\{T_k<\infty\}}\Big)^l\\
&= \sum_{l=0}^{\infty}\frac{(\alpha B)^l}{l!}\sum_{k=m+1}^{\infty}(k-m)^l\chi_{\{T_k<\infty,T_{k+1}=\infty\}},
\end{aligned}$$

and

$$E\left(\exp\left\{\alpha B\sum_{k=m+1}^{\infty}\chi_{\{T_k<\infty\}}\right\}\Big|\mathcal{F}_{T_{m+1}}\right)$$
$$=\sum_{l=0}^{\infty}\frac{(\alpha B)^l}{l!}\sum_{k=m+1}^{\infty}(k-m)^l E(\chi_{\{T_k<\infty,T_{k+1}=\infty\}}|\mathcal{F}_{T_{m+1}})$$
$$\leq\sum_{l=0}^{\infty}\frac{(\alpha B)^l}{l!}\sum_{k=m+1}^{\infty}(k-m)^l\gamma^{k-m-1}=\sum_{l=0}^{\infty}\frac{(\alpha B)^l}{l!}\sum_{k=1}^{\infty}k^l\gamma^{k-1}$$
$$=\sum_{k=1}^{\infty}\sum_{l=0}^{\infty}\frac{(\alpha kB)^l}{l!}\gamma^{k-1}=\frac{1}{\gamma}\sum_{k=1}^{\infty}(\gamma e^{\alpha B})^k.$$

Taking this estimate into (4.3.14), we get

$$\sum_{m=0}^{\infty}E(\exp\alpha(\varphi-\varphi_n)|\mathcal{F}_n)\chi_{X_n^{(m)}}\leq\sum_{m=0}^{\infty}\frac{1}{\gamma}\sum_{k=1}^{\infty}(\gamma e^{\alpha B})^k\chi_{X_n^{(m)}}$$
$$\leq\frac{e^{\alpha B}}{1-\gamma e^{\alpha B}}<\infty,$$

provided $e^{\alpha B}<\frac{1}{\gamma}$. The proof is finished. □

Remark For the *BMO*-norm estimate of φ, the condition $0\leq b_k(\omega)\leq B$ can be weaken as $|b_k(\omega)|\leq B$.

We are in the position to establish the *BLO* decomposition of *BMO*.

Theorem 4.3.9 *Each* $f\in\operatorname{Re}BMO$ *can be decomposed as* $f=g-h+\varphi$ *with* $g,h\in BLO$, $\varphi\in L^\infty$, *such that*

$$\|g\|_{BLO}+\|h\|_{BLO}+\|\varphi\|_\infty\leq C\|f\|_*. \tag{4.3.15}$$

Proof. From Theorem 4.1.2, for each $f\in BMO$, there exists $\alpha>0$, such that $\|E(\exp(\alpha|f-f_n|)|\mathcal{F}_n)\|_\infty\leq K_\alpha<\infty$. Now define

$$f^{(1)}=f-E_0(f),\ T_i=\inf\{n:|f_n^{(i)}|>\lambda\},\ \lambda>0\text{ being determined later},$$
$$f^{(i+1)}=f-f_{T_i},\qquad i=1,2,\cdots.$$

Since $f_n^{(i+1)}=f_n-f_{T_i\wedge n}$, we have $f_n^{(i+1)}=0$ on the set $\{n\leq T_i\}$, this means that $T_{i+1}>T_i$. Thus we get an increasing sequence $\{T_i\}_{i\geq1}$ of stopping times. We want to show that it is a γ-graded sequence provided λ is large enough. In fact, we have

$$E(\exp(\alpha|f^{(i+1)}|)|\mathcal{F}_{T_{i+1}})\geq\exp(\alpha E(|f^{(i+1)}|\ |\mathcal{F}_{T_{i+1}}))$$
$$\geq\exp(\alpha|E(f^{(i+1)}|\mathcal{F}_{T_{i+1}})|)$$
$$=\exp\left(\alpha|f_{T_{i+1}}^{(i+1)}|\right)\geq\exp(\alpha\lambda)\chi_{\{T_{i+1}<\infty\}}.$$

Meanwhile we have

$$\begin{aligned}K_\alpha &\geq E(\exp(\alpha|f - f_{T_i}|)|\mathcal{F}_{T_i})\\ &= E(E(\exp(\alpha|f - f_{T_i}|)\mathcal{F}_{T_{i+1}})|\mathcal{F}_{T_i})\\ &\geq E(e^{\alpha\lambda}\chi_{\{T_{i+1}<\infty\}}|\mathcal{F}_{T_i}),\end{aligned}$$

and hence we get

$$E(\chi_{\{T_{i+1}<\infty\}}|\mathcal{F}_{T_i}) \leq K_\alpha e^{-\alpha\lambda} = \gamma < 1, \quad \text{provided } \lambda \text{ large.}$$

Since $\{T_i\}_{i\geq 1}$ is a γ-graded sequence, $T_i \to \infty$, a.e., and $f_{T_i} \to f$, a.e. Denoting $T_0 = 0$, $f_{T_0} = E_0(f)$, we get f's decomposition

$$\begin{aligned}f - E_0(f) &= \sum_{i=1}^{\infty}(f_{T_i} - f_{T_{i-1}}) = \sum_{i=1}^{\infty} f_{T_i}^{(i)}\\ &= \sum_{i=1}^{\infty} f_{T_i}^{(i)+}\chi_{\{T_i<\infty\}} - \sum_{i=1}^{\infty} f_{T_i}^{(i)-}\chi_{\{T_i<\infty\}} + \sum_{i=1}^{\infty} f_{T_i}^{(i)}\chi_{\{T_i=\infty\}}\\ &= g - h + \varphi - E_0(f).\end{aligned}$$

Noticing that $|f_{T_i}^{(i)}\chi_{\{T_i=\infty\}}| \leq \lambda$, and $\{f_{T_i}^{(i)}\chi_{\{T_i=\infty\}} \neq 0\} \subset \{T_{i-1} < \infty, T_i = \infty\}$, and hence $\{f_{T_i}^{(i)}\chi_{\{T_i=\infty\}} \neq 0\}$ is disjoint, we see $\|\varphi\|_\infty \leq \|E_0(f)\|_\infty + \lambda$ Now we will examine the sequence $\{f_{T_i}^{(i)\pm}\}_{i\geq 1}$. Each $f_{T_i}^{(i)\pm}$ is nonnegative, measurable with respect to $\mathcal{F}_{T_i}$, uniformly bounded by $\lambda + \|f\|_{BMO_1}$. In fact the last claim follows from

$$\left|f_{T_i}^{(i)}\right| \leq \left|f_{T_i-1}^{(i)}\right| + \left|f_{T_i}^{(i)} - f_{T_i-1}^{(i)}\right| \leq \lambda + \|f^{(i)}\|_{BMO_1} \leq \lambda + \|f\|_{BMO_1}.$$

It remains to show that λ could be chosen to make (4.3.15) hold. From the proof of Lemma 4.1.1, we know that for the choice of α, like $\alpha = (2e\|f\|_{BMO_1})^{-1}$ for example, K_α could be an absolute constant, and hence for $\lambda = C\|f\|_{BMO_1}$ with C large enough, we would get $K_\alpha e^{-\alpha\lambda} = \gamma \leq \frac{1}{2}$, and

$$\|\varphi\|_\infty \leq \|E_0(f)\|_\infty + \lambda \leq C\|f\|_{BMO_1}.$$

$$\|g\|_{BLO} + \|h\|_{BLO} \leq \frac{4}{1-\gamma}(\|f\|_{BMO_1} + \lambda) \leq C\|f\|_{BMO_1}.$$

The proof of the theorem is thus finished. □

Remark We could give another proof of this theorem and formulate it more precisely.

Theorem 4.3.9′ *Each $f \in \mathrm{Re}\,BMO$ can be decomposed as $f = \varphi + g - h$, with $g = \sum_{\nu=0}^{\infty} E(\varepsilon_\nu^+|\mathcal{F}_\nu)$, $h = \sum_{\nu=0}^{\infty} E(\varepsilon_\nu^-|\mathcal{F}_\nu)$, where $\{\varepsilon_\nu\}_{\nu\geq 0}$ is a process satisfying $\left\||\varphi| + \sum_0^\infty |\varepsilon_\nu|\right\|_\infty \approx \|f\|_*$. Hence $\|g\|_{BLO} + \|h\|_{BLO} + \|\varphi\|_\infty \leq C\|f\|_*$.*

Proof. By making use of Theorem 4.2.8, we get φ and $\{\varepsilon_\nu\}_{\nu\geq 0}$ such that

$$\Big\| |\varphi| + \sum_0^\infty |\varepsilon_\nu| \Big\|_\infty \approx \|f\|_* \quad \text{and} \quad f = \varphi + \sum_{\nu=0}^\infty E(\varepsilon_\nu|\mathcal{F}_\nu).$$

Define $g = \sum_{\nu=0}^\infty E(\varepsilon_\nu^+|\mathcal{F}_\nu)$, $-h = f - g - \varphi$. Then

$$\begin{aligned} g_n &= \sum_{\nu=0}^n E(\varepsilon_\nu^+|\mathcal{F}_\nu) + \sum_{\nu=n+1}^\infty E(E(\varepsilon_\nu^+|\mathcal{F}_\nu) \mid \mathcal{F}_n) \\ &\leq \sum_{\nu=0}^\infty E(\varepsilon_\nu^+|\mathcal{F}_\nu) + \sum_{\nu=n+1}^\infty E(\varepsilon_\nu^+|\mathcal{F}_n) \leq g + \Big\| \sum_0^\infty \varepsilon_\nu^+ \Big\|_\infty, \end{aligned}$$

$$\begin{aligned} g_n - g_{n-1} = &\sum_0^n E(\varepsilon_\nu^+|\mathcal{F}_\nu) + \sum_{\nu=n+1}^\infty E(\varepsilon_\nu^+|\mathcal{F}_n) \\ &- \sum_{\nu=0}^{n-1} E(\varepsilon_\nu^+|\mathcal{F}_\nu) - \sum_{\nu=n}^\infty E(\varepsilon_\nu^+|\mathcal{F}_{n-1}), \end{aligned}$$

$$\begin{aligned} -\Big\| \sum_0^\infty \varepsilon_\nu^+ \Big\|_\infty &\leq -\sum_{\nu=n}^\infty E(\varepsilon_\nu^+|\mathcal{F}_{n-1}) \leq g_n - g_{n-1} \\ &\leq \sum_{\nu=n}^\infty E(\varepsilon_\nu^+|\mathcal{F}_n) \leq \Big\| \sum_0^\infty \varepsilon_\nu^+ \Big\|_\infty. \end{aligned}$$

This proves

$$\|g\|_{BLO} \leq \Big\| \sum_0^\infty \varepsilon_\nu^+ \Big\|_\infty \leq C\|f\|_*.$$

Analogously, we have

$$\|h\|_{BLO} \leq \Big\| \sum_0^\infty \varepsilon_\nu^- \Big\|_\infty \leq C\|f\|_*.$$

The proof is finished. □

In some cases, we want to require the $\log A_{\alpha',\beta'}$ property of g and h with α', β' as good as possible for $f \in \log A_{\alpha,\beta}$ fixed. In order to do that, we give a lemma first.

Lemma 4.3.10 *Let $\{T_k\}_{k\geq 1}$ be a γ_0-graded sequence of stopping times, and $\{A_k\}$ be a sequence of sets satisfying $A_k \subset \{T_k < \infty\}$ and A_k being $\mathcal{F}_{T_k}$-measurable, and*

$$E(\chi_{A_{k+1}} \mid \mathcal{F}_{T_k}) \leq \gamma_1, \quad \text{a.e.} \quad \forall k. \tag{4.3.16}$$

In addition, let $\{b_k\}_{k\geq 1}$ be a sequence of functions such that $0 \leq b_k \leq B$, b_k is measurable with respect to $\mathcal{F}_{T_k}$. Then there exist a $\frac{\gamma_1}{1-\gamma_0}$ graded (say γ_1 small enough such that $\frac{\gamma_1}{1-\gamma_0} < 1$) sequence $\{S_j\}_{j\geq 1}$ and a function sequence $\{c_j\}_{j\geq 1}$ satisfying $0 \leq c_j \leq B$, and c_j being measurable with respect to $\mathcal{F}_{S_j}$, such that

$$\varphi(\omega) = \sum_1^\infty b_k \chi_{A_k} = \sum_1^\infty c_j \chi_{\{S_j<\infty\}} = \theta(\omega). \tag{4.3.17}$$

Proof. If $\{A_k\}$ is a decreasing family, then a simple definition of $\{S_j\}$ and $\{c_j\}$ will do, that is $S_j = \begin{cases} T_j, & \omega \in A_j \\ \infty, & \omega \notin A_j \end{cases}$, and $c_j = b_j$, for all j. Naturally, in general case, $S_j(\omega) = T_{n_j(\omega)}(\omega)$ seems to be a right choice for some suitable definition of $n_j(\omega)$. Here is a right definition of $n_j(\omega)$.

$$n_0(\omega) = 0, \quad n_j(\omega) = \inf\{i > n_{j-1}(\omega) : \omega \in A_i\}, \quad j = 1, 2, \cdots. \tag{4.3.18}$$

Now we check that

$$S_j(\omega) = \begin{cases} T_{n_j(\omega)}(\omega), & n_j(\omega) < \infty, \\ \infty, & n_j(\omega) = \infty, \end{cases} \quad j = 1, 2, \cdots, \tag{4.3.19}$$

$$c_j(\omega) = \begin{cases} b_{n_j(\omega)}(\omega), & n_j(\omega) < \infty, \\ 0, & n_j(\omega) = \infty, \end{cases} \quad j = 1, 2, \cdots \tag{4.3.20}$$

are just what we want to find. At first, we show that

$$\{n_j(\omega) = k\} \in \mathcal{F}_{T_k}, \quad \forall j, \quad \forall k. \tag{4.3.21}$$

We prove it by induction with respect to j. Since $n_1(\omega) = \inf\{i : \omega \in A_i\}$,

$$\{n_1(\omega) = k\} = A_1^c \bigcap \cdots \bigcap A_{k-1}^c \bigcap A_k \in \mathcal{F}_{T_k},$$

the claim holds for $j = 1$. Suppose that $\{n_{j-1}(\omega) = l\} \in \mathcal{F}_{T_l}$, for all l. From

$$\{n_j(\omega) = k\} = \bigcup_{i=1}^{k-1} \left(\{n_{j-1}(\omega) = i\} \bigcap A_k \bigcap_{j=i+1}^{k-1} A_j^c \right),$$

we see that $\{n_j(\omega) = k\} \in \mathcal{F}_{T_k}$. Now we prove that $\{S_j\}_{j\geq 1}$ is an increasing sequence of stopping times. For all $k \geq 0$, we have

$$\{S_j(\omega) = k\} = \bigcap_{l=1}^\infty \{n_j(\omega) = l\} \bigcap \{T_l(\omega) = k\} \in \mathcal{F}_k,$$

and $S_j(\omega) \leq S_{j+1}(\omega)$ obviously. It proves the claim. Now we show

$$\{n_j(\omega) = k\} \in \mathcal{F}_{S_j}, \quad \forall j, \quad \forall k.$$

In fact, it follows from

$$\{n_j(\omega) = k\} \bigcap \{S_j(\omega) = n\} = \{n_j(\omega) = k\} \bigcap \{T_k(\omega) = n\} \in \mathcal{F}_n, \quad \forall n.$$

Now we can prove that $\{S_j\}_{j\geq 1}$ is $\frac{\gamma_1}{1-\gamma_0}$ graded. For all $k \geq 1$, we have

$$\begin{aligned} E(\chi_{\{S_{j+1}<\infty\}} \mid \mathcal{F}_{S_j})\chi_{\{n_j=k\}} &= E(\chi_{\{S_{j+1}<\infty, n_j=k\}} \mid \mathcal{F}_{S_j}) \\ &= E(\chi_{\{S_{j+1}<\infty, n_j=k\}} \mid \mathcal{F}_{T_k}) = E(\chi_{\{S_{j+1}<\infty\}} \mid \mathcal{F}_{T_k})\chi_{\{n_j=k\}} \\ &\leq E(\chi_{\cup_{k+1}^{\infty} A_i} \mid \mathcal{F}_{T_k})\chi_{\{n_j=k\}} \leq \sum_{i=k+1}^{\infty} E(\chi_{A_i} \mid \mathcal{F}_{T_k})\chi_{\{n_j=k\}}. \end{aligned}$$

Since

$$E(\chi_{A_{k+1}} \mid \mathcal{F}_{T_k}) \leq \gamma_1, \quad E(\chi_{A_{k+2}} \mid \mathcal{F}_{T_k}) \leq \gamma_1 E(\chi_{\{T_{k+1}<\infty\}} \mid \mathcal{F}_k) \leq \gamma_1\gamma_0,$$

we get

$$E(\chi_{\{S_{j+1}<\infty\}} \mid \mathcal{F}_{S_j})\chi_{\{n_j=k\}} \leq \gamma_1 \sum_{i=k+1}^{\infty} \gamma_0^{i-k-1} = \frac{\gamma_1}{1-\gamma_0}.$$

In addition, because of $\{n_j(\omega)=\infty\} = (\bigcup_k \{n_j(\omega)=k\})^c \in \mathcal{F}_{S_1}$, we also have

$$E(\chi_{\{S_{j+1}<\infty\}} \mid \mathcal{F}_{S_j})\chi_{\{S_j=\infty\}} = E(\chi_{\{S_{j+1}<\infty, S_j=\infty\}} \mid \mathcal{F}_{S_j}) = 0.$$

This proves

$$E(\chi_{\{S_{j+1}<\infty\}} \mid \mathcal{F}_{S_j}) \leq \frac{\gamma_1}{1-\gamma_0}, \quad \forall j \geq 1.$$

Now we turn to the check about $\{c_j(\omega)\}_{j\geq 1}$. Obviously, $0 \leq c_j \leq B$. In addition, c_j is measurable with respect to $\mathcal{F}_{S_j}$, since for all Borel sets Δ in $\mathbf{C}$, we have

$$\{c_j(\omega) \in \Delta\} \bigcap \{S_j = n\} = \bigcup_k \{n_j = k\} \bigcap \{T_k = n\} \bigcap \{b_k(\omega) \in \Delta\} \in \mathcal{F}_n.$$

Finally, it remains to prove $\varphi(\omega) = \theta(\omega)$, a.e. If $\omega \notin \bigcup_{i=1}^{\infty} A_i$, then $\varphi(\omega) = 0$. Meanwhile, for all $j \geq 1$, $n_j(\omega) = \infty$, $c_j(\omega) = 0$, and hence $\theta(\omega) = 0$. When $\omega \in A_{m_1} \bigcap \cdots \bigcap A_{m_j}$, with $m_1 < \cdots < m_j$, then we have

$$\begin{aligned} &n_1(\omega) = m_1,\ S_1(\omega) = T_{m_1}(\omega) < \infty, \ldots, n_j(\omega) = m_j,\ S_j(\omega) = T_{m_j}(\omega) < \infty, \\ &\text{but } \ n_{j+1}(\omega) = \infty, \ \ S_{j+1}(\omega) = \infty, \end{aligned}$$

and hence

$$\varphi(\omega) = b_{m_1} + \cdots + b_{m_j} = c_1 + \cdots + c_j = \theta(\omega).$$

Since $\left\{\omega : \omega \in \bigcap_{j=1}^{\infty} \bigcup_{i=j}^{\infty} A_j\right\}$ has measure zero (owing to $\sum_{i=j}^{\infty} |A_i| \leq \sum_{i=j}^{\infty} |\{T_i < \infty\}| \to 0$), we have (4.3.17) for a.e. ω. The proof of the lemma is complete. □

Theorem 4.3.11 *Let $f \in \log A_{\alpha,\beta} \bigcap BD$. Then for any $\varepsilon > 0$, there exists f's decomposition $f = g - h + \varphi$ with $g, h \in BLO$, such that for all $\tau > 0$, $g \in \log A_{\alpha-\varepsilon,\tau}$, $h \in \log A_{\beta-\varepsilon,\tau}$.*

Proof. Without loss of generality, assume $\beta \leq \alpha$. Then $E(\exp(\beta|f - f_n|)|\mathcal{F}_n) \leq K_\beta < \infty$. As shown in the proof of Theorem 4.3.9, for $\lambda > 0$ determined suitably and for $f^{(i)}$, T_i defined as follows

$$f^{(1)} = f - E_0(f),\ T_i = \inf\{n : |f_n^{(i)}| > \lambda\},\ f^{(i+1)} = f - f_{T_i},\quad i = 1, 2, \cdots,$$

we have following decomposition of f

$$f = \sum_1^\infty f_{T_i}^{(i)} \chi_{A_i} + \sum_1^\infty f_{T_i}^{(i)} \chi_{B_i} + \sum_1^\infty f_{T_i}^{(i)} \chi_{\{T_i = \infty\}} + E_0(f) = g - h + \varphi,$$

with

$$g = \sum_1^\infty f_{T_i}^{(i)} \chi_{A_i},\ A_i = \left\{\omega \in \{T_i < \infty\} : f_{T_i}^{(i)} > 0\right\},$$

$$h = -\sum_1^\infty f_{T_i}^{(i)} \chi_{B_i},\ B_i = \left\{\omega \in \{T_i < \infty\} : f_{T_i}^{(i)} \leq 0\right\},\ \varphi = f - g + h.$$

We have known that $\{T_i\}_{i \geq 1}$ is of $K_\beta e^{-\beta\lambda}$-graded. In addition, $A_i \in \mathcal{F}_{T_i}$, and

$$E(\chi_{A_i} \mid \mathcal{F}_{T_{i-1}}) \leq K_\alpha e^{-\alpha\lambda}.$$

The latter claim follows from

$$\begin{aligned} K_\alpha &\geq E(\exp(\alpha(f - f_{T_i})) \mid \mathcal{F}_{T_i}) \\ &= E(E(\exp(\alpha(f - f_{T_i})) \mid \mathcal{F}_{T_{i+1}}) \mid \mathcal{F}_{T_i}) \\ &\geq E(\exp(\alpha E(f - f_{T_i}) \mid \mathcal{F}_{T_{i+1}}) \mid \mathcal{F}_{T_i}) \\ &= E(\exp(\alpha f_{T_{i+1}}^{(i+1)}) \mid \mathcal{F}_{T_i}) \geq E(e^{\alpha\lambda} \chi_{A_{i+1}} \mid \mathcal{F}_{T_i}). \end{aligned}$$

In addition, $b_i(\omega) = \left|f_{T_i}^{(i)}\right|$ is measurable with respect to $\mathcal{F}_{T_i}$, and $b_i(\omega) \leq \lambda + \|f\|_{BMO_1}$. From Lemma 4.3.10, there exist γ-grad $\left(\gamma = \frac{K_\alpha e^{-\alpha\lambda}}{1 - K_\beta e^{-\beta\lambda}}\right)$ sequence $\{S_j\}_{j \geq 1}$ and corresponding $\{c_j\}_{j \geq 1}$ such that

$$g = \sum_{i=1}^\infty f_{T_i}^{(i)} \chi_{A_i} = \sum_1^\infty c_j \chi_{\{S_j < \infty\}}.$$

Now claim that g could be made in $\log A_{\alpha-\varepsilon,\tau}$, for all $\tau > 0$ and $\varepsilon > 0$ being arbitrarily small (but fixed), provided λ is large enough. Let $\varepsilon > 0$ be given. Choose first δ_1, $\delta_2 > 0$, such that $(\delta_1 + \delta_2)\alpha < \varepsilon$, and then λ large enough such that

$$\lambda + \|f\|_{BMO_1} \leq (1 + \delta_1)\lambda,\quad \gamma = \frac{K_\alpha e^{-\alpha\lambda}}{1 - K_\beta e^{-\beta\lambda}} \leq e^{-(1-\delta_2)\alpha\lambda}.$$

Noticing

$$\gamma e^{(\alpha-\varepsilon)(\lambda + \|f\|_{BMO_1})} \leq e^{(\alpha-\varepsilon)(1+\delta_1)\lambda} e^{-(1-\delta_2)\alpha\lambda},$$

$$(\alpha-\varepsilon)(1+\delta_1)\lambda-(1-\delta_2)\alpha\lambda=(-\varepsilon+(\delta_1+\delta_2)\alpha-\varepsilon\delta_1)\lambda<0,$$

from Lemma 4.3.8, we get

$$\sup_n \left\| E\left(e^{(\alpha-\varepsilon)(g-g_n)} \mid \mathcal{F}_n\right)\right\|_\infty \le K_{\alpha,\varepsilon}<\infty.$$

The same assertion about h follows similarly. Finally, since $g,\ h\in BLO$, for all $\tau>0$, $g\in \log A_{\alpha-\varepsilon,\tau}$, $h\in\log A_{\beta-\varepsilon,\tau}$. The proof of the Theorem is finished. □

Remark In Chapter 6 we will get Jone's famous factorization theorem of A_p weights in martingale setting from this theorem.

The decomposition in the theorem is a kind of Carleson's decomposition of BMO in the classical case. To see this more clearly, we consider the atom case. That is the case where all $\mathcal{F}_n$ are purely atomic. Let $(\Omega,\mathcal{F},\mu,\{\mathcal{F}_n\}_{n\ge0})$ be such one, $\mathcal{I}_n$ be the set of all atoms of $\mathcal{F}_n$, and $\mathcal{I}=\bigcup_n\mathcal{I}_n$.

Theorem 4.3.12 *Let $(\Omega,\mathcal{F},\mu,\{\mathcal{F}_n\}_{n\ge0})$ be as above, $\varphi\in BMO$. Then there exist $\psi,\{b_i\}\subset\mathbf{C},\{I_i\}\subset\mathcal{I}$, such that*

$$\varphi=\psi+\sum_i b_i\frac{1}{|I_i|}\chi_{I_i}, \tag{4.3.22}$$

$$\|\psi\|_\infty\le C\|\varphi\|_*, \tag{4.3.23}$$

$$\frac{1}{|I|}\sum_{I_i\subset I}|b_i|\le C\|\varphi\|_*, \quad \forall I\in\mathcal{I}. \tag{4.3.24}$$

Proof. We have got the decomposition as shown in Theorem 4.3.9,

$$\varphi=\psi+\sum_j\varphi_{T_j}^{(j)}\chi_{\{T_j<\infty\}},$$

with

$$\|\psi\|_\infty+\|\varphi_{T_j}^{(j)}\|_\infty\le\lambda+\|\varphi\|_*,\quad \varphi_{T_j}^{(j)}=\varphi_{T_j}-\varphi_{T_{j-1}},$$

where λ could be taken, for example as $2\|\varphi\|_*$. Each $\{T_j<\infty\}$ is a union of atoms, say $\{T_j<\infty\}=\bigcup_k I_{j,k}$. Denote $|I_{j,k}|\cdot\varphi_{T_j}^{(j)}|_{I_{j,k}}$, a complex number, by $b_{j,k}$. So the preceding decomposition could be rewritten as

$$\varphi=\psi+\sum_{j,k}b_{j,k}\frac{1}{|I_{j,k}|}\chi_{I_{j,k}}=\psi+\sum_i b_i\frac{1}{|I_i|}\chi_{I_i},$$

here the second equality follows from the absolute convergence of the series. We have shown that (4.3.22),(4.3.23) hold. Now prove (4.3.24). Let $I\in\mathcal{I}$ be given arbitrarily. Notice that if j is an index for which there would be $I_0 \in \mathcal{I}$ such that $I \subsetneq I_0 \subset \{T_j < \infty\}$ then for all $i \leq j$, there would exist such I_0 too. Now define

$$j_0(I) = \max\{j : \exists I_0 \in \mathcal{I}, \text{ such that } \quad I \subsetneq I_0 \subset \{T_j < \infty\}\}.$$

The case $j_0(I) = \infty$ means that for all j, there is a $I_0 \in \mathcal{I}$ such that $I \subsetneq I_0 \subset \{T_j < \infty\}$, so there would be no i, no atom $I_{i,k}$ of $\{T_i < \infty\}$ such that $I_{i,k} \subset I$, this implies $\sum_{I_i \subset I} |b_i| = 0$. So, it is enough to consider the case $j_0(I) < \infty$. For $i > j_0(I)$, for all atoms $I_{i,l}$ in $\{T_i < \infty\}$, we have either $I \bigcap I_{i,l} = \emptyset$, or $I_{i,l} \subset I$. So, for such i, $\{T_i < \infty\} \bigcap I \in \mathcal{F}_{T_i}$. Therefore, for $i > j_0(I) + 1$, we get

$$\begin{aligned}
|\{T_i < \infty\} \bigcap I| &\leq \frac{1}{\lambda} \int_{\{T_i < \infty\} \cap I} |\varphi_{T_i}^{(i)}| d\mu \\
&= \frac{1}{\lambda} \int_{\{T_i < \infty\} \cap I} |E(\varphi - \varphi_{T_{i-1}} \mid \mathcal{F}_{T_i})| d\mu \\
&\leq \frac{1}{\lambda} \int_{\{T_{i-1} < \infty\} \cap I} E(|\varphi - \varphi_{T_{i-1}} \mid |\mathcal{F}_{T_{i-1}}) d\mu \\
&\leq \frac{\|\varphi\|_*}{\lambda} |\{T_{i-1} < \infty\} \bigcap I| \\
&\leq \left(\frac{\|\varphi\|_*}{\lambda}\right)^{i-j_0} |\{T_{j_0} < \infty\} \bigcap I| \\
&\leq \left(\frac{\|\varphi\|_*}{\lambda}\right)^{i-j_0} |I|, \qquad (4.3.25)
\end{aligned}$$

for $i = j_0 + 1$, we have (with the choice of λ being $\lambda = 2\|\varphi\|_*$)

$$|\{T_{j_0+1} < \infty\} \bigcap I| \leq |I| = 2\frac{\|\varphi\|_*}{\lambda}|I|. \qquad (4.3.26)$$

Now suppose that all those i in the summation on the left-hand side of (4.3.24), are $\{I_{j,k}\}_{j,k}$ where $I_{j,k}$ is an atom in $\{T_j < \infty\}$, for each j. Then these j must make $j > j_0(I)$. In fact, on the contrary, we would have an atom $I_{j,k}$ in $\{T_j < \infty\}$, such that $I \subsetneq I_{j,k}$. In this case, no other $I_{j,l}$ could be included in I. So we get

$$\sum_{I_i \subset I} |I_i| \leq \sum_{j > j_0(I)} |\{T_j < \infty\} \bigcap I| \leq C \sum_{1}^{\infty} \left(\frac{\|\varphi\|_*}{\lambda}\right)^n |I| \leq C|I|,$$

$$\frac{1}{|I|} \sum_{I_i \subset I} |b_i| = \frac{1}{|I|} \sum_{I_i \subset I} |\varphi_{T_i}^{(i)}||I_i| \leq C(\lambda + \|\varphi\|_*) \frac{1}{|I|} \sum_{I_i \subset I} |I_i| \leq C\|\varphi\|_*.$$

This completes the proof of the theorem. □

4.4 Carleson measures and BMO

The classical Carleson measures and the relation between it and BMO have a martingale version as follows.

Definition 4.4.1 *Let ν be a nonnegative measure on $\Omega \times \mathbf{Z}^+$. ν is said to be a Carleson measure, if*

$$|||\nu||| = \sup |\{\tau < \infty\}|^{-1} |\{(\omega, k) : k \geq \tau(\omega), \tau(\omega) < \infty\}|_\nu < \infty, \tag{4.4.1}$$

where τ runs through all stopping times.

Theorem 4.4.2 *Let $g = (g_n)_{n\geq 0}$ be a martingale, and $d\nu = |\Delta_k g|^2 \delta_k d\mu$, where δ_k is Dirac measure centred at k. Then ν is a Carleson measure if and only if $g \in BMO$. Furthermore, in any case, we have $|||\nu||| = \|g\|^2_{BMO_2}$.*

Proof. Let $g = (g_n)_{n\geq 0}$ be a martingale, ν be generated by g as above, and τ be any stopping time. Then

$$\begin{aligned} |\{(\omega, k) : k \geq \tau(\omega), \tau(\omega) < \infty\}|_\nu &= E\Big(\sum_{k=0}^{\infty} |\Delta_k g|^2 \chi_{\{\tau(\omega)\leq k\}}\Big) \\ &= E\Big(E\Big(\sum_{k=\tau(\omega)}^{\infty} |\Delta_k g|^2 \,\Big|\, \mathcal{F}_\tau\Big) \chi_{\{\tau<\infty\}}\Big). \end{aligned} \tag{4.4.2}$$

Obviously, $g \in BMO$ implies $|||\nu||| \leq \|g\|^2_{BMO_2}$. Conversely, for any n, and any $F \in \mathcal{F}_n$, by defining

$$\tau = \begin{cases} n, & \omega \in F, \\ \infty, & \omega \notin F, \end{cases}$$

we get

$$\int_F \sum_{k=n}^{\infty} |\Delta_k g|^2 d\mu = |\{(\omega, k) : k \geq \tau(\omega), \tau(\omega) < \infty\}|_\nu \leq |||\nu|||\,|F|,$$

which implies $\|g\|^2_{BMO_2} \leq |||\nu|||$. This proves the theorem. □

The martingale version of classical Carleson's inequality concerning harmonic functions can be formulated as follows.

Theorem 4.4.3 *Let $d\nu = \nu_k \delta_k d\mu$, with the ν_k's being nonnegative random variables, be a Carleson measure, and $0 < p < \infty$. Then for all adapted processes $f = (f_n)_{n\geq 0}$, we have*

$$E\left(\sum_{k=0}^{\infty} \nu_k |f_k|^p\right) \leq |||\nu||| E((Mf)^p). \tag{4.4.3}$$

On the contrary, if (4.4.3) holds for some p, with $|||\nu|||$ replaced by a constant C_ν, then ν is a Carleson Measure and $|||\nu||| \leq C_\nu$.

Proof. Assume that ν is a Carleson measure, $f = (f_n)_{n\geq 0}$ is adapted. For any $\lambda > 0$, define the stopping time $\tau = \inf\{n : |f_n| > \lambda\}$. Then

$$\{(\omega, k) : |f_k| > \lambda\} \subset \{(\omega, k) : k \geq \tau(\omega), \tau(\omega) < \infty\}. \tag{4.4.4}$$

So, we get

$$\begin{aligned} E\Big(\sum_{k=0}^{\infty} \nu_k |f_k|^p\Big) &= p \int_0^{\infty} \lambda^{p-1} |\{(\omega, k) : |f_k| > \lambda\}|_\nu d\lambda \\ &\leq p \int_0^{\infty} \lambda^{p-1} |\{(\omega, k) : k \geq \tau(\omega), \tau(\omega) < \infty\}|_\nu d\lambda \\ &\leq p|||\nu||| \int_0^{\infty} \lambda^{p-1} |\{\tau < \infty\}| d\lambda = |||\nu||| E((Mf)^p). \end{aligned}$$

This proves (4.4.3). Assume that (4.4.3) holds for some p, with a constant C_ν. Let τ be any stopping time. Let $f_k = \chi_{\{\tau \leq k\}}$. Then $Mf = \chi_{\{\tau < \infty\}}$, and

$$E\Big(\sum_{k=0}^{\infty} \nu_k \chi_{\{\tau \leq k\}}\Big) = E\Big(\sum_{k=0}^{\infty} \nu_k |f_k|^p\Big) \leq C_\nu E((Mf)^p) = C_\nu |\{\tau < \infty\}|.$$

This proves $|||\nu||| \leq C_\nu$. The proof is finished. □

Remark In particular, when $d\nu = |\Delta_k g|^2 \delta_k d\mu$ with $g = (g_n)_{n\geq 0}$ is a martingale then $g \in BMO$, if and only if

$$\int_{\Omega\times Z^+} |f_k|^p d\nu \leq CE((Mf)^p), \quad \text{for all adapted } f = (f_n)_{n\geq 0}, \tag{4.4.5}$$

and

$$||g||^2_{BMO_2} = \inf\{C : C \text{ make (4.4.5) holds}\}. \tag{4.4.6}$$

Analogously, we have a bmo_2 version of the preceding assertions:

$$\int_{\Omega\times Z^+} |f_{k-1}|^p d\nu \leq CE((Mf)^p), \quad \text{for all adapted } f = (f_n)_{n\geq 0}, \tag{4.4.7}$$

if and only if $g \in bmo_2$, and in any case

$$||g||^2_{bmo_2} = \inf\{C : C \text{ make (4.4.7) holds}\}. \tag{4.4.8}$$

The proof is almost unchanged by noticing some slight modifications :

$$\{(\omega, k) : |f_{k-1}| > \lambda\} \subset \{(\omega, k) : k \geq \tau(\omega) + 1, \tau(\omega) < \infty\},$$

with the same τ defined by $\tau = \inf\{n : |f_n| > \lambda\}$, and

$$|\{(\omega, k) : |f_{k-1}| > \lambda\}|_\nu \leq E\Big(E\Big(\sum_{k=\tau+1}^{\infty} |\Delta_k g|^2 \mid \mathcal{F}_\tau\Big)\chi_{\{\tau<\infty\}}\Big).$$

4.5 Commutator theory related to BMO

As shown in the classical case that although a BMO function is not bounded in general, a commutator $[B,T]$ generated by a multiplication operator B defined by $b \in BMO$, and a suitable L^p-bounded operator T, could be L^p-bounded still. In martingale case, the situation is similar. We begin with the case where T is a kind of maximal operators.

Lemma 4.5.1 *Let $g = (g_n)_{n\geq 0}$ be a martingale, $1 < p < \infty$. Then $g \in BMO$, if and only if*

$$\Big\| \sup_n M_n f |g - g_{n-1}| \Big\|_p \leq C_g \|Mf\|_p, \quad \textit{for all adapted } f = (f_n)_{n\geq 0}. \tag{4.5.1}$$

Furthermore, $\|g\|_*$ is equivalent to the smallest C_g in (4.5.1).

Proof. First, we prove that (4.5.1) (even only for martingale f) implies $g \in BMO$, and $\|g\|_* \leq CC_g$. Let n and $F \in \mathcal{F}_n$ be fixed arbitrarily. Consider the martingale generated by χ_F. Then $f_k = \chi_F$, for $k \geq n$. From (4.5.1) we get

$$\begin{aligned}\int_F |g - g_{n-1}|^p d\mu &\leq E\left(\sup_k (M_k f |g - g_{k-1}|)^p\right) \\ &\leq C_g^p \|M\chi_F\|_p^p \leq C^p C_g^p \|\chi_F\|_p^p = C^p C_g^p |F|.\end{aligned}$$

This gives $\|g\|_* \leq CC_g$. Now assume that $g \in BMO$. By making use of g's decomposition as shown in Theorem 4.3.9′, we get $g = g^{(1)} - g^{(2)} + h$, $g^{(1)} = \sum_0^\infty E(\varepsilon_\nu^+ \mid \mathcal{F}_\nu)$, $g^{(2)} = \sum_0^\infty E(\varepsilon_\nu^- \mid \mathcal{F}_\nu)$, and $\big\| |h| + \sum_0^\infty |\varepsilon_\nu| \big\|_\infty \leq C\|g\|_*$. Without loss of generality, we could assume $g = \sum_0^\infty E(\varepsilon_\nu \mid \mathcal{F}_\nu)$ with $\varepsilon_\nu \geq 0$ and $\big\| \sum_0^\infty \varepsilon_\nu \big\|_\infty \leq C\|g\|_*$. Denote

$$\tilde{g}_n = \sum_0^n E(\varepsilon_\nu \mid \mathcal{F}_\nu), \quad g_n = E(g \mid \mathcal{F}_n), \quad \forall n.$$

Then we have

$$0 \leq E(g - \tilde{g}_{n-1} \mid \mathcal{F}_n) = E\Big(\sum_n^\infty E(\varepsilon_\nu \mid \mathcal{F}_\nu) \mid \mathcal{F}_n\Big) \leq C\|g\|_*, \tag{4.5.2}$$

$$0 \leq g_n - \tilde{g}_n = E(g - \tilde{g}_n \mid \mathcal{F}_n) = E\Big(\sum_{n+1}^\infty E(\varepsilon_\nu \mid \mathcal{F}_\nu) \mid \mathcal{F}_n\Big) \leq C\|g\|_*. \tag{4.5.3}$$

So, we have

$$|g - g_{n-1}| = |g - \tilde{g}_{n-1} + \tilde{g}_{n-1} - g_{n-1}| = g - \tilde{g}_{n-1} + O(\|g\|_*). \tag{4.5.4}$$

Now denoting $\alpha_n = (M_n f)^p$, $n \geq 0$, $\beta_n = (g - \tilde{g}_{n-1})^p$, $n \geq 1$, $\beta_0 = g$, and noticing that $\{\alpha_n\}_{n\geq 0}$ is a increasing process, meanwhile $\{\beta_n\}_{n\geq 0}$ is a decreasing one (so $\Delta_l \beta \leq 0$, $l \geq 1$, but $\Delta_0 \beta = g_0 \geq 0$), we get

$$\alpha_n \beta_n = \sum_{k=0}^n \Delta_k \alpha \sum_{l=0}^n \Delta_l \beta = \sum_{0\leq k<l\leq n} + \sum_{0\leq l\leq k\leq n}$$

$$= \sum_{l=1}^{n}\sum_{k=0}^{l-1} \Delta_k\alpha\Delta_l\beta + \sum_{k=0}^{n}\sum_{l=0}^{k} \Delta_l\beta\Delta_k\alpha$$

$$= \sum_{l=1}^{n} \alpha_{l-1}\Delta_l\beta + \sum_{k=0}^{n} \beta_k\Delta_k\alpha \le \sum_{0}^{\infty} \beta_k\Delta_k\alpha. \tag{4.5.5}$$

Notice that $E(\beta_n \mid \mathcal{F}_n) \le C\|g\|_*^p$. This can be seen from Garsia-Neveu's convexity lemma. In fact, since $(\tilde{g}_n)_{n\ge 0}$ is an increasing adapted process, and satisfying (4.5.2), the conditional convexity lemma gives $E((g-\tilde{g}_{n-1})^p \mid \mathcal{F}_n) \le C\|g\|_*^p$. Thus, we get

$$E\left(\sup_n \alpha_n\beta_n\right) \le E\left(\sum_{k=0}^{\infty} \beta_k\Delta_k\alpha\right) = E\left(\sum_{k=0}^{\infty} E(\beta_k \mid \mathcal{F}_k)\Delta_k\alpha\right)$$
$$\le C\|g\|_*^p E((Mf)^p).$$

This proves (4.5.1). The proof of the lemma is finished. □

Remark (4.5.1) could be improved slightly as follows: $g \in BMO$, $0 < p < \infty$, then

$$\left\|\sup_n M_n f \sup_{m\ge n} |g - g_{m-1}|\right\|_p \le C\|g\|_*\|Mf\|_p, \ \forall \text{ adapted } f = (f_n)_{n\ge 0}. \tag{4.5.6}$$

In fact, assuming again that $g = \sum_{\nu=0}^{\infty} E(\varepsilon_\nu \mid \mathcal{F}_\nu)$ with ε_ν's being nonnegative random varibles satisfying $\left\|\sum_{\nu=0}^{\infty}\varepsilon_\nu\right\|_\infty \le C\|g\|_*$, then we have

$$|g - g_{m-1}| = g - \tilde{g}_{m-1} + O(\|g\|_*), \quad \forall\, m,$$

and hence

$$\sup_{m\ge n} |g - g_{m-1}| \le g - \tilde{g}_{n-1} + O(\|g\|_*), \quad \forall\, n.$$

So, (4.5.6) follows from (4.5.1), which holds for $0 < p < \infty$ when $g \in BMO$ (for $0 < p \le 1$, the proof is unchanged, since $E((g-\tilde{g}_{n-1})^p|\mathcal{F}_n) \le C\|g\|_*^p$ holds for $0 < p < \infty$), immediately.

Now we can get the first result concerning commutators that arise from BMO martingales.

Theorem 4.5.2 *Let f, g be two martingales and $g \in BMO$, $1 < p < \infty$. Then*

$$\left\|\sup_n |E(fg \mid \mathcal{F}_n) - gf_n|\right\|_p \le C\|g\|_*\|f\|_p. \tag{4.5.7}$$

Proof. We have

$$\sup_n |E(fg \mid \mathcal{F}_n) - gf_n| \le \sup_n |E(fg \mid \mathcal{F}_n) - g_nf_n|$$
$$+ \sup_n |(g - g_{n-1})f_n| + \sup_n |f_n\Delta_n g|.$$

The second and third terms have the desirable majorant obviously. It remains the first term to be estimated. We have

$$\begin{aligned} E(fg \mid \mathcal{F}_n) - f_n g_n| &= E((f - f_n)(g - g_n)|\mathcal{F}_n) \\ &= E\left(\sum_{k=n+1}^{\infty} \Delta_k f \sum_{l=n+1}^{\infty} \Delta_l g \,\middle|\, \mathcal{F}_n \right) \\ &= E\left(\sum_{k=n+1}^{\infty} \Delta_k f \Delta_k g \,\middle|\, \mathcal{F}_n \right). \end{aligned}$$

Applying Fefferman's inequality written in conditional form (Proposition 2.3.1) we get

$$|E(fg \mid \mathcal{F}_n) - f_n g_n| \le C\|g\|_* E(S(f) \mid \mathcal{F}_n),$$

and hence

$$\left\| \sup_n |E(fg \mid \mathcal{F}_n) - f_n g_n| \right\|_p \le C\|g\|_* \|M(Sf)\|_p \le C\|g\|_* \|f\|_p.$$

This completes the proof of the theorem. □

Remark Let S be any measurable function from Ω to $\overline{\mathbf{Z}^+}$. Consider the operator α_S:

$$\text{martingale} \quad f = (f_n)_{n \ge 0} \to \alpha_S f = f_S.$$

Then α_S is a L^p-bounded linear operator for $1 < p < \infty$. Let B be the multiplication operator that arises from $g \in BMO$. Since

$$\sup_n |E(fg \mid \mathcal{F}_n) - g f_n| = \sup_S |(fg)_S - g f_S|,$$

then what we have just obtained says that the commutator $[B, \alpha_S] = \alpha_S B - B\alpha_S$ is a L^p-bounded operator, provided $g \in BMO$.

Next, we consider another kind of commutators which arise from B as above and a martingale transform. We will do more investigations about martingale transforms in the next chapter. But here, we want to use some results about it in advance. Let $v = (v_n)_{n \ge 0}$ be an adapted process such that $Mv \in L^p$, $0 < p < \infty$, in symbols $v \in V_p$. The following transform

$$T_v f = ((T_v f)_n)_{n \ge 0}, \quad T_v f = \sum_0^{\infty} v_{n-1} \Delta_n f, \quad \forall\, f = (f_n)_{n \ge 0}, \tag{4.5.8}$$

is called a martingale transform. (Notice that according to the usual convention, $v_{-1} = 0$, and hence $(Tf)_0 = 0$, for any $f = (f_n)_{n \ge 0}$.) We will use some boundedness results of T_v listed as follows:

$$v \in V_p,\ g \in BMO,\ 0 < p \le \infty, \text{ then } \|M(T_v g)\|_p \le C\|v\|_{V_p} \|g\|_*, \tag{4.5.9}$$

$v \in V_p$, $f \in H_q$, $0 < p \le \infty$, $1 < q < \infty$, then

$$\|T_v f\|_r \le C\|v\|_{V_p}\|f\|_q, \quad \frac{1}{r} = \frac{1}{p} + \frac{1}{q} < 1, \tag{4.5.10}$$

In addition we need a boundedness result of the operator

$$(f,g) \to d(f,g) = \sum_0^\infty E(\Delta_k f \Delta_k g \mid \mathcal{F}_{k-1}), \quad f_0 \text{ or } g_0 = 0, \tag{4.5.11}$$

the proof of which is similar to the proof of Fefferman's duality inequality. We formulate it in the following theorem.

Theorem 4.5.3 *Let $g = (g_n)_{n \ge 0}$, $g_0 = 0$, be in BMO, $1 < p < \infty$. Then*

$$\|d(f,g)\|_p \le C\|g\|_* \|f\|_p. \tag{4.5.12}$$

Proof. Let $g = (g_n)_{n\ge 0} \in BMO$ with $g_0 = 0$, and $f = (f_n)_{n \ge 0} \in L^p$, $1 < p < \infty$. Denote $\theta_n = \sum_1^{n+1} E(\Delta_k f \Delta_k g | \mathcal{F}_{k-1})$, $n \ge 1$, $\theta_0 = 0$. We have

$$\begin{aligned} E(|d(f,g) - \theta_{n-1}||\mathcal{F}_n) &= E\Big(\Big|\sum_{k=n+1}^\infty E(\Delta_k f \Delta_k g|\mathcal{F}_{k-1})\Big|\Big|\mathcal{F}_n\Big) \\ &\le E\Big(\sum_{k=n+1}^\infty |\Delta_k f||\Delta_k g|\Big|\mathcal{F}_n\Big) \\ &\le E\Big(\sum_{k=n+1}^\infty |\Delta_k g|^2 S_k(h)\Big|\mathcal{F}_n\Big)^{\frac12} E\Big(\sum_{k=n+1}^\infty |\Delta_k f|^2 S_k(h)^{-1}\Big|\mathcal{F}_n\Big)^{\frac12} \\ &= (AB)^{\frac12}, \end{aligned} \tag{4.5.13}$$

where $h = f - f_n$ and

$$S_k(h) = \left(\sum_{n+1}^k |\Delta_j h|^2\right)^{\frac12} = \left(\sum_{n+1}^k |\Delta_j f|^2\right)^{\frac12}, \quad k \ge n+1, \quad S_n(h) = 0.$$

Now we estimate A and B. We have

$$B \le 2E\left(\sum_{n+1}^\infty (S_k(h) - S_{k-1}(h))\Big|\mathcal{F}_n\right) \le 2E(S(h)|\mathcal{F}_n),$$

$$\begin{aligned} A &\le E\left(\sum_{k=n+1}^\infty |\Delta_k g|^2 \sum_{j=n+1}^k (S_j(h) - S_{j-1}(h))\Big|\mathcal{F}_n\right), \\ &= E\left(\sum_{j=n+1}^\infty E\left(\sum_{k=j}^\infty |\Delta_k g|^2\Big|\mathcal{F}_j\right)(S_j(h) - S_{j-1}(h))\Big|\mathcal{F}_n\right) \\ &\le \|g\|_*^2 E(S(h)|\mathcal{F}_n). \end{aligned}$$

Thus we get (noticing $S(h) \le S(f)$)

$$d(f,g)^{\#}_{\theta} \le \sqrt{2}\|g\|_* M(S(f)), \tag{4.5.14}$$

and hence from Theorem 3.5.7′, for $1 < p < \infty$,

$$E(|d(f,g)|^p) \le CE(d(f,g)^{\#}_{\theta}) \le C\|g\|_*^p E(M(S(f))^p) \le C\|g\|_*^p E(|f|^p).$$

The proof is thus finished. □

Now we turn to the boundedness of the commutator $[B, T_v]$. First, we want to express the multiplication operator in a more precise form. Let $g = (g_n)_{n\ge 0}$ and $f = (f_n)_{n\ge 0}$ be two L^1-bounded martingales. Then

$$g_n f_n = \sum_{l=1}^{n} g_{l-1}\Delta_l f + \sum_{k=1}^{n} f_{k-1}\Delta_k g + \sum_{k=0}^{n} \Delta_k f \Delta_k g, \tag{4.5.15}$$

$$\begin{aligned} gf &= \sum_{l=1}^{\infty} g_{l-1}\Delta_l f + \sum_{k=1}^{\infty} f_{k-1}\Delta_k g + \sum_{k=0}^{\infty} \Delta_k f \Delta_k g \\ &= T_g f + T_f g + \mathcal{D}(f,g). \end{aligned} \tag{4.5.16}$$

We only consider the case $g \in BMO$, $g_0 = 0$ and $f \in L^q$, $1 < q < \infty$. In this case, each series in (4.5.16) converges in $L^{q-\varepsilon}$, and so

$$(gf)_n = \sum_{k=1}^{n} g_{k-1}\Delta_k f + \sum_{k=1}^{n} f_{k-1}\Delta_k g + E(\mathcal{D}(f,g)|\mathcal{F}_n), \quad \forall n. \tag{4.5.17}$$

Now the commutator $[B, T_v]$ could be written as follows. We have

$$T_v(gf) = \sum_{k=1}^{\infty} v_{k-1}g_{k-1}\Delta_k f + \sum_{k=1}^{\infty} v_{k-1}f_{k-1}\Delta_k g + T_v(\mathcal{D}(f,g)),$$

$$\begin{aligned} T_v(\mathcal{D}(f,g)) &= \sum_{k=1}^{\infty} v_{k-1}[\Delta_k f \Delta_k g - E(\Delta_k f \Delta_k g|\mathcal{F}_{k-1})] \\ &\quad + T_v\left(\sum_{l=1}^{\infty} E(\Delta_l f \Delta_l g|\mathcal{F}_{l-1})\right), \end{aligned}$$

$$\begin{aligned} gT_v f &= \sum_{k=1}^{\infty} g_{k-1}\Delta_k(T_v f) + \sum_{k=1}^{\infty} (T_v f)_{k-1}\Delta_k g + \mathcal{D}(g, T_v f) \\ &= \sum_{k=1}^{\infty} v_{k-1}g_{k-1}\Delta_k f + \sum_{k=1}^{\infty} (T_v f)_{k-1}\Delta_k g \end{aligned}$$

$$+\sum_{k=1}^{\infty} v_{k-1}E(\Delta_k f \Delta_k g|\mathcal{F}_{k-1})$$
$$+\sum_{k-1}^{\infty} v_{k-1}[\Delta_k f \Delta_k g - E(\Delta_k f \Delta_k g|\mathcal{F}_{k-1})].$$

So, we have

$$\begin{aligned}
[B,T_v]f &= T_v(gf) - gT_v f \\
&= \sum_{k=1}^{\infty} v_{k-1} f_{k-1}\Delta_k g - \sum_{k=1}^{\infty}(T_v f)_{k-1}\Delta_k g \\
&\quad + T_v\left(\sum_{l=1}^{\infty} E(\Delta_l f \Delta_l g|\mathcal{F}_{l-1})\right) - \sum_{k=1}^{\infty} E(v_{k-1}\Delta_k f \Delta_k g|\mathcal{F}_{k-1}) \\
&= \sum_{k=1}^{\infty} v_{k-1} f_{k-1}\Delta_k g - \sum_{k=1}^{\infty}(T_v f)_{k-1}\Delta_k g \\
&\quad + T_v(d(f,g)) - d(T_v f, g).
\end{aligned} \tag{4.5.18}$$

Now we are in the position to get $[B,T_v]$'s boundedness.

Theorem 4.5.4 *Let* $0 < p \le \infty$, $1 < q < \infty$, $\frac{1}{r} = \frac{1}{p} + \frac{1}{q} < 1$, $v \in V_p$, $f \in L^q$, $g \in BMO$. *Then*

$$\|[B,T_v]f\|_r \le C\|g\|_*\|v\|_{V_p}\|f\|_q. \tag{4.5.19}$$

Proof. Since $(v_k f_k)_{k\ge 0} \in V_r$, and $T_v f \in H_r \subset V_r$ (owing to (4.5.10)), we see that the first and second terms of the last part of (4.5.18) belong to L^r with the required norm estimates owing to (4.5.9). Now consider the third and fourth terms in (4.5.18). $d(f,g) \in H_q$ owing to the Thecrem 4.5.3 and $q > 1$. Then by means of (4.5.10), $T_v(d(f,g)) \in H_r$. Similarly, $T_v f \in H_r$ owing to (4.5.10), and then Theorem 4.5.3 gives $d(T_v f, g) \in H_r$ since $r > 1$. And all the required norm estimate are implied in the arguments. The assertion has been proved. □

Remark How could we conclude g's behavior from the given boundedness of $[B,T_v]$ for fixed v seems to be a bit difficult. If the question is proposed as follows, then the answer is easy. Let $g = (g_n)_{n\ge 0}$ be a martingale such that

$$\|[B,T_v]f\|_2 \le C_g\|v\|_{V_p}\|f\|_q,$$
$$\frac{1}{p} + \frac{1}{q} = \frac{1}{2}, \quad 1 < q < \infty, \quad \forall v \in V_p,\ \forall f \in L^q. \tag{4.5.20}$$

Then $\|g\|_{bmo_2} \le CC_g$. In fact, for n and $F \in \mathcal{F}_n$ fixed, set

$$v_k = \begin{cases} 0, & k < n, \\ \chi_F, & k \ge n, \end{cases} \qquad f = \chi_F.$$

Notice that

$$\Delta_k f = \begin{cases} E(\chi_F|\mathcal{F}_k) - E(\chi_F|\mathcal{F}_{k-1}), & k \le n, \\ 0, & k > n. \end{cases}$$

Then we have

$$T_v f = \sum_1^\infty v_{k-1}\Delta_k f = 0,$$

$$d(f,g) = \sum_{k=1}^n E(\Delta_k f \Delta_k g|\mathcal{F}_{k-1}), \quad \Delta_k(d(f,g)) = 0, \quad k \ge n,$$

and so we see that the second, third and fourth term in (4.5.18) are vanishing. It remains the first term to be considered. Noting

$$v_k f_k = \begin{cases} 0, & k < n, \\ \chi_F, & k \ge n, \end{cases}$$

we get

$$[B, T_v]f = \sum_{k=n+1}^\infty v_{k-1} f_{k-1} \Delta_k g = (g - g_n)\chi_F.$$

From (4.5.20), we get

$$\left(\int_F |g - g_n|^2 d\mu\right)^{\frac{1}{2}} \le C_g \|v\|_{V_p} \|f\|_q \le C_g |F|^{\frac{1}{p}} |F|^{\frac{1}{q}} = C_g' |F|^{\frac{1}{2}}.$$

This means $\|g\|_{bmo_2} \le C_g$. The proof of the assertion is finished.

Lemma 4.5.5 *Let $1 < p < \infty$, $f \in L^p$, $g \in L^{p'}$, $v \in V_\infty$. Then*

$$\|fT_v g - gT_v f\|_{H_1} \le C\|v\|_{V_\infty} \|f\|_p \|g\|_{p'}. \tag{4.5.21}$$

Proof. Assume $f \in L^\infty$. Then $fT_v g - gT_v f \in L^{1+\varepsilon} \subset H_1$. So

$$\begin{aligned} \|fT_v g - gT_v f\|_{H_1} &= \sup_{b \in BMO, \|b\|_* \le 1} |E(b(fT_v g - gT_v f))| \\ &= \sup_b |E(f[bT_v g - T_v(bg)])| \le C\|v\|_{V_\infty}\|f\|_p\|g\|_{p'}. \end{aligned}$$

For $f \in L^p$ arbitrarily, choosing $\{f^{(n)}\} \subset L^\infty$ and $f^{(n)} \to f$ in L^p. Then $\{f^{(n)}T_v g - gT_v f^{(n)}\}$ is a Cauchy sequence in H_1, so its limit $fT_v g - gT_v f$ is in H_1, and (4.5.21) holds. The proof is finished. □

4.6 Distance of $f \in BMO$ to L^∞

As well known, $L^\infty \subset BMO$ in any case. We want to know if L^∞ is dense in BMO, or when it is not the case, if L^∞ is closed in BMO. The answer is negative. First, we investigate these problems and then determine the distance of $f \in BMO$ to L^∞ quantitatively.

Lemma 4.6.1 *Let $g \in BMO$, $\{\tau_n\}_{n\geq 1}$ be a sequence of stopping times satisfying $\tau_n \to \infty$, a.e. Then $\{g^{(\tau_n)}\}_{n\geq 1}$ converges to g in the weak topology $\sigma(BMO, H_1)$.*

Proof. Let $g \in BMO$, $f \in H_1$. We have $\langle f, g^{(\tau_n)}\rangle = \langle f^{(\tau_n)}, g\rangle$. Since

$$E(S(f^{(\tau_n)} - f)) = E\Big(\Big(\sum_{\tau_n}^{\infty} |\Delta_k f|^2\Big)^{\frac{1}{2}}\Big) \to 0,$$

we get $\langle f^{(\tau_n)}, g\rangle \to \langle f, g\rangle$. So $\langle f, g^{(\tau_n)}\rangle \to \langle f, g\rangle$. This proves the assertion. □

Corollary 4.6.2 *Let B be the unit ball of BMO. Then $B \bigcap L^\infty$ is dense in B under the weak topology $\sigma(BMO, H_1)$.*

Proof. Let $g \in B$. Define $\tau_n = \inf\{m : |g_m| > n\}$. Then $\tau_n \to \infty$, a.e. Otherwise we would have a set E of positive measure and a positive number K such that $|\tau_n| \leq K$ on E. This implies

$$M_K g \mid_E \geq M_{\tau_n} g \mid_E \geq n, \quad \forall n.$$

This is not possible. And furthermore $|g^{(\tau_n)}| \leq n + \|g\|_*$. That is $g^{(\tau_n)} \in B \cap L^\infty$. Since $g^{(\tau_n)} \to g$ in $\sigma(BMO, H_1)$ owing to the preceding lemma, we get the required assertion. □

Corollary 4.6.3 *Let $f = (f_n)_{n\geq 0} \in L^1$. Then inspite of $f \in H_1$ or not, we have*

$$C\|f\|_{H_1} \leq |||f|||_1 = \sup_{g \in B\cap L^\infty} |E(fg)| \leq C\|f\|_{H_1}. \tag{4.6.1}$$

Proof. Assume $f \in H_1$. Then there exists $g \in BMO$ such that

$$|\langle f, g\rangle| = \|f\|_{H_1}, \text{ and } \|g\|_* = 1,$$

owing to the Hahn-Banach Theorem. Since $B \bigcap L^\infty$ is dense in B under the weak topology $\sigma(BMO, H_1)$,

$$\sup_{g\in B} |\langle f, g\rangle| = \sup_{g\in B\cap L^\infty} |\langle f, g\rangle|, \quad \forall f \in H_1.$$

This proves the assertion when $f \in H_1$. Now let $f \in L^1 - H_1$. Consider the stopped martingale $f^{(N)}$ at N. Then $|||f^{(N)}|||_1 \geq C\|f^{(N)}\|_{H_1} \to \infty$. Noticing that the norm $|||\cdot|||_1$ is decreasing under the action of stopping times. So $|||f|||_1 \geq |||f^{(N)}|||_1 \to \infty$. This finishes the proof of the corollary. □

Now turn to the main theorem.

Theorem 4.6.4 *Assume $L^\infty \neq BMO$. Then L^∞ is neither dense nor closed in BMO.*

Proof. The argument is by contradiction. Suppose that L^∞ were dense in BMO. We would like to prove $H_1 = L^1$, and get a contradiction to the assumption $L^\infty \neq BMO$. Since H_1 is dense in L^1, it is enough to prove that any sequence $\{f^{(n)}\} \subset H_1$ which converges to 0 in L^1 must converge to 0 in H_1. Assume that $\{f^{(n)}\}$ is such a sequence in H_1. Without loss of generality, we can assume that $\|f^{(n)}\|_{H_1}$ is bounded, otherwise we could replace it by $\{\max(1, \|f^{(n)}\|_{H_1})^{-1} f^{(n)}\}$. Since $\{f^{(n)}\}$ converges to 0 in L^1, it converges also to 0 under the weak topology $\sigma(L^1, L^\infty)$. We claim that $\{f^{(n)}\}$ converges to 0 under the weak topology $\sigma(H_1, BMO)$. In fact, let $g \in B$, $\{g^{(k)}\} \subset B \bigcap L^\infty$ such that $\|g - g^{(k)}\|_* \leq \varepsilon_k \to 0$ owing to the density of L^∞ in BMO. Then for $\varepsilon > 0$ given arbitrarily, at first choose k such that ε_k small enough, say $\varepsilon_k \leq (2C \sup_n \|f^{(n)}\|_{H_1})^{-1}\varepsilon$, then choose n_0 such that $n \geq n_0$ implies $|\langle f^{(n)}, g^{(k)}\rangle| \leq \frac{\varepsilon}{2}$. Thus we have, when $n \geq n_0$,

$$|\langle f^{(n)}, g\rangle| \leq |\langle f^{(n)}, g - g^{(k)}\rangle| + |\langle f^{(n)}, g^{(k)}\rangle| \leq C \sup_n \|f^{(n)}\|_{H_1}\varepsilon_k + \frac{\varepsilon}{2} \leq \varepsilon.$$

Now applying a fact about strong convergence in H_1, see Theorem 2.6.4, we see that $\{f^{(n)}\}$ converges to 0 in H_1 because of the convergence of $\{f^{(n)}\}$ in $\sigma(H_1, BMO)$ and in L^1. The assertion $H_1 = L^1$ has been proved. The contradiction gives the proof of the assertion that L^∞ is not dense in BMO.

Now suppose that L^∞ was closed in BMO. This implies that the identity embedding I from L^∞ to BMO has a closed region. From the well known Inverse Operator Theorem, we see that I^{-1} is bounded from $BMO \bigcap L^\infty$ to L^∞. In particular

$$\|f\|_\infty \leq C\|f\|_*, \quad \forall f \in B \cap L^\infty.$$

So, $\|\cdot\|_\infty$ and $\|\cdot\|_*$ are equivalent on $B \bigcap L^\infty$. Applying Corollary 4.6.3, we get

$$\|f\|_{H_1} \approx |||f|||_1 = \sup_{g \in B \cap L^\infty, \|g\|_* \leq 1} |E(fg)| \approx \sup_{g \in B \cap L^\infty, \|g\|_\infty \leq 1} |E(fg)| \approx \|f\|_1.$$

This says that $H_1 = L^1$. This contradiction proves that L^∞ is not closed in BMO. The proof is finished. □

Now we want to know how to measure the distance of a $f \in BMO$ to L^∞ quantitatively. One half of the result depends on a regularity assumption on $(\Omega, \mathcal{F}, \mu, \{\mathcal{F}_n\}_{n \geq 0})$. This reguarity is just one will be treated in Chapter 7. That is

$$E(f \mid \mathcal{F}_n) \leq dE(f \mid f_{n-1}), \quad n = 1, 2, \cdots, \ \forall f \in L^1_+. \tag{4.6.2}$$

What we want to use to measure the distance of $f \in BMO$ to L^∞ is the following quantity

$$\alpha_f = \sup\{\alpha : \sup_n \|E(\exp(\alpha|f - f_n|) \mid \mathcal{F}_n)\|_\infty \leq K_\alpha < \infty\}. \tag{4.6.3}$$

Intuitively, the larger α_f is, the nearer f is from L^∞. This suggests that α_f might be a suitable quantity to measure the distance of f to L^∞. The main step to approach the problem had been done in BMO's decomposition in §4.3. There we have got the decomposition $f = \sum_i f_{T_i}^{(i)} \chi_{\{T_i<\infty\}} + \sum_i f_{T_i}^{(i)} \chi_{\{T_i=\infty\}}$. Basing on it. we want to separate a L^∞-part from $\sum_i f_{T_i}^{(i)} \chi_{\{T_i<\infty\}}$, such that the remained part has it's BMO norm dominated by $\frac{c}{\alpha_f}$. To do that, we formulate two lemmas at first.

Lemma 4.6.5 *Assume that* (4.6.2) *holds. Let* R, S *be two stopping times such that* $R \leq S$, *and*

$$E(\chi_{\{S<\infty\}} \mid \mathcal{F}_R) \leq \gamma^m, \quad 0<\gamma<1, \ m \in \mathbf{Z}^+. \tag{4.6.4}$$

Then there exists a sequence $\{T_j\}_{j=0}^m$ *of stopping times satisfying* $R = T_0 \leq T_1 \leq \cdots \leq T_m = S$, *and*

$$E(\chi_{\{T_{j+1}<\infty\}} \mid \mathcal{F}_{T_j}) \leq d\gamma, \quad j=0,\cdots,m-1. \tag{4.6.5}$$

Proof. Set $f = \chi_{\{S<\infty\}}$, $f_n = E(f \mid \mathcal{F}_n)$ For $j = 0,\cdots,m-1$, define $T_j = \inf\{n : f_n > \gamma^{m-j}\}$, $T_0 = R$, $T_m = S$. Then $\{T_j\}_1^{m-1}$ is an increasing sequence of stopping times. Obviously we also have $T_0 < T_1, T_{m-1} \leq T_m$. In fact, since $f_R = E(\chi_{\{S<\infty\}} \mid \mathcal{F}_R) \leq \gamma^m < \gamma^{m-1}$, $T_1 > R$ follows. Meanwhile, on the set $\{S<\infty\}$, $f_S = \chi_{\{S<\infty\}} = 1 > \gamma^{m-(m-1)}$, so $T_{m-1} \leq S$. Notice that $T_j \neq 0$, a.e., $\forall j = 1,\cdots,m-1$, this follows from $R < T_1 \leq T_j$. So, $f_{T_j-1} \leq \gamma^{m-j}, j = 1,\cdots,m-1$. Now we prove that $\{T_j\}_{j=0}^m$ is $d\gamma$-graded. We have

$$\gamma^{m-1} E(\chi_{\{T_1<\infty\}} \mid \mathcal{F}_R) \leq E(f_{T_1} \mid \mathcal{F}_R) = f_R \leq \gamma^m,$$

$$E(\chi_{\{T_1<\infty\}} \mid \mathcal{F}_R) \leq \gamma.$$

And for $j = 1,\cdots,m-1$, we have

$$\gamma^{m-(j+1)} E(\chi_{\{T_{j+1}<\infty\}} \mid \mathcal{F}_{T_j}) \leq E(f_{T_{j+1}} \mid \mathcal{F}_{T_j}) = f_{T_j} \leq d f_{T_j-1} \leq d\gamma^{m-j}$$

$$E(\chi_{\{T_{j+1}<\infty\}} \mid \mathcal{F}_{T_j}) \leq d\gamma.$$

This ends the proof of the lemma. □

Lemma 4.6.6 *Assume that* (4.6.2) *holds. Let* $\{T_i\}_{i=1}^\infty$ *be a* γ^m*-graded sequence of stopping times,* $0<\gamma<1$, $m \in \mathbf{Z}^+$, *and* $d\gamma < 1$. *Let* $\{b_i(\omega)\}_{i\geq 1}$ *be a sequence such that each* b_i *is* f_{T_i}*-measurable and* $|b_i(\omega)| \leq B$. *Then there exist a* $d\gamma$*-graded sequence* $\{S_j\}_{j\geq 0}$ *of stopping times and a sequence* $\{c_j\}_{j\geq 0}$ *with* c_j *being* $\mathcal{F}_{S_j}$*-measurable and* $|c_j| \leq B$, *such that*

$$\left\| \sum_{i=1}^\infty b_i \chi_{\{T_j<\infty\}} - \frac{1}{m} \sum_{j=0}^\infty c_j \chi_{\{S_j<\infty\}} \right\|_\infty \leq B, \tag{4.6.6}$$

$$\left\| \frac{1}{m} \sum_{j=0}^\infty c_j \chi_{\{S_j<\infty\}} \right\|_* \leq \frac{1}{m} \frac{2B}{1-d\gamma}. \tag{4.6.7}$$

Proof. According to Lemma 4.6.5, there exists $\{S_{i,j}\}$ such that $T_i = S_{i,0} \le S_{i,1} \le \cdots \le S_{i,m} = T_{i+1}$, and

$$E(\chi_{\{S_{i,j+1}<\infty\}} \mid \mathcal{F}_{S_{i,j}}) \le d\gamma, \quad \forall i, j.$$

Rearrange $\{S_{i,j}\}$ as $\{S_j\}_{j=0}^{\infty}$ such that

$$T_i = S_{(i-1)m} \le \cdots \le S_{(i-1)m+m-1} \le S_{im} = T_{i+1}, \quad i = 1, 2, \cdots.$$

Define

$$c_j = b_i, \quad \text{for } (i-1)m \le j < im, \ i = 1, 2, \cdots.$$

Notice that c_j is measurable with respect to $\mathcal{F}_{T_i} = \mathcal{F}_{S_{(i-1)m}} \subset \mathcal{F}_{S_j}$, and $|c_j| \le B$. Now prove (4.6.6) and (4.6.7). Denote

$$X_\infty^{(k)} = \{\omega : T_1 < \infty, \cdots, T_k < \infty, T_{k+1} = \infty\},$$

$$X_\infty^{(0)} = \{\omega : T_1 = \infty\}, \ X_\infty^{(\infty)} = \{T_k < \infty, \ \forall\, k\}.$$

Notice that $|X_\infty^{(\infty)}| = 0$, and $X_\infty^{(0)}$ is outside of our consideration. We have

$$\varphi(\omega) = \sum_{i=1}^{\infty} b_i \chi_{\{T_i<\infty\}} = \sum_{k=1}^{\infty}\sum_{i=1}^{k} b_i \chi_{X_\infty^{(k)}}.$$

Assume $\omega \in X_\infty^{(k)}$. Then $S_{(k-1)m} < \infty$, $S_{km} = \infty$. Assume for some $j_0 < m$, we have $S_{(k-1)m+j} < \infty$, $\forall j = 1, \cdots, j_0$, but $S_{(k-1)m+j_0+1} = \infty$. Then we have

$$\frac{1}{m}\theta(\omega) = \frac{1}{m}\sum_{j=0}^{\infty} c_j \chi_{(S_j<\infty)} = \frac{1}{m}\sum_{j=0}^{km-1} c_j + \frac{j_0 - m}{m} b_k = \sum_{i=1}^{k} b_i + \frac{j_0 - m}{m} b_k.$$

So we get

$$\left|\varphi(\omega) - \frac{1}{m}\theta(\omega)\right| \le \left|\frac{j_0 - m}{m} b_k\right| \le B.$$

And by means of Lemma 4.3.8., we get

$$\left\|\frac{1}{m}\theta\right\|_* \le \frac{1}{m}\frac{2B}{1 - d\gamma}.$$

The proof is finished. □

We are in the position to prove the distance theorem.

Theorem 4.6.7 *Let $f \in BMO$. Then with the BMO_1 norm we have*

$$\operatorname{dist}(f, L^\infty) \ge \frac{1}{e\alpha_f}. \tag{4.6.8}$$

Furthermore, when (4.6.2) *holds, we have*

$$\operatorname{dist}(f, L^\infty) \le \frac{C_d}{\alpha_f}. \tag{4.6.9}$$

Proof. We prove (4.6.8) by contradiction. Suppose that for some $f \in BMO$, we would have $\text{dist}(f, L^\infty) \leq \frac{C_f}{\alpha_f}$, $C_f < \frac{1}{e}$. Then there would be $g \in L^\infty$ such that

$$\|h\|_{BMO_1} = \|f - g\|_{BMO_1} \leq \frac{C_f + \varepsilon}{\alpha_f} < \frac{1}{e\alpha_f}.$$

Let $\alpha > \alpha_f$ be such that $(C_f + \varepsilon)\frac{\alpha}{\alpha_f} < \frac{1}{e}$. Then we would have

$$\|\alpha h\|_{BMO_1} \leq (C_f + \varepsilon)\frac{\alpha}{\alpha_f} < \frac{1}{e}.$$

And so from Theorem 4.1.2, we would have

$$E(\exp(\alpha|f - f_n|) \mid \mathcal{F}_n) \leq \exp 2\alpha\|g\|_\infty E(\exp(\alpha|h - h_n|) \mid \mathcal{F}_n) \leq K_\alpha < \infty.$$

This would contradict to the definition of α_f. (4.6.8) is thus proved.

Now let $f \in BMO$, and $\alpha < \alpha_f$. Then $E(\exp(\alpha|f - f_n|) \mid \mathcal{F}_n) \leq K_\alpha < \infty$. We have got f's following decomposition

$$f = \varphi + \psi + E_0(f), \quad \varphi = \sum_{i=1}^\infty f_{T_i}^{(i)} \chi_{\{T_i < \infty\}}, \quad \psi = \sum_{i=1}^\infty f_{T_i}^{(i)} \chi_{\{T_i = \infty\}},$$

with $\{T_i\}_1^\infty$ a $K_\alpha e^{-\alpha\lambda}$-graded sequence of stopping times, and $\{b_i\}_1^\infty = \{f_{T_i}^{(i)}\}_1^\infty$ a sequence associated to $\{T_i\}_1^\infty$ in the sense: b_i is $\mathcal{F}_{T_i}$-measurable and $|b_i| \leq B = \lambda + \|f\|_*$, λ determined suitably. Now assume that $\varepsilon > 0$, $\delta > 0$ are arbitrary and $\lambda \geq \lambda_0$ is large enough such that

$$\|f\|_* \leq \varepsilon\lambda, \quad K_\alpha e^{-\alpha\lambda} \leq e^{-(1-\delta)\alpha\lambda}.$$

For any $\beta > \frac{\log d}{1-\delta}$, denote $\gamma = e^{-(1-\delta)\beta}$. Notice that $d\gamma < 1$. Assume $\lambda \geq \lambda_0$ being chosen such that $\frac{\alpha\lambda}{\beta} = m \in \mathbf{Z}^+$ as well. Then $\{T_i\}_1^\infty$ is a γ^m-graded sequence. Thus we get a pair $(\{T_i\}_1^\infty, \{b_i\}_1^\infty)$ satisfying the conditions in Lemma 4.6.6, and hence by that lemma, we get

$$\theta = \sum_{j=0}^\infty c_j \chi_{\{S_j < \infty\}} \in BMO,$$

and f's decomposition

$$f = E_0(f) + \psi + \varphi - \frac{1}{m}\theta + \frac{1}{m}\theta = g + \frac{1}{m}\theta.$$

Noticing

$$\|g\|_\infty \leq \|E_0(f)\|_\infty + \lambda + (1 + \varepsilon)\lambda,$$

and

$$\left\|\frac{1}{m}\theta\right\|_* \leq \frac{(1+\varepsilon)\lambda}{m}\frac{2}{1 - de^{-(1-\delta)\beta}} = \frac{2(1+\varepsilon)\beta}{1 - de^{-(1-\delta)\beta}}\frac{1}{\alpha},$$

we get

$$\operatorname{dist}(f, L^\infty) \le \inf_{\varepsilon,\delta,\alpha,\beta} \frac{2(1+\varepsilon)\beta}{\alpha(1-de^{-(1-\delta)\beta})} = \inf_{\beta>\log d} \frac{2\beta}{1-de^{-\beta}}\frac{1}{\alpha_f} = C_d\frac{1}{\alpha_f}.$$

This proves (4.6.9). The theorem is proved. □

Remark Roughly speaking, C_d in (4.6.9) is of order $\log d$ for large d. For example, let $de^{-\beta}$ be like $\frac{1}{2}$, then β is like $\log d$, so C_d is like $\log d$ too. We could not expect that C_d can be replaced by some C independent of d as shown by following example. Consider $(\Omega, \mathcal{F}, \mu, \{\mathcal{F}_n\}_{n\ge 0})$ with $\Omega = \mathbf{Z}^+$, $\mathcal{F} = \{\text{all subsets of } \Omega\}$, $\mu(\{n\}) = (1-p)p^n$, and $\mathcal{F}_n$ generated by atoms $\{0\}, \{1\}, \cdots, \{n-1\}, \{n, n+1, \cdots\}$. Notice that each $\mathcal{F}_n$-atom either is a $\mathcal{F}_{n-1}$-atom, or comes from spliting a $\mathcal{F}_{n-1}$-atom in two parts the measures of which are $(1-p)p^n$, p^{n+1} respectively. It is easy to show that (4.6.2) holds with $d = \frac{1}{p}$. Now consider a martingale $f = (f_n)_{n\ge 0}$, with $f(\omega) = \omega$. Notice that

$$f_n(\omega) = \begin{cases} \omega, & \omega \le n-1, \\ c, & \omega \ge n, \end{cases} \qquad \forall n,$$

with c satisfying $c\sum_{k=n}^{\infty}(1-p)p^k = \sum_{k=n}^{\infty} k(1-p)p^k$, $c = n + \frac{p}{1-p}$, and hence

$$\Delta_n f(\omega) = \begin{cases} 0, & \omega \le n-2, \\ -\frac{p}{1-p}, & \omega = n-1, \\ 1, & \omega \ge n, \end{cases} \qquad \forall n \ge 1,\ \Delta_0 f = \frac{p}{1-p}.$$

Thus, for $\alpha > 0$, we have

$$E(\exp(\alpha|f-f_n|) \mid \mathcal{F}_n) = \begin{cases} 1, & \omega \le n-1, \\ \sum\limits_{k=n}^{\infty} e^{\alpha|k-n-\frac{p}{1-p}|}(1-p)p^k \le & \\ p^n \sum\limits_{k=0}^{\infty} e^{\frac{\alpha p}{1-p}}(1-p)(e^\alpha p)^k \le C < \infty, & \omega \ge n, \end{cases}$$

proved $\alpha < \log\frac{1}{p}$. This means $f \in BMO$, and $\alpha_f = \log\frac{1}{p}$. We claim that $\operatorname{dist}(f, L^\infty) \ge 1$. In fact, suppose that there would be some $g \in L^\infty$ such that $\|f-g\|_{BMO_1} \le 1-\varepsilon$, then we would have $|\Delta_n(f-g)| \le 1-\varepsilon$. But $\Delta_n f = 1$ on $\{\omega \ge n\}$, so, $\Delta_n g \ge \varepsilon$ on $\{\omega \ge n\}$, and hence $g - g_0 \ge n\varepsilon$ on $\{\omega \ge n\}$. This contradicts to $g \in L^\infty$. Thus we get

$$\operatorname{dist}(f, L^\infty) \ge 1 = \frac{\alpha_f}{\alpha_f},$$

so C_d in (4.6.9) must be such that

$$C_d \ge \alpha_f = \log\frac{1}{p} \approx \log d.$$

This proves the assertion. The same example shows as well that H_∞^S is not dense in BMO in general. In fact, for the $f \in BMO$ constructed as above, any g which makes $\|f - g\|_{BMO_1} \leq 1 - \varepsilon$, we have

$$\Delta_n g \geq \varepsilon \text{ on } \{\omega \geq n\}, \quad S_n(g) \geq \sqrt{n}\varepsilon \text{ on } \{\omega \geq n\},$$

and hence, $g \notin H_\infty^S$. This gives the assertion.

Corollary 4.6.8 *Let $\overline{L}^\infty$ be the closure of L^∞ in BMO. Then $\overline{L}^\infty \subset \{f \in BMO : \alpha_f = \infty\}$. And under the condition (4.6.2), we have the inverse inclusion, and hence $\overline{L}^\infty = \{f : \alpha_f = \infty\}$.*

Proof. These two assertions follow from (4.6.8) and (4.6.9) respectively. □

Notes to Chapter 4

§4.1. Lemma 4.1.1 is a version of Garsia-Neveu's convexity lemma, the proof of which in the book is taken from Meyer [1], but the idea is due to Stroock [1]. In the BMO case, we face a martingale $f = (f_n)_{n \geq 0}$ satisfying $E(|f - f_{n-1}| \mid \mathcal{F}_n) \leq C$, which is neither nonnegative, nor increasing, so the stopping time argument which has been effective in the convexity lemma case would to be modified slightly. Introducing two stopping times is enough to deal with this kind of problems, as we have seen in the proof of Lemma 4.1.1, or in Chapter 2. Theorem 4.1.2 is a martingale version of John-Nirenberg's theorem the statement of which is as follows. Let $f(x) \in L^1_{loc}(\mathbf{R}^n)$. f is said to be in BMO, if $\|f\|_* = \sup\limits_I \frac{1}{|I|} \int_I |f - f_I| dx < \infty$, where I is a cube with its edges parallel to the axes, and $f_I = \frac{1}{|I|} \int_I f(x) dx$. J-N's theorem says that for all $f \in BMO$, for all I, we have

$$|\{x \in I : |f(x) - f_I| > \lambda\}| \leq B|I| e^{-\frac{b}{\|f\|_*}\lambda}, \quad \forall \lambda > 0,$$

with B, $b > 0$ independent of f and I. It could be improved slightly by replacing $|f - f_I|$ by $M(f - f_I)$, where M is classical Hardy-Littlewood's maximal operator. Theorems 4.1.3, 4.1.4 are taken from Garsia [1].

§4.2. Theorem 4.2.1, and Corollaries 4.2.2, 4.2.3, 4.2.4 are well known in the classical case. Theorem 4.2.5 is due to Meyer [1]. Theorem 4.2.6, Lemma 4.2.7 and Theorem 4.2.8 are taken from Garsia [1], but the idea of Theorem 4.2.8 is due to C. Herz. In Meyer [2], there has been another way to formulate Theorem 4.2.8 and Corollary 4.2.10 as follow. (According to what Meyer said, the idea to do this is due to C. Herz and D. Lepingle.) Let $M = (M_t)_{t \geq 0} \in BMO$, $E(|M_\infty - M_t| \mid \mathcal{F}_t) \leq 1$, $|\Delta M| \leq a$. Choose $c > 1$. Define a sequence of stopping times

$$T_{n+1} = \inf\{t > T_n : |M_t - M_{T_n}| > c\}, \quad n \geq 0,\ T_{-1} = 0,$$

and

$$H_n = E(\chi_{\{T_{n+1} < \infty\}} | \mathcal{F}_{T_n}), \quad n \geq 0,$$

$$K_n = (1 - H_n)^{-1}\chi_{\{T_n<\infty, T_{n+1}=\infty\}}(M_{T_n} - M_{T_{n-1}}), \quad n \geq 0.$$

Notice that for any $\omega \in \Omega$, only one n makes $K_n(\omega) \neq 0$, that is the last n such that $T_n(\omega) < \infty$, and that $|M_{T_n-1} - M_{T_{n-1}}| \leq c$, and so $|M_{T_n} - M_{T_{n-1}}| \leq c + a$. As regards to H_n, we have

$$c\chi_{\{T_{n+1}<\infty\}} \leq |M_{T_{n+1}} - M_{T_n}|,$$

$$\begin{aligned} cE(\chi_{\{T_{n+1}<\infty\}}|\mathcal{F}_{T_n}) &\leq E(|M_{T_{n+1}} - M_{T_n}| \mid \mathcal{F}_{T_n}) \\ &= E(|E(M_\infty - M_{T_n} \mid \mathcal{F}_{T_{n+1}}) \mid \mathcal{F}_{T_n}) \\ &\leq E(|M_\infty - M_{T_n}| \mid \mathcal{F}_{T_n}) \leq 1. \end{aligned}$$

So, $H_n \leq \frac{1}{c}$, and hence

$$|K_n| \leq \frac{c(c+a)}{c-1}.$$

Thus, $U = \{U_t\}_{t\geq 0}$, with $U_t = \sum_0^\infty K_n\chi_{\{t\geq T_n\}}$, is a nonadapted process satisfying $\int_{0-}^\infty |dU_t| \leq \frac{c(c+a)}{c-1}$. Since

$$\begin{aligned} E(K_n \mid \mathcal{F}_{T_n}) &= E(\chi_{\{T_n<\infty, T_{n+1}=\infty\}}|\mathcal{F}_{T_n})(1 - H_n)^{-1}(M_{T_n} - M_{T_{n-1}}) \\ &= (M_{T_n} - M_{T_{n-1}})\chi_{\{T_n<\infty\}}, \end{aligned}$$

the adapted process

$$\overline{U} = (\overline{U}_t)_{t\geq 0}, \ \overline{U}_t = \sum_0^\infty E(K_n \mid \mathcal{F}_{T_n})\chi_{\{t\geq T_n\}}$$

satisfies $\overline{U}_\infty = M_L$, with $L =$ last T_n such that $T_n < \infty$. Now setting

$$\overline{V} = (\overline{V}_t)_{0\leq t\leq\infty}, \quad \text{with } \overline{V}_t = \overline{U}_t, \ t < \infty,$$

and

$$\overline{V}_\infty = \overline{U}_\infty + M_\infty - M_L = M_\infty,$$

then we get following expression of $M \in BMO$

$$M_\infty = M_\infty - M_L + \overline{U}_\infty = M_\infty - M_L + \textstyle\sum_0^\infty E(K_n \mid \mathcal{F}_{T_n}). \qquad (*)$$

Noticing $|M_\infty - M_L| \leq c$ (because $L =$ last T_n such that $T_n < \infty$, we have $T_{n+1} = \infty$ and $|M_\infty - M_L| = |M_\infty - M_{T_n}| \leq c$), and $\|\sum_0^\infty |K_n|\|_\infty \leq \frac{c(c+a)}{c-1}$, we see that $(*)$ is just a continuous version of (4.2.15). Basing on $(*)$, a continuous version of (4.2.19) has been formulated in Meyer [2]. In Meyer [2] and in Dellacherie-Meyer-Yor [1], there have been some applications of this formula about H_1-BMO duality. Theorem 4.2.9 is taken from Garsia [1].

§4.3. The classical BLO was introduced by Coifman-Rochberg [1] : $f \in L^1_{\text{loc}}$ is said to be in BLO, if $f_I \leq \inf_{x\in I} f(x) + C$, for all I. The relation between $\log A_{\alpha,\beta}$ and

A_p weights was well known. The concept of γ-graded sequence is due to Varopoulos [1]. For Lemma 4.3.8, he estimated φ's BMO norm in the case $b_k = 1$, for all k, and Long [3] did such estimates for generalized γ-graded function. Theorem 4.3.9 is essentially due to Varopoulos [1]. Lemma 4.3.10 and Theorem 4.3.11 are due to Long-Peng [1]. As regards to Theorem 4.3.12, Carleson [1] established it in classical case. It says that $f \in BMO(\mathbf{R}^n)$, if and only if (for $\varphi(x)$ fixed suitably, and $\varphi_t(x) = t^{-n}\varphi(t^{-1}x)$.)

$$f(x) = g(x) + \int_{R^{n+1}} \varphi_t(x-y)d\mu(y,t),$$

where $g \in L^\infty$, and μ is a Carleson measure, and

$$\|g\|_\infty + |||\mu||| \approx \|f\|_*.$$

The so-called Carleson measure is a nonnegative measure on $\mathbf{R}^{n+1}_+$ satisfying

$$|||\mu||| = \sup_I \frac{\hat{I}_\mu}{|I|} < \infty, \quad \forall I \subset \mathbf{R}^n,$$

where $\hat{I}$ is the tent based on I, the definition of which is

$$\widehat{I} = \{(y,t) \in \mathbf{R}^{n+1}_+ : y \in I,\ 0 < t \le l(I)\},$$

or more standard, for ball B replacing I,

$$\widehat{B} = \{(y,t) \in \mathbf{R}^{n+1}_+ : B(y,t) \subset B\},$$

where $B(y,t)$ is the ball with its center y and radius t. This decomposition is called Carleson's decomposition. In the dyadic case, this decomposition is reduced as follows (see Garnett-Jones [2]): Let $f \in BMO_d$, and Q_0 be any dyadic cube. Then there exist a sequence $\{Q_k\}$ of dyadic cubes $Q_k \subset Q_0$, and a sequence $\{c_k\}$ of complex numbers such that

$$f(x) = f_{Q_0} + \sum_1^\infty c_k \chi_{Q_k} + g, \quad \|g\|_\infty \le C\|f\|_*,$$

$$\sum_{k:Q_k \subset Q_0} |c_k||Q_k| \le C\|f\|_*|Q|, \quad \forall \text{ dyadic } Q.$$

Chao [2] got such decompositions for q-regular martingales. Here it is established in the atom case without any regularity.

§4.4. The classical Carleson measure was introduced by L. Carleson as we just showed. The Carleson's inequality is also due to Carleson. Let $u(x,t)$ be a harmonic function on $\mathbf{R}^{n+1}_+$, and $u^*(x) = \sup_{(y,t)\in\Gamma(x)} |u(y,t)|$, where $\Gamma(x)$ is the cone vertexed at x. Then the Carleson's inequality reads

$$\int_{R^{n+1}_+} |u(y,t)|^p d\nu \le C|||\nu||| \int_{R^n} u^{*p}(x)dx, \quad 0 < p < \infty,$$

provided ν is a Carleson measure. As regards to the relation between Carleson measure and BMO function, it was Fefferman-Stein [1] who discovered that $\varphi \in BMO$ if and only if $|\Delta\Phi(x,t)|^2 tdxdt$ is Carleson measure, where $\Phi(x,t)$ is the Possion extension of φ. In martingale setting, under the correspondences between harmonic functions and martingales, derivative and martingale differences, tents and $\{(\omega, k) : k \geq \tau(\omega), \tau(\omega) < \infty\}$, integral averages and conditional expectations, etc., we could consider all these problems concerning Carleson measure and BMO in martingale setting, like what we have done in this section. Notice that owing to the two possibilities of correspondence between average and expectation (that is the correspondence between f_I and f_n or f_I and f_{n-1}), we get bmo$_2$ version of BMO theory in this topic.

§4.5. The commutator between a multiplication operator B defined by a $b \in BMO$, and an operator T of a kind of maximal operator was discussed first in classical case by Coifman-Rochberg-Weiss [1]. They defined two operators

$$M(b,f)(x) = \sup_{I \ni x} |(b(x) - b_I) f_I|, \quad N(b,f)(x) = \sup_{I \ni x} |(bf)_I - b(x) f_I|,$$

and proved that for $1 < p < \infty$, $M(b,\cdot)$ is L^p-bounded if and only if $b \in BMO$, and that $b \in BMO$ implies N's L^p-boundedness. In martingale setting, Meyer [3] considered the corresponding problems as shown in Lemma 4.5.1, and Theorem 4.5.2. The result stated in Lemma 4.5.1 is slightly stronger than the classical one since the operator in Lemma 4.5.1 is larger than $M(b,f)$. In the classical case, the commutator $[B,T]$ with T a singular integral operator has been investigated widely, by A.P. Calderón, R. Coifman and Y. Meyer $\cdots$. In the martingale case, first Meyer [3], then Lepingle [1], studied the same problems with T being stochastic integrals which are defined by predictable process bounded by 1. Here we consider T as a martingale transform with its multiplier $v = (v_n)_{n\geq 0}$ satisfying $Mv \in L^p, 0 < p \leq \infty$. Theorem 4.5.4 and Corollary 4.5.5 are parallel to those in Lepingle [1].

§4.6. Theorem 4.6.4 and its proof are due to Dellacherie-Mayer-Yor [1], there they used the formula of H_1-BMO duality in Meyer [2] to do this, here we used Theorem 4.2.8 and Corollary 4.2.10 instead. In the classical case, to determine $\mathrm{dist}(f, L^\infty)$ quantitatively for $f \in BMO$ was considered first by Carnett-Jones [1]. In martingale setting, under the "continuous path" hypothesis, it was Varopoulos [1] who established the same result. Emey [1] showed by examples that if we want to get an estimate $\frac{c_1}{\alpha_f} \leq \mathrm{dist}(f, L^\infty) \leq \frac{c_2}{\alpha_f}$ with universal constants, then the "continuous path" hypothesis could not be taken off. The example in the remark after Theorem 4.6.7 is taken from Emey [1], but it is due to I.V. Pavlov.

5 Martingale Transforms

Let $(\Omega, \mathcal{F}, \mu, \{\mathcal{F}_n\}_{n\geq 0})$ be a probability space with a sequence $\{\mathcal{F}_n\}_{n\geq 0}$ of sub-σ-fields satisfying usual conditions as before. Let $f\cdot = (f_n)_{n\geq 0}$ be a martingale and $v = (v_n)_{n\geq 0}$ be an adapted process. For the simplicity, assume $f_0 = 0$ in general in this chapter, and as usual a_{-1} is meant as zero for any process $(a_n)_{n\geq 0}$. The following transform

$$g_n = \sum_{k=1}^{n} v_{k-1}\Delta_k f, \quad n \geq 1, \quad g_0 = 0,$$

where $f = (f_n)_{n\geq 0}$ is a martingale or submartingale, is called a martingale transform. When $g_n \in L^1$, for all n, $g = (g_n)_{n\geq 0}$ is also a martingale provided f is; or a submartingale provided f is and v is nonnegative. Such transforms, particularly in the case in which v may take only 0 and 1 as possible values, have a long history and an interesting gambling interpretation. And in preceding chapters, we have already met a few examples of such transforms such as $\sum_0^\infty \pm\Delta_n f$ and stopped martingales $\sum_0^\infty \chi_{\{n\leq\tau\}}\Delta_n f$ with τ any stopping time. In this chapter, we want to study such transforms in more detail. We will see in §5.1 that such transforms may be used in many problems as a powerful technique, and see that they have very good boundedness behavior on various spaces of martingales, especially on BMO in §5.2; in §5.3 we will introduce some related transforms and combine them to martingale transforms to get some further results; in §5.4. we will give some necessary conditions for the given boundedness of such transforms. §5.5 will be devoted to Burkholder's famous works on the best constants in martingale transforms.

5.1 Martingale transforms as a technique

First, we want to know under what conditions a martingale transform converges pointwise. Here are some mild conditions.

Theorem 5.1.1 *Let $f = (f_n)_{n\geq 0}(f_0 = 0)$ be a L^1-bounded martingale (in symbols $f \in L^1$ as before) and g be f's martingale transform with the multiplier $v = (v_n)_{n\geq 0}$*

$$g_n = \sum_1^n v_{k-1}\Delta_k, \quad n \geq 1, \quad g_0 = 0. \tag{5.1.1}$$

Then $g = (g_n)_{n\geq 0}$ converges a.e. on the set $\{Mv(x) = \sup |v_n(x)| < \infty\}$.

Proof. We first consider the case $Mv \leq 1$. When $f \in L^2, g \in L^2$ too, and hence g converges a.e. Suppose that f is a uniformly bounded submartingale. Without loss of generality, we can assume $f \geq 0$. Writing $\Delta_n f$ as d_n, we have

$$E(f_{n-1}d_n) = E(f_{n-1}E(d_n|\mathcal{F}_{n-1})) \geq 0,$$

$$E(f_n^2) \geq E(f_{n-1}^2) + E(d_n^2) \geq \sum_0^n E(d_k^2), \quad \forall n. \tag{5.1.2}$$

Consider the martingale $\tilde{f} = (\tilde{f}_n)_{n\geq 0}$, $\tilde{f}_n = \sum_0^n \tilde{d}_k$ with $\tilde{d}_k = d_k - E(d_k|\mathcal{F}_{k-1})$. Then $\tilde{f} \in L^2$ because of (5.1.2) and

$$E(\tilde{d}_n^2) = E(d_n^2) - E(E(d_n|\mathcal{F}_{n-1})^2) \leq E(d_n^2).$$

And hence both $\sum_0^n E(d_k|\mathcal{F}_{k-1}) = f_n - \tilde{f}_n$ and $\tilde{g}_n = \sum_1^n v_{k-1}\tilde{d}_k$ converge a.e. Finally g converges a.e. owing to

$$g_n = \tilde{g}_n + \sum_1^n v_{k-1}E(d_k|\mathcal{F}_{k-1})$$

and $E(d_k|\mathcal{F}_{k-1}) \geq 0$, $|v_k| \leq 1$. For general $f \in L^1$, assuming $f \geq 0$, considering $-f^{(N)} = (-f_n^{(N)})_{n\geq 0}$ with $f_n^{(N)} = \min(f_n, N)$, we get a sequence of uniformly bounded submartingales. For each $N, f^{(N)}$'s martingale transform $g^{(N)}$ converges a.e. But $g^{(N)} = g$ on the set $\{x : Mf(x) \leq N\}$. From this follows that g converges a.e. by letting $N \to \infty$. Now it remains to get rid of the restriction $Mv \leq 1$. Let $v_n^{(N)} = v_n\chi_{\{|v_n|\leq N\}}$, and consider f's martingale transform $g^{(N)}$ with $v^{(N)}$ replacing v. Then $g^{(N)}$ converges a.e. Since $g^{(N)} = g$ on the set $\{x : Mv(x) \leq N\}$, by making $N \to \infty$, g converges a.e. on $\{x : Mv(x) < \infty\}$. The proof of the theorem is finished. □

Remark The assertion of the theorem remains being true for L^1-bounded submartingales owing to Doob's decomposition.

The preceding convergence theorem about martingale transforms can be used to the convergence problem of martingales.

Theorem 5.1.2 *Let $f = (f_n)_{n\geq 0}$ be a martingale. Then f converges a.e. on the set $\{x : \sigma(f)(x) < \infty\}$.*

Proof. Let $f = (f_n)_{n\geq 0}(f_0 = 0)$ be any martingale, and

$$\sigma_n(f) = \left(\sum_1^n E(|\Delta_k f|^2|\mathcal{F}_{k-1})\right)^{\frac{1}{2}}.$$

We do not need assume $\sigma_n(f) < \infty$, a.e. For $n \geq 1$, set $v_{n-1} = \sigma_n(f)^2 + 1$. Notice that there is an exception to the usual convention about a_{-1} for any process $(a_n)_{n\geq 0}$, here $v_{-1} = 1$. Consider the martingale transform

$$g_n = \sum_1^n v_{k-1}^{-1}\Delta_k f, \quad n \geq 1, \quad g_0 = 0. \tag{5.1.3}$$

Then we have

$$\sigma_n(g)^2 = \sum_1^n v_{k-1}^{-2} E(|\Delta_k f|^2|\mathcal{F}_{k-1}) = \sum_1^n v_{k-1}^{-2}(v_{k-1} - v_{k-2})$$
$$\leq \sum_1^n \int_{v_{k-2}}^{v_{k-1}} \frac{dt}{t^2} \leq \int^{\infty} \frac{dt}{t^2} = 1.$$

So, g as a L^2 martingale converges a.e. But on the set $\{x : \sigma(f) < \infty\}$,

$$f_n = \sum_1^n v_{k-1}\Delta_k g = v_{n-1}g_n + \sum_1^{n-1} g_k(v_{k-1} - v_k).$$

Since on this set, v_k tends to a finite limit monotonously and g_k converges, so f converges a.e. on the set. The proof of the theorem is finished. □

Remark It is well known that for any martingale $f = (f_n)_{n\geq 0}$ with $E(Md) < \infty$, $Md = \sup|d_n|$, $d_n = \Delta_n f$,

$$\{x : f \text{ converges }\} = \{x : Mf < \infty\} = \{x : S(f) < \infty\} \text{ (modulo measure 0)}.$$

The second equality may be seen as follows. For any N, consider the stopping time $\tau = \inf\{n : S_n(f) > N\}$ and the stopped martingale $f^{(\tau)} = (f_{n\wedge\tau})_{n\geq 0}$. Then we have

$$S(f^{(\tau)}) \leq S_{\tau-1}(f) + Md, \quad E(S(f^{(\tau)})) < \infty,$$

and so

$$E(M_\tau f) = E(M(f^{(\tau)}) \leq CE(S(f^{(\tau)})) < \infty, \quad M_\tau f < \infty, \text{ a.e.}$$

Since on the set $\{x : \tau = \infty\}$, $Mf = M_\tau f$, we get

$$\{x : S(f) \leq N\} = \{\tau = \infty\} = \{\tau = \infty\} \bigcap \{M_\tau f < \infty\} \subset \{x : Mf < \infty\}.$$

This proves $\{S(f) < \infty\} \subset \{Mf < \infty\}$. A similar argument gives the inverse inclusion. The first equality follows from a similar stopping time argument. For

$N > 0$, consider the stopped martingale $f^{(\tau)}$ with $\tau = \inf\{n : |f_n| > N\}$. Then $E(|f_n^{(\tau)}|) \le N + E(Md) < \infty$, and $f^{(\tau)}$ converges a.e. But on the set $\{\tau = \infty\} = \{Mf \le N\}$, $f = f^{(\tau)}$, so f converges a.e. on $\{Mf \le N\}$. Making $N \to \infty$, we get $\{Mf < \infty\} \subset \{f \text{ converges}\}$. The inverse inclusion is obvious. This theorem tells us that $\{x : \sigma(f) < \infty\} \subset \{x : f \text{ converges}\}$ without any restriction on $Md = \sup_n |\Delta_n f|$. In addition, in the proof of this theorem we see that any martingale $f = (f_n)_{n\ge 0}$ with $\sigma(f) < \infty$, a.e., is a martingale transform of an L^2-bounded martingale $g = (g_n)_{n\ge 0}$ with the multiplier $v = (v_n)_{n\ge 0}$, $v_n = \sigma_{n+1}(f) + 1$, $n \ge 0$. This result has following extension. For any $0 < p_1, p_2 < \infty$ and $f \in h_{p_1}$ given, there exist $g \in h_{p_2}$ and v such that f is g's martingale transform. And the same assertion holds for the spaces $\mathcal{P}_{p_1}$, $\mathcal{P}_{p_2}$, too.

Theorem 5.1.3 *Let $0 < p_1, p_2 < \infty$ and $f = (f_n)_{n\ge 0} \in h_{p_1}$, $f_0 = 0$, be given. Then there exist $g = (g_n)_{n\ge 0} \in h_{p_2}$ satisfying*

$$C_{p_1,p_2}\|f\|_{h_{p_1}}^{p_1} \le \|g\|_{h_{p_2}}^{p_2} \le C_{p_1,p_2}\|f\|_{h_{p_1}}^{p_1}, \tag{5.1.4}$$

and an adapted nonnegative and increasing process $v = (v_n)_{n\ge 0}$ satisfying

$$E((Mv)^\alpha)(= E(v_\infty^\alpha)) < \infty, \quad \alpha = \frac{p_1 p_2}{p_2 - p_1}, \tag{5.1.5}$$

such that f is g's martingale transform

$$f_n = \sum_1^n v_{k-1}\Delta_k g, \quad n \ge 1, \quad f_0 = 0. \tag{5.1.6}$$

Conversely, in the case $p_1 \le p_2$, each martingale transform (5.1.6) with v an adapted process satisfying (5.1.5) and $g = (g_n)_{n\ge 0}$ satisfying $\|g\|_{h_{p_2}} < \infty$, is in h_{p_1}, and

$$\|f\|_{h_{p_1}} \le \|Mv\|_\alpha \|g\|_{h_{p_2}}. \tag{5.1.7}$$

Proof. For given $f \in h_{p_1}$, define g as follows

$$g_n = \sum_1^n \sigma_k(f)^{-\frac{p_1}{\alpha}} \Delta_k f, \quad n \ge 1, \quad g_0 = 0. \tag{5.1.8}$$

We have

$$\begin{aligned}\sigma_n(g)^2 &= \sum_1^n \sigma_k(f)^{-\frac{2p_1}{\alpha}} E(|\Delta_k f|^2|\mathcal{F}_{k-1}) \\ &= \sum_1^n (\sigma_k(f)^2 - \sigma_{k-1}(f)^2)\sigma_k(f)^{\frac{2p_1}{p_2}-2}\end{aligned}$$

By making use of the elementary inequalities (2.2.1–2.2.2), i.e. $(B \ge A)$

$$B^p - A^p \le (B^2 - A^2)B^{p-2} \le \frac{2}{p}(B^p - A^p), \quad 0 < p \le 2, \tag{5.1.9}$$

$$\frac{2}{p}(B^p - A^p) \le (B^2 - A^2)B^{p-2} \le B^p - A^p, \quad p \ge 2,$$

we get

$$\sigma_n(g)^2 \approx \sum_1^n \left(\sigma_k(f)^{\frac{2p_1}{p_2}} - \sigma_{k-1}(f)^{\frac{2p_1}{p_2}}\right) = \sigma_n(f)^{\frac{2p_1}{p_2}}.$$

This proves (5.1.4). (5.1.6) with $v_k = \sigma_{k+1}(f)^{\frac{p_1}{\alpha}}$ holds obviously, and $E(v_\infty^\alpha) = E\left(\sigma(f)^{p_1}\right) < \infty$ gives (5.1.5).

Now suppose that $p_1 \le p_2$, $v = (v_n)_{n\ge 0}$ is an adapted process and $g \in h_{p_2}$. For f given in (5.1.6), we have

$$\sigma_n(f)^2 = \sum_1^n |v_{k-1}|^2(\sigma_k(g)^2 - \sigma_{k-1}(g)^2) \le (M_{n-1}v)^2\sigma_n(g)^2,$$

and hence

$$\begin{aligned}\|\sigma_n(f)\|_{p_1}^{p_1} &\le E((M_{n-1}v)^{p_1}\sigma_n(g)^{p_1}) \\ &\le E((M_{n-1}v)^\alpha)^{\frac{p_2-p_1}{p_2}} E(\sigma_n(g)^{p_2})^{\frac{p_1}{p_2}}.\end{aligned}$$

This gives (5.1.7). The proof of theorem is thus finished. □

What happens when $p_2 = \infty$? We have following

Theorem 5.1.4 *Let $0 < p_1 < p_2 = \infty$, $f \in h_{p_1}$, $f_0 = 0$. Then there exists an adapted increasing process $v = (v_n)_{n\ge 0}$ satisfying $\|v_\infty\|_{p_1} \le C\|f\|_{h_{p_1}}$, and $g = (g_n)_{n\ge 0} \in bmo_2$ satisfying $\|g\|_{bmo_2} \le 1$, such that*

$$f_n = \sum_1^n v_{k-1}\Delta_k g, \quad n \ge 1, \quad g_0 = 1.$$

Conversely, g's martingale transform f with $v = (v_n)_{n\ge 0}$ an adapted process satisfying $\|Mv\|_{p_1} < \infty$, and $\|g\|_{bmo_2} < \infty$, must be in h_{p_1}, and

$$\|f\|_{h_{p_1}} \le C\|g\|_{bmo_2}\|Mv\|_{p_1}. \tag{5.1.10}$$

Proof. The idea is same as in the case $p_2 < \infty$. We have some modification owing to the fact that (5.1.9) is not useful for $p = 0$. Take $p_0 < \min(p_1, 1)$, define

$$v_n = \max_{m\le n} E(\sigma(f)^{p_0}|\mathcal{F}_m)^{\frac{1}{p_0}}, \quad n \ge 0, \tag{5.1.11}$$

$$g_n = \sum_1^n v_{k-1}^{-1}\Delta_k f, \quad n \ge 1, \quad g_0 = 0.$$

Then we have

$$E(|g_N - g_n|^2|\mathcal{F}_n) = \sum_{n+1}^{N} E(v_{k-1}^{-2}|\Delta_k f|^2|\mathcal{F}_n)$$
$$= \sum_{n+1}^{N} E(v_{k-1}^{-2}(\sigma_k(f)^2 - \sigma_{k-1}(f)^2)|\mathcal{F}_n).$$

By means of the fact

$$1 \le E(\sigma(f)^{p_0}|\mathcal{F}_{k-1})E(\sigma(f)^{-p_0})|\mathcal{F}_{k-1}) \le v_{k-1}^{p_0} E(\sigma(f)^{-p_0}|\mathcal{F}_{k-1}),$$

$$v_{k-1}^{-2} \le E(\sigma(f)^{-p_0}|\mathcal{F}_{k-1})^{\frac{2}{p_0}} \le E(\sigma(f)^{-2}|\mathcal{F}_{k-1}), \quad \forall k, \tag{5.1.12}$$

we get

$$E(|g_N - g_n|^2\mathcal{F}_n) \le \sum_{n+1}^{N} E([\sigma_k(f)^2 - \sigma_{k-1}(f)^2]E(\sigma(f)^{-2}|\mathcal{F}_{k-1})|\mathcal{F}_n)$$
$$= \sum_{n+1}^{N} E(\sigma(f)^{-2}(\sigma_k(f)^2 - \sigma_{k-1}(f)^2)|\mathcal{F}_n) \le 1.$$

This proves $g \in bmo_2$ and $||g||_{bmo_2} \le 1$. In addition, since $p_0 < p_1$, we have

$$||v_\infty||_{p_1} \le E((M(\sigma(f)^{p_0})^{\frac{p_1}{p_0}})^{\frac{1}{p_1}} \le C||\sigma(f)||_{p_1}.$$

This completes the proof of first part of the theorem. As regards the second part of the theorem, we postpone its proof to next section (see Theorem 5.2.4′). The proof is finished. □

Remark As an application of Theorem 5.1.3, we give another proof of the fact "$h_p \subset L_0^p$ for $0 < p \le 2$ and $L_0^p \subset h_p$ for $2 \le p < \infty$". In fact, from

$$g_n = \sum_{1}^{n} \sigma_k(f)^{\frac{p-2}{2}} \Delta_k f, \quad f_n = \sum_{1}^{n} \sigma_k(f)^{\frac{2-p}{2}} \Delta_k g,$$

we can get

$$Mg \le CMf\sigma(f)^{\frac{p-2}{2}}, \quad \text{for } p \ge 2,$$

$$Mf \le CMg\sigma(f)^{\frac{2-p}{2}}, \quad \text{for } p \le 2,$$

and the desired assertion owing to the fact $||\sigma(f)||_p^p \approx ||g||_2^2$.

Now we turn to same representation problem for the spaces $\mathcal{P}_p$.

Theorem 5.1.5 *Let $0 < p_1 < p_2 < \infty$, $\alpha = \frac{p_2 p_1}{p_2 - p_1}$, $f \in \mathcal{P}_1$, and $\lambda_f = (\lambda_n(f))_{n \geq 0}$ be f's optimal majorant. Then $g = (g_n)_{n \geq 0}$ with*

$$g_n = \sum_1^n \lambda_{k-1}(f)^{-\frac{p_1}{\alpha}} \Delta_k f, \quad n \geq 1, \quad g_0 = 0, \tag{5.1.13}$$

satisfies

$$\left(\frac{1}{2}\right)^{p_2} \|f\|_{\mathcal{P}_{p_1}}^{p_1} \leq \|g\|_{\mathcal{P}_{p_2}}^{p_2} \leq \left(\frac{p_2}{p_1}\right)^{p_2} \|f\|_{\mathcal{P}_{p_1}}^{p_1}. \tag{5.1.14}$$

And hence f is a martingale transform of $g \in \mathcal{P}_{p_2}$,

$$f_n = \sum_1^n v_{k-1} \Delta_k g, \quad n \geq 1, \ f_0 = 0, \ v_k = \lambda_k(f)^{\frac{p_1}{\alpha}}, \tag{5.1.15}$$

with $g \in \mathcal{P}_{p_2}$ satisfying (5.1.14) and $E(v_\infty^\alpha) \leq C\|f\|_{\mathcal{P}_{p_1}}^{p_1}$. When $0 < p_2 < p_1 < \infty$, each $f \in \mathcal{P}_{p_1}$ is a martingale transform of $g \in \mathcal{P}_{p_2}$ with a decreasing multiplier $v = (v_k)_{k \geq 0}$, $v_k = \lambda_k(f)^{\frac{p_1}{\alpha}} (\alpha < 0)$, where g and v satisfying

$$\|g\|_{\mathcal{P}_{p_2}}^{p_2} \leq C\|f\|_{\mathcal{P}_{p_1}}^{p_1}, \quad E(v_\infty^\alpha) \leq C\|f\|_{\mathcal{P}_{p_1}}^{p_1}.$$

Conversely, in the case $p_1 \leq p_2$, each martingale transform (5.1.15) with $v = (v_n)_{n \geq 0}$ an adapted process satisfying $\|Mv\|_\alpha < \infty$ and $g \in \mathcal{P}_{p_2}$, must be in $\mathcal{P}_{p_1}$, and

$$\|f\|_{\mathcal{P}_{p_1}} \leq C\|Mv\|_\alpha \|g\|_{\mathcal{P}_{p_2}}. \tag{5.1.16}$$

Proof. Denote $\beta = 1 - \frac{p_1}{p_2} = \frac{p_1}{\alpha}$ in the case $p_1 < p_2$. For g in (5.1.13), we have

$$\begin{aligned}
|g_n| &= |\lambda_{n-1}(f)^{-\beta} f_n + \sum_1^{n-1} f_k(\lambda_{k-1}(f)^{-\beta} - \lambda_k(f)^{-\beta})| \\
&\leq \lambda_{n-1}(f)^{1-\beta} + \sum_1^{n-1} \lambda_{k-1}(f)(\lambda_{k-1}(f)^{-\beta} - \lambda_k(f)^{-\beta}) \\
&= \sum_1^n \lambda_{k-1}(f)^{-\beta}(\lambda_{k-1}(f) - \lambda_{k-2}(f)) \leq \int_0^{\lambda_{n-1}(f)} t^{-\beta} dt \\
&= \frac{p_2}{p_1} \lambda_{n-1}(f)^{\frac{p_1}{p_2}}.
\end{aligned}$$

This proves $g \in \mathcal{P}_{p_2}$, and $\|g\|_{\mathcal{P}_{p_2}}^{p_2} \leq (\frac{p_2}{p_1})^{p_2} \|f\|_{\mathcal{P}_{p_1}}^{p_1}$. And from

$$|f_n| = \left| \sum_1^n \lambda_{k-1}(f)^\beta \Delta_k g \right| \leq 2\lambda_{n-1}(g)\lambda_{n-1}(f)^\beta,$$

we get

$$\|f\|_{\mathcal{P}_{p_1}}^{p_1} \leq 2^{p_1} E(\lambda_\infty(g)^{p_2})^{\frac{p_1}{p_2}} E(\lambda_\infty(f)^{p_1})^{1-\frac{p_1}{p_2}}, \quad \|f\|_{\mathcal{P}_{p_1}}^{p_1} \leq 2^{p_2} \|g\|_{\mathcal{P}_{p_2}}^{p_2}$$

This proves (5.1.14). When $p_2 < p_1$, we have

$$g = \sum_1^\infty \lambda_{k-1}(f)^{\frac{p_1}{p_2}-1}\Delta_k f,$$

$$f = \sum_1^\infty \lambda_{k-1}(f)^{1-\frac{p_1}{p_2}}\Delta_k g = \sum_1^\infty v_{k-1}\Delta_k g.$$

The same argument can give $\|g\|_{\mathcal{P}_{p_2}}^{p_2} \leq C\|f\|_{\mathcal{P}_{p_1}}^{p_1}$ (but not its inverse). Here $(v_k)_{k\geq 0}$ is an adapted, decreasing process satisfying

$$E(v_\infty^\alpha) = E(\lambda_\infty(f)^{p_1}) = \|f\|_{\mathcal{P}_{p_1}}^{p_1}.$$

Conversely, in the case $p_1 < p_2$, we will see in next section (Theorem 5.2.5) that each $f = \sum_1^\infty v_{k-1}\Delta_k g \in \mathcal{P}_{p_1}$ and $\|f\|_{\mathcal{P}_{p_1}} \leq C\|Mv\|_\alpha \|g\|_{\mathcal{P}_{p_2}}$ since $\frac{1}{p_1} = \frac{1}{\alpha} + \frac{1}{p_2}$. The proof of the theorem is finished. □

Now consider the case $p_2 = \infty$.

Theorem 5.1.6 *Let $0 < p_1 < p_2 = \infty$. Then each $f \in \mathcal{P}_{p_1}$ is a martingale transform of $g \in BMO$ with an adapted, nonnegative, increasing process $v = (v_k)_{k\geq 0}$ as the multiplier, and*

$$\|g\|_*(= \|g\|_{BMO}) \leq C, \quad \|v_\infty\|_{p_1} \leq C\|f\|_{\mathcal{P}_{p_1}}. \tag{5.1.17}$$

On the contrary, each martingale transform f of $g \in BMO$ with an adapted process $v = (v_k)_{k\geq 0}$, $\|Mv\|_{p_1} < \infty$, as multiplier, must be in $\mathcal{P}_{p_1}$ and satisfies

$$\|f\|_{\mathcal{P}_{P_1}} \leq C\|g\|_*\|Mv\|_{p_1}. \tag{5.1.18}$$

Proof. Take $p_0 < \min(p_1, 1)$, and define

$$v_k = \sup_{m\leq k} E(\lambda_\infty(f)^{p_0}|\mathcal{F}_m)^{\frac{1}{p_0}}, \quad \forall k, \tag{5.1.19}$$

$$g_n = \sum_1^n v_{k-1}^{-1}\Delta_k f, \quad n \geq 1, \quad g_0 = 0.$$

We have

$$\begin{aligned}
g_N - g_{n-1} &= \sum_n^N v_{k-1}^{-1}\Delta_k f \\
&= v_{N-1}^{-1}(f_N - f_{n-1}) + \sum_n^{N-1}(f_k - f_{n-1})(v_{k-1}^{-1} - v_k^{-1}), \\
|g_N - g_{n-1}| &\leq 2v_{N-1}^{-1}\lambda_{N-1}(f) + 2\sum_n^{N-1}\lambda_{k-1}(f)(v_{k-1}^{-1} - v_k^{-1}) \\
&= 2v_{n-1}^{-1}\lambda_{n-1}(f) + 2\sum_n^{N-1} v_k^{-1}(\lambda_k(f) - \lambda_{k-1}(f)).
\end{aligned}$$

As in the proof of Theorem 5.1.4, we have

$$1 \le E(\lambda_\infty(f)^{p_0}|\mathcal{F}_k)E(\lambda_\infty(f)^{-p_0}|\mathcal{F}_k) \le v_k^{p_0}E(\lambda_\infty(f)^{-p_0}|\mathcal{F}_k),$$

$$v_k^{-1} \le E(\lambda_\infty(f)^{-p_0}|\mathcal{F}_k)^{\frac{1}{p_0}} \le E(\lambda_\infty(f)^{-1}|\mathcal{F}_k), \quad \forall k. \tag{5.1.20}$$

So, we get

$$\begin{aligned} E(|g_N - g_{n-1}|\,|\mathcal{F}_n) &\le 2E(\lambda_{n-1}(f)E(\lambda_\infty(f)^{-1}|\mathcal{F}_{n-1})|\mathcal{F}_n) \\ &\quad +2E\left(\sum_n^{N-1}[\lambda_k(f) - \lambda_{k-1}(f)]E(\lambda_\infty(f)^{-1}|\mathcal{F}_k)\Big|\mathcal{F}_n\right) \\ &\le 2 + 2E\left(\sum_n^{N-1}\lambda_\infty(f)^{-1}(\lambda_k(f) - \lambda_{k-1}(f))\Big|\mathcal{F}_n\right) \le 4. \end{aligned}$$

In addition, we have also $E(v_\infty^{p_1}) \le CE(\lambda_\infty(f)^{p_1})$. This proves the first part of the theorem.

Now suppose that $Mv \in L^p, g \in BMO$. We will see in next section (Theorem 5.2.4) that $\|MT_v(g)\|_{p_1} \le C\|g\|_*\|Mv\|_{p_1}$, $f = T_v g$. But

$$|(T_v g)_n| \le M_{n-1}(T_v g) + M_{n-1}v\|g\|_*,$$

$$\|T_v g\|_{\mathcal{P}_{p_1}} \le C\|\; MT_v g\|_{p_1} + C\|g\|_*\|Mv\|_{p_1} \le C\|g\|_*\|Mv\|_{p_1}.$$

This proves the second part of the theorem. The proof is finished. □

As other applications of martingale transform we have following theorems.

Theorem 5.1.7 *Let $0 < p < \infty$. Then $\mathcal{P}_p \subset h_p$, and the imbedding is continuous.*

Proof. Let $f \in \mathcal{P}_p$, and $\lambda_f = (\lambda_k(f))_{k\ge 0}$ be its optimal majorant. Take $q > \max(2,p)$, and define (denote $1 - \frac{p}{q} = \beta$)

$$g_n = \sum_1^n \lambda_{k-1}(f)^{-\beta}\Delta_k f, \quad n \ge 1, \quad g_0 = 0.$$

As shown in Theorem 5.1.5, $g = (g_n)_{n\ge 0} \in \mathcal{P}_q$, $\|g\|_{\mathcal{P}_q}^q \approx \|f\|_{\mathcal{P}_p}^p$. So, from $f_n = \sum_1^n \lambda_{k-1}(f)^\beta \Delta_k g$, we get

$$\sigma_n(f)^2 = \sum_1^n \lambda_{k-1}(f)^{2\beta}E(|\Delta_k g|^2|\mathcal{F}_{k-1}) \le \lambda_{n-1}(f)^{2\beta}\sigma_n(g)^2,$$

$$\begin{aligned} E(\sigma_n(f)^p) &\le E(\lambda_{n-1}(f)^p)^{1-\frac{p}{q}}E(\sigma_n(g)^q)^{\frac{p}{q}} \\ &\le CE(\lambda_\infty(f)^p)^{1-\frac{p}{q}}E((Mg)^q)^{\frac{p}{q}} \le CE(\lambda_\infty(f)^p). \end{aligned}$$

This proves the theorem. □

Another application is characterizing local martingales, a new concept to which we do not want to pay more attention in this book. $f = (f_n)_{n\geq 0}$ is called a local martingale if there exists an increasing sequence $\{\tau_k\}_{k\geq 1}$ of stopping times such that $\lim_{k\to\infty} \tau_k = \infty$, almost everywhere, and $(f_{n\wedge\tau_k}\chi_{\{\tau_k>0\}})_{n\geq 0}$ are all uniformly integrable martingales. A stopping time τ is called to reduce a process $f = (f_n)_{n\geq 0}$, if $(f_{n\wedge\tau}\chi_{\{\tau>0\}})_{n\geq 0}$ is a uniformly integrable martingale. For a local martingale $f = (f_n)_{n\geq 0}$, each reduction sequence of stopping times in the definition is called f's fundamental sequence. We can characterize local martingales in such a way: each local martingale is a martingale transform. For doing this, we give a lemma about finiteness of generalized conditional expectation introduced in §2.1, at first.

Lemma 5.1.8 *Let $(\Omega, \mathcal{F}, \mu)$ be a probability space, $\mathcal{B} \subset \mathcal{F}$ be a sub-σ-field, and $\mathcal{B}, \mathcal{F}$ be complete. Let f be nonnegative $\mathcal{F}$-measurable, and $g = E(f|\mathcal{B})$ be f's generalized conditional expectation. Then $g < \infty$, a.e. if and only if the measure $\nu : \mathcal{B} \to \overline{\mathbf{R}}^+$ defined by $\nu(A) = \int_A f d\mu$, for all $A \in \mathcal{B}$ is σ-finite.*

Proof. Suppose $g < \infty$, a.e. Denote $\Omega_k = \{x : 2^k \leq g < 2^{k+1}\}$. Then $\{\Omega_k\}_k$ is a $\mathcal{B}$-measurable partition of Ω. Since

$$\nu(\Omega_k) = \int_{\Omega_k} f d\mu = \int_{\Omega_k} g d\mu < 2^{k+1},$$

this proves ν's σ-finiteness. On the contrary, suppose $\Omega = \bigcup \Omega_k$, with $\nu(\Omega_k) < \infty$, for all k. Write $f = \sum_k f_k = \sum_k f\chi_{\Omega_k}$. Then $f_k \in L^1, E(f_k|\mathcal{B}) < \infty$ a.e. But $g\chi_{\Omega_k} = E(f_k|\mathcal{B})$, so $g < \infty$, a.e. The proof of the lemma is finished. □

Theorem 5.1.9 *Let $f = (f_n)_{n\geq 0}$ with $f_0 = 0$ be an adapted process. Then the following two conditions are equivalent*

(a) *f is a local martingale,*

(b) *f is a martingale transform.*

Proof. (a) ⇒ (b). Let $f = (f_n)_{n\geq 0}$, $f_0 = 0$, be a local martingale and $\{\tau_k\}_k$ be a fundamental sequence of f. Then $f_{n\wedge\tau_k}\chi_{\{\tau_k>0\}}$ is integrable for all n, for all k. In particular,

$$E(|f_{n+1}|\chi_{\{\tau_k>n\}}) = E(|f_{(n+1)\wedge\tau_k}|\chi_{\{\tau_k>n\}}) < \infty.$$

So,

$$E(|f_{n+1}||\mathcal{F}_n)\chi_{\{\tau_k>n\}} = E(|f_{n+1}|\chi_{\{\tau_k>n\}}|\mathcal{F}_n) < \infty, \text{ a.e.}$$

Making $k \to \infty$, we see $E(|f_{n+1}||\mathcal{F}_n) < \infty$ a.e. By means of Lemma 5.1.8, the measure $A \to \nu(A) = \int_A |f_{n+1}| d\mu$, for all $A \in \mathcal{F}_n$, is σ-finite. So, in order to prove

$$E(\Delta_{n+1} f|\mathcal{F}_n) = 0, \quad \forall n, \tag{5.1.21}$$

it is enough to prove that for all $A \in \mathcal{F}_n, \nu(A) < \infty$, we have

$$\int_A f_{n+1} d\mu = \int_A f_n d\mu. \tag{5.1.22}$$

Since $(|f_{n\wedge\tau_k}|\chi_{\{\tau_k>0\}})_{n\geq 0}$ is a submartingale, for all $A \in \mathcal{F}_n$, $\nu(A) < \infty$, we have

$$\begin{aligned}\int_{A\cap\{\tau_k>n\}} |f_n| d\mu &= \int_{A\cap\{\tau_k>n\}} |f_{n\wedge\tau_k}|\chi_{\{\tau_k>0\}} d\mu \\ &\leq \int_{A\cap\{\tau_k>n\}} |f_{(n+1)\wedge\tau_k}|\chi_{\{\tau_k>0\}} d\mu \\ &\leq \int_{A\cap\{\tau_k>n\}} |f_{n+1}| d\mu.\end{aligned}$$

Making $k \to \infty$, we get $\int_A |f_n| d\mu \leq \int_A |f_{n+1}| d\mu < \infty$. By Lebesgue's dominated convergence theorem, from (since $(f_{n\wedge\tau_k}\chi_{\{\tau_k>0\}})_{n\geq 0}$ is martingale)

$$\int_{A\cap\{\tau_k>n\}} f_n d\mu = \int_{A\cap\{\tau_k>n\}} f_{n+1} d\mu,$$

we get (5.1.22), and hence (5.1.21).

Now define

$$v_{n-1} = E(|\Delta_n f|\,|\mathcal{F}_{n-1}), \quad n \geq 1, \quad w_n = v_n^{-1}, \quad n \geq 0, \tag{5.1.23}$$

$$g_n = \sum_1^n w_{k-1}\Delta_k f, \quad n \geq 1, \quad g_0 = 0. \tag{5.1.24}$$

Then $v_n < \infty$, a.e. for all n, but not w_n. But $\Delta_{n+1} f = 0$, where $w_n = \infty$, so g_n have meaning for all n. And we have always

$$E(|\Delta_n g|\,|\mathcal{F}_{n-1}) = w_{n-1}E(|\Delta_n f|\,|\mathcal{F}_{n-1}) \leq 1, \quad \forall n,$$
$$E(\Delta_n g|\mathcal{F}_{n-1}) = w_{n-1}E(\Delta_n f|\mathcal{F}_{n-1}) = 0, \quad \forall n.$$

So, $g = (g_n)_{n\geq 0}$ is a martingale, and f is g's martingale transform.

(b)$\Rightarrow$(a). Let f be g's martingale transform with the multiplier v. According to the assumption, $f_0 = g_0 = 0$, for all $k \in \mathbf{Z}^+$, define the stopping time

$$T_k = \inf\{n : |v_n| > k\}, \quad \tau_k = T_k \wedge k.$$

Then $\{\tau_k\}_{k\geq 1}$ is increasing to infinity (otherwise we would have $T_k(x) \leq N$, for all k, for all $x \in E, |E| > 0$, and hence $M_N v = \infty$ on E, a contradiction). Now prove that $(f_{n\wedge\tau_k}\chi_{\{\tau_k>0\}})_{n\geq 0}$ is uniformly integrable martingale, for all k. Notice that

$$f_{n\wedge\tau_k}\chi_{\{\tau_k>0\}} = \sum_{m=1}^n v_{m-1}\chi_{\{m\leq\tau_k\}}\Delta_m g, \quad |v_{m-1}|\chi_{\{m\leq\tau_k\}} \leq k, \quad \forall m \geq 1.$$

This implies that $f_{n\wedge\tau_k}\chi_{\{\tau_k>0\}}$ is uniformly integrable with respect to n owing to $n \wedge \tau_k \leq k$, for all n. In addition, for all m,

$$E(v_{m-1}\chi_{\{m\leq\tau_k\}}\Delta_m g|\mathcal{F}_{m-1}) = v_{m-1}\chi_{\{m\leq\tau_k\}}E(\Delta_m g|\mathcal{F}_{m-1}) = 0.$$

So f is a local martingale. The proof of the theorem is complete. □

5.2 Boundedness of martingale transforms as operators

We introduce a new space as follows. Let $0 < p \leq \infty$, denote the following space of adapted processes by V_p

$$V_p = \{v = (v_n)_{n\geq 0} : \|v\|_{V_p} = \|Mv\|_{p<\infty}\}. \tag{5.2.1}$$

With $v \in V_p$ as a multiplier, we get a martingale transform $T_v : f = (f_n)_{n\geq 0}$, $(f_0 = 0) \to T_v f = (g_n)_{n\geq 0}$,

$$g_n = \sum_1^n v_{k-1}\Delta_k f, \quad n \geq 1, \quad g_0 = 0. \tag{5.2.2}$$

We want to investigate the boundedness properties of T_v acting on various spaces of martingales such as H_p^S, H_p^*, h_p, Λ_α, and λ_α, where H_p^S denotes the Hardy spaces defined by square function, and H_p^* by maximal function. We remind some facts about Lipshitz spaces Λ_α and λ_α, $0 \leq \alpha \leq 1$. Their definition are

$$\Lambda_\alpha = \{f = (f_n)_{n\geq 0} : \|f\|_{\Lambda_\alpha} = \sup_n \|\omega_n^{-\alpha} E(|f - f_{n-1}|^2|\mathcal{F}_n)^{\frac{1}{2}}\|_\infty < \infty\},$$

$$\lambda_\alpha = \{f = (f_n)_{n\geq 0} : \|f\|_{\lambda_\alpha} = \sup_n \|\omega_n^{-\alpha} E(|f - f_n|^2|\mathcal{F}_n)^{\frac{1}{2}}\|_\infty < \infty\},$$

where $\omega_n = \sum |I|\chi_I$ with the summation over all $\mathcal{F}_n$-atoms I. Remember that $\Lambda_0 = BMO$, $\lambda_0 = bmo_2$, and the Banach dual of H_p and h_p are respectively Λ_α and λ_α when $\alpha = \frac{1}{p} - 1 \geq 0$.

For the sake of simplicity, we adopt the concepts of type and weak type concerning operator's boundedness. Let A, B be quasi-normed spaces, and T be a quasi-linear operator defined on A. T is said to be of type (A, B), if T is bounded from A to B (i.e. $\|Tf\|_B \leq C\|f\|_A$, for all $f \in A$); is said to be of weak type (A, B), for $B = L^p$, if T is bounded from A to WL^p (i.e. $|\{|Tf| > \lambda\}| \leq (\frac{C}{\lambda}\|f\|_A)^p$, for all f and $\lambda > 0$); for $B = H_p^*$ or H_p^S, or h_p, if MT, or ST, or σT is bounded from A to WL^p (or say, T is bounded from A to WH_p^*, or WH_p^S, or Wh_p). A linear space is called a quasi-normed space, if it is endowed with a quasi-norm $\|\cdot\|$ which is different from the usual norms by the following generalized trigonometric inequality

$$\|f + g\| \leq C(\|f\| + \|g\|), \quad \forall f, g. \tag{5.2.3}$$

An operator T defined on a linear space and valued in a linear space consisting of complex valued functions is called quasi-linear, if

$$|T(\alpha f)| = |\alpha|\,|Tf|, \quad |T(f+g)| \leq k(|Tf| + |Tg|). \tag{5.2.4}$$

When $k = 1$, it is called sublinear.

We have the following two elementary but important results in this field.

Theorem 5.2.1 *Let $0 < p, q \leq \infty$, $v \in V_p$. Then T_v is of type (H_q^S, H_r^S) and of type (h_q, h_r) with the bound $\|T_v\| \leq \|v\|_{V_p}$ where r satisfies*

$$\frac{1}{r} = \frac{1}{p} + \frac{1}{q}. \tag{5.2.5}$$

Proof. This follows from following pointwise estimates

$$S(T_v f)(x) \le Mv(x)S(f)(x),$$

$$\sigma(T_v f)(x) \le Mv(x)\sigma(f)(x).$$

The theorem is proved. □

Notice that T_v is selfadjoint in the sense, for all f, g good enough

$$E(gT_v f) = \sum_1^\infty E(v_{k-1}\Delta_k f \Delta_k g) = E(fT_v g). \tag{5.2.6}$$

So we can get by duality

Theorem 5.2.2 *Let* $0 \le \alpha \le 1$, $\frac{1}{1+\alpha} < p \le \infty$, $\beta = \alpha - \frac{1}{p} = -\frac{1}{s}$, *and* $v \in V_p$. *Then when* $\frac{1}{p} \le \alpha$, T_v *is of type* $(\Lambda_\alpha, \Lambda_\beta)$ *and* $(\lambda_\alpha, \lambda_\beta)$; *when* $\alpha < \frac{1}{p} < 1+\alpha, T_v$ *is of type* (Λ_α, L^s) *and* (λ_α, h_s). *And in all cases, we have* $||T_v|| \le ||v||_{V_p}$.

Proof. Let $r = \frac{1}{1+\alpha}$. When $\frac{1}{p} \le \alpha$, there is q, $0 < q \le 1$ such that $\frac{1}{p} + \frac{1}{q} = 1 + \alpha = \frac{1}{r}$. From T_v's type (H_q^S, H_r^S) we get its type $(\Lambda_\alpha, \Lambda_\beta)$ with the same bound. And analogously we get T_v's type $(\lambda_\alpha, \lambda_\beta)$. When $\alpha < \frac{1}{p} < 1+\alpha$, there is q, $1 < q < \infty$, such that $\frac{1}{p} + \frac{1}{q} = 1 + \alpha = \frac{1}{r}$. Noticing $1 - \frac{1}{q} = \frac{1}{s}$, by the same duality argument, we get what we want. The theorem is thus proved. □

Remark In particular, when $\alpha = 0, 1 < p < \infty$, we get

$$||T_v f||_p \le C||f||_* ||v||_{V_p}, \tag{5.2.7}$$

even

$$||T_v f||_p \le C||T_v f||_{h_p} \le C||f||_{bmo_2} ||v||_{V_p}, \quad 1 < p \le 2. \tag{5.2.8}$$

This gives a remarkable property of BMO or bmo_2 martingales, i.e. they behave like L^∞ martingales in one kind of multiplication operators.

The proof in theorem 5.2.2 is not able to tell us what happens when $p \le \frac{1}{1+\alpha}$. In order to state the situation in this case we need some kind of extrapolation lemmas, a very useful technique we will meet frequently in what follows.

Lemma 5.2.3 *Let* $0 < p_0 \le r_0 \le \infty, T$ *be a linear operator defined on* V_∞ *and valued in the space of all martingales. Suppose that* T *is of weak type* $(V_{p_0}, H_{r_0}^*)$, *or weak type* $(V_{p_0}, H_{r_0}^S)$, *or weak type* (V_{p_0}, h_{r_0}), *with the bound* $||T||$, *and* T *is commutable with stopping times in the following sense*

$$M(T(v - v^{(\tau-1)}))\chi_{\{\tau=\infty\}} = 0, \text{ a.e.}, \forall \tau, \ \forall v \in V_\infty, \tag{5.2.9}$$

or

$$S(T(v - v^{(\tau-1)}))\chi_{\{\tau=\infty\}} = 0, \quad \text{a.e.}, \forall\tau, \ \forall v \in V_\infty, \tag{5.2.10}$$

or

$$\sigma(T(v - v^{(\tau-1)}))\chi_{\{\tau=\infty\}} = 0, \quad \text{a.e.}, \forall\tau, \ \forall v \in V_\infty, \tag{5.2.11}$$

where the process $v^{(\tau-1)}$ *is defined as usual*

$$v^{(\tau-1)} = (v_{n\wedge(\tau-1)})_{n\geq 0}, \ v_{n\wedge(\tau-1)} = v_0\chi_{\{\tau=1\}} + \cdots + v_n\chi_{\{\tau\geq n+1\}}, \forall n. \tag{5.2.12}$$

Then for all (p, r) *satisfying*

$$\frac{1}{p} - \frac{1}{r} = \frac{1}{p_0} - \frac{1}{r_0}, \quad 0 < p \leq p_0, \tag{5.2.13}$$

T *is of type* (V_p, H_r^*), *or type* (V_p, H_r^S), *or type* (v_p, h_r), *with bound* $\leq C\|T\|$.

Proof. Let $v \in V_\infty$ be given. Without loss of generality, we can assume $\|v\|_{V_p} = 1$, and $\|T\| = 1$, otherwise we consider $v/\|v\|_{V_p}$ and $T/\|T\|$ instead. For $\lambda > 0$, set $\delta = \lambda^{\frac{r}{p}}$, and define a stopping time $\tau = \inf\{n : |v_n| > \delta\}$. Then

$$\{\tau < \infty\} = \{Mv > \delta\}, \quad M(v^{(\tau-1)}) \leq \delta.$$

It is easy to verify that $v^{\tau-1}$ is really an adapted process. Decompose v as $v = v - v^{(\tau-1)} + v^{(\tau-1)}$. By means of the condition (5.2.9), we have

$$\{M(Tv) > \lambda\} \subset \{\tau < \infty\} \bigcup \{M(Tv^{(\tau-1)}) > \lambda\},$$

$$\begin{aligned}
|\{M(Tv) > \lambda\}| &\leq |\{\tau < \infty\}| + C\lambda^{-r_0}\Big(\int_{\{\tau<\infty\}} + \int_{\{\tau=\infty\}} M(v^{(\tau-1)})^{p_0} d\mu\Big)^{\frac{r_0}{p_0}} \\
&\leq |\{\tau < \infty\}| + C\lambda^{\frac{rr_0}{p} - r_0}|\{\tau < \infty\}|^{\frac{r_0}{p_0}} \\
&\quad + C\lambda^{-r_0}\Big(\int_{\{Mv\leq\delta\}} (Mv)^{p_0} d\mu\Big)^{\frac{r_0}{p_0}} \\
&= \sum_{i=1}^{3} I_i.
\end{aligned} \tag{5.2.14}$$

Now estimate each $I_i, i = 1, 2, 3$. We have

$$\begin{aligned}
\int_0^\infty \lambda^{r-1} I_1 d\lambda &\leq \int_0^\infty \lambda^{r-1}|\{Mv > \lambda^{\frac{r}{p}}\}| d\lambda \\
&= C\int_0^\infty \lambda^{p-1}|\{Mv > \lambda\}| d\lambda = C.
\end{aligned}$$

Noticing that

$$\begin{aligned}
\lambda^{\frac{rr_0}{p} - r_0}|\{Mv > \delta\}|^{\frac{r_0}{p_0} - 1} &= \lambda^{\frac{r}{p}(r_0 - p(\frac{r_0}{p_0} - 1)) - r_0}(\delta^p|\{Mv > \delta\}|)^{\frac{r_0}{p_0} - 1} \\
&\leq \lambda^{rr_0(\frac{1}{p} - \frac{1}{p_0} + \frac{1}{r_0}) - r_0} \leq 1,
\end{aligned}$$

we get

$$\int_0^\infty \lambda^{r-1} I_2 d\lambda \le C \int_0^\infty \lambda^{r-1} |\{\tau < \infty\}| d\lambda \le C.$$

For the remained term, setting $a = rp^{-1}(p - p_0) - 1$, we have

$$\begin{aligned}\int_0^\infty \lambda^{r-1} I_3 d\lambda \le \sup_{\lambda>0} &\left\{\lambda^{r-1-r_0-a}\Big(\int_{\{Mv\le\delta\}} (Mv)^{p_0} d\mu\Big)^{\frac{r_0}{p_0}-1}\right\} \\ &\times \int_0^\infty \lambda^a \int_{\{Mv\le\delta\}} (Mv)^{p_0} d\mu d\lambda = J_1 J_2. \end{aligned} \tag{5.2.15}$$

For J's estimates, we have

$$\begin{aligned} J_1 &\le \sup_{\lambda>0}\{\lambda^{r-1-r_0-a}\delta^{(p_0-p)(\frac{r_0}{p_0}-1)}\} \\ &= \sup_{\lambda>0} \lambda^{r-1-r_0-a+\frac{r}{p}(p_0-p)(\frac{r_0}{p_0}-1)} \\ &= \sup_{\lambda>0} \lambda^{r-r_0+\frac{rr_0}{pp_0}(p_0-p)} = 1, \\ J_2 &= \int_\Omega \int_{(Mv)^{\frac{p}{r}}}^\infty \lambda^a d\lambda (Mv)^{p_0} d\mu \\ &\le C \int_\Omega (Mv)^{\frac{p}{r}(a+1)+p_0} d\mu = C \int_\Omega (Mv)^p d\mu = C. \end{aligned}$$

Combining these three estimates, we get (assuming $\|T\| = 1$)

$$\int_0^\infty \lambda^{r-1} |\{M(Tv) > \lambda\}| d\lambda \le C, \quad \forall v \in V_\infty, \quad \|v\|_{V_p} = 1.$$

Since V_∞ is dense in V_p we get the inequality for all $v \in V_p$. And by linearity, we get

$$\|Tv\|_{H_r^*} \le C\|T\|\|v\|_{V_p}, \quad \forall v \in V_p. \tag{5.2.16}$$

The other two assertions follow in same way. The proof of the lemma is finished. □

Now we can give a supplement to Theorem 5.2.2.

Theorem 5.2.4 *Let $0 \le \alpha \le 1$, $0 < p < \frac{1}{\alpha}$, $\frac{1}{r} = \frac{1}{p} - \alpha$. Then T_v is of type (Λ_α, H_r^*) with the bound $\le C\|v\|_{V_p}$.*

Proof. We have got the result when $\frac{1}{p} < 1 + \alpha$ in Theorem 5.2.2. Now consider the case $\frac{1}{p} \ge 1 + \alpha$. We want to use Lemma 5.2.3. Let $f \in \Lambda_\alpha$ be given and consider $T_v(f)$ as an operator T defined on V_∞. It has been known that T is of type (V_{p_0}, H_{r_0}) for $0 \le \alpha \le 1$, for some (p_0, r_0), $\frac{1}{\alpha} > p_0 > \frac{1}{1+\alpha}$, $\frac{1}{r_0} = \frac{1}{p_0} - \alpha < 1$, with the bound $\le C\|f\|_{\Lambda_\alpha}$. That is to say

$$\|M(T_v f)\|_{r_0} \le C\|f\|_{\Lambda_\alpha}\|v\|_{V_{p_0}}, \quad \forall v \in V_\infty, \quad \forall f \in \Lambda_\alpha.$$

It remains to verify (5.2.9). Let $v \in V_\infty$ and τ be any stopping time. Then

$$(v - v^{(\tau-1)})_{n-1} = v_{n-1} - (v_0\chi_{\{\tau=1\}} + \cdots + v_{n-1}\chi_{\{\tau\geq n\}}),$$

$$(T_{v-v^{(\tau-1)}}f)_n\chi_{\{\tau=\infty\}} = \sum_1^n (v - v^{(\tau-1)})_{k-1}\Delta_k f\chi_{\{\tau=\infty\}} = 0, \text{ a.e.}$$

So, (5.2.9) holds. Now applying Lemma 5.2.3, we get

$$\|T_v f\|_{H_r^*} \leq C\|f\|_{\Lambda_\alpha}\|v\|_{V_p}, \quad \forall v \in V_p, \quad \forall f \in \Lambda_\alpha,$$

for all (p, r), $0 < p \leq p_0$, $\frac{1}{p} - \frac{1}{r} = \frac{1}{p_0} - \frac{1}{r_0} = \alpha$. The theorem is thus proved. □

We have also the (λ_α, h_r) version of Theorem 5.2.4.

Theorem 5.2.4' *Let $0 \leq \alpha \leq 1$, $0 < p < \frac{1}{\alpha}$, $\frac{1}{r} = \frac{1}{p} - \alpha$. Then T_v is of type (λ_α, h_r) with the bound $\leq C\|v\|_{V_p}$.*

Proof. The proof is just the imitation of that in Theorem 5.2.4. In order to apply Lemma 5.2.3, it should be verified that T_v for fixed f satisfies the condition (5.2.11) and that T_v is of type (V_{p_0}, h_{p_0}) for some (p_0, r_0), $\frac{1}{\alpha} > p_0 > \frac{1}{1+\alpha}$. The former is easy, the latter has been known in Theorem 5.2.2. The proof is finished. □

Remark In particular, from these two theorems, we get, for $\alpha = 0$, $0 < p < \infty$,

$$\|T_v f\|_{H_p^*} \leq C\|f\|_*\|v\|_{V_p}, \quad \forall v \in V_p, \quad \forall f \in BMO, \tag{5.2.17}$$

$$\|T_v f\|_{h_p} \leq C\|f\|_{bmo_2}\|v\|_{V_p}, \quad \forall v \in V_p, \quad \forall f \in bmo_2. \tag{5.2.18}$$

The latter supplies the proof of the second part of Theorem 5.1.4.

Is T_v of type (H_q^S, H_r^*)? We could give an affirmtive answer for $q \geq 1$. For the space $\mathcal{P}_q$, we could do better by proving T_v's $(\mathcal{P}_q, \mathcal{P}_r)$-boundedness for $0 < q < \infty$.

Theorem 5.2.5 *Let $0 < p \leq \infty$, $0 < q < \infty$. Then T_v is of type $(\mathcal{P}_q, \mathcal{P}_r)$ with the bound $\leq C\|v\|_{V_p}$, where $\frac{1}{r} = \frac{1}{p} + \frac{1}{q}$.*

Proof. Let $g \in \mathcal{P}_q$, $\lambda_g = (\lambda_n(g))_{n\geq 0}$ be g's optimal majorant. Then we have

$$|\Delta_n g| \leq 2\lambda_{n-1}(g), \quad |\Delta_n(T_v g)| \leq 2M_{n-1}v\lambda_{n-1}(g) = \rho_{n-1},$$

$$|(T_v g)_n| \leq M_{n-1}(T_v g) + \rho_{n-1}. \tag{5.2.19}$$

So, $T_v g$ is predictable. Applying a well known fact (see Th.3.5.4): for any martingale $h = (h_n)_{n\geq 0}$ with $\Delta_n h$ satisfying $|\Delta_n h| \leq D_{n-1}$, where $D = (D_n)_{n\geq 0}$ is a nonnegative increasing and adapted process, both $(Mh, S(h) + D_\infty)$ and $(S(h), Mh + D_\infty)$ satisfy so-called good λ-inequality, we get

$$\|S(g)\|_q \leq C(\|Mg\|_q + \|\lambda_\infty(g)\|_q) \leq C\|g\|_{\mathcal{P}_q}, \tag{5.2.20}$$

and

$$\begin{aligned}||M(T_v g)||_r &\le C(||S(T_v g)||_r + ||\rho_\infty||_r) \\ &\le C||v||_{V_p}(||S(g)||_q + ||\lambda_\infty(g)||_q) \\ &\le C||v||_{V_p}||g||_{\mathcal{P}_q}.\end{aligned}$$

Finally, from (5.2.19), we get

$$||T_v g||_{\mathcal{P}_r} \le C(||M(T_v g)||_r + ||\rho_\infty||_r) \le C||v||_{V_p}||g||_{\mathcal{P}_q}. \tag{5.2.21}$$

This ends the proof of the theorem. □

Remark (5.2.21) supplies the second part of Theorem 5.1.5.

Theorem 5.2.6 *Let $0 < p \le \infty$, $1 \le q < \infty$, $v \in V_p$, $\frac{1}{r} = \frac{1}{p} + \frac{1}{q}$. Then T_v is of type (H_q, H_r^*) with the bound $\le C||v||_{V_p}$.*

Proof. By making use of Davis' decomposition we get $g = g^{(1)} + g^{(2)}$ with

$$||g^{(1)}||_{\mathcal{A}_q} = \Big\|\sum_n |\Delta_n g^{(1)}|\Big\|_q \le C||g||_{H_q},$$

$$|\Delta_n g^{(2)}| \le CM_{n-1}g, \quad ||g^{(2)}||_{\mathcal{P}_q} \le C||g||_{H_q}.$$

Then we have

$$\begin{aligned}||M(T_v g)||_r &\le C(||M(T_v g^{(1)})||_r + ||M(T_v g^{(2)})||_r) \\ &\le C||v||_{V_p}(||g^{(1)}||_{\mathcal{A}_q} + ||g^{(2)}||_{\mathcal{P}_q}) \le C||v||_{V_p}||g||_{H_q}.\end{aligned} \tag{5.2.22}$$

This proves the theorem. □

Remark When $p = \infty$, we have a result stronger than (5.2.22) as follows.

Proposition 5.2.7 *Let $1 \le q < \infty$, $g \in H_q$ be fixed. Consider the martingale transform $T_v(g)$ as an operator T acting on V_∞. Then with $\sigma(\lambda) = \sup_{v:||v||_{V_\infty}\le 1} |\{M(T_v g) > \lambda\}|$, we have*

$$\int_0^\infty \lambda^{q-1}\sigma(\lambda)d\lambda \le C^q||g||_{H_q}^q. \tag{5.2.23}$$

Proof. Making g's Davis decomposition as above. We have

$$\begin{aligned}\sigma(2\lambda) &\le \sup_{v:||v||_{V_\infty}\le 1} |\{M(T_v g^{(1)}) > \lambda\}| + \sup_{v:||v||_{V_\infty}\le 1} |\{M(T_v g^{(2)}) > \lambda\}| \\ &= \sigma_1(\lambda) + \sigma_2(\lambda),\end{aligned}$$

$$\sigma_1(\lambda) \le \sup_{v:\|v\|_{V_\infty}\le 1} \left|\left\{\|Mv\|_\infty \sum_n |\Delta_n g^{(1)}| > \lambda\right\}\right|$$
$$\le \left|\left\{\sum_n |\Delta_n g^{(1)}| > \lambda\right\}\right|,$$

$$\int_0^\infty \lambda^{q-1}\sigma_1(\lambda)d\lambda \le C^q\|g^{(1)}\|^q_{\mathcal{A}_q} \le C^q\|g\|^q_{H_q}.$$

Now estimate $\sigma_2(\lambda)$. Notice that $T_v g^{(2)}$ is predictable, and $S_n(T_v g^{(2)}) \le S_{n-1}(T_v g^{(2)}) + \rho_{n-1}$ with $\rho_n = CM_n v M_n g$, thus we have following good λ-inequality: there is $\alpha > 1$, for all $\beta > 0$, there is C_β satisfying $\lim_{\beta\to 0} C_\beta = 0$ such that

$$|\{M(T_v g^{(2)}) > \alpha\lambda\}| \le$$
$$C_\beta|\{M(T_v g^{(2)}) > \lambda\}| + |\{S(T_v g^{(2)}) > \beta\lambda\}| + |\{CMvMg > \beta\lambda\}|.$$

Taking "sup" over all v, $\|v\|_{V_\infty} \le 1$, we get

$$\sigma_2(\alpha\lambda) \le C_\beta\sigma_2(\lambda) + |\{S(g^{(2)}) > \beta\lambda\}| + |\{Mg > C\beta\lambda\}|.$$

Chosing β snall enough, we get (5.2.23). This ends the proof. □

There is a counterpart of the extrapolation Lemma 5.2.3 dealing with the case $p > r$, basing on a kind of inequality like (5.2.23). But this counterpart is much less powerful than Lemma 5.2.3 because of the difficulty to verify the implied conditions.

Lemma 5.2.8 *Let T be a linear operator defined on V_∞ and valued in the space of all martingales, and satisfying* (5.2.9). *In addition, suppose that for $0 < p_0 \le \infty$, $p_0 > r_0$,*

$$\int_0^\infty \lambda^{r_0-1}\sigma_T(\lambda)d\lambda \le C_T^{r_0}, \tag{5.2.24}$$

where

$$\sigma_T(\lambda) = \sup_{v:\|v\|_{V_{p_0}}\le 1} |\{M(Tv) > \lambda\}|, \quad \lambda > 0. \tag{5.2.25}$$

Then for all (p, r) satisfying (5.2.13), *T is of type (V_p, H_r^*) with the bound $\le CC_T$.*

Proof. Let $v \in V_\infty$, with $\|v\|_{V_p} = 1$. We can assume $C_T = 1$, because T/C_T satisfies (5.2.24) with $C_{T/C_T} = 1$:

$$\int_0^\infty \lambda^{r_0-1}\sigma_{T/C_T}(\lambda)d\lambda = \int_0^\infty \lambda^{r_0-1}\sigma_T(C_T\lambda)d\lambda$$
$$= C_T^{-r_0}\int_0^\infty \lambda^{r_0-1}\sigma_T(\lambda)d\lambda \le 1.$$

For $\lambda > 0$, define $\tau = \inf\{n : |v_n| > \delta\}$ with $\delta = \lambda^{\frac{r}{p}}$, and decompose v as $v - v^{(\tau-1)} + v^{(\tau-1)}$. We have

$$\int_0^\infty \lambda^{r-1}|\{M(Tv) > \lambda\}|d\lambda \leq$$
$$\int_0^\infty \lambda^{r-1}|\{\tau < \infty\}|d\lambda + \int_0^\infty \lambda^{r-1}|\{M(Tv^{(\tau-1)}) > \lambda\}|d\lambda.$$

The first term $\leq C$, we have seen it in Lemma 5.2.3. For the second term, noticing

$$\|v^{(\tau-1)}\|_{V_{p_0}} \leq \delta^{\frac{p_0-p}{p_0}} = \lambda^{\frac{r}{pp_0}(p_0-p)},$$

we get

$$|\{M(Tv^{(\tau-1)}) > \lambda\}| =$$
$$|\{M(T(\lambda^{\frac{r}{pp_0}(p-p_0)}v^{(\tau-1)})) > \lambda^{1+\frac{r}{pp_0}(p-p_0)}\}| \leq \sigma_T(\lambda^{\frac{r}{r_0}}).$$

And hence

$$\int_0^\infty \lambda^{r-1}|\{M(Tv^{(\tau-1)}) > \lambda\}|d\lambda \leq \int_0^\infty \lambda^{r-1}\sigma_T(\lambda^{\frac{r}{r_0}})d\lambda$$
$$= C\int_0^\infty \lambda^{r_0-1}\sigma_T(\lambda)d\lambda \leq C.$$

This completes the proof of the Lemma. □

Remark In general, it is not easy to verify (5.2.24). But Proposition 5.2.7 tells us that $T_v(g)$ for $g \in H_q$, $q \geq 1$, fixed, satisfies (5.2.24) with $p_0 = \infty$, $r_0 = q$, $C_T = C\|g\|_{H_q}$. So, Theorem 5.2.6 could be deduced by this lemma.

What boundedness can we say about T_v's action on L^1? A natural answer is weak boundedness. It is really true.

Theorem 5.2.9 *Let $0 < p \leq \infty$, $v \in V_p$, $\frac{1}{r_0} = 1 + \frac{1}{p}$, Then T_v is of type $(L^1, WH^*_{r_0})$, and $(L^1, WH^S_{r_0})$, i.e. for all L^1-bounded martingales $f = (f_n)_{n\geq 0}$,*

$$|\{M(T_v f) > \lambda\}| \leq \left(\frac{c}{\lambda}\|f\|_1\right)^{r_0}, \quad \forall\lambda > 0, \tag{5.2.26}$$

$$|\{S(T_v f) > \lambda\}| \leq \left(\frac{c}{\lambda}\|f\|_1\right)^{r_0}, \quad \forall\lambda > 0. \tag{5.2.27}$$

Proof. Without loss of generality, we can assume $\|v\|_{V_p} = \|f\|_1 = 1$. According to Gundy's decomposition, for $\delta = \lambda^{r_0}$, we get $f = \sum_1^3 f^{(i)}$, with $f^{(i)}$ satisfying

$$\|f^{(1)}\|_1 \leq C\|f\|_1, \quad |A| = |\{\sup_n |\Delta_n f^{(1)}| \neq 0\}| \leq \frac{c}{\delta}\|f\|_1,$$

$$f^{(2)} \in \mathcal{A}_1, \quad \|f^{(2)}\|_{\mathcal{A}_1} \leq C\|f\|_1,$$

$$f^{(3)} \in L^2, \quad \|f^{(3)}\|_2^2 \le c\delta\|f\|_1.$$

Now, from

$$\begin{aligned}\{M(T_v f) > 2\lambda\} &\subset \{M(T_v f^{(1)}) \ne 0\}|\bigcup\{M(T_v f^{(2)}) > \lambda\}\bigcup\{M(T_v f^{(3)}) > \lambda\} \\ &\subset A\bigcup\{M(T_v f^{(2)}) > \lambda\}\bigcup\{M(T_v f^{(3)}) > \lambda\},\end{aligned}$$

we get (let r be such that $1/r = 1/2 + 1/p$)

$$\begin{aligned}|\{M(T_v f) > 2\lambda\}| &\le |A| + \lambda^{-r_0}\|M(T_v f^{(2)})\|_{r_0}^{r_0} + C\lambda^{-r}\|f^{(3)}\|_2^r \\ &\le C\lambda^{-r_0} + \lambda^{-r_0}\Big\|Mv\sum_n |\Delta_n f^{(2)}|\Big\|_{r_0}^{r_0} + C\delta^{\frac{r}{2}}\lambda^{-r}\|f\|_1^{\frac{r}{2}} \\ &\le C\lambda^{-r_0}.\end{aligned}$$

And analogously,

$$\begin{aligned}|\{S(T_v f) > 2\lambda\}| &\le |A| + \lambda^{-r_0}\|S(T_v f^{(2)})\|_{r_0}^{r_0} + C\lambda^{-r}\|f^{(3)}\|_2^r \\ &\le C\lambda^{-r_0} + \lambda^{-r_0}\Big\|Mv\sum_n |\Delta_n f^{(2)}|\Big\|_{r_0}^{r_0} + C\lambda^{-r_0} \\ &\le C\lambda^{-r_0}.\end{aligned}$$

This proves the theorem. □

As the end of this section, we list all the boundedness results obtained in this section. In terms of bilinear operator language, a martingale transform acting as an operator

$$T : (v, f) \to T_v f,$$

is of type

$$(V_p, H_q^S; H_r^S) \text{ and } (V_p, h_q; h_r), \ 0 < p, \ q \le \infty, \ \frac{1}{r} = \frac{1}{p} + \frac{1}{q},$$

$$(V_p, \Lambda_\alpha; \Lambda_\beta) \text{ and } (V_p, \lambda_\alpha; \lambda_\beta), \ 0 < p \le \infty, \ 0 \le \alpha \le 1, \ \beta = \alpha - \frac{1}{p} \ge 0,$$

$$(V_p, \Lambda_\alpha; H_r^*) \text{ and } (V_p, \lambda_\alpha; h_r), \ 0 < p < \frac{1}{\alpha}, \ 0 \le \alpha \le 1, \ \frac{1}{r} = \frac{1}{p} - \alpha,$$

$$(V_p, H_q; H_r^*), \ 0 < p \le \infty, \ 1 \le q < \infty, \ \frac{1}{r} = \frac{1}{p} + \frac{1}{q},$$

$$(V_p, \mathcal{P}_q; \mathcal{P}_r), \ 0 < p \le \infty, \ 0 < q < \infty, \ \frac{1}{r} = \frac{1}{p} + \frac{1}{q},$$

$$(V_p, L^1; WH_r^*) \text{ and } (V_p, L^1; WH_r^S), \ 0 < p \le \infty, \ \frac{1}{r} = 1 + \frac{1}{p}.$$

5.3 Paraproduct operator and some related ones

In this section we want to make a further investigation about martingale transforms when the implied multiplier $v = (v_n)_{n\geq 0}$ is also a martingale. In this case, it is called paraproduct operator in martingale setting, and $T_f(g)$ is rewritten as $\mathcal{C}(f,g)$. The fact that both f and g are martingales makes us to be able to interchange the role of f and g when we need. Following elementary equality

$$\sum_1^n f_{k-1}\Delta_k g = g_n f_{n-1} - g_0 f_0 - \sum_1^n g_k \Delta_k f, \tag{5.3.1}$$

shows that it is natural to introduce another related operator

$$\mathcal{B}(f,g) = \sum_1^\infty f_k \Delta_k g, \quad \forall f = (f_n)_{n\geq 0},\ g = (g_n)_{n\geq 0},\ f_0 = g_0 = 0, \tag{5.3.2}$$

and the difference of these two

$$\mathcal{D}(f,g) = \sum_1^\infty \Delta_k f \Delta_k g, \quad \forall f = (f_n)_{n\geq 0},\ g = (g_n)_{n\geq 0},\ f_0 = g_0 = 0. \tag{5.3.3}$$

Notice that for all $f, g \in L^1$, $\mathcal{B}(f,g)$ and $\mathcal{D}(f,g)$ have meaning pointwise, because of Theorem 5.1.1 and (5.3.1). $\mathcal{B}$ and $\mathcal{D}$ are operators mapping martingales into measurable functions. Notice again that $\mathcal{D}$ is symmetric and $\mathcal{B}$ is selfadjoint with respect to f for fixed g. The latter can be seen as follows

$$E(\mathcal{B}(f,g)h) = \sum_k E(f_k \Delta_k g h_k) = E(\mathcal{B}(h,g)f), \quad \forall \text{ good } f, g, h. \tag{5.3.4}$$

At first study $\mathcal{D}$'s boundedness on various spaces of martingales such as H_p^S, $\mathcal{P}_p$, Λ_α, $\mathcal{A}_1$ and L^1. Only following cases are worth to be considered, i.e.:

$$f \in H_p^S,\ g \in H_q^S \text{ or } L^1; \quad f \in L^1,\ g \in L^1;$$

$$f \in \Lambda_\alpha,\ g \in H_q (q \geq 1) \text{ or } \Lambda_\beta, \text{ or } \mathcal{P}_q (0 < q < 1), \text{ or } \mathcal{A}_1, \text{ or } L^1.$$

Theorem 5.3.1 *Let $0 < p, q \leq \infty$, and $1/r = 1/p + 1/q$. Then $\mathcal{D}$ is of type $(H_p^S, H_q^S; L^r)$, i.e.*

$$\|\mathcal{D}(f,g)\|_r \leq \|f\|_{H_p^S} \|g\|_{H_q^S}, \quad \forall f, g. \tag{5.3.5}$$

Proof. This follows from the pointwise estimate $|\mathcal{D}(f,g)| \leq S(f)S(g)$. □

Theorem 5.3.2 *Let $0 < p \leq \infty$, $1/r_0 = 1/p + 1$. Then $\mathcal{D}$ is of type $(H_p^S, L^1; WL^{r_0})$, i.e.*

$$|\{\mathcal{D}(f,g) > \lambda\}| \leq \left(\frac{c}{\lambda}\|f\|_{H_p^S}\|g\|_1\right)^{r_0}, \quad \forall \lambda > 0. \tag{5.3.6}$$

Proof. Let $f \in H_p^S$, $g \in L^1$, with $f_0 = g_0 = 0$, $\|f\|_{H_p^S} = \|g\|_1 = 1$. For $\delta = \lambda^{r_0}$, make Gundy's decomposition of g, $g = \sum_1^3 g^{(i)}$. Set $r = (1/p + 1/2)^{-1}$. We get

$$\begin{aligned}|\{|\mathcal{D}(f,g)| > 2\lambda\}| &\leq |A| + |\{|\mathcal{D}(f,g^{(2)})| > \lambda\}| + |\{|\mathcal{D}(f,g^{(3)})| > \lambda\}| \\ &\leq |A| + \left|\left\{S(f)\sum_n |\Delta_n g^{(2)}| > \lambda\right\}\right| + C\lambda^{-r}\|g^{(3)}\|_2^r \\ &\leq C\lambda^{-r_0}\|g\|_1 + C\lambda^{-r_0}\|S(f)\|_p^{r_0}\|g^{(2)}\|_{\mathcal{A}_1}^{r_0} + C\lambda^{\frac{rr_0}{2}-r}\|g\|_1^{\frac{r}{2}} \\ &\leq C\lambda^{-r_0}.\end{aligned}$$

This proves the theorem. □

Theorem 5.3.3 *$\mathcal{D}$ is of type $(L^1, L^1; WL^{\frac{1}{2}})$, i.e.*

$$|\{|\mathcal{D}(f,g) > \lambda\}| \leq \left(\frac{C}{\lambda}\|f\|_1\|g\|_1\right)^{\frac{1}{2}}, \quad \forall \lambda > 0. \tag{5.3.7}$$

Proof. At first, make Gundy's decomposition for f and $\delta = \lambda^{\frac{1}{2}}$ (assuming $\|f\|_1 = \|g\|_1 = 1$). We have

$$\begin{aligned}&|\{|\mathcal{D}(f,g)| > \lambda\}| \leq \\ &\qquad |A| + |\{|\mathcal{D}(f^{(3)},g)| > \lambda\}| + |\{|\mathcal{D}(f^{(2)},g)| > \lambda\}| = \sum_1^3 J_i.\end{aligned}$$

We have known (see (5.3.6)) that with r_0 such that $1/r_0 = 1 + 1/2 = 3/2$,

$$|\{|\mathcal{D}(f^{(3)},g)| > \lambda\}| \leq \left(\frac{C}{\lambda}\|f^{(3)}\|_2\|g\|_1\right)^{r_0} \leq C(\lambda^{\frac{1}{4}-1})^{\frac{2}{3}} = C\lambda^{-\frac{1}{2}}.$$

So, except J_3, we have desired order $\lambda^{-\frac{1}{2}}$. In order to estimate J_3, make Gundy's decomposition again for g. We can assume $\|f^{(2)}\|_{\mathcal{A}_1} = 1$ (since $\|f^{(2)}\|_{\mathcal{A}_1} \leq C\|f\|_1 \leq C$). For $\delta = \lambda^{\frac{1}{2}}$, we get

$$\begin{aligned}|\{|\mathcal{D}(f^{(2)},g)| > 2\lambda\}| &\leq C\lambda^{-\frac{1}{2}} + |\{|\mathcal{D}(f^{(2)},g^{(2)})| > \lambda\}| \\ &\leq C\lambda^{-\frac{1}{2}} + \left|\left\{\sum_k |\Delta_k f^{(2)}| \sum_k |\Delta_k g^{(2)}| > \lambda\right\}\right| \\ &\leq C\lambda^{-\frac{1}{2}} + C\lambda^{-\frac{1}{2}}\|f^{(2)}\|_{\mathcal{A}_1}^{\frac{1}{2}}\|g^{(2)}\|_{\mathcal{A}_1}^{\frac{1}{2}} \leq C\lambda^{-\frac{1}{2}}.\end{aligned}$$

This proves the theorem. □

Now we turn to the case $f \in \Lambda_\alpha$. We need following lemmas.

Lemma 5.3.4 *Let $0 < a < \infty, 0 \leq \alpha \leq 1, a\alpha < 1$. Then the fractional maximal operator $M_{a,\alpha}$ defined by*

$$M_{a,\alpha}f(x) = \sup_n \omega_n^\alpha E(|f|^a|\mathcal{F}_n)^{\frac{1}{a}}, \quad \forall f = (f_n)_{n\geq 0}, \tag{5.3.8}$$

is of type $(L^{\tilde{a}}, WL^{q_0})$, *and of type* (L^p, L^q) *where*

$$\alpha = \frac{1}{p} - \frac{1}{q} = \frac{1}{a} - \frac{1}{q_0}, \quad a < p < q \le \infty.$$

And the bound ≤ 1.

Proof. Let $f = (f_0)_{n \ge 0}$ and $\lambda > 0$ be given. Define

$$\tau = \inf\{n : \omega_n^\alpha E(|f|^a|\mathcal{F}_n)^{\frac{1}{a}} > \lambda\}.$$

According to the definition of ω_n and the fact that $E(|f|^a|\mathcal{F}_n)$ is constant over any $\mathcal{F}_n$-atoms, so $\{\tau = n\} = \bigcup_j I_j^{(n)}$ with $I_j^{(n)}$ being $\mathcal{F}_n$-atoms. And we have $\omega_n^\alpha E(|f|^a|\mathcal{F}_n)^{\frac{1}{a}} > \lambda$ over each $I_j^{(n)}$. Thus, we get

$$|I_j^{(n)}|^{1-a\alpha} < \lambda^{-a} \int_{I_j^{(n)}} E(|f|^a|\mathcal{F}_n) d\mu = \lambda^{-a} \int_{I_j^{(n)}} |f|^a d\mu,$$

$$|\{\tau < \infty\}|^{1-a\alpha} = \Big(\sum_n |\{\tau = n\}|\Big)^{1-a\alpha} \le \sum_{n,j} |I_j^{(n)}|^{1-a\alpha} \le \lambda^{-a} \|f\|_a^a.$$

It is just

$$|\{M_{a,\alpha} f > \lambda\}| \le \Big(\frac{1}{\lambda}\|f\|_a\Big)^{q_0}.$$

Now let $f \in L^{1/\alpha}$. Since $a < \frac{1}{\alpha}$, we have

$$\omega_n^\alpha E(|f|^a|\mathcal{F}_n)^{\frac{1}{a}} \le \omega_n^\alpha E(|f|^{\frac{1}{\alpha}}|\mathcal{F}_n)^\alpha \le \|f\|_{\frac{1}{\alpha}}.$$

By Marcinkiewicz' interpolation theorem, we get

$$\|M_{a,\alpha} f\|_q \le \|f\|_p.$$

This ends the proof of the lemma. □

Another lemma is a version of Lemma 4.5.3.

Lemma 5.3.5 *Let* $0 \le \alpha < 1$, $f \in \Lambda_\alpha$. *Then for all* $g = (g_n)_{n \ge 0}$, *there exists an adapted process* $\theta = \theta_{f,g} = (\theta_n)_{n \ge 0}$, *such that*

$$E(|\mathcal{D}(f,g) - \theta_{n-1}|\mathcal{F}_n) \le \sqrt{2}\|f\|_{\Lambda_\alpha} \omega_n^\alpha E(S(g - g_{n-1})|\mathcal{F}_n), \quad \forall n. \tag{5.3.9}$$

Proof. Denote $h = g - g_{n-1}$, and set $\theta_n = \sum_1^n \Delta_k f \Delta_k g$, $n \ge 1$, $\theta_0 = 0$. We have

$$E(|\mathcal{D}(f,g) - \theta_{n-1}||\mathcal{F}_n)$$

$$\le E\Big(\Big(\sum_n^\infty |\Delta_k f|^2 S_k(h)\Big)^{\frac{1}{2}} \Big(\sum_n^\infty |\Delta_k g|^2 S_k(h)^{-1}\Big)^{\frac{1}{2}} \Big|\mathcal{F}_n\Big)$$

$$\le E\Big(\sum_n^\infty |\Delta_k f|^2 S_k(h)\Big|\mathcal{F}_n\Big)^{\frac{1}{2}} E\Big(\sum_n^\infty |\Delta_k g|^2 S_k(h)^{-1}\Big|\mathcal{F}_n\Big)^{\frac{1}{2}} = (AB)^{\frac{1}{2}},$$

where $h = g - g_{n-1}$ and

$$S_k(h) = \Big(\sum_n^k |\Delta_j h|^2\Big)^{\frac{1}{2}}, \ k \geq n; \quad S_{n-1}(h) = 0.$$

Noticing $\Delta_k h = \Delta_k g$, for all $k \geq n$, and

$$|\Delta_k g|^2 = S_k(h)^2 - S_{k-1}(h)^2 \leq 2S_k(h)(S_k(h) - S_{k-1}(h)), \quad \forall k \geq n,$$

we get

$$B \leq 2E\Big(\sum_n^\infty (S_k(h) - S_{k-1}(h))\Big|\mathcal{F}_n\Big) \leq 2E(S(h)|\mathcal{F}_n). \tag{5.3.10}$$

For A, we have

$$\begin{aligned}
A &\leq E\Big(\sum_n^\infty |\Delta_k f|^2 \sum_{j=n}^k (S_j(h) - S_{j-1}(h))\Big|\mathcal{F}_n\Big) \\
&= E\Big(\sum_{j=n}^\infty E\Big(\sum_{k=j}^\infty |\Delta_k f|^2\Big|\mathcal{F}_j\Big)(S_j(h) - S_{j-1}(h))\Big|\mathcal{F}_n\Big) \\
&\leq \|f\|_{\Lambda_\alpha}^2 E\Big(\sum_{j=n}^\infty \omega_j^{2\alpha}(S_j(h) - S_{j-1}(h))\Big|\mathcal{F}_n\Big).
\end{aligned}$$

Noticing that ω_k is decreasing with respect to k (since each $\mathcal{F}_{k+1}$-atom $I^{(k+1)}$ is contained in some $\mathcal{F}_k$-atom $I^{(k)}$), we get

$$A \leq \|f\|_{\Lambda_\alpha}^2 \omega_n^{2\alpha} E(S(h)|\mathcal{F}_n). \tag{5.3.11}$$

Combining (5.3.10) and (5.3.11), (5.3.9) follows. The proof of the lemma is finished. □

Remark For $f \in \lambda_\alpha$, we have following version of (5.3.9)

$$E(|\mathcal{D}(f,g) - \theta_n||\mathcal{F}_n) \leq \sqrt{2}\|f\|_{\lambda_\alpha}\omega_n^\alpha E(\sigma(g - g_n)|\mathcal{F}_n), \ \forall n. \tag{5.3.12}$$

This can be seen by denoting $h = g - g_n$, replacing $S_k(h)$ by

$$\sigma_k(h) = \Big(\sum_{j=n+1}^k E(|\Delta_j g|^2|\mathcal{F}_{j-1})\Big)^{\frac{1}{2}}, \tag{5.3.13}$$

and noticing

$$\begin{aligned}
E(|\mathcal{D}(f,g) - \theta_n||\mathcal{F}_n) \leq{}& E\Big(\sum_{j=n+1}^\infty E\Big(\sum_{k=j}^\infty |\Delta_k f|^2\Big|\mathcal{F}_{j-1}\Big)(\sigma_j(h) - \sigma_{j-1}(h))\Big|\mathcal{F}_n\Big)^{\frac{1}{2}} \\
&\times E\Big(\sum_{k=n+1}^\infty E(|\Delta_k g|^2|\mathcal{F}_{k-1})\sigma_k(g)^{-1}\Big|\mathcal{F}_n\Big)^{\frac{1}{2}}.
\end{aligned}$$

Theorem 5.3.6 *Let $0 \le \alpha < 1$, $1 < p < \frac{1}{\alpha}$. Then with r satisfying*

$$\alpha = \frac{1}{p} - \frac{1}{r}, \quad p \le r < \infty, \tag{5.3.14}$$

$\mathcal{D}$ is of type $(\Lambda_\alpha, H_p; L^r)$, i.e.

$$\|\mathcal{D}(f,g)\|_r \le C\|f\|_{\Lambda_\alpha}\|g\|_{H_p}. \tag{5.3.15}$$

Proof. For $f \in \Lambda_\alpha, g \in H_p$ given, set $\theta_n = \sum_1^n \Delta_k f \Delta_k g$, $\theta_0 = 0$. Taking "sup" over n on both sides of (5.3.9), we get

$$\mathcal{D}(f,g)_\theta^\# \le C\|f\|_{\Lambda_\alpha} M_{1,\alpha}(S(g)), \tag{5.3.16}$$

where $h_\theta^\#$ is the generalized sharp function defined (in §3.5) by

$$h_\theta^\# = \sup_n E(|h - \theta_{n-1}(h)||\mathcal{F}_n),$$

$\theta_h = \theta = (\theta_n(h))_{n\ge 0}$ adapted. From Theorem 3.5.7′, there exists constant c depending only on r such that

$$\|Mh\|_r \le C\|h_\theta^\#\|_r, \quad 0 < r < \infty. \tag{5.3.17}$$

Applying (5.3.17) to $h = \mathcal{D}(f,g)$, and Lemma 5.3.4 to $f = S(g)$, we get

$$\|\mathcal{D}(f,g)\|_r \le C\|\mathcal{D}(f,g)_\theta^\#\|_r \le C\|f\|_{\Lambda_\alpha}\|M_{1,\alpha}(S(g))\|_r \le C\|f\|_{\Lambda_\alpha}\|S(g)\|_p.$$

This completes the proof of the theorem. □

When $g \in L^p$, $\frac{1}{\alpha} \le p \le \infty$, or $g \in \Lambda_\beta$, we can say that $\mathcal{D}(f,g)$ is almost in $\Lambda_{\alpha-\frac{1}{p}}$ or $\Lambda_{\alpha+\beta}$ in following sense.

Theorem 5.3.7 *Let $0 \le \alpha < 1$, $\frac{1}{\alpha} \le p \le \infty$, $f \in \Lambda_\alpha$, $g \in L^p$ or $g \in \Lambda_\beta$, then there exists an adapted process $\theta = \theta_{f,g} = (\theta_n)_{n\ge 0}$, such that*

$$E(|\mathcal{D}(f,g) - \theta_{n-1}||\mathcal{F}_n) \le C\|f\|_{\Lambda_\alpha}\|g\|_p \omega_n^{\alpha-\frac{1}{p}}, \tag{5.3.18}$$

or

$$E(|\mathcal{D}(f,g) - \theta_{n-1}||\mathcal{F}_n) \le C\|f\|_{\Lambda_\alpha}\|g\|_{\Lambda_\beta}\omega_n^{\alpha+\beta}. \tag{5.3.19}$$

Proof. From (5.3.9) we have

$$\begin{aligned} E(|\mathcal{D}(f,g) - \theta_{n-1}||\mathcal{F}_n) &\le C\|f\|_{\Lambda_\alpha}\omega_n^\alpha E(S(g)^p|\mathcal{F}_n)^{\frac{1}{p}} \\ &\le C\|f\|_{\Lambda_\alpha}\omega_n^{\alpha-\frac{1}{p}}(\omega_n E(|g|^p|\mathcal{F}_n))^{\frac{1}{p}} \\ &\le C\|f\|_{\Lambda_\alpha}\|g\|_p\omega_n^{\alpha-\frac{1}{p}}, \end{aligned}$$

or in the latter case,

$$\begin{aligned}E(|\mathcal{D}(f,g)-\theta_{n-1}||\mathcal{F}_n) &\le C\|f\|_{\Lambda_\alpha}\omega_n^\alpha E(S(g-g_{n-1})^2|\mathcal{F}_n)^{\frac{1}{2}}\\ &\le C\|f\|_{\Lambda_\alpha}\omega_n^\alpha E(|g-g_{n-1}|^2|\mathcal{F}_n)^{\frac{1}{2}}\\ &\le C\|f\|_{\Lambda_\alpha}\|g\|_{\Lambda_\beta}\omega_n^{\alpha+\beta}.\end{aligned}$$

This ends the proof. □

Is there any possibility to extend (5.3.15) to $p \le 1$? We can get $\mathcal{D}$'s type $(\Lambda_\alpha, H_1; L^{r_0})$ and $(\Lambda_\alpha, \mathcal{P}_p; L^r)$ with r_0, r such that $\frac{1}{r_0} = 1-\alpha$ and $\frac{1}{r} = \frac{1}{p} - \alpha$, $0 < p < 1$. This benefits from following extrapolation lemma, a version of Lemma 5.2.3.

Lemma 5.3.8 *Let T be a linear operator defined on*

$$L_0^\infty = \{\textit{martingale } f = (f_n)_{n\ge 0} : f \in L^\infty, f_0 = 0\}, \tag{5.3.20}$$

valued in the space of all measurable functions. Suppose that T is of weak type (H_p^*, L^{r_0}) for $0 < p_0 \le r_0 \le \infty$ with the bound $\|T\|$, and T is commutable with stopping times in following sense

$$T(f - f^{(\tau)})\chi_{\{\tau=\infty\}} = 0, \text{ a.e.}, \quad \forall f \in L_0^\infty,\ \forall \tau. \tag{5.3.21}$$

Then for all (p,r), $0 < p \le p_0$, $\frac{1}{p} - \frac{1}{r} = \frac{1}{p_0} - \frac{1}{r_0}$, T is of the type $(\mathcal{P}_p, L^r)$ with the bound $\le C\|T\|$.

Proof. Let $f \in \mathcal{P}_p$, with $\lambda_f = (\lambda_n(f))_{n\ge 0}$ as its optimal majorant. Assume $\|f\|_{\mathcal{P}_p} = \|T\| = 1$. For $\lambda > 0$, set $\delta = \lambda^{\frac{r}{p}}$, and define $\tau = \inf\{n : \lambda_n(f) > \delta\}$. Then $\{\tau < \infty\} = \{\lambda_\infty(f) > \delta\}$ and $M_\tau f \le \delta$. From

$$\begin{aligned}|\{|Tf| > \lambda\}| &\le |\{\tau<\infty\}| + C\lambda^{-r_0}\Big(\int_{\{\tau<\infty\}} + \int_{\{\tau=\infty\}} (Mf^{(\tau)})^{p_0} d\mu\Big)^{\frac{r_0}{p_0}}\\ &\le |\{\tau<\infty\}| + C\lambda^{\frac{rr_0}{p}-r_0}|\{\tau<\infty\}|^{\frac{r_0}{p_0}}\\ &\quad + C\lambda^{-r_0}\Big(\int_{\{\lambda_\infty(f)\le\delta\}} \lambda_\infty(f)^{p_0} d\mu\Big)^{\frac{r_0}{p_0}},\end{aligned}$$

the desired inequality

$$\|Tf\|_r \le C\|T\|\,\|\lambda_\infty(f)\|_p = C\|T\|\,\|f\|_{\mathcal{P}_p}$$

follows. The proof is finished. □

Remark The difference between this lemma and Lemma 5.2.3 lies in that we have to replace (5.2.9) by (5.3.21) when we face an operator acting on a space of martingales but not on a space of adapted process, and hence we have to replace H_p^* by $\mathcal{P}_p$ in the conclusion.

Theorem 5.3.9 *Let $0 \leq \alpha < 1, 0 < p < 1$. Then $\mathcal{D}$ is of type $(\Lambda_\alpha, H_1; L^{r_0})$ and $(\Lambda_\alpha, \mathcal{P}_p; L^r)$, with r_0, r satisfying*

$$\frac{1}{r_0} = 1 - \alpha, \quad \frac{1}{r} = \frac{1}{p} - \alpha.$$

Proof. Consider $\mathcal{D}$ as an operator T_f acting on g for fixed $f \in \Lambda_\alpha$. Then T_f is of type (H_{p_0}, L^{r_0}), $1 < p_0 < \infty$, $\alpha = 1/p_0 - 1/r_0$ with the bound $\leq C\|f\|_{\Lambda_\alpha}$. And hence T_f is of type $(\mathcal{P}_p, L^r)$, $0 < p \leq 1$, $1/p - 1/r = \alpha$, with same bound. Now for the assertion in the case $p = 1$, it remains to apply Davis' decomposition and T_f's type $(\Lambda_\alpha, \mathcal{A}_1; L^{r_0})$, which we will prove in next theorem. As such, we get

$$\begin{aligned} \|\mathcal{D}(f,g)\|_{r_0} &\leq \|\mathcal{D}(f,g^{(1)})\|_{r_0} + \|\mathcal{D}(f,g^{(2)})\|_{r_0} \\ &\leq C\|f\|_{\Lambda_\alpha}(\|g^{(1)}\|_{\mathcal{A}_1} + \|g^{(2)}\|_{\mathcal{P}_1}) \\ &\leq C\|f\|_{\Lambda_\alpha}\|g\|_{H_1}. \end{aligned} \tag{5.3.22}$$

The proof is finished. □

Now consider the case $g \in \mathcal{A}_1$.

Theorem 5.3.10 *Let $0 \leq \alpha \leq 1$, $\frac{1}{r_0} = 1 - \alpha$. Then $\mathcal{D}$ is of type $(\Lambda_\alpha, \mathcal{A}_1; L^{r_0})$, with the bound ≤ 1, i.e.*

$$\|\mathcal{D}(f,g)\|_{r_0} \leq \|f\|_{\Lambda_\alpha}\|g\|_{\mathcal{A}_1}. \tag{5.3.23}$$

Proof. We have $|\Delta_k f| \leq \|f\|_{\Lambda_\alpha}\omega_k^\alpha$, for all k, a.e. So

$$\|\mathcal{D}(f,g)\|_{r_0} \leq \sum_1^\infty \|\Delta_k f \Delta_k g\|_{r_0} \leq \|f\|_{\Lambda_\alpha} \sum_1^\infty \|\omega_k^\alpha \Delta_k g\|_{r_0}. \tag{5.3.24}$$

But for each k, we have

$$\|\omega_k^\alpha \Delta_k g\|_{r_0} = \Big(\sum_j \int_{I_j^{(k)}} \omega_k^{\alpha r_0} |\Delta_k g|^{r_0} d\mu\Big)^{\frac{1}{r_0}},$$

and on each $I_j^{(k)}$,

$$|\Delta_k g| = |I_j^{(k)}|^{-1} \int_{I_j^{(k)}} |\Delta_k g| d\mu, \tag{5.3.25}$$

so we get

$$\begin{aligned} \int_{I_j^{(k)}} \omega_k^{\alpha r_0} |\Delta_k g|^{r_0} d\mu &= |I_j^{(k)}|^{\alpha r_0 + 1 - r_0} \Big(\int_{I_j^{(k)}} |\Delta_k g| d\mu\Big)^{r_0} \\ &= \Big(\int_{I_j^{(k)}} |\Delta_k g| d\mu\Big)^{r_0}, \end{aligned}$$

and hence

$$\|\omega_k^\alpha \Delta_k g\|_{r_0} = \Big(\sum_j \Big(\int_{I_j^{(k)}} |\Delta_k g| d\mu\Big)^{r_0}\Big)^{\frac{1}{r_0}} \le \sum_j \int_{I_j^{(k)}} |\Delta_k g| d\mu \le \|\Delta_k g\|_1.$$

From (5.3.24), we finally get

$$\|\mathcal{D}(f,g)\|_{r_0} \le \|f\|_{\Lambda_\alpha} \sum_1^\infty \|\Delta_k g\|_1 = \|f\|_{\Lambda_\alpha} \|g\|_{\mathcal{A}_1}.$$

Noticing that the case $\alpha = 1 (r_0 = \infty)$ is easier to deal with owing to (5.3.25). This completes the proof of the theorem. □

Theorem 5.3.11 *Let $0 \le \alpha < 1$, $\frac{1}{r_0} = 1 - \alpha$. Then $\mathcal{D}$ is of type $(\Lambda_\alpha, L^1; WL^{r_0})$, with the bound $\le C$.*

Proof. This follows from Gundy's decomposition and the results obtained for $g \in \mathcal{A}_1$ and $g \in L^p$, $p > 1$. □

Now we list $\mathcal{D}$'s boundedness properties as follows. $\mathcal{D}$ is of type

$$(H_p^S, H_q^S; L^r), \quad 0 < p, q \le \infty, \quad \frac{1}{r} = \frac{1}{p} + \frac{1}{q},$$

$$(H_p^S, L^1; WL^{r_0}), \quad 0 < p \le \infty, \quad \frac{1}{r_0} = \frac{1}{p} + 1,$$

$$(L^1, L^1; WL^{\frac{1}{2}}),$$

$$(\Lambda_\alpha, H_p; L^r), \quad 1 \le p < \frac{1}{\alpha}, \; 0 \le \alpha < 1, \; \frac{1}{r} = \frac{1}{p} - \alpha,$$

$$(\Lambda_\alpha, \mathcal{P}_p; L^r), \quad 0 < p < \frac{1}{\alpha}, \; 0 \le \alpha < 1, \; \frac{1}{r} = \frac{1}{p} - \alpha,$$

$$(\Lambda_\alpha, \mathcal{A}_1; L^{r_0}), \quad 0 \le \alpha < 1, \; \frac{1}{r_0} = 1 - \alpha,$$

$$(\Lambda_\alpha, L^1; WL^{r_0}), \quad 0 \le \alpha < 1, \; \frac{1}{r_0} = 1 - \alpha.$$

This list together with the one in §5.2, shows that the boundedness properties of $\mathcal{C}$ and $\mathcal{D}$ have some remarkable differences, although they are similar in most cases. For example, we have no smoothness results for $\mathcal{C}$ when f is smooth, and no H_r^* results for $\mathcal{D}$ either. Is it possible to get these result? Now we want answer it negatively. By the way, we want also to show that some other related boundedness results are the best possible ones by making use of (5.3.1). From (5.3.1) and Theorem 5.1.1, we have

$$fg = \mathcal{C}(f,g) + \mathcal{B}(g,f) = \mathcal{C}(g,f) + \mathcal{B}(f,g), \; \forall f,g \in L^1, \; f_0 = g_0 = 0. \quad (5.3.26)$$

This is a very useful fact in following argument.

Theorem 5.3.12 *$\mathcal{C}$ is of type $(L^\infty, BMO; BMO)$, but may be not of type $(BMO, BMO; BMO)$. $\mathcal{D}$ is of type $(H_1, BMO; L^1)$, but may be not of type $(H_1, BMO; H_1)$. $\mathcal{B}$ is of type $(L^\infty, BMO; BMO)$ and $(H_1, BMO; L^1)$, but may be not of type $(BMO, BMO; BMO)$, either of type $(H_1, BMO; H_1)$.*

Proof. The main step of the proof is to construct a probability space $(\Omega, \mathcal{F}, \mu, \{\mathcal{F}_n\}_{n\geq 0})$ and a $g \in BMO$, such that $\mathcal{C}(g, g) \notin BMO$. Consider dyadic martingales on (0, 1]. Denote $I_k = (2^{-(k+1)}, 2^{-k}]$, $k = 0, 1, \cdots$, set

$$g(x) = \sum_{k=0}^{\infty} k\chi_{I_k}. \tag{5.3.27}$$

We want to show that it is a typical example of $f \in BMO - L^\infty$. For any dyadic interval I of which 0 is not one end-point, we have $\int_I |g - g_I| dx = 0$. Let $J = J_n = (0, 2^{-n}]$. We have

$$\begin{aligned} g_J &= \frac{1}{|J|}\int_J g\,dx = 2^n \sum_{k=n}^{\infty} k2^{-k-1} \\ &= 2^n(n\{2^{-n-1} + 2^{-n-2} + \cdots\} + \{2^{-n-2} + 2^{-n-3} + \cdots\} \\ &\quad + \{2^{-n-3} + 2^{-n-4} + \cdots\} + \cdots) \\ &= 2^n(n2^{-n} + 2^{-n-1} + 2^{-n-2} + \cdots) = n + 1, \end{aligned} \tag{5.3.28}$$

and

$$\begin{aligned} \frac{1}{|J|}\int_J |g - g_J| dx &= 2^n \sum_{k=n}^{\infty} |k - n - 1| 2^{-k-1} \\ &= 2^n\left(2^{-n-1} + \sum_{n+1}^{\infty}(k - n - 1)2^{-k-1}\right) \\ &= 2^n\left(2^{-n-1} + 2^{-n-1}\sum_{0}^{\infty} j2^{-j-1}\right) = 1. \end{aligned}$$

This proves that $g \in BMO$, and $\|g\|_* \leq C$.

Now we estimate $\mathcal{C}(g, g)$. We have (noticing $I_k \in \mathcal{F}_{k+1}$)

$$g_{n-1} = \sum_{0}^{n-2} k\chi_{I_k} + n\chi_{(0, 2^{-n-1}]},$$

$$g_n = \sum_{0}^{n-2} k\chi_{I_k} + (n-1)\chi_{(2^{-n}, 2^{-n+1}]} + (n+1)\chi_{(0, 2^{-n}]},$$

$$\Delta_n g = \chi_{(0, 2^{-n}]} - \chi_{(2^{-n}, 2^{-n+1}]},$$

and so

$$\Delta_n \mathcal{C}(g, g) = g_{n-1}\Delta_n g \notin L^\infty.$$

This proves $\mathcal{C}(g,g) \notin BMO$. So, in this case, $\mathcal{C}$ is not of type $(BMO,\ BMO;\ BMO)$.

In the same case, $\mathcal{B}$ is not of type $(BMO, BMO; BMO)$, since $\mathcal{D}$ is and $\mathcal{C}$ is not. In addition, $\mathcal{B}$ is not of type $(H_1, BMO; H_1)$ either, otherwise $\mathcal{B}$ would be of type $(BMO,\ BMO;\ BMO)$ by duality, a contradiction. As a result, $\mathcal{D}$ is not of type $(H_1, BMO; H_1)$, since $\mathcal{C}$ is and $\mathcal{B}$ is not. This completes the proof of the theorem. □

Remark The assertions in the theorem remain to be true even when $g \in L^\infty$. This can be seen as follows. Suppose that $\mathcal{C}$ was of type $(BMO, L^\infty; BMO)$. From (5.3.26) and the fact $\mathcal{B}(g,f) \in BMO$, $fg = \mathcal{C}(f,g) + \mathcal{B}(g,f) \in BMO$, for all $f \in BMO$, $g \in L^\infty$. This is obviously a contradiction. So $\mathcal{C}$ is not of type $(BMO, L^\infty; BMO)$. Since $\mathcal{D}$ is, $\mathcal{B}$ is not. By duality, $\mathcal{B}$ is not of type $(H_1, L^\infty; H_1)$. Since $\mathcal{C}$ is of type $(H_1, L^\infty; H_1)$, $\mathcal{D}$ is not. This proves the assertions.

This theorem and the remark tell us that f's smoothness (for example $f \in BMO$) has nothing beneficial to the operator $\mathcal{C}$, and $\mathcal{D}$'s L^r-result could not be improved to H_r-result. There are some other similar phenomena as shown in following theorems.

Theorem 5.3.13 *Let $1 \leq p < \infty$. Then both $\mathcal{C}$ and $\mathcal{B}$ may not be of type $(BMO, L^p; WL^p)$.*

Proof. Suppose that $\mathcal{C}$ was of type $(BMO, L^p; WL^p)$. Since $\mathcal{B}$ is of type $(L^p, BMO; L^p)$, from (5.3.26), fg would be in WL^p for all $f \in BMO, g \in L^p$. This is not true in general. For example, choose $(\Omega, \mathcal{F}, \mu, \{\mathcal{F}_n\}_{n\geq 0})$ such that $BMO \neq L^\infty$. Choose $f \in BMO - L^\infty$. For any $N > 0, |E| = |\{|f| > N\}| > 0$. Set $g = |E|^{-\frac{1}{p}}\chi_E$. Then $||g||_p = 1$, and

$$||fg||_{WL^p} = \sup_{t>0} t^{\frac{1}{p}}(fg)^*(t) \geq \sup_{t>0} t^{\frac{1}{p}} N|E|^{-\frac{1}{p}}\chi_{(0,|E|)}(t) = N^\dagger.$$

This shows that $\mathcal{C}$ could not be of type $(BMO, L^p; WL^p)$. The same proof gives the assertion about $\mathcal{B}$. The proof is thus finished. □

Theorem 5.3.14 *Suppose that $(\Omega, \mathcal{F}, \mu, \{\mathcal{F}_n\}_{n\geq 0})$ be such that $H_r^S = H_r^* = H_r$, $0 < r < \infty$, $L^{r_1} \neq L^{r_2}$, when $r_1 \neq r_2$, and $BMO \neq L^\infty$. Let $1 \leq p, q < \infty, 1/r = 1/p + 1/q \geq 1$. Then $\mathcal{D}$ is not of type $(H_p, H_q; H_r)$.*

Proof. Suppose that $\mathcal{D}$ was. Since $\mathcal{C}$ is, $\mathcal{B}$ would be of type $(H_p, H_q; H_r)$. By duality, $\mathcal{B}$ would be of type $(\Lambda_\alpha, H_q; H_p')$, where $\alpha = 1/r - 1$, $H_p' = L^{p'}$ when $p > 1$, or BMO, when $p = 1$. Since $\mathcal{C}$ is always of type $(H_q, \Lambda_\alpha; H_p')$, from (5.3.26), we would have

$$fg \in H_p', \quad \forall f \in H_q,\ \forall g \in \Lambda_\alpha. \tag{5.3.29}$$

Notice that $q < p'$ when $\alpha > 0$, and $q = p'$ when $\alpha = 0$. When $\alpha = 0$, we have seen that (5.3.29) is not true. When $\alpha > 0$, it is even easy to show that (5.3.29) is not

† Here * denotes the rearrangement function.

true either by taking $g = 1 \in \Lambda_\alpha$. This proves the theorem. □

Theorem 5.3.15 *Each of $\mathcal{C}, \mathcal{B}$ and $\mathcal{D}$ may be not of type $(L^\infty, L^\infty; L^\infty)$.*

Proof. Obviously, $\mathcal{D}$ may be not of type $(L^\infty, L^\infty; L^\infty)$, since $\mathcal{D}(f,f) = S(f)^2$ and $S(f) \notin L^\infty$ for some $f \in L^\infty$ in many cases. From

$$f^2 = \mathcal{C}(f,f) + \mathcal{B}(f,f), \quad \forall f \in L^\infty,$$

we see that $\mathcal{C}(f,f)$ and $\mathcal{B}(f,f)$ must be in or not in L^∞ simultaneously. But the equality

$$\mathcal{D}(f,f) = \mathcal{B}(f,f) - \mathcal{C}(f,f)$$

shows that only the case $\mathcal{C}(f,f) \notin L^\infty$ and $\mathcal{B}(f,f) \notin L^\infty$ can happen. The proof is finished. □

Remark As a result, from the theorem, we see that $\mathcal{B}$ may be not of type $(L^1, L^\infty; L^1)$, and $\mathcal{C}$ may be not of type $(L^\infty, L^1; L^1)$.

5.4 Some necessary conditions

Considering the martingale transform $T_v(f)$ as an operator on $v \in V_p$, we see that when $f \in \Lambda_\alpha$, $0 \le \alpha \le 1$, then $T_v(f) \in L^r$, with $\frac{1}{r} = \frac{1}{p} - \alpha$. Our question is, if $T_v(f)$ is of type (V_p, L^r), $p \le r$, $r \ge 1$, what can we say about f's behavior? A necessary condition could be obtained simply. In fact, for all n, and any $A^{(n)} \in \mathcal{F}_n$, setting $v = (v_k)_{k\ge 0}$, with $v_k = 0$, $k \le n-1$, $= \chi_{A^{(n)}}$, $k \ge n$, then $\|v\|_{V_p} = |A^{(n)}|^{1/p}$, and $T_v(f) = \sum_{n+1}^\infty v_{k-1}\Delta_k f = (f - f_n)\chi_{A^{(n)}}$. From T_v's type (V_p, L^r), we get

$$\left(\int_{A^{(n)}} |f - f_n|^r d\mu\right)^{\frac{1}{r}} \le C\|T_v\||A^{(n)}|^{\frac{1}{p}},$$

$$\left(|A^{(n)}|^{-1}\int_{A^{(n)}} |f - f_n|^r d\mu\right)^{\frac{1}{r}} \le C\|T_v\||A^{(n)}|^{\alpha}.$$

By taking $A^{(n)} = I^{(n)}(\mathcal{F}_n\text{-atoms})$ or $A^{(n)} \ni x$, $|A^{(n)}| \to 0$, for $x \notin \bigcup I_j^{(n)}$, we get

$$E(|f - f_n|^r|\mathcal{F}_n)^{\frac{1}{r}} \le C\|T_v\|\omega_n^\alpha, \quad \forall n.$$

This means $f \in {}_r\lambda_\alpha$. This result is almost best possible, since when $r \le 2$, $f \in {}_2\lambda_\alpha$ is already a sufficient condition. When we consider operators $\mathcal{C}_g(\cdot) = \mathcal{C}(\cdot, g)$, $\mathcal{B}_g$ or $\mathcal{D}_g$, the same question could be asked, but the same simple method works no longer well. This section will be devoted to this topic. Again, we will benefit from the extrapolation lemma, i.e. Lemma 5.3.8 without the restriction $p_0 \le r_0$.

Lemma 5.4.1 *Lemma* 5.3.8 *holds without any restriction about the order between* p_0 *and* r_0.

Proof. By making use of following lemma about $\mathcal{P}_p$'s interpolation theory, it is enough to prove that T is of weak type $(\mathcal{P}_p, L^r)$ under the condition of Lemma 5.3.8 but without the restriction $p_0 \le r_0$. It is easier. With same concepts, we have

$$\begin{aligned}|\{|Tf|>\lambda\}| &\le |\{\tau<\infty\}| + |\{|Tf^{(\tau)}|>\lambda\}|\\ &\le C\lambda^{-r}\|\lambda_\infty(f)\|_p^p + C\lambda^{-r_0}(E(Mf^{(\tau)})^{p_0})^{\frac{r_0}{p_0}}\\ &\le C\lambda^{-r} + C\lambda^{\frac{rr_0}{pp_0}(p_0-p)-r_0} \le C\lambda^{-r}.\end{aligned}$$

This proves the lemma. □

Lemma 5.4.2 *Let* $0<p_1,\ p_2<\infty,\ 0<\theta<1,\ p$ *be such that* $\frac{1}{p}=\frac{1-\theta}{p_1}+\frac{\theta}{p_2}$. *Then for* $\mathcal{P}_r$*'s real interpolation spaces we have*

$$(\mathcal{P}_{p_1}, \mathcal{P}_{p_2})_{\theta,p} = \mathcal{P}_p. \tag{5.4.1}$$

Proof. By the reiteration theorem (see Bergh-Löfström [1]), it is enough to prove $(\mathcal{P}_{p_0}, L_0^\infty)_{\theta,p} = \mathcal{P}_p$, with $p=\frac{p_0}{1-\theta}$. We want to establish the following estimate for the K-functional $K(t,f)=K(t,f,\mathcal{P}_{p_0},L_0^\infty)$

$$K(t,f)^{p_0} \le C\int_0^{t^{p_0}} \lambda_\infty^{*p_0}(\tau)d\tau, \quad \forall f\text{'s admissible majorant } \lambda=(\lambda_n), \forall t>0, \tag{5.4.2}$$

(Here $*$ denotes again the rearrangement function.)

$$\int_0^{t^{p_0}} \lambda_\infty^{*p_0}(\tau)d\tau \le CK(t,f)^{p_0}, \quad \exists f\text{'s admissible majorant } \lambda=(\lambda_n), \forall t>0. \tag{5.4.3}$$

Once it is proved, we will have

$$C\|f\|_{\mathcal{P}_p} \le \|f\|_{(\mathcal{P}_{p_0},L_0^\infty)_{\theta,p}} \le C\|f\|_{\mathcal{P}_p}.$$

Now aim at (5.4.2). Let $f\in\mathcal{P}_{p_0}+L_0^\infty$ be given, and $\lambda=(\lambda_n)_{n\ge0}$ be nonnegative, increasing and adapted, and satisfying $|f_n|\le\lambda_{n-1}$, for all n. Let $t>0$. In the case $\lambda_\infty^*(t^{p_0})=0$, the inequality is easy to be seen, since $\int_0^{t^{p_0}}\lambda_\infty^{*p_0}(\tau)d\tau \ge \|f\|_{\mathcal{P}_{p_0}}^{p_0}$. When $\lambda_\infty^*(t^{p_0})>0$, define $\tau=\inf\{n:\lambda_n>\lambda_\infty^*(t^{p_0})\}$. Then we have

$$|\{\tau<\infty\}| = |\{\lambda_\infty>\lambda_\infty^*(t^{p_0})\}| \le t^{p_0}, \quad Mf^{(\tau)} \le \lambda_{\tau-1} \le \lambda_\infty^*(t^{p_0}). \tag{5.4.4}$$

Make f's Calderón-Zygmund Decomposition $f=f-f^{(\tau)}+f^{(\tau)}$. Noticing that the martingale $((f-f^{(\tau)})_n)_{n\ge0}$, where

$$(f-f^{(\tau)})_n = f_n - f_{n\wedge\tau} = (f_n-f_\tau)\chi_{\{\tau<n\}},$$

has an adapted, nonnegative increasing process $(2\lambda_n \chi_{\{\tau<n+1\}})_{n\geq 0}$ as its admissible majorant, we see

$$\|f - f^{(\tau)}\|_{\mathcal{P}_{p_0}}^{p_0} \leq \int_{\{\tau<\infty\}} (2\lambda_\infty)^{p_0} d\mu \leq C \int_0^{t^{p_0}} \lambda_\infty^{*p_0}(\tau) d\tau. \tag{5.4.5}$$

On the other hand, $f^{(\tau)} \in L_0^\infty$, and

$$t^{p_0}\|f^{(\tau)}\|_{L_0^\infty}^{p_0} \leq t^{p_0}\lambda_\infty^{*p_0}(t^{p_0}) \leq \int_0^{t^{p_0}} \lambda_\infty^{*p_0}(\tau) d\tau. \tag{5.4.6}$$

Combining (5.4.5) and (5.4.6), we get (5.4.2). Now aim at (5.4.3). Let $f \in \mathcal{P}_{p_0} + L_0^\infty$. It is easy to prove $\int_0^{t^{p_0}} (Mf)^{*p_0}(\tau)d\tau \leq CK(t,f)^{p_0}$ for all t. So $K(t,f) \neq 0$, for all $t > 0$, if $f \not\equiv 0$. Thus for each $t > 0$, there is f'decomposition $f = f^{(0)} + f^{(1)}$ such that

$$\|f^{(0)}\|_{\mathcal{P}_{p_0}} + t\|f^{(1)}\|_{L_0^\infty} \leq 2K(t,f).$$

Let $\tilde{\lambda}^{(i)} = \tilde{\lambda}_{f^{(i)}}$ be $f^{(i)}$'s optimal majorant. Define $\tilde{\lambda}^{(t)} = \tilde{\lambda}^{(0)} + \tilde{\lambda}^{(1)}$. We have

$$\begin{aligned}\int_0^{t^{p_0}} \tilde{\lambda}_\infty^{(t)*p_0}(\tau)d\tau &\leq C\Big\{\int_0^{t^{p_0}} \tilde{\lambda}_\infty^{(0)*p_0}(\tau)d\tau + \int_0^{t^{p_0}} \tilde{\lambda}_\infty^{(1)*p_0}(0)d\tau\Big\} \\ &\leq C(\|f^{(0)}\|_{\mathcal{P}_{p_0}}^{p_0} + t^{p_0}\|f^{(1)}\|_{L_0^\infty}^{p_0}) \leq CK(t,f)^{p_0}.\end{aligned}$$

Now let $\{t_j\}$ be the set of all rational numbers in $(0,\infty)$, and define $\lambda = \inf_j\{\tilde{\lambda}^{(t_j)}\}$. Then we get $\lambda = (\lambda_n)$, a suitable admissible majorant of f which satisfies

$$\int_0^{t^{p_0}} \lambda_\infty^{*p_0}(\tau)d\tau \leq CK(t,f)^{p_0}, \quad \forall t > 0.$$

Thus (5.4.3) is proved. And the proof of the lemma is finished. □

The last lemma we need is about C_g's adjoint C_g'.

Lemma 5.4.3 *Let $g \in L^p$, $1 < p < \infty$. Then C_g' can be represented as follows*

$$C_g'(h) = gh - g_0h_0 - \sum_1^\infty [g_kh_k - E(g_kh_k|\mathcal{F}_{k-1})], \ \forall h = (h_n)_{n\geq 0} \in L^\infty. \tag{5.4.7}$$

Proof. The series in (5.4.7) converges in L^1. In fact, we see that

$$\begin{aligned}&\sum_1^\infty [g_kh_k - E(g_kh_k|\mathcal{F}_{k-1})] \\ &\qquad = \sum_1^\infty g_k\Delta_kh + \sum_1^\infty h_{k-1}\Delta_kg - \sum_1^\infty E(\Delta_kg\Delta_kh|\mathcal{F}_{k-1}),\end{aligned}$$

and the first two converge in L^p, and the last one converges in L^1 because of

$$\sum_1^\infty E(E(|\Delta_k g|\,|\Delta_k h|\,|\mathcal{F}_{k-1})) \le E(S(g)^p)^{\frac{1}{p}} E(S(h)^{p'})^{\frac{1}{p'}} \le C\|h\|_\infty \|g\|_p.$$

Meanwhile

$$\mathcal{C}_g(f) = gf - g_0 h_0 - \sum_1^\infty g_k \Delta_k f,\ g \in L^p,\ f \in L^\infty,$$

with the series converging in L^p. So we can interchange order between integration and summation in following argument to get

$$\begin{aligned} E(\mathcal{C}_g(f)h) &= E(ghf) - E(g_0 h_0 f_0) - \sum_1^\infty E(hg_k \Delta_k f) \\ &= E(ghf) - E(g_0 h_0 f_0) - \sum_1^\infty E([h_k g_k - E(g_k h_k|\mathcal{F}_{k-1})]\Delta_k f) \\ &= E(\mathcal{C}_g'(h)f). \end{aligned}$$

This ends the proof of the lemma. □

Remark $\mathcal{C}_g'$ satisfies the condition $\mathcal{C}_g'(h-h^{(\tau)})\chi_{\{\tau=\infty\}} = 0$. In fact, for any $h \in L_0^\infty$ and any stopping time τ, we have $(h - h^{(\tau)})_k = \sum_{j=0}^k \chi_{\{j>\tau\}}\Delta_j h$, so

$$g(h - h^{(\tau)})\chi_{\{\tau=\infty\}} = 0 = g_k(h - h^{(\tau)})_k \chi_{\{\tau=\infty\}},\quad \forall k,$$

$$E(g_k(h - h^{(\tau)})_k|\mathcal{F}_{k-1})\chi_{\{\tau=\infty\}} = \sum_{j=0}^k E(g_k \Delta_j h|\mathcal{F}_{k-1})\chi_{\{j>\tau\}}\chi_{\{\tau=\infty\}} = 0.$$

Now we are in the position to get our main results.

Theorem 5.4.4 *Let $1 < p < \infty$, T be any one of $\mathcal{C}_g, \mathcal{B}_g$ or $\mathcal{D}_g$. Suppose that T is of type (L^p, L^p), then $g \in bmo_1$, $\|b\|_{bmo_1} \le C\|T\|$.*

Proof. Consider the case $T = \mathcal{B}_g$ or $\mathcal{C}_g$. Applying Lemma (5.4.1) to T', we see that T' is of type $(\mathcal{P}_1, L^1)$, and hence T is of type $(L^\infty, \mathcal{P}_1')$. Notice that $\mathcal{P}_1' \cap L_u^1 = bmo_1$. Since $g - g_0 = T(1) \in L^p \cap \mathcal{P}_1'$, this implies $g \in bmo_1$, and

$$\|g\|_{bmo_1} \le C\|g\|_{\mathcal{P}_1'} \le C\|T\|.$$

Now consider the case $T = \mathcal{D}_g$. First consider the operator sequence $\{\mathcal{D}_{g_n}\}$. Notice that $\mathcal{D}_{g_n}(f) = \mathcal{D}_g(f_n)$, so $\|\mathcal{D}_{g_n}\| \le \|\mathcal{D}_g\|$. Lemma 5.4.1 can be used to $\mathcal{D}$, too. Thus, the sequence of linear functionals on $\mathcal{P}_1$ defined by

$$ln(f) = E(g_n f) = E\Big(\sum_1^n \Delta_k g \Delta_k f\Big) = E(\mathcal{D}_{g_n}(f)),\quad f \in \mathcal{P}_1,$$

is uniformly bounded: $|l n(f)| \leq \|\mathcal{D}_g\| \|f\|_{\mathcal{P}_1}$. So $g_n \in \mathcal{P}'_1$ uniformly. In addition, $g_n \in L^{p'}$ uniformly, for n. In fact, for all $f \in L^p$, we have

$$|E(g_n f)| = |E(\mathcal{D}_{g_n}(f))| < \infty.$$

By the uniform boundedness principle, $g_n \in L^{p'}$ uniformly. Thus we get

$$E(|g_n - g_k| \,|\mathcal{F}_k) \leq C\|\mathcal{D}_g\|, \quad \forall n,\ \forall k \leq n,$$

and Fatou's lemma gives the desired result, i.e. $\|g\|_{bmo_1} \leq C\|\mathcal{D}_g\|$. This ends the proof. □

Theorem 5.4.5 *Let $0 < p \leq r < \infty$, $1 \leq r$, T be any one of C_g, $\mathcal{B}_g$, or $\mathcal{D}_g$. Suppose that T is of type (H_p^*, L^r). Then $g \in {}_1\lambda_\alpha$ with $\alpha = \frac{1}{p} - \frac{1}{r}$, and $\|g\|_{{}_1\lambda_\alpha} \leq C\|T\|$.*

Proof. In the case $T = \mathcal{B}_g$ or C_g, applying Lemma 5.4.1 to T', we see that T is of type $(\mathcal{P}_{p_0}, L^1)$ with $\frac{1}{p_0} - 1 = \alpha$, and hence T is of type $(L^\infty, \mathcal{P}'_{p_0})$. Noticing that $\mathcal{P}'_{p_0} \cap L^1_u = {}_1\lambda_\alpha$, we get $\|g\|_{{}_1\lambda_\alpha} \leq C\|T\|$. In the case $T = \mathcal{D}_g$, the proof is unchanged. The proof is finished. □

Finally consider the case $1 < r < p$.

Theorem 5.4.6 *Let $1 < r < p < \infty$, T be any one of C_g, $\mathcal{B}_g$, or $\mathcal{D}_g$. Suppose that T is of type (L^p, L^r) with the bound $\|T\|$. Then $g \in {}_1k_q$ with q satisfying $1/q = 1/r - 1/p$, and $\|g\|_{{}_1k_q} \leq C\|T\|$.*

Proof. The same argument gives that $\mathcal{B}_g$ or C_g is of type $(L^\infty, \mathcal{P}'_{p_0})$, $p_0 = q'$, so $g \in \mathcal{P}'_{p_0} \cap L^r \subset \mathcal{P}'_{p_0} \cap L^1_u = {}_1k_q$. In the case $T = \mathcal{D}_g$, $g_n \in {}_1k_q$ uniformly, so $g \in {}_1k_q$. Here the fact $\mathcal{P}'_r \cap L^1_u = {}_1k_{r'}$, $1 < r < \infty$, is referred to §2.8. The proof is finished. □

5.5 Best constants in some martingale inequalities

In the early in 1980's D. Burkholder discovered a technical method to get best constants implied in some martingale inequalities. This method is based on the idea of differential subordination and boundary value problems. The latter is the problem: Let $S \subset H \times H$ (H a real or complex Hilbert space) be a biconvex set, $S_0 \subset S$ be a nonempty subset, and $F(x, y)$ be a function on S_0, is there a biconcave (or biconvex) function $u(x, y)$ defined on S, such that $F \leq u$ (or $F \geq u$) on S_0 ? If so, find the least (or biggest) one. The fomer is the concept: Let $f = (f_n)_{n \geq 0}, g = (g_n)_{n \geq 0}$ be two H-valued martingales with respect to an usual

$(\Omega, \mathcal{F}, \mu, \{\mathcal{F}_n\}_{n\geq 0})$. $\Delta_n f$ and $\Delta_n g$ is denoted by d_n, e_n respectively. g is said to be differentially subordinate to f, if

$$|e_n| \leq |d_n|, \text{ a.e. } \forall n \geq 0. \tag{5.5.1}$$

For such f, g, how could we establish some inequalities with best constants between them? After D. Burkholder, say considering the L^p-inequality, we can rewrite the inequality what we want as following form

$$E(v(f,g)) = E(|g|^p - c_p|f|^p) \leq 0.$$

The problem is that for which c_p as small as possible, is there a biconcave function $u(x,y)$ (It means that as a function of $t \in \mathbf{R}$, $u(x_0+tx_1, y_0+ty_1)$ is concave.) defined on $S = H \times H$ such that for all x, y, $v(x,y) \leq u(x,y)$, and when $|y| \leq |x|$, $u(x,y) \leq 0$? Once it is solvable, then the desired inequality (with c_p may be best one) follows immediately from

$$E(v(f_n, g_n)) \leq E(u(f_n, g_n)) \leq E(u(f_{n-1}, g_{n-1})) \leq \cdots \leq E(u(f_0, g_0)) \leq 0.$$

In this section, we want to introduce several sharp inequalities due to D. Burkholder, for differentially subordinate H-valued martingales, such as L^p-type one, weak type one and exponential one. Obviously, they could be applied to martingale transforms provided the implied multiplier $v = (v_n)_{n\geq 0} \in V_\infty$, and to square functions as well.

In what follows, the inner product and the norm in H are denoted by $\langle\, ,\rangle$, $|\cdot|$ respectively, and $(\, ,)$ denotes $\mathrm{Re}\langle\, ,\rangle$. First, we state a differential formula which is easy to be verified: for all $x, h \in H$, for all $t \in \mathbf{R}$,

$$\frac{d}{dt}|x+th| = |x+th|^{-1}(x+th, h), \text{ provided } |x+th| \neq 0. \tag{5.5.2}$$

Now we begin with the L^p-type inequality. Let $1 < p < \infty$, p^* denotes $\max(p, p')$ with p' conjugate index of p.

Theorem 5.5.1 *Let $1 < p < \infty$, and f, g be two H-valued martingales satisfying* (5.5.1). *Then*

$$\|g\|_p \leq (p^* - 1)\|f\|_p. \tag{5.5.3}$$

In the nontrivial case $0 < \|f\|_p < \infty$, there is the equality in (5.5.3) *if and only if $p = 2$ and equalities in* (5.5.1) *hold. And the constant $p^* - 1$ is best possible.*

Proof. Assume $0 < \|f\|_p < \infty$. We can assume that $|f_n| \neq 0$, $|g_n| \neq 0$, for all n, for all $\omega \in \Omega$. Otherwise we could assume that f and g take their values in a proper subspace $H_0 \subset H$ by enlarging H, and consider $f + a$ and $g + a$ for some $a \in H_0^\perp$, and then let $a \to 0$. Define two functions v and u on $H \times H$ by

$$v(x,y) = |y|^p - (p^* - 1)^p|x|^p, \tag{5.5.4}$$

$$u(x,y) = \alpha_p(|y| - (p^* - 1)|x|)(|x| + |y|)^{p-1}, \tag{5.5.5}$$

with $\alpha_p = p(1-\frac{1}{p^*})^{p-1}$. We claim that

$$v(x,y) \le u(x,y), \quad \forall x,y, \tag{5.5.6}$$

$$E(u(f_n,g_n)) \le E(u(f_{n-1},g_{n-1})), \quad \forall n \ge 1, \tag{5.5.7}$$

$$E(u(f_0,g_0)) \le 0. \tag{5.5.8}$$

Once they are proved, (5.5.3) follows immediately. First, notice that for $\varphi(p) = (p-1)^{p-1}p^{2-p}$, we have

$$\varphi(p) \le 1 \text{ for } 1 < p \le 2; \text{ and } \varphi(p) \ge 1, \text{ for } 2 \le p < \infty. \tag{5.5.9}$$

To show (5.5.9), it is enough to consider $\psi(p) = \ln\varphi(p)(= \log_e \varphi(p))$. We have

$$\psi(1) = \psi(2) = 0, \psi(\infty) = \infty, \text{ and } \psi''(p) = \frac{1}{p}\Big(\frac{1}{p-1} - \frac{2}{p}\Big).$$

From the facts $\psi''(p) \ge 0$ on $(1,2]$, $\psi''(p) \le 0$ on $[2,\infty)$ and the boundary conditions as above, we get that $\psi(p) \le 0$ on $(1,2]$ and $\psi(p) \ge 0$ on $[2,\infty)$. (5.5.9) is thus proved. Now prove (5.5.6). When $|x|+|y| = 0$, it is obvious. When $|x|+|y| > 0$, owing to the homogeneity, we can assume $|x|+|y| = 1$, and (5.5.6) is reduced to

$$F(t) = \alpha_p(1-p^*t) - (1-t)^p + (p^*-1)^p t^p \ge 0, \quad 0 \le t \le 1. \tag{5.5.10}$$

Now prove (5.5.10). We have

$$F'(t) = -\alpha_p p^* + p(1-t)^{p-1} + p(p^*-1)^p t^{p-1},$$

$$F''(t) = -p(p-1)(1-t)^{p-2} + p(p-1)(p^*-1)^p t^{p-2}.$$

Both $F'(t)$ and $F''(t)$ have only one zero on $(0,1)$. The unique zero of $F'(t)$ is just $1/p^*$, which is also a zero of $F(t)$. The zero of $F''(t)$ is some t_0 which is in the interval $(0,1/p^*)$ when $p > 2$, and in $(1/p^*,1)$ when $p < 2$. Now consider two cases according to $p \ge 2$ or $p \le 2$ separately. When $p \ge 2$, we have

$$F(0) = \alpha_p - 1 = \varphi(p) - 1 \ge 0, \text{ and } F\Big(\frac{1}{p^*}\Big) = 0.$$

If there would be some $t_1 \in (0,1/p^*)$ such that $F(t_1) < 0$, then there would be some $t_2 \in (0,1/p^*)$ such that $F'(t_2) = 0$, this would contradict to the uniqueness of zero of F'. So $F(t) \ge 0$ on $[0,1/p^*]$. Since $F''(t) > 0$ on $[1/p^*,1]$ and $F(1/p^*) = F'(1/p^*) = 0$, we get $F(t) \ge 0$ on $[1/p^*,1]$. The proof of (5.5.6) in the case $p < 2$ is similar. $F(1) = 1-\varphi(p) \ge 0$, $F(1/p^*) = 0$, together with the uniqueness of zero of F' implies $F(t) \ge 0$ on $[\frac{1}{p^*},1]$. And $F(\frac{1}{p^*}) = F'(\frac{1}{p^*}) = 0$ together with $F''(t) > 0$ on $(0,\frac{1}{p^*})$ implies $F(t) \ge 0$ on same interval. The proof of (5.5.6) is finished. (5.5.8) is easy to see. Since $|g_0| \le |f_0|$, we have

$$\begin{aligned} &u(f_0,g_0) \le 2^{p-1}\alpha_p(2-p^*)|f_0|^p, \\ &E(u(f_0,g_0)) \le 2^{p-1}\alpha_p(2-p^*)E(|f_0|^p) \le 0. \end{aligned} \tag{5.5.11}$$

Next step is to prove (5.5.7). To do that we define a real variable function: For x, y, h, $k \in H$ fixed with $|k| \le |h|$, set

$$G(t) = u(x + th, y + tk), \quad \forall t \in \mathbf{R}. \tag{5.5.12}$$

In our case, x, y, h, k make $M(t) = |x + th|$ and $N(t) = |y + k|$ never to be zero for all $t \in \mathbf{R}$, so $G(t)$ is infinitely differential. We want to show that $G(t)$ is concave on $\mathbf{R}$. First consider the case $p \ge 2$. We have

$$G(t) = \alpha_p((M(t) + N(t))^p - pM(t)(M(t) + N(t))^{p-1}),$$

$$G'(t) = \alpha_p(p(M + N)^{p-1}N' - p(p-1)M(M + N)^{p-2}(M' + N')),$$

$$\begin{aligned} G''(t) = \alpha_p\{&p(M + N)^{p-1}N'' \\ &+p(p-1)(M + N)^{p-2}(M' + N')N' \\ &-p(p-1)(M + N)^{p-2}(M' + N')M' \\ &-p(p-1)(M + N)^{p-2}(M'' + N'')M \\ &-p(p-1)(p-2)(M + N)^{p-3}(M' + N')^2M\} = \sum_{i=1}^{5} I_i, \end{aligned}$$

$$\begin{aligned} I_1 = &-\alpha_p p(p-2)(M + N)^{p-1}N'' \\ &+\alpha_p p(p-1)(M + N)^{p-2}(M + N)N'' = I_{1,1} + I_{1,2}, \end{aligned}$$

$$\begin{aligned} I_{1,2} + I_2 + I_3 + I_4 &= -\alpha_p p(p-1)(M + N)^{p-2}\{-(M + N)N'' \\ &\quad -(M' + N')N' + (M' + N')M' + (M'' + N'')M\} \\ &= -\alpha_p p(p-1)(M + N)^{p-2}(MM'' + (M')^2 - NN'' - (N')^2) \\ &= -\alpha_p p(p-1)(M + N)^{p-2}(|h|^2 - |k|^2) \le 0. \end{aligned}$$

Since $N(t)$ is convex, $N'' \ge 0$, and hence $I_{1,1} \le 0$. In addition, $I_5 \le 0$. The proof of the concavity of $G(t)$ when $p \ge 2$ is finished. The case of $p \le 2$ can be handled by symmetry, since

$$(p-1)G(t) = -\alpha_p(M(t) - (p-1)N(t))(N(t) + M(t))^{p-1} = -\alpha_p H(t).$$

As we have shown, noticing the interchangment of $M(t)$ and $N(t)$, we have

$$\begin{aligned} H''(t) = &-(p(p-2)(M + N)^{p-1}M'' + p(p-1)(M + N)^{p-2}(|k|^2 - |h|^2) \\ &+p(p-1)(p-2)(M + N)^{p-3}(N' + M')^2N) \ge 0. \end{aligned}$$

This completes the proof of G's concavity. Thus, we get

$$G(1) \le G(0) + G'(0). \tag{5.5.13}$$

Since $G'(0) = (\varphi(x, y), h) + (\psi(x, y), k)$, with

$$\varphi(x, y) = \alpha_p((p - p^*)|y| - p(p^* - 1)|x|)(|x| + |y|)^{p-2}\frac{x}{|x|},$$

$$\psi(x,y) = \alpha_p(p|y| + (p+p^* - pp^*)|x|)(|x|+|y|)^{p-2}\frac{y}{|y|},$$

we get

$$u(x+h, y+k) \le u(x,y) + (\varphi(x,y), h) + (\psi(x,y), k). \tag{5.5.14}$$

Applying (5.5.14) to the case $x = f_{n-1}$, $y = g_{n-1}$, $h = d_n$, $k = e_n$ (As we have shown, d_n and e_n could be assumed to be in $H_0 \subset H$, and f_{n-1}, g_{n-1} have their non-vanishing projections in $H_0^\perp$, so $x + th$, $y + tk$ could never be zero.) we get

$$u(f_n, g_n) \le u(f_{n-1}, g_{n-1}) + (\varphi(f_{n-1}, g_{n-1}), d_n) + (\psi(f_{n-1}, g_{n-1}), e_n),$$

and hence (5.5.7). Up to now, we have proved (5.5.3).

Now prove that when $p \ne 2$ and $0 < ||f||_p < \infty$, then (5.5.3) is strict. Let m be the first integer such that $||f_m||_p > 0$. Then $|g_m| = |e_m| \le |d_m| = |f_m|$, so by (5.5.11), we get

$$E(u(f_n, g_n)) \le E(u(f_m, g_m)) \le 2^{p-1}\alpha_p(2-p^*)||f_m||_p^p < 0, \quad \forall n \ge m,$$

$$||g_n||_p^p \le (p^*-1)^p||f_n||_p^p + 2^{p-1}\alpha_p(2-p^*)||f_m||_p^p, \quad \forall n \ge m.$$

This proves the strictness of (5.5.3).

It remains to prove that $p^* - 1$ is best possible. By duality, it is enough to consider the case $p \ge 2$. Let $x > 0$, $\delta > 0$ be given arbitrarily. Let q be the unique root of the equality of $y : x^p = y^p - py^{p-1}$. It is easy to see that $q > p$, and $\lim_{x\to 0} q = p$. Denote $\theta = 1 - \frac{1}{q} = \frac{1}{q'}$. For $n \ge 1$, set $\pi_n = (x/x + (n-1)\delta)^q$, and define a family of σ-fields on $(0, 1]$ as follows:

$\mathcal{F}_n$ is generated by the atom $(0, \pi_n]$ and Borel sets on $(\pi_n, 1]$.

Define $g = (g_n)_{n\ge 1}$ with

$$g_n(t) = (x + (n-1)\delta)\chi_{(0,\pi_n]} + \theta x t^{\theta-1}\chi_{(\pi_n,1]}.$$

Then g is a martingale. It is adapted obviously. And $E(\Delta_n g|\mathcal{F}_{n-1}) = 0$ follows from $E(g_n) = x$, for all n, and $\mathrm{supp}\Delta_n g \subset (0, \pi_{n-1}]$,

$$\Delta_n g = \delta\chi_{(0,\pi_n]} + (\theta x t^{\theta-1} - x - (n-2)\delta)\chi_{(\pi_n,\pi_{n-1}]}.$$

Define g's martingale transform f by the multiplier $v = (1, -1, 1, -1\cdots)$. We want to get the estimate

$$\sup_n |f_n - x| \le \delta + x(1-\theta)t^{\theta-1}.$$

Once it is proved, then

$$||\sup_n |f_n - x|\,||_{L^p} \le \delta + 1.$$

But $||g||_p$ is easy to estimate:

$$\lim_{n\to\infty} ||g_n||_p^p = \lim_{n\to\infty}\left(|x+(n-1)\delta|^p\pi_n + \int_{\pi_n}^1 (\theta x t^{\theta-1})^p dt\right) = (q-1)^p.$$

Thus, we would have

$$\lim_{x\to 0}\lim_{\delta\to 0}\|g\|_p = (p-1)\lim_{x\to 0}\lim_{\delta\to 0}\|Mf\|_p.$$

Since g is f's martingale transform, this would prove the best possibility of the constant p^*-1 in (5.5.3).

Now return to estimate $\sup_k |f_k - x|$. For n fixed, consider the interval (π_{n+1}, π_n). Since $\Delta_1 g = x$, $\Delta_k g = \delta$, $2 \le k \le n$, on this interval, so

$$\left\| \max_{k\le n} |(f_k - x)\chi_{(\pi_{n+1},\pi_n)}| \right\|_\infty \le \delta.$$

But for $k \ge n+2$, $\Delta_k g$ are all supported in $(0, \pi_{n+1}]$, so $f_k = f_{n+1}$ on $(\pi_{n+1}, \pi_n]$, for all $k \ge n+2$. Now on (π_{n+1}, π_n)

$$\begin{aligned} |f_{n+1} - f_n| &= g_n - g_{n+1} = x + (n-1)\delta - \theta x t^{\theta-1} \\ &= x\pi_n^{\theta-1} - \theta x t^{\theta-1} \le x(1-\theta)t^{\theta-1}. \end{aligned}$$

This proves the assertion, and the proof of the theorem is thus finished. □

Now we turn to the exponential type inequality. More precisely, let $\Phi(t)$ be an increasing convex function on $[0,\infty)$ satisfying $\Phi(0) = \Phi'(0) = 0$, Φ' is strictly convex (so Φ'' is strictly increasing), and $\int_0^\infty \Phi(t)e^{-t}dt < \infty$.

Theorem 5.5.2 *Let f, g be two H-valued martingales as above, with $\|f\|_\infty \le 1$. Then for Φ as above, we have*

$$\sup_n E(\Phi(|g_n|)) < \frac{1}{2}\int_0^\infty \Phi(t)e^{-t}dt. \tag{5.5.15}$$

And the inequality is best possible.

Proof. Denote $S = \{(x,y) \in H\times H : |x| \le 1\}$. Since $\|f\|_\infty \le 1$, we can assume $(f_n(\omega), g_n(\omega)) \in S$, for all ω, for all n. We want to define a function $u(x,y)$ from S to $\mathbf{R}$ satisfying

$$E(\Phi(|g_n|)) \le E(u(f_n, g_n)), \tag{5.5.16}$$

$$E(u(f_n, g_n)) \le E(u(f_{n-1}, g_{n-1})), \quad \forall n, \tag{5.5.17}$$

$$E(u(f_0, g_0)) \le \frac{1}{2}\int_0^\infty \Phi(t)e^{-t}dt. \tag{5.5.18}$$

Once it is done, (5.5.15) without the strictness follows immediately. To define $u(x,y)$, let $A(t)$, $B(t)$ be two functions on $[1,\infty)$ given by $B(t) = \Phi(t-1)$, and

$$A(t) = e^t\int_t^\infty B(s)e^{-s}\,ds = e^{t-1}\int_{t-1}^\infty \Phi(s)e^{-s}\,ds.$$

Set $F(t) = A(t) - B(t)$. Noticing $A' = A - B$, we have

$$F(t) \geq 0, \tag{5.5.19}$$

$$F'(t) = A(t) - B(t) - B'(t) \geq 0, \tag{5.5.20}$$

$$F''(t) = A(t) - B(t) - B'(t) - B''(t) \geq 0, \tag{5.5.21}$$

$$tF'(t) - F(t) \geq 0. \tag{5.5.22}$$

Now define

$$u(x,y) = \begin{cases} (1 + |y|^2 - |x|^2)\dfrac{A(1)}{2}, & |x| + |y| \leq 1, \\ (1 - |x|)A(|x| + |y|) + |x|B(|x| + |y|), & |x| + |y| > 1. \end{cases} \tag{5.5.23}$$

Then $u(x,y)$ is continuous on S, since two formulas of u have same limit $y|A(1)$ at (x,y) such that $|x| + |y| = 1$. Now for $x, y, h, k \in H$, such that $|x| \leq 1$, $|x + h| \leq 1$, and $|k| \leq |h|$, for all $t \in \mathbf{R}$ define $G(t) = u(x+th, y+tk)$. We claim that $G'(t)$ exists everywhere and is continuous on $(0,1)$. In fact, denoting $M(t) = |x+th|$, $N(t) = |y+tk|$, then the possible exceptional set $\{M(t) = 0\} \bigcup \{N(t) = 0\} \bigcup \{M(t) + N(t) = 1\}$ is a finite set, since when $h \neq 0$ $M(t)$ is strictly convex, and when $h = 0$, $M(t)$ is constant, and analogously for $N(t)$. Consider these exceptional points. Let t_0 be such that $M(t_0) + N(t_0) = 1$, and t is in a small neighbourhood of t_0, then

$$\lim_{t \to t_0} \frac{G(t) - G(t_0)}{t - t_0} = \lim_{\varepsilon_t \to 0} G'(t_0 + \varepsilon_t),$$

where $t_0 + \varepsilon_t$ is never exceptional, and $G'(t_0 + \varepsilon_t)$ is represented by one of following formulas

$$G'(t) = ((y + tk, k) - (x + th, h))A(1), \quad M(t) + N(t) < 1, \tag{5.5.24}$$

$$G'(t) = (A - B)N' - (A - B - B')(M' + N')M,$$
$$M(t) > 0, \quad N(t) > 0, \quad M(t) + N(t) > 1. \tag{5.5.25}$$

It is easy to see that at such t_0, both (5.5.24) and (5.5.25) have the same limit $((y + t_0k, k) - (x + t_0h, h))A(1)$. This proves that G is differentiable at such t_0 and G' is continuous at such a t_0. Let t_0 be such that $M(t_0) + N(t_0) \neq 1$, but $M(t_0) = 0$ or $N(t_0) = 0$, then same argument (notice that when $N(t_0) = 0$, only consider the case $M(t_0) + N(t_0) < 1$) shows that G is differentiable and G' is continuous at t_0. The assertion about $G'(t)$ is thus proved. Since $G''(t)$ exists outside an exceptional finite set E, so it is enough to prove $G''(t) \leq 0$ outside E for G's concavity. First, we have $G''(t) = -(|h|^2 - |k|^2)A(1) \leq 0$ on $\{t : M(t) + N(t) < 1\}$. Then on the set $\{t : M(t) > 0, N(t) > 0, M(t) + N(t) > 1\}$, we have

$$\begin{aligned} G''(t) &= (F(M + N)N' - F'(M + N)(M' + N')M)' \\ &= -M(M' + N')^2 F''(M + N) \\ &\quad -N''((M + N)F'(M + N) - F(M + N)) \\ &\quad -(|h|^2 - |k|^2|)F'(M + N) \leq 0. \end{aligned} \tag{5.5.26}$$

The concavity of $G(t)$ is thus proved. As a result, we get

$$G(1) \le G(0) + G'(0), \tag{5.5.27}$$

with

$$G'(0) = (\varphi(x,y), h) + (\psi(x,y), k),$$

where

$$\varphi(x,y) = \begin{cases} -A(1)x, & |x| + |y| \le 1, \\ (B(|x|+|y|) + B'(|x|+|y|) - A(|x|+|y|))x, & |x| + |y| > 1, \end{cases}$$

$$\psi(x,y) = \begin{cases} A(1)y, & |x| + |y| \le 1, \\ \{(B(|x|+|y|) + B'(|x|+|y|) - A(|x|+|y|))|x| \\ \quad +A(|x|+|y|) - B(|x|+|y|)\}\frac{y}{|y|}, & |x| + |y| > 1. \end{cases}$$

Applying (5.5.27) to $x = f_{n-1}$, $y = g_{n-1}$, $h = d_n$, $k = e_n$, we get (5.5.17). And applying (5.5.27) to $x = y = 0$, $h = d_0$, $k = e_0$, we get

$$u(f_0, g_0) = u(d_0, e_0) \le u(0,0) = \frac{1}{2}\int_0^\infty \Phi(t)e^{-t}dt,$$

since $\varphi(0,0) = \psi(0,0) = 0$. This gives (5.5.18). Now aim at (5.5.16), i.e.

$$\Phi(|y|) \le u(x,y), \quad |x| \le 1. \tag{5.5.28}$$

When $|x| + |y| \le 1$, owing to $|y| \le 1$ and $|x| \le 1 - |y|$, we have

$$\Phi(|y|) \le |y|\Phi(1) \le |y|A(1) \le (1 + |y|^2 - |x|^2)\frac{A(1)}{2} = u(x,y).$$

Now consider the case $|x| \le 1$ and $|x| + |y| > 1$. We have the following boundary majorant $\Phi(|y|) \le u(x,y)$ when $|x| = 1$. For $0 < |x| < 1$, writing $x = -\frac{1-|x|}{2}\frac{x}{|x|} + \frac{1+|x|}{2}\frac{x}{|x|}$, by making use of the concavity of $u(x,y)$, we get

$$\begin{aligned} u(x,y) &\ge \frac{1-|x|}{2}u\left(-\frac{x}{|x|}, y\right) + \frac{1+|x|}{2}u\left(\frac{x}{|x|}, y\right) \\ &\ge \left(\frac{1-|x|}{2} + \frac{1+|x|}{2}\right)\Phi(|y|) = \Phi(|y|). \end{aligned}$$

This completes the proof of (5.5.16), and hence of (5.5.15) without the strictness.

Now prove the strictness of (5.5.15). Let x, y, h, k be such that $|x| + |y| \le 1 < |x+h| + |y+k|$. Then $h \ne 0$, and $M(t) = |x + th|$ is strictly convex, and hence $M'(t) + N'(t) \ne 0$ outside of a finite set. So from (5.5.26), we know that $G''(t) < 0$ outside of a finite set, and $G'(t)$ is decreasing strictly, and

$$G(1) < G(0) + G'(0) \tag{5.5.29}$$

holds strictly. Now suppose that $\|g\|_\infty > 1$, then either $|\{|f_0| + |g_0| > 1\}| > 0$, or $|\{|f_n| + |g_n| > 1 \geq |f_{n-1}| + |g_{n-1}|\}| > 0$ for some n, must hold. Thus, (5.5.29) may be used, and at least one inequality in the following inequality chain must be strict

$$E(\Phi(|g_n|)) \leq E(u(f_n, g_n)) \leq E(u(f_{n-1}, g_{n-1})) \leq \cdots$$
$$\leq E(u(f_0, g_0) \leq \frac{1}{2}\int_0^\infty \Phi(t)e^{-t}dt.$$

Meanwhile, when $\|g\|_\infty \leq 1$, we have

$$E(\Phi(|g_n|)) \leq \Phi(1) < \frac{1}{2}\Phi'(1) < \frac{1}{2}\int_0^\infty \Phi'(t)e^{-t}dt,$$

here $\Phi'(1) < \int_0^\infty \Phi'(t)e^{-t}dt$ follows from $1 = \int_0^\infty te^{-t}dt$ and Jensen's inequality, and $\Phi(1) = \int_0^1 \Phi'(t)dt < \frac{1}{2}\Phi'(1)$ can be seen from an area interpretation of both hand sides, by noticing the convexity of $\Phi'(t)$.

Now prove that the constant $\frac{1}{2}\int_0^\infty \Phi'(t)e^{-t}dt$ is best possible. Let $0 < \delta < 2$, $\gamma = \frac{1}{2+\delta}$, $\beta = \frac{2-\delta}{2+\delta}$. Divide $(0, 1]$ by points $\beta^k\gamma$, $k = 0, 1, 2, \cdots$. Define a family of σ-fields on $(0, 1]$ by: $\mathcal{F}_n$ is generated by atoms $(0, \beta^{n-1}\gamma]$, $(\beta^{n-1}\gamma, \beta^{n-2}\gamma]$, $\cdots$, $(\gamma, 1]$, $n \geq 1$, and $\mathcal{F}_0$ is trivial. Define a martingale $g = (g_n)_{n\geq 0}$, $e_n = \Delta_n g$, by

$$e_0 = \frac{1}{2}\chi_{(0,1]}, \quad e_1 = \frac{1+\delta}{2}\chi_{(0,\gamma]} - \frac{1}{2}\chi_{(\gamma,1]},$$

$$e_n = \delta\chi_{(0,\beta^{n-1}\gamma]} - (1 - \frac{\delta}{2})\chi_{(\beta^{n-1}\gamma,\beta^{n-2}\gamma]}, \quad n \geq 2.$$

Consider g's martingale transform f by $v = (1, -1, 1, -1, \cdots)$. Then it is easy to see that $\|f\|_\infty = 1$. For g's estimate, notice that on $(\beta^n\gamma, \beta^{n-1}\gamma]$, we have

$$e_0 = \frac{1}{2}, \quad e_1 = \frac{1+\delta}{2}, \quad e_k = \delta,\ 2 \leq k \leq n, \quad e_{n+1} = -\left(1 - \frac{\delta}{2}\right).$$

And hence (write $\alpha = \frac{1}{\delta}\ln\frac{1}{\beta} = \ln(1 + \delta(1 - \frac{\delta}{2})^{-1})^{\frac{1}{\delta}}$)

$$g_{n+1} = \left(1 + \frac{2n-1}{2}\delta\right)\chi_{(0,\beta^n\gamma]} + \sum_{k=1}^{n} k\delta\chi_{(\beta^k\gamma,\beta^{k-1}\gamma]},$$

$$E(\Phi(|g|)) \geq \gamma(1-\beta)\sum_1^\infty \Phi(k\delta)\beta^{k-1} = \gamma(1-\beta)\sum_1^\infty \Phi(k\delta)e^{-\alpha(k-1)\delta}$$
$$\geq \gamma(1-\beta)\sum_1^\infty \int_{(k-1)\delta}^{k\delta} \frac{\Phi(t)e^{-\alpha t}}{\delta}dt = 2\int_0^\infty \frac{\Phi(t)e^{-\alpha t}dt}{(2+\delta)^2}.$$

Since $\lim_{\delta\to 0}\alpha = 1$, we get $\lim_{\delta\to 0} E(\Phi(|g|)) \geq \frac{1}{2}\int_0^\infty \Phi(t)e^{-t}dt$. This ends the proof. □

We have following simple corollaries.

Corollary 5.5.3 *Let $2 < p < \infty, \Phi(t) = t^p$. Then for f, g, with g differentially subordinate to f, and $||f||_\infty \leq 1$, we have*

$$||g||_p^p < \frac{1}{2}\Gamma(p+1). \tag{5.5.30}$$

And the constant is best possible.

Corollary 5.5.4 *Let $0 < \alpha < 1, \Phi(t) = e^{\alpha t} - 1 - \alpha t$. Then for f, g as in Corollary 5.5.3, we have*

$$\sup_n E(e^{\alpha|g_n|}) + \alpha(1 - ||g||_1) < \left(1 - \frac{\alpha^2}{2}\right)(1-\alpha)^{-1}. \tag{5.5.31}$$

And the constant is best possible.

Proof. Set $p > 1$ such that $p\alpha < 1$, then $\Phi(t)^p$ satisfies the conditions in Theorem 5.5.2. So, we get

$$\sup_n E(\Phi^p(|g_n|)) < \frac{1}{2}\int_0^\infty \Phi(t)^p e^{-t} dt < \infty.$$

This means that $(\Phi(|g_n|))_{n\geq 0}$ is a L^p-bounded submartingale, and hence

$$\sup_n E(\Phi(|g_n|)) = E(\Phi(|g_\infty|)).$$

But applying Theorem 5.5.2 to $\Phi(t)$, we get

$$E(e^{\alpha|g_\infty|} - 1 - \alpha|g_\infty|) < \frac{1}{2}\int_0^\infty \Phi(t)e^{-t}dt = \frac{1}{2}\frac{\alpha^2}{1-\alpha}.$$

Thus we get

$$\begin{aligned}\sup_n E(e^{\alpha|g_n|}) + \alpha(1 - ||g||_1) &= E(e^{\alpha|g_\infty|}) + \alpha(1 - ||g||_1)\\ &< \left(1 - \frac{\alpha^2}{2}\right)(1-\alpha)^{-1}.\end{aligned}$$

This proves the corollary. □

Now we turn to the weak type inequality. In order to illustrate the idea, we consider the weak type (1, 1) at first, then consider general so-called Φ-weak type inequality.

Theorem 5.5.5 *Let $f = (f_n)_{n\geq 0}$, $g = (g_n)_{n\geq 0}$ be two H-valued martingales with g differentially subordinate to f. Then*

$$|\{|f_n| + |g_n| > \lambda\}| \leq \frac{2}{\lambda}||f||_1, \quad \forall \lambda > 0. \tag{5.5.32}$$

And the constant 2 is best possible.

Proof. Define a function $u(x,y)$ on $H \times H$ by

$$u(x,y) = \begin{cases} 1 + |x|^2 - |y|^2, & \text{when } |x| + |y| < 1; \\ 2|x|, & \text{when } |x| + |y| \geq 1. \end{cases} \tag{5.5.33}$$

Denote $w(x,y) = 1 + |x|^2 - |y|^2$, $v(x,y) = 2|x|$. Obviously, u is continuous and

$$u(x,y) = 2|x|, \quad |x| + |y| \geq 1, \tag{5.5.34}$$

$$u(x,y) \geq 1, \quad |y| \leq |x|, \tag{5.5.35}$$

$$2|x| + 1 \geq u(x,y) \geq v(x,y), \quad \forall x, y. \tag{5.5.36}$$

For x, y, h, $k \in H$, with $|k| \leq |h|$, define

$$G_u(t) = u(x + th, y + tk) \text{ (analogously for } G_w(t), G_v(t)), \quad \forall t \in \mathbf{R}. \tag{5.5.37}$$

Obviously, $G_w(t)$ and $G_v(t)$ are convex, we claim that $G_u(t)$ is convex too Suppose that $G_u(t)$ were not convex. Then there would be $t_1 < t_2$ such that the chord connecting two points $(t_i, G_u(t_i))$, $i = 1, 2$, is strictly under the arc with the same endpoints. Owing to the convexity of $G_w(t)$ and $G_v(t)$, there would be some t_0 in (t_1, t_2), say $t_0 = \alpha t_1 + (1-\alpha)t_2$, such that $|x + t_0 h| + |y + t_0 k| = 1$. We claim that this is impossible. In fact, such t_0 would satisfy following two opposite facts:

$$G_u(t_0) > \alpha G_u(t_1) + (1-\alpha) G_u(t_2),$$

$$G_u(t_0) = G_v(t_0) \leq \alpha G_v(t_1) + (1-\alpha) G_v(t_2) \leq \alpha G_u(t_1) + (1-\alpha) G_u(t_2).$$

The contradiction proves the convexity of $G_u(t)$. As a result, we have

$$G_u(1) \geq G_u(0) + G'_u(0), \tag{5.5.38}$$

provided $G'_u(0)$ exists. We want to deduce the following, from (5.5.38),

$$u(x+h, y+k) \geq u(x,y) + (\varphi(x,y), h) + (\psi(x,y), k), \tag{5.5.39}$$

where

$$\varphi(x,y) = \begin{cases} 2x, & |x| + |y| < 1 \text{ or } x = 0; \\ \dfrac{2x}{|x|}, & |x| + |y| \geq 1, \ x \neq 0, \end{cases} \tag{5.5.40}$$

$$\psi(x,y) = \begin{cases} 2y, & |x| + |y| < 1; \\ 0, & |x| + |y| \geq 1. \end{cases} \tag{5.5.41}$$

Consider four cases. Case (1): $|x| + |y| < 1$. In this case $G'_u(0) = (\varphi(x,y), h) + (\psi(x,y), k)$, the conclusion follows. Case (2): $|x| + |y| > 1$, $|x| \neq 0$. It is similar to (1). Case (3): $|x| + |y| \geq 1$, $x \neq 0$. Consider $\{x_j\} \subset H$, $|x_j| > |x|$, $|x_j| \to x$. Applying (5.5.38) to (x_j, y) then taking the limit by the continuity of φ and ψ at (x,y) with

$x \neq 0$, we get the conclusion. Case (4): $|x| + |y| \geq 1$, $x = 0$. In this case $|y| \geq 1$. The right-hand side of (5.5.39) is 0 and u is nonnegative. We have proved (5.5.39) in all cases.

Now we are in the position to get (5.5.32). Without loss of generality, we can assume $\lambda = 1$. Noticing that $u(x, y) \leq 2|x| + 1$ (hence $2|x| + 1 - u(x, y) \geq 0$) we have

$$\begin{aligned}|\{|f_n| + |g_n| > 1\}| &\leq |\{2|f_n| \geq u(f_n, g_n)\}| \\ &= |\{2|f_n| - u(f_n, g_n) + 1 \geq 1\}| \\ &\leq E(2|f_n| - u(f_n, g_n) + 1).\end{aligned}$$

The problem is thus reduced to prove

$$E(u(f_n, g_n)) \geq 1, \quad \forall n \geq 0.$$

From (5.5.39), we have $E(u(f_n, g_n)) \geq E(u(f_0, g_0))$. Since $|g_0| \leq |f_0|$, $u(f_0, g_0) \geq 1$. This ends the proof of (5.5.32).

The best possibility of the constant 2 can be shown by following two examples. One is a pair of dyadic martingales on (0, 1] with f, g defined respectively by

$$d_0 = 1,\ d_1 = \chi_{(0,\frac{1}{2}]} - \chi_{(\frac{1}{2},1]},\ d_2 = 2\chi_{(0,\frac{1}{4}]} - 2\chi_{(\frac{1}{4},\frac{1}{2}]},\ d_n = 0,\ n \geq 3,$$

$$e_0 = d_0,\ e_1 = -d_1,\ e_n = d_n,\ n \geq 2.$$

Then $|g| = 2$, and $\|f\|_1 = 1$, and hence

$$|\{|g| \geq 2\}| = 2\Big(\frac{1}{2}\|f\|_1\Big).$$

Another example is taken as follows. Let $\Omega = \{-\frac{1}{2}, \frac{3}{2}\}$, $\mu(\{-\frac{1}{2}\}) = \frac{3}{4}$, $\mu(\{\frac{3}{2}\}) = \frac{1}{4}$, $e_0 = d_0 = \frac{1}{2}$, $e_1 = -d_1$, where $d_1(\omega) = \omega$, $e_n = d_n = 0$, $n \geq 2$. Then $|g| = 1$, $\|f\|_1 = 1/2$. So $|\{|g| \geq 1\}| = 1 = 2\|f\|_1$. The proof of the theorem is thus finished. □

A simple stopping time argument can strengthen this theorem as follows.

Theorem 5.5.5′ *With the same conditions as in Theorem* 5.5.5, *we have*

$$|\{\sup_k(|f_k| + |g_k|) > \lambda\}| \leq \frac{2}{\lambda}\|f\|_1, \quad \forall \lambda > 0. \tag{5.5.42}$$

Proof. Only consider $\lambda = 1$. Define the stopping time

$$\tau^{(n)} = \tau \wedge n, \quad \tau = \inf\{k : |f_k| + |g_k| > 1\}, \tag{5.5.43}$$

and consider the stopped martingales $f^{(\tau^{(n)})}$, $g^{(\tau^{(n)})}$. Noticing that

$$|\Delta_k g^{(\tau^{(n)})}| = |\chi_{\{k \leq \tau^{(n)}\}} e_k| \leq |\chi_{\{k \leq \tau^{(n)}\}} d_k| = |\Delta_k f^{(\tau^{(n)})}|,$$

and that

$$|\{\sup_{k\leq n}(|f_k|+|g_k|)>1\}|\leq|\{|f_n^{(\tau^{(n)})}|+|g_n^{(\tau^{(n)})}|>1\},$$

from Theorem 5.5.5, we get

$$|\{\sup_{k\leq n}(|f_k|+|g_k|)>1\}|\leq 2\|f_n^{(\tau^{(n)})}\|_1\leq 2\|f_n\|_1\leq 2\|f\|_1.$$

Letting $n\to\infty$, this gives (5.5.42). The theorem is proved. □

Now consider so-called the Φ-weak type. Let $\Phi(t)$ be an increasing convex function on $\mathbf{R}^+$ such that $\Phi(0)=\Phi'(0)=0$ and $\Phi'(t)$ are strictly concave. Then $\Phi''(t)$ is strictly decreasing. So, with the same notations as in Theorem 5.5.2, we have $F(t)=A(t)-B(t)\geq 0$, $F'(t)=A(t)-B(t)-B'(t)\geq 0$, but

$$F''(t)=A(t)-B(t)-B'(t)-B''(t)\leq 0,\quad tF'(t)-F(t)\leq 0. \tag{5.5.44}$$

Theorem 5.5.6 *Let $\Phi(t)$ be as above, f, g be two H-valued martingales with g differentially subordinate to f. Then*

$$|\{g^*>\lambda\}|\leq\Big(\frac{1}{2}\int_0^\infty\Phi(t)e^{-t}dt\Big)^{-1}\sup_n E\Big(\Phi(\frac{|f_n|}{\lambda})\Big),\quad\forall\lambda>0. \tag{5.5.45}$$

The inequality is strict in nontrivial case. And the constant is best possible.

Proof. Define functions u, v, w as follows

$$w(x,y)=\begin{cases}(1+|x|^2-|y|^2)\dfrac{A(1)}{2}, & |x|+|y|\leq 1,|y|\leq 1;\\ (1-|y|)A(|x|+|y|)+|y|B(|x|+|y|), & |x|+|y|>1,|y|\leq 1,\end{cases}$$

$$v(x,y)=\Phi(|x|),\quad\forall x,y,$$

$$u(x,y)=\begin{cases}w(x,y), & \text{when } |y|\leq 1;\\ v(x,y), & \text{when } |y|>1.\end{cases}$$

Obviously, $u(x,y)$ is continuous, and following elementary estimate holds:

$$0\leq u(x,y)-v(x,y)\leq\frac{A(1)}{2},\quad\forall x,y. \tag{5.5.46}$$

The left-hand side one of (5.5.46) can be seen as follows. When $|x|+|y|\leq 1$, then $(1+|x|^2-|y|^2)\frac{A(1)}{2}\geq|x|A(1)\geq|x|\Phi(1)\geq\Phi(|x|)$. Consider the case $|x|+|y|>1$. Denote $|x|=t$, $|y|=s$. Fixing t and considering

$$H(s,t)=(1-s)A(s+t)+sB(s+t)=A(s+t)-sF(s+t)$$

as a function of $s\in[0,1]$, we have

$$H(1,t)=\Phi(t),\quad H(0,t)>\Phi(t),$$

the latter of which follows from

$$t = \int_{t-1}^{\infty} s e^{t-1} e^{-s} ds,$$

$$\Phi(t) < e^{t-1} \int_{t-1}^{\infty} \Phi(s) e^{-s} ds = A(t).$$

Since $H_s'(s,t) = -sF'(s+t) < 0$, on $(0, 1]$, we get $H(s,t) \geq \Phi(t)$, for all $s \in [0,1]$.

Now prove the inequality on the right-hand side of (6.5.46). Fixing $|x| = t$, and considering $u(x,y)$ as a function $K_t(s)$ of $|y| = s \in [0,1]$. Since $K_t'(s) \leq 0$ on $[0, 1]$, $K_t(s)$ is decreasing. In order to establish $K_t(s) \leq \frac{A(1)}{2} + \Phi(t)$, it is enough to prove $K_t(0) \leq \frac{A(1)}{2} + \Phi(t)$. It is reduced to prove

$$A(1) \leq 2\Phi(1), \ A'(t) \leq \Phi'(t) \text{ on } [1,\infty), \text{ and } \Phi(t) \geq \Phi(1)t^2 \text{ on } [0,1).$$

In fact, once they are proved we would have

$$\begin{aligned} K_t(0) &= \begin{cases} (1+|x|^2)\dfrac{A(1)}{2}, & |x| \leq 1 \\ A(|x|), & |x| > 1 \end{cases} \\ &\leq \begin{cases} \dfrac{A(1)}{2} + \Phi(1)t^2, & t \leq 1 \\ A(t), & t \geq 1 \end{cases} \leq \frac{A(1)}{2} + \Phi(t). \end{aligned}$$

Now aim at the three inequalities waited to be proved as above. $A(1) \leq 2\Phi(1)$ and $A'(t) \leq \Phi'(t)$ come respectively from (noticing the concavity of Φ')

$$\frac{1}{2}\int_0^{\infty} \Phi(s)e^{-s} ds = \frac{1}{2}\int_0^{\infty} \Phi'(s)e^{-s} ds \leq \frac{1}{2}\Phi'(1) \leq \int_0^1 \Phi'(s)\, ds = \Phi(1),$$

$$A'(t) = A(t) - B(t) = e^t \int_t^{\infty} B'(s)e^{-s} ds \leq B'(t+1) = \Phi'(t).$$

The last one can be seen (assuming $\Phi(1) = 1$) as follows. Let $t \in (0,1)$ be given, when $\Phi'(t) \geq 2t$, we have $\Phi'(s) \geq \frac{s}{t}\Phi'(t) \geq 2s$ for $0 \leq s \leq t$, so $\Phi(t) = \int_0^t \Phi'(s)ds \geq 2\int_0^t s\, ds = t^2$; and when $\Phi'(t) < 2t$, we have $\Phi'(s) \leq \frac{s}{t}\Phi'(t) \leq 2s$, for $t \leq s \leq 1$, so $\Phi(t) = 1 - \int_t^1 \Phi'(s)ds \geq 1 - 2\int_t^1 s\, ds = t^2$. The second part of (6.5.46) is proved. This completes the proof of (6.5.46).

Now define $G_u(t), G_w(t), G_v(t)$, for all $x, y, h, k \in H$, with $|k| \leq |h|$. Of course, $G_w(t)$ is meaningful only when $|x+tk| \leq 1$. Now we claim that $G_u(t)$ is convex. We can show it in the same way as in Theorem 5.5.5. In fact, suppose that $G_u(t)$ was not convex, then there would be $t_1 < t_2$ such that the chord is strictly under the arc between the points $(t_i, G_u(t_i))$, $i = 1,2$. Then there would be some t_0 in (t_1,t_2) such that $|y + t_0 k| = 1$, otherwise it would contradict to the convexity of $G_v(t)$ when $|y+tk| > 1$, for all $t \in (t_1,t_2)$, or to the convexity of $G_w(t)$ when $|y+tk| < 1$, for all $t \in (t_1, t_2)$. The convexity of $G_w(t)$ comes from the facts: $G'_w(t)$ exists and is continuous and $G''_w(t) \geq 0$ as shown by (noticing (5.5.44))

$$\begin{aligned} G''_w(t) = &-N(M'+N')F''(M+N) - M''((M+N)F'(M+N) \\ &-F(M+N)) - (|k|^2 - |h|^2)F'(M+N) \geq 0. \end{aligned}$$

But such t_0 satisfies two opposite properties

$$G_u(t_0) > \alpha G_u(t_1) + (1-\alpha)G_u(t_2),$$

$$G_u(t_0) = G_v(t_0) \leq \alpha G_v(t_1) + (1-\alpha)G_v(t_2) \leq \alpha G_u(t_1) + (1-\alpha)G_u(t_2).$$

This contradiction gives the convexity of $G_u(t)$. Thus we get

$$G_u(1) \geq G_u(0) + G'_u(0)$$

provided $G'_u(0)$ exists. We want to deduce, from it,

$$u(x+h, y+k) \geq u(x,y) + (\varphi(x,y), h) + (\psi(x,y), k), \tag{5.5.47}$$

where

$$\varphi(x,y) = \begin{cases} A(1)x, & |x|+|y| \leq 1, \\ \{F(|x|+|y|) - F'(|x|+|y|)|y|\}\dfrac{x}{|x|}, & |x|+|y| > 1,\ |y| < 1, \\ \Phi'(|x|)\dfrac{x}{|x|}, & |y| \geq 1, \end{cases}$$

$$\psi(x,y) = \begin{cases} -A(1)y, & |x|+|y| \leq 1, \\ -F'(|x|+|y|)y, & |x|+|y| > 1,\ |y| < 1, \\ 0, & |y| \geq 1. \end{cases}$$

Notice that $G'_u(0) = (\varphi(x,y), h) + (\psi(x,y), k)$ possibly except $\{|x| = 0\} \bigcup \{|x|+|y| = 1\} \bigcup \{|y| = 1\}$. We have seen in the proof of Theorem 5.5.2 that (x,y) such that $|x|+|y| = 1$ is really not exceptional. Now it is easy to show that (x,y) such that $|y| = 1$ is not exceptional either. Let $|y| = 1$. When $|y+k| > 1$, we have

$$\begin{aligned} u(x+h, y+k) = \Phi(|x+h|) &\geq \Phi(|x|) + \Phi'(|x|)\left(\frac{x}{|x|}, h\right) \\ &= u(x,y) + (\varphi(x,y), h) + (\psi(x,y), k). \end{aligned}$$

And when $|y+k| \leq 1$, no trouble occurs, since only $w(x,y)$ is concerned. As regards the restriction $x \neq 0$, we could get it rid of by assuming $f = (f_n)_{n \geq 0}$ never taking value 0 as above. (5.5.47) is thus proved. Now return to the proof of (5.5.45). We have

$$\begin{aligned} |\{|g_n| > 1\}| &\leq \left|\left\{\Phi(|f_n|) - u(f_n, g_n) + \frac{A(1)}{2} \geq \frac{A(1)}{2}\right\}\right| \\ &\leq \left(\frac{A(1)}{2}\right)^{-1} E\left(\Phi(|f_n|) - u(f_n, g_n) + \frac{A(1)}{2}\right) \\ &\leq \left(\frac{A(1)}{2}\right)^{-1} E(\Phi(|f_n|)), \end{aligned}$$

since from (5.5.47), we have

$$E(u(f_n,g_n)) \geq E(u(f_{n-1},g_{n-1})) \geq E(u(f_0,g_0)) \geq E(u(0,0)) = \frac{A(1)}{2}.$$

The rest remains unchanged to get (5.5.45). When f is nontrivial, then at least one inequality in the following chain is strict

$$E(u(f_n,g_n)) \geq E(u(f_{n-1},g_{n-1})) \geq \cdots \geq E(u(0,0)) = \frac{A(1)}{2}.$$

And the example in Theorem 5.5.2 shows that (5.5.45) is sharp. Since in that example, $\lim_{\delta\to 0} E(\Phi(|f|)) = \frac{1}{2}A(1)$, and $|\{g^* > 1-\varepsilon_\delta\}$ may be as near to 1 as possible when δ is small. The theorem is proved. □

Now we give some applications of preceding theorems to square functions.

Theorem 5.5.7 *Let $1 < p < \infty$. Then for all H-valued martingales f,*

$$(p^*-1)^{-1}||S(f)||_p \leq ||f||_p \leq (p^*-1)||S(f)||_p. \tag{5.5.48}$$

In particular,

$$||f||_p \geq (p-1)||S(f)||_p, \quad 1 < p \leq 2. \tag{5.5.49}$$

$$||f||_p \leq (p-1)||S(f)||_p, \quad 2 \leq p < \infty. \tag{5.5.50}$$

The constant $p-1$ is best possible. When $0 < ||f||_p < \infty$, then equality holds if and only if $p = 2$.

Proof. Let $K = l^2(H) = \{x = (x_j)_{j\geq 0} : |x|_K = (\sum_0^\infty |x_j|^2)^{\frac{1}{2}} < \infty\}$. Then K is a Hilbert space. Consider two K-valued martingales

$$F = (F_n)_{n\geq 0}, \quad D_k = (d_k, 0, 0, \cdots),$$

$$G = (G_n)_n \geq 0, \quad E_k = (0, \cdots, d_k, 0, \cdots) \text{ (d_k is in k-th term).}$$

Then $|D_k|_K = |E_k|_K$, that is to say F and G is differentially subordinate each other. We have

$$F_n = \sum_{k=0}^n D_k = (f_n, 0, \cdots), \quad G_n = \sum_{k=0}^n E_k = (d_0, d_1, \cdots, d_n, 0, \cdots).$$

Therefore $|F_n|_K = |f_n|$, $|G_n|_K = S_n(f)$, $||F||_p = ||f||_p$, $||G||_p = ||S(f)||_p$. From Theorem 5.5.1, we get (5.5.48). And the assertion about equality follows from that theorem, too.

The sharpness of (5.5.49) and (5.5.50) can be seen by following example. For the simplicity, we prefer an example with continuous parameter.

Let $1 < p < r < \infty$. Define $\{\mathcal{F}_t\}_{0 \leq t < 1}$ on [0, 1) by $\mathcal{F}_t$: generalized by the atom [t, 1) and Borel sets on [0, t]. Set $f(s) = (1-s)^{-\frac{1}{r}}$, $H(t) = \frac{1}{1-t}\int_t^1 f(s)ds$, and

$$f_t(s) = H(t)\chi_{\{t \leq s\}} + f(s)\chi_{\{t > s\}}, \quad 0 \leq t, s < 1.$$

Then $(f_t)_{0 \leq t < 1}$ is a martingale, $H(s) = \sup_t f_t(s) = \frac{r}{r-1}f(s)$ and $S(f)(s) = H(s) - f(s) = \frac{1}{r-1}f(s)$. We have

$$\|H\|_p = \frac{r}{r-1}\|f\|_p, \quad \|S(f)\|_p = \frac{1}{r-1}\|f\|_p.$$

Letting $r \to p$, and $p > 2$, we see that (5.5.50) is sharp. If (5.5.49) were not sharp, then (5.5.50) were so by duality. This completes the proof of the theorem. □

Remark We have another sharp inequality

$$\|Mf\|_p \leq p\|S(f)\|_p, \quad 2 \leq p < \infty. \tag{5.5.51}$$

This follows by combining (5.5.50) and Doob's maximal inequality $\|Mf\|_p \leq p'\|f\|_p$, $1 < p < \infty$. And the sharpness is shown by the same example.

Theorem 5.5.8 *For any H-valued martingale $f = (f_n)_{n \geq 0}$, we have*

$$|\{|f_n| + S_n(f) > \lambda\}| \leq \frac{2}{\lambda}\|f\|_1, \quad \forall \lambda > 0. \tag{5.5.52}$$

Proof. This follows from Theorem 5.5.5 and the argument in the proof of Theorem 5.5.7. □

Here is the last application in this section concerning a version of square function. Let $T = \{T_j\}_{j \geq 0}$ be an increasing sequence of stopping times. For any H-valued martingale f, define

$$S_n(T, f) = \Big(|f_{T_0 \wedge n}|^2 + \sum_j |f_{T_j \wedge n} - f_{T_{j-1} \wedge n}|^2\Big)^{\frac{1}{2}}, \tag{5.5.53}$$

$$S(T, f) = \Big(|f_{T_0}|^2 + \sum_j |f_{T_j} - f_{T_{j-1}}|^2\Big)^{\frac{1}{2}}. \tag{5.5.54}$$

Theorem 5.5.9 *Let f, g be two H-valued martingales with g differentially subordinate to f. Then*

$$|\{|f_n| + S_n(T, g) > \lambda\}| \leq \frac{2}{\lambda}\|f_n\|_1, \quad \forall \lambda > 0. \tag{5.5.55}$$

Proof. Set $K = l^2(H)$. Consider K-valued martingales

$$F = (F_n)_{n\geq 0} \quad \text{with } D_k(\omega) = (d_k(\omega), 0, \cdots),$$
$$G = (G_n)_{n\geq 0} \quad \text{with}$$
$$E_k(\omega) = (e_k(\omega), 0, \cdots), \quad k \leq T_0(\omega)(e_k(\omega) \text{ is in 0-th term}),$$
$$E_k(\omega) = (0, \cdots, e_k(\omega), 0, \cdots), \quad T_{j-1}(\omega) \leq k \leq T_j(\omega)$$
$$(e_k(\omega) \text{ is in } j\text{-th term}).$$

Then

$$|E_k(\omega)|_K = |e_k(\omega)| \leq |d_k(\omega)| = |D_k(\omega)|_K,$$

$$F_n = (f_n, 0, \cdots), \quad |F_n|_K = |f_n|,$$

$$G_n(\omega) = (g_{T_0\wedge n}(\omega), g_{T_1\wedge n}(\omega) - g_{T_0\wedge n}(\omega), \cdots, g_{T_j\wedge n}(\omega) - g_{T_{j-1}\wedge n}(\omega), \cdots),$$

$$|G_n(\omega)|_K = S_n(T, g)(\omega).$$

So, from Theorem 5.5.5, we get (5.5.55). The theorem is proved. □

Notes to Chapter 5

Martingale transforms have a long history, an intimate relation with some concepts in analysis (for example with conjugate harmonic functions), and a lot of applications in Probability, in Analysis, even in Geometry (related to Banach spaces). So, naturally, many people have been attracted to this field. What we have introduced in this chapter can not be complete because of the limitation of personal ability and interest.

In §5.1, we tried to show that martingale transforms could be used to reduce some problems concerning one kind of objects to another. Except Theorem 5.1.1, all other theorems are of this kind. Theorem 5.1.1 is taken from Burkholder's first famous paper on martingale transforms. Theorem 5.1.9 is taken from A. Shiryayer's book "Probability". Theorems 5.1.2–5.1.7 are essentially due to A. Garsia and taken from his book, but there are some improvements which are due to Chao-Long [2]. For example, Theorems 5.1.3, 5.1.4, 5.1.5, 5.1.6, have given a satisfactory characterization about interchanging the space h_{p_1} to h_{p_2}, or $\mathcal{P}_{p_1}$ to $\mathcal{P}_{p_2}$, via martingale transforms, for full scope of index: $0 < p_1, p_2 < \infty$, or $0 < p_1 < p_2 = \infty$.

§§5.2, 5.3, 5.4 are devoted to summarizing the joint works due to J. A. Chao and the author, which can be referred to Chao-Long [1] and its sequels. We did it in order to fill the gap in the operator theory in martingale setting. Among other things, we tried to show that the space BMO could take a better role than L^∞ in operator theory.

All results in §5.5, except the example in Theorem 5.5.7, and the idea to get it are due to D. Burkholder. Burkholder's idea could be used to conjugate harmonic functions, to stochastic integrals, to a generalization of martingale (so-called very weak martingale), and to martingales taken its values in Banach spaces. All of these can be referred to Burkholder's papers listed in the Reference. The example taken in Theorem 5.5.7 is due to L. Dubins and D. Gilat. Some cases about the inequalities concerning $S(f)$ have not yet been covered by the results in this section. For example, (5.5.52) is not able to give the best constant implied in $|\{S(f) > \lambda\}| \leq \frac{c}{\lambda}\|f\|_1$, which is $e^{\frac{1}{2}}$ as shown by D.C.Cox; and how about (5.5.49) for $2 \leq p < \infty$, and (5.5.50) for $1 < p \leq 2$ are still open. Another interesting problem is what happens when the differential subordination is replaced by $S(g) \leq S(f)$, or $S_n(g) \leq S_n(f)$, for all n.

6 Weight Theory and Weighted Φ-inequalities

In this chapter, we will introduce the weight theory and weighted Φ-inequalities in martingale setting. The former is the main content of this chapter which contains A_p condition and its extension b_λ (§6.1–6.2), inverse Hölder inequality (§6.3), and the relation between A_p weights and BMO martingales (§6.4–6.5). The last section §6.6 will be devoted to the weighted Φ-inequalities. The assumptions imposed on $(\Omega, \mathcal{F}, \mu, \{\mathcal{F}_n\}_{n\geq 0})$ are usual.

6.1 A_p conditions and its extensions b_λ

The weights we consider in what follows are mainly martingales generated by $z \in L^1$ which are strictly positive. But it is convenient to consider weights as strictly positive processes $z = (z_n)_{0\leq n\leq\infty}$, rather than martingales. The so-called strict positivity of process $z = (z_n)_{o\leq n\leq\infty}$ means $z_n > 0$, a.e. for all n.

Definition 6.1.1 *Let $\lambda \in \overline{\mathbf{R}}$, $z = (z_n)_{0\leq n\leq\infty}$ be a weight. $z = (z_n)$ is said to satisfy a b_λ condition, in symbols $z \in b_\lambda$, if there exists constant K such that*

$$\frac{1}{K} z_n \leq E(z_\infty^\lambda|\mathcal{F}_n)^{\frac{1}{\lambda}} \leq K z_n, \text{ a.e.,} \quad \forall n,\ \lambda \neq 0,\ \lambda \neq \pm\infty, \tag{6.1.1.i}$$

$$\frac{1}{K} z_n \leq \exp E(\log z_\infty|\mathcal{F}_n) \leq K z_n, \text{ a.e.,} \quad \forall n,\ \lambda = 0, \tag{6.1.1.ii}$$

$$\frac{1}{K} z_n \leq z_\infty \leq K z_n, \text{ a.e.,} \ \forall n,\ \lambda = \pm\infty. \tag{6.1.1.iii}$$

The conditions $b_\lambda^-(K)$ and $b^+(K)$ denote two parts of b_λ shown by the left and right hand sides of preceding inequalities respectively.

Definition 6.1.2 *$z = (z_n)_{0\le n\le\infty}$ is said to satisfy an S condition, in symbols $z \in S$, if there exists K such that*

$$\frac{1}{K} z_{n-1} \le z_n \le K z_{n-1}, \text{ a.e. }, \forall n \ge 1. \tag{6.1.2}$$

The conditions S^- and S^+ have similar meaning.

Definition 6.1.3 *$z = (z_n)_{0\le n\le\infty}$ is said to satisfy a B_λ condition, if $z \in b_\lambda \bigcap S$.*

For weights $z = (z_n)_{0\le n\le\infty}$ in b_λ, $z^\lambda = (z_n^\lambda)_{0\le n\le\infty}$ are naturally associated weights. Notice that, even when $z = (z_n)$ is a martingale, $z^\lambda = (z_n^\lambda)$ is no longer a martingale. This is one of the reasons to consider weights as processes but not only as martingales.

Proposition 6.1.4 *Let $\lambda \ne 0$, $\lambda \ne \pm\infty$. Then for $\lambda > 0$, $z \in b_\lambda^\pm(K)$ if and only if $z^\lambda \in b_1^\pm(K^\lambda)$; for $\lambda < 0$, $z \in b_\lambda^\pm(K)$, if and only if $z^\lambda \in b_1^\mp(K^{-\lambda})$.*

The proof is obvious.

Proposition 6.1.5 *Let $z \in b_\lambda(K)$, T be any stopping time. then the stopped process $z^{(T)} = (z_{n\wedge T})_{0\le n\le\infty} \in b_\lambda(K^2)$.*

Proof. Since $z \in b_\lambda(K)$ if and only if $z^\lambda \in b_1(K^{|\lambda|})$, it is enough to consider the case $\lambda = 1$. Notice that in (6.1.1.i–6.1.1.iii), n could be replaced by any stopping time T. Now denote $z^{(T)}$ by $\tilde{z}$. Then $\tilde{z}_\infty = z_T$. From the b_1 condition of z we have

$$\frac{1}{K}\tilde{z}_\infty = \frac{1}{K} z_T \le E(z_\infty|\mathcal{F}_T),$$

$$E\left(\frac{1}{K}\tilde{z}_\infty\middle|\mathcal{F}_n\right) \le E(E(z_\infty|\mathcal{F}_T)\mathcal{F}_n) = E(z_\infty|\mathcal{F}_{T\wedge n}) \le K z_{T\wedge n} = K\tilde{z}_n.$$

This means $\tilde{z} \in b_1^+(K^2)$. Similarly from

$$K\tilde{z}_\infty = K z_T \ge E(z_\infty|\mathcal{F}_T),$$

we get

$$E(K\tilde{z}_\infty|\mathcal{F}_n) \ge E(z_\infty|\mathcal{F}_{T\wedge n}) \ge \frac{1}{K} z_{T\wedge n} = \frac{1}{K}\tilde{z}_n,$$

i.e., $\tilde{z} \in b_1^-(K^2)$. The proof is finished. □

The most interesting ones of the conditions b_λ are b_λ^- for $\lambda < 0$, which are just so-called A_p conditions. For $\lambda < 0$, set $p = 1 - \frac{1}{\lambda}$. Then the b_λ^- conditions

$$\frac{1}{K} z_n \le E\left(z_\infty^{-\frac{1}{p-1}}\middle|\mathcal{F}_n\right)^{-(p-1)}$$

could be rewritten

$$z_n E\left(z_\infty^{-\frac{1}{p-1}}\middle|\mathcal{F}_n\right)^{p-1} \le K, \quad \text{a.e.}, \forall n. \tag{6.1.3}$$

Definition 6.1.6 *Let $1 \le p \le \infty, z = (z_n)_{0 \le n \le \infty}$ be a weight. z is said to satisfy an $A_p(K)$ condition for $1 < p < \infty$, if (6.1.3) holds; to satisfy $A_1(K)$ if*

$$z_n \le K z_\infty, \quad \text{a.e., } \forall n; \tag{6.1.4}$$

denote $A_p = \bigcup_K A_p(K), 1 \le p < \infty$. z is said to satisfy A_∞, if $z \in A_p$ for some p (write $A_\infty = \bigcup_p A_p$).

Remark The larger p is the weaker the condition A_p is, i.e.

$$A_1 \subset A_p \subset A_q \subset A_\infty, \quad 1 \le p \le q \le \infty. \tag{6.1.5}$$

This can be seen easily.

Now consider the special weight case furthermore. Let $z_\infty \in L^1$ be strictly positive, $z = (z_n)_{0 \le n \le \infty}$ be the martingale generated by z_∞. Notice that the strict positivity of z_∞ implies that so z_n are for all n, since for positive martingale $z = (z_n)_{0 \le n \le \infty}$,

$$\{z_n = 0\} \subset \{z_\infty = 0\}, \quad \forall n.$$

Assume $E(z_\infty) = 1$. Then

$$d\hat{\mu} = z_\infty d\mu, \tag{6.1.6}$$

is another probability measure on $(\Omega, \mathcal{F}, \mu)$. We claim that both μ and $\hat{\mu}$ have the same sets of measure 0. It is enough to prove that $\hat{\mu}(F) = 0$ for some $F \in \mathcal{F}$ implies $\mu(F) = 0$. In fact $\hat{\mu}(F) = \int_F z_\infty d\mu = 0$ implies $z_\infty = 0$ a.e. on F. Since z_∞ is strictly positive, $\mu(F) = 0$ must hold. Thus, associated with such strictly positive $z_\infty \in L^1$, we have a pair $(\mu, \hat{\mu})$ of probability measures. We call z or z_∞ as a measure change. For $(\Omega, \mathcal{F}, \hat{\mu})$, we could consider all concepts like for $(\Omega, \mathcal{F}, \mu)$, such as $\widehat{L^p}$, conditional expectation $\widehat{E}$, and martingale, etc. What is the relation between E and $\widehat{E}$? Here we are at

Proposition 6.1.7 *When the following conditional expectations have meaning, we have*

$$\widehat{E}(f|\mathcal{F}_n) = \frac{1}{z_n} E(f z_\infty|\mathcal{F}_n), \tag{6.1.7}$$

$$E(f|\mathcal{F}_n) = z_n \widehat{E}(f z_\infty^{-1}|\mathcal{F}_n). \tag{6.1.8}$$

Proof. For any $F \in \mathcal{F}_n$, we have

$$\begin{aligned} \int_F \widehat{E}(f|\mathcal{F}_n) d\hat{\mu} &= \int_F f d\hat{\mu} = \int_F f z_\infty d\mu \\ &= \int_F E(f z_\infty|\mathcal{F}_n) d\mu = \int_F z_n^{-1} E(f z_\infty|\mathcal{F}_n) d\hat{\mu}. \end{aligned}$$

This proves (6.1.7). Here we have used the fact: for all $F \in \mathcal{F}_n$

$$\int_F d\hat{\mu} = \int_F z_\infty d\mu = \int_F E(z_\infty|\mathcal{F}_n) d\mu = \int_F z_n d\mu.$$

(6.1.8) can be proved similarly or followed by (6.1.7). The proof is finished. □

Remarks

1. (6.1.7) and (6.1.8) are same things. In fact, considering $(\Omega, \mathcal{F}, \hat{\mu})$ as underlying space and $d\mu = z_\infty^{-1} d\hat{\mu}$, and noticing

$$\widehat{E}(z_\infty^{-1}|\mathcal{F}_n) = z_n^{-1} E(z_\infty z_\infty^{-1}|\mathcal{F}_n) = z_n^{-1},$$

then (6.1.8) is just (6.1.7), on the new underlying space $(\Omega, \mathcal{F}, \hat{\mu})$.

2. In (6.1.7), (6.1.8), n could be replaced by any stopping time T. In fact, considering (6.1.7), we have

$$\begin{aligned}
\widehat{E}(f|\mathcal{F}_T) &= \sum_0^\infty \widehat{E}(f|\mathcal{F}_n)\chi_{\{T=n\}} + \widehat{E}(f|\mathcal{F}_T)\chi_{\{T=\infty\}} \\
&= \sum_0^\infty \widehat{E}(f|\mathcal{F}_n)\chi_{\{T=n\}} + f\chi_{\{T=\infty\}} \\
&= \sum_0^\infty z_n^{-1} E(f z_\infty|\mathcal{F}_n)\chi_{\{T=n\}} + z_\infty^{-1} f z_\infty \chi_{\{T=\infty\}} \\
&= z_T^{-1} E(f z_\infty|\mathcal{F}_T)\chi_{\{T<\infty\}} + z_\infty^{-1} E(f z_\infty|\mathcal{F}_T)\chi_{\{T=\infty\}} \\
&= z_T^{-1} E(f z_\infty|\mathcal{F}_T).
\end{aligned}$$

We have seen that for a special weight z in the sense $z = (z_n)_{0\le n\le\infty}$ is a measure change (i.e a strictly positive martingale with respect to $(\Omega, \mathcal{F}, \mu, \{\mathcal{F}_n\}_{n\ge 0})$ with $E(z_\infty) = 1$), $z^{-1} = (z_n^{-1})_{0\le n\le\infty}$ is also a special weight with respect to $(\Omega, \mathcal{F}, \hat{\mu}, \{\mathcal{F}_n\}_{n\ge 0})$. It is natural to ask what the condition $z^{-1} \in \widehat{A}_p(K)$ is like. Here is the answer.

Proposition 6.1.8 *Let $1 < p < \infty$. Then $z^{-1} \in \widehat{A}_p(K)$, in symbols $z \in \widetilde{A}_p(K)$, if and only if $z \in b_{p'}^+(K^{\frac{1}{p}})$ with p' being the conjugate index to p.*

Proof. Assume $z^{-1} \in \widehat{A}_p(K)$. Then

$$z_n^{-1}\widehat{E}\left(z_\infty^{\frac{1}{p-1}}\Big|\mathcal{F}_n\right)^{p-1} \le K, \quad \text{a.e.,} \quad \forall n.$$

It is just

$$z_n^{-1} z_n^{1-p} E\left(z_\infty^{\frac{1}{p-1}+1}\Big|\mathcal{F}_n\right)^{p-1} \le K,$$

$$E\left(z_\infty^{p'}|\mathcal{F}_n\right)^{\frac{1}{p'}} \le K^{\frac{1}{p}} z_n \quad \text{a.e.,} \quad \forall n. \tag{6.1.9}$$

So $z \in b_{p'}^+(K^{\frac{1}{p}})$. The reciprocal argument works well. The proof is finished. □

Remark For a positive martingale $z = (z_n)_{0 \le n \le \infty}$, following kind of inequality is called inverse Hölder's inequality

$$E(z_\infty^r | \mathcal{F}_n)^{\frac{1}{r}} \le C z_n, \quad \forall n, \tag{6.1.10}$$

where $r > 1$ is some index.

Definition 6.1.9 *Let $z = (z_n)_{0 \le n \le \infty}$ be a special weight. z is said to satisfy $\widetilde{A}_p$ for $1 \le p \le \infty$, if $z^{-1} \in \widehat{A}_p(K)$ for some K; is said to satisfy $\widetilde{A}_\infty$, if $z = (z_n)_{0 \le n \le \infty}$ satisfies* (6.1.10) *for some $r > 1$, some C.*

6.2 The b_λ conditions in the case $\lambda < 0$ or $\lambda > 1$

In this section we do some further studies of b_λ in the case $\lambda < 0$ or $\lambda > 1$. In these cases, for special weights $z = (z_n)_{0 \le n \le \infty}$, only the conditions $b_\lambda^-, \lambda < 0$ and $b_q^+, q > 1$, are non-trivial. So, even for general weights (not necessarily being martingales), we consider only b_λ^- for $\lambda < 0$ equivalently A_p, and b_q^+ for $q > 1$. (Essentially only b_λ^-, since $z \in b_q^+$ if and only if $z^{-1} \in \widehat{A}_{q'}$.)

Let λ, p be such that

$$-\infty < \lambda < 0, \quad 1 < p < \infty, \quad \lambda = -\frac{1}{p-1} \quad \left(p = 1 - \frac{1}{\lambda}\right). \tag{6.2.1}$$

We have following simple propositions about the condition A_p.

Proposition 6.2.1 *Let $1 \le p < \infty$. Then weight $z = (z_n)_{0 \le n \le \infty} \in A_p(K)$, if and only if the operator family $\{T_n\}$ with*

$$T_n : f \to z_n^{\frac{1}{p}} E\left(f z_\infty^{-\frac{1}{p}} | \mathcal{F}_n\right), \quad n = 0, 1, \cdots, \tag{6.2.2}$$

is bounded on L^p uniformly. More precisely, $z \in A_p(K)$ implies

$$\sup_n \|T_n\| \le K^{\frac{1}{p}};$$

and $\sup \|T_n\| < \infty$ *implies* $z \in A_p(\sup_n \|T_n\|^p)$.

Proof. Assume $z \in A_p(K)$. then

$$\begin{aligned} z_n |E(f z_\infty^{-\frac{1}{p}} | \mathcal{F}_n)|^p &\le z_n (E(|f|^p | \mathcal{F}_n)^{\frac{1}{p}} E(z_\infty^{-\frac{1}{p-1}} | \mathcal{F}_n)^{\frac{1}{p'}})^p \\ &\le K E(|f|^p | \mathcal{F}_n), \end{aligned}$$

$$\sup_n \|T_n\| \le K^{\frac{1}{p}}.$$

On the contrary, when $p = 1$, from

$$E(f z_n z_\infty^{-1}) = E(z_n E(f z_\infty^{-1} | \mathcal{F}_n)) \le \|T_n\| E(f), \quad \forall f \in L_+^1,$$

we get $z_n \le \|T_n\| z_\infty$, a.e., for all n; and when $1 < p < \infty$, from

$$E\big(z_n E(f z_\infty^{-\frac{1}{p}} | \mathcal{F}_n)^p\big) \le \|T_n\|^p E(f^p), \quad \forall f \in L_+^p,$$

by setting $f = z_\infty^{-\frac{1}{p(p-1)}} \chi_{\{z_\infty > \varepsilon\}} \chi_A$, for any $\varepsilon > 0$, any $A \in \mathcal{F}_n$, we get

$$\int_A z_n E\left(z_\infty^{-\frac{1}{p-1}} \chi_{\{z_\infty > \varepsilon\}} \Big| \mathcal{F}_n \right)^p d\mu \le \|T_n\|^p \int_A z_\infty^{-\frac{1}{p-1}} \chi_{\{z_\infty > \varepsilon\}} d\mu,$$

$$z_n E\left(z_\infty^{-\frac{1}{p-1}} \chi_{\{z_\infty > \varepsilon\}} \Big| \mathcal{F}_n \right)^p \le \|T_n\|^p E\left(z_\infty^{-\frac{1}{p-1}} \chi_{\{z_\infty > \varepsilon\}} \Big| \mathcal{F}_n \right),$$

$$z_n E\left(z_\infty^{-\frac{1}{p-1}} \Big| \mathcal{F}_n \right)^{p-1} \le \|T_n\|^p.$$

This proves $z \in A_p(\sup_n \|T_n\|^p)$. The proof is finished. □

Remark Since in the definition of $A_p(K)$, n could be replaced by any stopping time τ, we get that $z \in A_p(K)$, if and only if the operator family $\{T_\tau\}$ with

$$T_\tau : f \to z_\tau^{\frac{1}{p}} E(f z_\infty^{-\frac{1}{p}} | \mathcal{F}_\tau), \tag{6.2.3}$$

is bounded on L^p uniformly.

Proposition 6.2.2 *Let $1 \le p < \infty$, $z = (z_n)_{0 \le n \le \infty}$ be a special weight. Then $z \in A_p(K)$, if and only if for all nonnegative f and stopping times τ, we have*

$$E(f|\mathcal{F}_\tau)^p \le K \widehat{E}(f^p | \mathcal{F}_\tau), \tag{6.2.4}$$

and if and only if $\{E_{\mathcal{F}_\tau}\} = \{E_\tau\}$ is bounded on $\widehat{L}^p$ uniformly with $\sup_\tau \|E_\tau\| \le K^{\frac{1}{p}}$.

Proof. Assume $z \in A_p(K)$. Then

$$\begin{aligned} E(f|\mathcal{F}_\tau) &= E(f z_\infty^{\frac{1}{p}} z_\infty^{-\frac{1}{p}} | \mathcal{F}_\tau) \le E(f^p z_\infty | \mathcal{F}_\tau)^{\frac{1}{p}} E(z_\infty^{-\frac{1}{p-1}} | \mathcal{F}_\tau)^{\frac{1}{p'}} \\ &\le K^{\frac{1}{p}} z_\tau^{-\frac{1}{p}} E(f^p z_\infty | \mathcal{F}_\tau)^{\frac{1}{p}} = K^{\frac{1}{p}} \widehat{E}(f^p | \mathcal{F}_\tau)^{\frac{1}{p}}. \end{aligned}$$

Now assume that (6.2.4) holds. Then, obviously, $\{E_\tau\} = \{E_{\mathcal{F}_\tau}\}$ is bounded on $\widehat{L}^p$ uniformly. Finally assume

$$\widehat{E}(|E(f|\mathcal{F}_\tau)|^p) \le \|T_\tau\|^p \widehat{E}(|f|^p), \quad \forall f.$$

Just as in the proof of the Proposition 6.2.1, by setting $f = z_\infty^{-\frac{1}{p-1}} \chi_{\{\chi_\infty > \varepsilon\}} \chi_A$, $A \in \mathcal{F}_\tau$, we have

$$\int_A z_\tau E(z_\infty^{-\frac{1}{p-1}} \chi_{\{z_\infty > \varepsilon\}} | \mathcal{F}_\tau)^p d\mu \le \|E_\tau\|^p \int_A z_\infty^{-\frac{1}{p-1}} \chi_{\{z_\infty > \varepsilon\}} d\mu.$$

From it we get

$$z_\tau E(z_\infty^{-\frac{1}{p-1}}|\mathcal{F}_\tau)^{p-1} \le \|E_\tau\|^p, \quad z \in A_p(\sup_\tau \|E_\tau\|^p).$$

The proof is finished. □

Parallelly, we get

Proposition 6.2.3 *Let $1 < p \le \infty$, $z = (z_n)_{0 \le n \le \infty}$ be a special weight. Then $z \in b_q^+(K)$, if and only if $\widehat{E}(f|\mathcal{F}_\tau)^{q'} \le K^{q'} E(f^{q'}|\mathcal{F}_\tau)$, and if and only if $\{\widehat{E}_\tau\} = \{\widehat{E}_{\mathcal{F}_\tau}\}$ is bounded on $L^{q'}$ uniformly, with $\sup_\tau \|E_\tau\| \le K$.*

Proof. We know that

$$z \in b_q^+(K) \Longleftrightarrow z^{-1} \in \widehat{A}_{q'}(K^{q'}), \quad 1 < q \le \infty.$$

Now consider $(\Omega, \mathcal{F}, \hat{\mu})$ as the underlying space, and $d\mu = z_\infty^{-1} d\hat{\mu}$ as a measure change. Then from Proposition 6.2.2, we see that $z_\infty^{-1} \in \widehat{A}_{q'}(K^{q'})$, if and only if $\widehat{E}(f|\mathcal{F}_\tau)^{q'} \le K^{q'} E(f^{q'}|\mathcal{F}_\tau)$, if and only if $\{\widehat{E}_\tau\}$ is bounded on $L^{q'}$ uniformly, with $\sup \|\widehat{E}_\tau\| \le K$. This completes the proof of the Propostion. □

There is another characterization of A_p in terms of the weak boundedness of the maximal operator M. We postpone it to §6.6.

6.3 Gehring's lemma, and inverse Hölder's inequality

In classical analysis, two kinds of inequalities about the integral means over cubes of nonnegative locally integrable function concern the topic, one is Gehring's, another is called inverse Hölder's inequality. Here we want to establish a unified inequality which tells a deeper property of b_λ. First we give a lemma.

Lemma 6.3.1 *Let U be nonnegative measurable on the finite measure space $(\Omega, \mathcal{F}, \mu)$, and constants $K \ge 0, \beta > 0, \varepsilon (0 < \varepsilon \le 1)$ be such that*

$$\int_{\{U > \lambda\}} U d\mu \le K \lambda^\varepsilon \int_{\{U > \beta\lambda\}} U^{1-\varepsilon} d\mu, \quad \forall \lambda \ge C_0 E(U). \tag{6.3.1}$$

Then there exist $r > 1$ and C depending only on K, β, ε such that

$$E(U^r)^{\frac{1}{r}} \le C C_0^{\frac{r}{r-1}} E(U). \tag{6.3.2}$$

Proof. β could be assumed to be less than 1 obviously. In addition, $E(U)$ could be assumed to be 1, otherwise we could consider tU (with $t = E(U)^{-1}$) which satisfies

$$\int_{\{tU > \lambda\}} tU d\mu = t \int_{\{U > \frac{\lambda}{t}\}} U d\mu \le tK \left(\frac{\lambda}{t}\right)^\varepsilon \int_{\{U > \frac{\beta\lambda}{t}\}} U^{1-\varepsilon} d\mu$$

$$= K\lambda^{\varepsilon} \int_{\{tU>\beta\lambda\}} (tU)^{1-\varepsilon} d\mu, \quad \forall \lambda \geq C_0 E(tU).$$

So it is enough to prove that there exist r and C such that for U satisfying (6.3.1) and $E(U) = 1$, we have

$$E(U^r)^{\frac{1}{r}} \leq C C_0^{\frac{r-1}{r}}. \tag{6.3.3}$$

Assume (6.3.1) holds, and U is essentially bounded. Integrating two sides of (6.3.1) multiplied by $a\lambda^{a-1}$(a determined later) with respect to λ on $[C_0, \infty)$, we get

$$\int_{\{U>C_0\}} U \int_{C_0}^{U} a\lambda^{a-1} d\lambda d\mu = \int_{\{U>c_0\}} (U^{1+a} - UC_0^a) d\mu,$$

and

$$\begin{aligned} K &\int_{\{U>\beta C_0\}} U^{1-\varepsilon} \int_{C_0}^{\frac{U}{\beta}} a\lambda^{a-1+\varepsilon} d\lambda d\mu \\ &= \frac{Ka}{a+\varepsilon} \int_{\{U>\beta C_0\}} U^{1-\varepsilon} \left(\left(\frac{U}{\beta}\right)^{a+\varepsilon} - C_0^{a+\varepsilon} \right) d\mu \\ &\leq \frac{Ka}{a+\varepsilon} \frac{1}{\beta^{a+\varepsilon}} \int_{\{U>\beta C_0\}} U^{1+a} d\mu \\ &= k \int_{\{U>C_0\}} U^{1+a} d\mu + k \int_{\{\beta C_0 < U \leq C_0\}} U^{1+a} d\mu. \end{aligned}$$

Noticing

$$k = \frac{Ka}{a+\varepsilon} \frac{1}{\beta^{a+\varepsilon}} \to 0 \quad \text{as} \quad a \to 0,$$

k could be taken less than 1 provided a small enough. Since U is bounded, $E(U^{1+\varepsilon}) < \infty$. Thus we get

$$(1-k) \int_{\{U>C_0\}} U^{1+a} d\mu \leq \int_{\{U>C_0\}} C_0^a U d\mu + k \int_{\{U \leq C_0\}} U^{1+a} d\mu,$$

$$\begin{aligned} (1-k)E(U^{1+a}) &\leq \int_{\{U>C_0\}} C_0^a U d\mu + k \int_{\{U \leq C_0\}} U^{1+a} d\mu \\ &\quad + (1-k) \int_{\{U \leq C_0\}} U^{1+a} d\mu \\ &\leq C_0^a \int_{\{U>C_0\}} U d\mu + \int_{\{U \leq C_0\}} U^{1+a} d\mu \leq C_0^a. \end{aligned}$$

By setting $r = 1+a$, we get (6.3.3) under the additional assumption: U is essentially bounded.

Now get rid of the assumption. Suppose U is not bounded. Then for m being large arbitrarily, there exists $\omega \in \Omega$ such that $U(\omega) = m$. Consider a new measure

$$\mu' = \mu\chi_{\{U<m\}} + j\varepsilon_\omega, \quad j = \frac{1}{m}\int_{\{U\geq m\}} U d\mu,$$

where ε_ω is a point meassure centered at ω. We claim that U besides being bounded, also satisfies (6.2.1) and $E(U) = 1$, with respect to this new measure μ'. Obviously,

$$\int_\Omega U d\mu' = \int_{\{U<m\}} U d\mu + U(\omega)j = \int_\Omega U d\mu = 1;$$

meanwhile when $\lambda < m$, we have

$$\int_{\{U>\lambda\}} U d\mu' = \int_{\{U>\lambda\}} U d\mu, \int_{\{U>\beta\lambda\}} U^{1-\varepsilon} d\mu' \geq \int_{\{U>\beta\lambda\}} U^{1-\varepsilon} d\mu,$$

the latter of which follows from

$$\begin{aligned}\int_{\{U>\beta\lambda\}} U^{1-\varepsilon} d\mu' &= \int_{\{U\geq m\}} + \int_{\{\beta\lambda<U<m\}} = U(\omega)^{1-\varepsilon} j + \int_{\{\beta\lambda<U<m\}} \\ &= m^{-\varepsilon}\int_{\{U\geq m\}} U d\mu + \int_{\{\beta\lambda<U<m\}} U^{1-\varepsilon} d\mu \\ &\geq \int_{\{U>\beta\lambda\}} U^{1-\varepsilon} d\mu,\end{aligned}$$

and when $\lambda \geq m$, $\int_{\{U>\lambda\}} U d\mu' = 0$, so

$$\int_{\{U>\lambda\}} U d\mu' \leq K\lambda^\varepsilon \int_{\{U>\beta\lambda\}} U^{1-\varepsilon} d\mu', \quad \forall \lambda \geq C_0.$$

The proof of the assertions about U's conditions with respect to μ' is complete. Thus we get

$$\left(\int_\Omega U^r d\mu'\right)^{\frac{1}{r}} \leq C C_0^{\frac{r-1}{r}}.$$

Letting $m \to \infty$, we get (6.3.3). The proof of the lemma is finished. □

We have another version of the lemma.

Lemma 6.3.1′ *Let z be nonnegative measurable on the finite measure space $(\Omega, \mathcal{F}, \mu), q > 1$ and $K, h (0 < h < 1)$ be such that*

$$\int_{\{z>\nu\}} z^q d\mu \leq K\nu^{q-1}\int_{\{z>h\nu\}} z d\mu, \quad \forall \nu \geq C_0 E(z^q)^{\frac{1}{q}}. \tag{6.3.4}$$

Then there exists $p (> q)$ and C such that

$$\|z\|_p \leq C C_0^\delta \|z\|_q, \quad \delta = \frac{1}{q} - \frac{1}{p}. \tag{6.3.5}$$

Proof. Let $U = z^q$, $\lambda = \nu^q, \beta = h^q$, then (6.3.4) becomes

$$\int_{\{U>\lambda\}} U d\mu \le K\lambda^{\frac{q-1}{q}} \int_{\{U>\beta\lambda\}} U^{\frac{1}{q}} d\mu.$$

Applying Lemma 6.3.1 with $\varepsilon = 1 - \frac{1}{q}$, and $p = rq$, we get

$$E(z^p) = E(U^r) \le CC_0^{r-1}E(U)^r,$$

$$E(z^p)^{\frac{1}{p}} \le CC_0^{\frac{(r-1)}{p}} E(z^q)^{\frac{1}{q}} = CC_0^{\delta}E(z^q)^{\frac{1}{q}}.$$

The proof is finished. □

Remark When $\varepsilon \ne 1$, Lemma 6.3.1 follows from Lemma 6.3.1′, too. Let $z = U^{\frac{1}{q}}$, $q = \frac{1}{1-\varepsilon}$, $\nu = \lambda^{\frac{1}{q}}$, $h = \beta^{\frac{1}{q}}$ for U, ε, λ given, then (6.3.1) implies (6.3.4), and (6.3.5) implies (6.3.2) with $r = \frac{p}{q}$.

We are in the position to get the main theorem in this section.

Theorem 6.3.2 *Let $z = (z_n)_{0\le n\le\infty}$ be a weight, and $z \in S^+ \bigcap b_\lambda^- \bigcap b_\nu$ with $\nu >$ $\max(\lambda, 0)$. Then there exists $\varepsilon > 0$ such that $z \in b_{\nu+\varepsilon}^+$.*

Proof. (a) Assume $0 < \lambda < \nu$. Considering $z^\lambda = (z_n^\lambda)_{0\le n\le\infty}$ instead, the problem reduces to the case $\lambda = 1, \nu = q > 1$. This is just the case Gehring's lemma tells about. First, consider the case $z_0 = 1$, a.e. Since $z \in b_q^-$, we have

$$1 = z_0^q \le C_0^q E(z_\infty^q|\mathcal{F}_0), \quad 1 \le C_0 E(z_\infty^q)^{\frac{1}{q}}.$$

For any $\lambda \ge C_0 E(z_\infty^q)^{\frac{1}{q}}$, define stopping time $\tau = \inf\{n : z_n > \lambda\}$. Then τ never takes the value 0, so from $z \in b_q^+ \bigcap S^+$, we get

$$E(z_\infty^q|\mathcal{F}_\tau) \le Kz_\tau^q \le K_1 z_{\tau-1}^q.$$

Thus we get

$$\int_{\{z_\infty>\lambda\}} z_\infty^q d\mu = \int_{\{\tau<\infty\}} E(z_\infty^q|\mathcal{F}_\tau)d\mu \le K_1\lambda^q|\{\tau<\infty\}|.$$

Meanwhile, from $z \in b_1^-$, we have

$$\begin{aligned}\lambda|\{\tau<\infty\}| &\le \int_{\{\tau<\infty\}} z_\tau d\mu \le K_2 \int_{\{\tau<\infty\}} z_\infty d\mu \\ &= K_2\left\{\int_{\{\tau<\infty, z_\infty>h\lambda\}} z_\infty d\mu + \int_{\{\tau<\infty, z_\infty\le h\lambda\}} z_\infty d\mu\right\} \\ &\le K_2 \int_{\{z_\infty>h\lambda\}} z_\infty d\mu + K_2 h\lambda|\{\tau<\infty\}|.\end{aligned}$$

When h makes $K_2 h < 1$, we get

$$\lambda|\{\tau < \infty\}| \leq \frac{K_2}{1 - K_2 h} \int_{\{z_\infty > h\lambda\}} z_\infty \, d\mu.$$

From preceding two estimates, we get

$$\int_{\{z_\infty > \lambda\}} z_\infty^q \, d\mu \leq K_1 \frac{K_2}{1 - K_2 h} \lambda^{q-1} \int_{\{\tau_\infty > h\lambda\}} z_\infty \, d\mu, \quad \forall \lambda \geq C_0 E(z_\infty^q)^{\frac{1}{q}}.$$

By means of Lemma 6.3.1′ , we see that there exists $p(> q)$ and C depending only on the constants implied in the assumptions such that $\|z_\infty\|_p \leq C\|z_\infty\|_q$.

We could prove $z \in b_p^+$ from preceding arguments. Consider new space $(\Omega', \mathcal{F}', \mu', \{\mathcal{F}'_m\}_{m\geq 0})$ with

$$\Omega' = A \in \mathcal{F}_n, \quad \mathcal{F}' = \mathcal{F} \bigcap A, \quad \mathcal{F}'_m = \mathcal{F}_{m+n} \bigcap A, \; m \geq 0, \quad \mu' = |A|^{-1} \mu|_{\mathcal{F}'},$$

and new process $z' = (z'_m)_{m\geq 0}$, with

$$z'_m = \frac{z_{m+n}}{z_n} \chi_A, \quad \forall m \geq 0.$$

Then $z' \in b_1^- \bigcap b_q \bigcap S^+$ with respect to the new underlying space, and with the same constants. Since $z'_0 = 1$, a.e. on A, we get

$$E'((z'_\infty)^p)^{\frac{1}{p}} \leq C E'((z'_\infty)^q)^{\frac{1}{q}},$$

$$\left(\frac{1}{|A|} \int_A \left(\frac{z_\infty}{z_n} \right)^p d\mu \right)^{\frac{1}{p}} \leq C \left(\frac{1}{|A|} \int_A \left(\frac{z_\infty}{z_n} \right)^q d\mu \right)^{\frac{1}{q}} \leq C,$$

$$E(z_\infty^p | \mathcal{F}_n)^{\frac{1}{p}} \leq C z_n.$$

This completes the proof in the case $0 < \lambda < \nu$.

(b) Assume $\lambda < 0 < \nu$. Considering $z^\nu = (z_n^\nu)_{0\leq n \leq \infty}$, the problem reduces to the case $\lambda < 0, \nu = 1$. Noticing $b_\lambda^- = A_p$, with $p = 1 - \frac{1}{\lambda}$, this is just the case of inverse Hölder's inequality considered by R. Coifman-C. Fefferman. First, assume $z_0 = 1$, a.e. again. Since $z \in A_p(K)$, for all nonnegative U_∞ , and for all stopping times τ, we have (see (6.2.4))

$$z_\tau U_\tau^p \leq K E(z_\infty U_\infty^p | \mathcal{F}_\tau).$$

Set $U_\infty = \chi_{\{z_\infty \leq \beta z_\tau\}}$, $\beta < 1$ determined later. We get

$$z_\tau E(\chi_{\{z_\infty \leq \beta z_\tau\}} | \mathcal{F}_\tau)^p \leq K E(z_\infty \chi_{\{z_\infty \leq \beta z_\tau\}} | \mathcal{F}_\tau) \leq K \beta z_\tau.$$

Since $0 < z_\tau < \infty$, a.e. we get

$$E(\chi_{\{z_\infty \leq \beta z_\tau\}} | \mathcal{F}_\tau)^p \leq K\beta.$$

When $K\beta < 1$, we see that for some $a > 0$,

$$E(\chi_{\{z_\infty > \beta z_\tau\}} | \mathcal{F}_\tau) \geq a, \quad \text{a.e.}$$

$$E(\chi_{\{z_\infty > \beta z_\tau\}}\chi_{\{\tau<\infty\}}|\mathcal{F}_\tau) \geq a\chi_{\{\tau<\infty\}}. \tag{6.3.6}$$

Since $z \in b_1^-$, $1 = z_0 \leq C_0E(z_\infty|\mathcal{F}_0)$, $1 \leq C_0E(z_\infty)$. For all $\lambda \geq C_0E(z_\infty)$, define the stopping time $\tau = \inf\{n : z_n > \lambda\}$. Then as in the proof in (a), we have

$$z_\tau \geq \lambda \quad \text{on} \quad \{\tau < \infty\}, \quad \text{and} \quad z_\tau \leq C\lambda. \tag{6.3.7}$$

By making use of the condition b_1^+, we get

$$\begin{aligned} E(z_\infty\chi_{\{z_\infty>\lambda\}}) &= E(E(z_\infty\chi_{\{\tau<\infty\}}|\mathcal{F}_\tau)) \leq KE(z_\tau\chi_{\{\tau<\infty\}}) \\ &\leq KC\lambda|\{\tau<\infty\}| \leq \frac{KC\lambda}{a}|\{z_\infty > \beta z_\tau, \tau < \infty\}|. \end{aligned}$$

From Lemma 6.3.1 ($\varepsilon = 1$ case), there exists $r > 1$ such that

$$E(z_\infty^r)^{\frac{1}{r}} \leq CE(z_\infty).$$

By the same technique as in the proof in (a), we get

$$E(z_\infty^r|\mathcal{F}_n)^{\frac{1}{r}} \leq Cz_n.$$

The proof of the assertion $z \in b_r^+ = b_{1+\varepsilon}^+$ is finished.

(c) Assume $\lambda = 0 < \nu$. It follows from the case $0 < \lambda < \nu$, owing to the fact: b_0^- is stronger than b_λ^-, for all $\lambda > 0$. The fact that b_0^- is stronger than b_1^- can be seen from

$$z_n \leq K \exp E(\log z_\infty|\mathcal{F}_n) \leq KE(z_\infty|\mathcal{F}_n).$$

Analogously, $z \in b_0^-$ implies $z^\alpha \in b_1^-$, for all $\alpha > 0$,

$$z_n^\alpha \leq K^\alpha \exp(\alpha E(\log z_\infty|\mathcal{F}_n)) \leq K^\alpha E(z_\infty^\alpha|\mathcal{F}_n).$$

So, we have

$$z \in b_0^- \Longrightarrow z^\lambda \in b_1^- \Longrightarrow z \in b_\lambda^-, \quad \forall \lambda > 0.$$

The proof of the theorem is thus finished. □

Corollary 6.3.3 *Let $z = (z_n)_{0\leq n\leq\infty}$ be a weight in $S^-\bigcap b_\lambda \bigcap b_\theta^+$ with $\lambda < \min(\theta, 0)$. Then there exists $q, 1 < q < p = 1 - \frac{1}{\lambda}$, such that $z \in A_q$.*

Proof. Consider $z^{-1} = (z_n^{-1})_{0\leq n\leq\infty}$. Then

$$z^{-1} \in S^+\bigcap b_{-\theta}^-\bigcap b_{-\lambda}.$$

Since $-\theta < -\lambda$, we see from Theorem 6.3.4 that $z^{-1} \in b^+_{\frac{1}{p-1}+\varepsilon}$, $\varepsilon > 0$. That is $z \in b^-_{-(\frac{1}{p-1}+\varepsilon)} = A_q$, with $q = 1 + (\frac{1}{p-1} + \varepsilon)^{-1} < p$. The proof of the corollary is finished. □

Remark When $z = (z_n)_{0\leq n\leq\infty}$ is a special weight, then $z \in b_1^+ \bigcap b_\lambda^+$, for all $\lambda < 0$ naturally. In this case, the condition $z \in S^- \bigcap A_p$ is sufficient for the existence of $q < p$ such that $z \in A_q$. The condition $z \in S^-$ here is superfluous in the classical case. We collect some assertions for special weights in the following corollaries.

Corollary 6.3.4 *Assume $z \in S$. Then the following assertions are equivalent*

(a) $z \in A_p = b_\lambda^-$, *for some* $p > 1$, $\lambda = -\frac{1}{p-1}$,

(b) $z \in b_\mu^+$, *for some* $\mu > 1$,

(c) $z^{-1} \in \widehat{A}_p = \hat{b}_\lambda^-$, *for some* $p > 1$, $\lambda = -\frac{1}{p-1}$,

(d) $z^{-1} \in \hat{b}_\mu^+$, *for some* $\mu > 1$.

Proof.

(a) $\Longrightarrow$ (b). Since $z \in S^+ \bigcap b_\lambda^- \bigcap b_1, \lambda < 0$, we have $z \in b_{1+\varepsilon}^+$.

(b) $\Longleftrightarrow$ (c). It follows from: $z^{-1} \in \widehat{A}_p$ if and only if $z \in b_{p'}^+, 1 < p < \infty$, without any added condition.

(c) $\Longrightarrow$ (d). Same as (a) $\Longrightarrow$ (b), since $z^{-1} \in S^+$ if and only if $z \in S^-$.

(d) $\Longleftrightarrow$ (a). Same as (b) $\Longleftrightarrow$ (c). The proof is finished. □

Corollary 6.3.5 *Let $z \in A_p \bigcap S$. Then there exists $\varepsilon > 0$ such that both, $z^{1+\varepsilon}$ and the martingale generated by $z_\infty^{1+\varepsilon}$ are in A_p.*

Proof. Denote $U_\infty = z_\infty^{-\frac{1}{p-1}}, U_n = E(U_\infty|\mathcal{F}_n)$. Then it is obvious that $z \in A_p$ if and only if $U \in A_{p'}$. And since

$$1 = E(z_\infty^{\frac{1}{p}} z_\infty^{-\frac{1}{p}}|\mathcal{F}_n)^p \le E(z_\infty|\mathcal{F}_n)E(z_\infty^{-\frac{1}{p-1}})^{p-1} = z_n U_n^{p-1} \le K, \tag{6.3.8}$$

$z \in S$ is equivalent to $U \in S$. Thus, besides $z \in A_p \bigcap S$ we also have $U \in A_{p'} \bigcap S$. By making use of Corollary 6.3.4 we see

$$z \in b_{1+\alpha}^+, U \in b_{1+\beta}^+, \quad \text{for some} \quad \alpha, \beta > 0.$$

Thus for $\varepsilon = \min(\alpha, \beta)$, we have

$$z_n \le E(z_\infty^{1+\varepsilon}|\mathcal{F}_n)^{\frac{1}{1+\varepsilon}} \le C z_n, \quad U_n \le E(U_\infty^{1+\varepsilon}|\mathcal{F}_n)^{\frac{1}{1+\varepsilon}} \le C U_n.$$

From this we get the assertion that both $z^{1+\varepsilon} = (z_n^{1+\varepsilon})_{0 \le n \le \infty}$ and the martingale generated by $z_\infty^{1+\varepsilon}$ are in $A_p \bigcap S$. The proof is finished. □

Now we discuss what happens without the condition $z \in S$. But as a compensation we take a regularity assumption occurring in Chapter 7 which we have met several times , that is

$$\text{(R)} \quad f_n = E(f|\mathcal{F}_n) \le dE(f|\mathcal{F}_{n-1}), \quad n \ge 1, \quad \forall f \in L_+^1,$$

where $d \ge 1$ is a constant.

Proposition 6.3.6 *Assume* (R) *holds. Let z be a special weight in $b_q^+ (q > 1)$. Then $z \in b_{q+\varepsilon}^+$ for some $\varepsilon > 0$.*

Proof. (R) implies $z \in S^+$. Since $Z \in S^+ \bigcap b_1^- \bigcap b_q$, the assertion follows. □

Proposition 6.3.7 *Assume* (R) *holds. Let z be a special weight in $A_p (p > 1)$. Then $z \in A_{p-\varepsilon}$, for some $\varepsilon > 0$.*

Proof. Denote $U_\infty = z_\infty^{-\frac{1}{p-1}}, U_n = E(U_\infty|\mathcal{F}_n)$. Then $U_n \le dU_{n-1}, n \ge 1$, from (R). By means of (6.3.8), we have

$$z_{n-1} \le CU_{n-1}^{1-p} \le C_d U_n^{1-p} \le C_d z_n, \quad n \ge 1.$$

So $z \in S^-$. And hence $z \in A_{p-\varepsilon}$, for some $\varepsilon > 0$. □

Proposition 6.3.8 *Assume* (R) *holds. Let z be a special weight in A_p. Then inverse Hölder's inequality holds*

$$E(z_\infty^{1+\varepsilon}|\mathcal{F}_n)^{\frac{1}{1+\varepsilon}} \le K z_n, \quad \forall n.$$

Proof. Since $z \in S^+ \bigcap b_\lambda^- \bigcap b_1, \lambda = -\frac{1}{p-1}, z \in b_{1+\varepsilon}^+$. The proof is complete. □

Proposition 6.3.9 *Assume* (R) *holds. Let z be a special weight in A_∞. Then $z \in \widetilde{A}_\infty$ (i.e., $z^{-1} \in \widehat{A}_\infty$). But the inverse may not be true.*

Proof. Assume $z \in A_\infty$. Then $z \in A_p$ for some $p < \infty$. So $z \in b_{1+\varepsilon}^+$, i.e., $z \in \widetilde{A}_{(1+\varepsilon)'} \subset \widetilde{A}_\infty$. This proves the first assertion.

Consider a counterexample. Let $\Omega = (0, 1]$, $d\mu = dx$, $\mathcal{F} = \bigvee_{n\ge 1} \mathcal{F}_n$, $\mathcal{F}_1$ be trivial, $\mathcal{F}_n$ be generated by atoms

$$\left\{ \left(0, \frac{1}{2^{n-1}}\right], \ I_k = \left(\frac{1}{2^k}, \frac{1}{2^{k-1}}\right], \quad k = 1, \cdots, n-1 \right\}.$$

Let $\{c_k\}_1^\infty$ be a number sequence with $c_k > 0, c_k$ being decreasing. Denote

$$z_\infty = \sum_1^\infty c_k \chi_{I_k}, \quad z_n = E(z_\infty|\mathcal{F}_n), \quad n \ge 1.$$

We have

$$z_{n+1} = \sum_1^n c_k \chi_{I_k} + 2^n \sum_{k \ge n+1} c_k |I_k| \chi_{(0,2^{-n}]},$$

$$E(z_\infty^2|\mathcal{F}_n) = \sum_1^{n-1} c_k^2 \chi_{I_k} + 2^{n-1} \sum_{k\ge n} c_k^2 |I_k| \chi_{(0,2^{-n+1}]}.$$

For $a = 1, 2$, we have (owing to the monotonity of c_k)

$$c_n^a \le 2^n \sum_{k\ge n} c_k^a 2^{-k} \le c_n^a + 2^n c_{n+1}^a \sum_{k \ge n+1} 2^{-k} \le 2c_n^a.$$

This means

$$2^{n-1}\sum_{k\geq n} c_k^2 2^{-k} \leq c_n^2 \leq \left(2^n \sum_{k\geq n} c_k 2^{-k}\right)^2,$$

and hence

$$E(z_\infty^2|\mathcal{F}_n)^{\frac{1}{2}} \leq 2E(z_\infty|\mathcal{F}_n),$$

$z \in b_2^+$, i.e., $z \in \widetilde{A}_2 \subset \widetilde{A}_\infty$. But $z \notin A_\infty$ provided $\{c_k\}$ chosen suitably, say, $\lim\limits_{n\to\infty} \frac{c_n}{c_{n+1}} = \infty$, since in this case on $(0, 2^{-n}]$,

$$\frac{z_n}{z_{n+1}} \geq \frac{1}{2}\frac{c_n}{c_{n+1}} \to \infty.$$

(Suppose $z \in A_\infty$, then together with (R), we would have $z \in S$, as shown in the proof of Proposition 6.3.7.) This proves the assrtion. □

Proposition 6.3.10 *Assume* (R) *holds. Let z be a special weight in A_p. Then there exists $\varepsilon > 0$ such that $z^{1+\varepsilon}$ and the martingales $(E(z_\infty^{1+\varepsilon}|\mathcal{F}_n))_{n\geq 0}$ are in A_p too.*

Proof. It follows from the fact: the condition A_p together (R) implies the condition S. □

As the end of this section, we take an example to show that the inverse Hölder's inequality is not true in general.

Example For any $p > 1$ given, there exists weight $z \in A_p$, but $z \notin A_{p-\varepsilon}$, for all $\varepsilon > 0$. Let $\Omega = (0, 1]$, $d\mu = dx$, $\mathcal{F} = \bigvee_{n\geq 1} \mathcal{F}_n$, $\mathcal{F}_1$ be trivial, $\mathcal{F}_n$ be generated by atoms

$$\left\{\left(0, \frac{1}{n!}\right], \; I_k = \left(\frac{1}{(k+1)!}, \frac{1}{k!}\right], \; k = 1, \cdots, n-1\right\}.$$

At first, consider the case $p = 2$. Set $z = \sum_1^\infty z_k \chi_{I_k}$, with

$$z_k = b\frac{2^k}{k!}, \quad b^{-1} = \sum_1^\infty \frac{2^k}{k!}|I_k| < \infty,$$

and $U = z^{-1}$. For all $r > 1$, we have

$$E(U^r) = b^{-r}\sum_1^\infty (k!)^r 2^{-kr}\frac{k}{(k+1)!} \geq \frac{1}{2}b^{-r}\sum_1^\infty (k!)^{r-1}2^{-kr} = \infty.$$

This shows that for all $\varepsilon > 0, z \notin A_{2-\varepsilon}$, otherwise we would have $E(U^{(1-\varepsilon)^{-1}}) < \infty$. We claim $z \in A_2$. Notice $I_k \in \mathcal{F}_n, k \leq n$, and $(0, \frac{1}{n!}] \in \mathcal{F}_n$. We have

$$\frac{1}{|I_k|}\int_{I_k} z dx \frac{1}{|I_k|}\int_{I_k} U dx = b\frac{2^k}{k!}b^{-1}\frac{k!}{2^k} = 1, \quad \forall k \leq n.$$

So it remains to prove

$$n!\int_0^{\frac{1}{n!}} z\,dx\, n!\int_0^{\frac{1}{n!}} U\,dx \le C.$$

We have

$$b^{-1}\int_0^{\frac{1}{n!}} z dx = \sum_{k\ge n}\frac{2^k}{k!}|I_k| \le \sum_{k\ge n}\frac{2^k}{(k!)^2} \le \frac{2^{n+1}}{(n!)^2},$$

$$b\int_0^{\frac{1}{n!}} U dx = \sum_{k\ge n}\frac{k!}{2^k}|I_k| \le \sum_{k\ge n} 2^{-k} \le 2^{-n+1},$$

and hence

$$n!\int_0^{\frac{1}{n!}} z\,dx\, n!\int_0^{\frac{1}{n!}} U\,dx \le 4.$$

This proves the assertion. This example could be modified to adapt the general $p > 1$ case. Denote

$$z = \sum_1^\infty z_k\chi_{I_k}, \quad z_k = b\Big(\frac{2^k}{k!}\Big)^{p-1}, \quad b^{-1} = \sum_1^\infty \Big(\frac{2^k}{k!}\Big)^{p-1}|I_k|.$$

In this case, for all $r > 1$, we have

$$E(z^{-\frac{r}{p-1}}) = b^{\frac{r}{p-1}}\sum_1^\infty \Big(\frac{2^k}{k!}\Big)^{-r}|I_k| \ge C\sum_1^\infty (k!)^{r-1}2^{-kr} = \infty.$$

So $z \notin A_{p-\varepsilon}$, for all $\varepsilon > 0$. But

$$\frac{1}{|I_k|}\int_{I_K} z\,dx\Big(\frac{1}{|I_k|}\int_{I_k} z^{-\frac{1}{p-1}}dx\Big)^{p-1} \le C, \quad \forall k,$$

$$n!\int_0^{\frac{1}{n!}} z\,dx\Big(n!\int_0^{\frac{1}{n!}} z^{-\frac{1}{p-1}}dx\Big)^{p-1} \le C, \quad \forall n,$$

so $z \in A_p$.

6.4 A_p weights and BMO martingales

For a real maritingale $f = (f_n)_{n\ge 0} \in BMO$, some associated weights $z = (z_n)_{0\le n\le\infty} \in A_p$ could be constructed in various ways. For example, choose a suitable $\lambda > 0$, define $z_\infty = e^{\lambda f_\infty}$ and consider the martingale $z = (z_n)_{0\le n\le\infty}$ generated by z_∞; or simply consider $z = (e^{\lambda f_n})_{0\le n\le\infty}$; or a little more complicated consider a stochastic exponent. (But we do not want to introduce the concept of stochastic exponents in this book.) On the contrary, from $z \in A_p$, under the condition $z \in S$, we could construct an associated $f \in BMO$. In this section, we will discuss these problems. First consider the special weights case.

Proposition 6.4.1 *Let $z = e^f$ be strictly positive in L^1. Then $z \in A_p, 1 < p < \infty$, if and only if $f \in \log A_{1,(p-1)^{-1}}$.*

Proof. Assume $z \in A_p$. Then $f_n = E(f_\infty|\mathcal{F}_n)$ has meaning. In fact, since $\log u$ is concave on $(0,\infty)$, we have

$$E(\log z_\infty|\mathcal{F}_n) \le \log E(z_\infty|\mathcal{F}_n) < \infty, \quad \text{a.e.,}$$

$$-\frac{1}{p-1}E(\log z_\infty|\mathcal{F}_n) \le \log E\left(z_\infty^{-\frac{1}{p-1}}|\mathcal{F}_n\right) < \infty, \quad \text{a.e.}$$

So,

$$f_n = E(f|\mathcal{F}_n) \ne \infty, \quad \text{a.e.}$$

Since the condition $\log A_{1,(p-1)^{-1}}$ could be formulated as

$$E(e^f|\mathcal{F}_n)E(e^{-\frac{1}{p-1}f}|\mathcal{F}_n) \le K < \infty, \quad \text{a.e.,} \quad \forall n,$$

which is just the condition A_p. This proves the assertion. □

Theorem 6.4.2 *For all $f \in BMO$, there exist $\alpha, \beta > 0$ such that*

$$z_1 = e^{\alpha f} \in A_{p_1} \bigcap S, \quad p_1 = 1 + \frac{\alpha}{\beta}, \tag{6.4.1.i}$$

$$z_2 = e^{-\beta f} \in A_{p_2} \bigcap S, \quad p_2 = 1 + \frac{\alpha}{\beta}. \tag{6.4.1.ii}$$

On the contrary, if $\lambda \ne 0$ makes

$$e^{-\lambda f} \in A_p \bigcap S, \quad \textit{for some} \quad p > 1, \tag{6.4.2}$$

then $f \in BMO$.

Proof. We know that $\operatorname{Re} BMO = \bigcup_{\alpha,\beta>0} \left(\log A_{\alpha,\beta} \bigcap BD\right)$. So for any $f \in \operatorname{Re} BMO$, there exist $\alpha, \beta > 0$ such that $f \in \log A_{\alpha,\beta}$, i.e.

$$\alpha f \in \log A_{1,\frac{\beta}{\alpha}}, \quad -\beta f \in \log A_{1,\frac{\alpha}{\beta}}.$$

From Proposition 6.4.1, we see $e^{\alpha f} \in A_{p_1}, e^{-\beta f} \in A_{p_2}$. As for the S property, say $e^{\alpha f} \in S^+$, we have

$$E(e^{\alpha f}|\mathcal{F}_n) \le Ke^{\alpha f_n} \le Ke^{\alpha f_{n-1}} \le KE(e^{\alpha f}|\mathcal{F}_{n-1}), \quad \text{a.e.,} \quad \forall n \ge 1.$$

Now assume that there exists $\lambda \ne 0$ such that $e^{\lambda f} \in A_p \bigcap S$. then $\lambda f \in \log A_{1,(p-1)^{-1}}$, and

$$e^{\lambda f_{n-1}} \le E(e^{\lambda f}|\mathcal{F}_{n-1}) \le KE(e^{\lambda f}|\mathcal{F}_n) \le Ke^{\lambda f_n} = e^{\lambda f_n + K},$$

and hence $\lambda(f_{n-1} - f_n) \le k$. Analogously, we also have $\lambda(f_n - f_{n-1}) \le K$. So $f \in BD$. Since $f \in \log A_{\alpha,\beta}$ implies

$$E(\exp(\varepsilon|f - f_n|)|\mathcal{F}_n) \le K.$$

This proves $f \in BMO$. The proof is finished. □

Now consider the $p = 1$ case. We first give some properties of A_1.

Proposition 6.4.3 *Let z be a special weight in $A_1 \bigcap S$. Then there exist $g, h \in L^1_+$ satisfying $Mg < \infty$, $0 < a \le h \le b < \infty$, a.e., and $\delta(0 < \delta < 1)$ such that*

$$z = h(Mg)^\delta. \tag{6.4.3}$$

Proof. Since $z \in A_1 \bigcap S$, then there exists $\varepsilon > 0$ such that

$$E(z^{1+\varepsilon}|\mathcal{F}_n)^{\frac{1}{1+\varepsilon}} \le KE(z|\mathcal{F}_n) \le Kz, \quad \text{a.e,} \quad \forall n.$$

Thus, we have

$$z^{1+\varepsilon} \le M(z^{1+\varepsilon}) \le Kz^{1+\varepsilon}.$$

Denote $\delta = \frac{1}{1+\varepsilon}$, $g = z^{1+\varepsilon}$, $h = z(Mg)^{-\delta}$. We get

$$z = z(Mg)^{-\delta}(Mg)^\delta = h(Mg)^\delta,$$

with $\frac{1}{K} \le h \le 1$, and $Mg \le Kz^{1+\varepsilon} < \infty$, a.e. The proof is finished. □

Remark In the classical case, the reciprocal of the proposition is true, too. That is to say, if $g \in L^1_+$ such that $Mg < \infty$, a.e, then $(Mg)^\delta \in A_1$, for all $\delta < 1$. We do not know how about in the martingal setting. We will discuss this problem under an added condition. We give the conditional version of weak type (1,1) of M at first.

Lemma 6.4.4 *Let $f = (f_n)_{n \ge 0}$ be a martingale in L^1. Then for all $\lambda > 0$, we have*

$$E_0(\chi_{\{Mf>\lambda\}}) \le \frac{1}{\lambda}E_0(|f_\infty|), \quad \text{a.e.}, \tag{6.4.4}$$

and for all $\delta, 0 < \delta < 1$, for all $F \in \mathcal{F}_0$, we have

$$\frac{1}{|F|}\int_F (Mf)^\delta d\mu \le C_\delta\Big(\frac{1}{|F|}\int_F |f_\infty| d\mu\Big)^\delta. \tag{6.4.5}$$

Proof. For $\lambda > 0$, define a stopping time $\tau = \inf\{n : |f_n| > \lambda\}$. Then

$$\chi_{\{Mf>\lambda\}} \le \frac{1}{\lambda}|f_\tau|\chi_{\{Mf>\lambda\}} = \frac{1}{\lambda}\sum_{n=0}^\infty |f_n|\chi_{\{\tau=n\}},$$

$$\begin{aligned} E_0(\chi_{\{Mf>\lambda\}}) &\le \frac{1}{\lambda}\sum_0^\infty E_0(|f_n|\chi_{\{\tau=n\}}) \le \frac{1}{\lambda}\sum_{n=0}^\infty E_0(E(|f_\infty|\chi_{\{\tau=n\}}|\mathcal{F}_n)) \\ &\le \frac{1}{\lambda}E_0(|f_\infty|\chi_{\{Mf>\lambda\}}) \le \frac{1}{\lambda}E_0(|f_\infty|). \end{aligned}$$

(6.4.4) has been proved. From it we also have, for all $F \in \mathcal{F}_0$,

$$|F \bigcap \{Mf > \lambda\}| = \int_F \chi_{\{Mf>\lambda\}} d\mu \leq \frac{1}{\lambda} \int_F |f_\infty| d\mu.$$

Now for $0 < \delta < 1$, we get

$$\begin{aligned}\int_F (Mf)^\delta d\mu &= \delta \int_0^\infty \lambda^{\delta-1} |\{Mf\chi_F > \lambda\}| d\lambda \\ &= \delta \int_0^\infty \lambda^{\delta-1} |\{Mf > \lambda\} \bigcap F| d\lambda \\ &\leq \delta \int_0^\alpha \lambda^{\delta-1} |F| d\lambda + \delta \int_\alpha^\infty \lambda^{\delta-2} \int_F |f_\infty| d\mu d\lambda \\ &= 2\Big(\frac{\delta}{1-\delta}\Big)^\delta \frac{1}{|F|^{\delta-1}} \Big(\int_F |f_\infty| d\mu\Big)^\delta,\end{aligned}$$

where we have chosen

$$\alpha = \frac{\delta}{1-\delta} \frac{1}{|F|} \int_F |f_\infty| d\mu.$$

The proof of the lemma is finished. □

Now we add a condition: there exists sequence $\{\mathcal{I}\}_{n\geq 0}$ of set families, such that $\mathcal{I}_n \subset \mathcal{F}_n$, for all n, and for all $f \in L^1_+$, we have

$$f_n(\omega) \leq C \sup_{F:\omega \in F \in \mathcal{I}_n} \frac{1}{|F|} \int_F |f| d\mu, \text{ a.e.,} \tag{6.4.6.i}$$

$$\sup_{F:\omega\in F\in \bigcup_n \mathcal{I}_n} \frac{1}{|F|} \int_F |f| d\mu \leq CMf(\omega), \text{ a.e.} \tag{6.4.6.ii}$$

Proposition 6.4.5 *Let* $(\Omega, \mathcal{F}, \mu, \{\mathcal{F}_n\}_{n\geq 0})$ *be such that* (6.4.6.i), (6.4.6.ii) *hold. Assume* $g \in L^1_+$ *satisfying* $Mg < \infty$, a.e. *Then for all* $\delta, 0 < \delta < 1, (Mg)^\delta \in A_1$, *and*

$$M((Mg)^\delta) \leq C_\delta (Mg)^\delta, \quad \text{a.e.} \tag{6.4.7}$$

Proof. For n fixed, consider a new family $\{\mathcal{F}'_m\}_{m\geq 0}$ with $\mathcal{F}'_m = \mathcal{F}_{n+m}$. By making use of (6.4.5), (6.4.6.i), (6.4.6.ii) we get

$$\begin{aligned}E\Big((\sup_{m\geq n} g_m)^\delta | \mathcal{F}_n\Big) &\leq C \sup_{\omega\in F\in \mathcal{I}_n} \frac{1}{|F|} \int_F (\sup_{m\geq n} g_m)^\delta d\mu \\ &\leq C_\delta \sup_{\omega \in F\in\mathcal{I}_n} \Big(\frac{1}{|F|}\int_F g d\mu\Big)^\delta \leq c_\delta (Mg)^\delta, \quad \forall n.\end{aligned}$$

□

Remark If the $\mathcal{F}'_n s$ are all generated by atoms, then the conditions (6.4.6.i) and (6.4.6.ii) are true by chosing $\mathcal{I}_n$ as the family of all $\mathcal{F}_n$-atoms.

Now consider the relation between A_1 and BLO.

Theorem 6.4.6 *$f \in BLO$ if and only if for some $\lambda > 0$, $z = e^{\lambda f} \in A_1 \bigcap S$.*

Proof. Assume $f \in BLO$. Then there exists $\lambda > 0$ such that

$$E(\exp \lambda(f - f_n)|\mathcal{F}_n) \leq K_1.$$

Thus,

$$E(e^{\lambda f}|\mathcal{F}_n) \leq K_1 e^{\lambda f_n} \leq K \exp(\lambda(f + \|f\|_{BLO})) = K e^{\lambda f}.$$

This means $z = e^{\lambda f} \in A_1$. As for $z \in S$, we have (say $z \in S^+$)

$$E(e^{\lambda f}|\mathcal{F}_n) \leq K_1 e^{\lambda f_n} \leq K_1 e^{\lambda f_{n-1}} \leq K E(e^{\lambda f}|\mathcal{F}_{n-1}).$$

Assume $z = e^{\lambda f} \in A_1 \bigcap S$. Then

$$e^{\lambda f_n} \leq E(e^{\lambda f}|\mathcal{F}_n) \leq K e^{\lambda f} = e^{\lambda f + K}, \quad f_n \leq f + \frac{K}{\lambda}, \quad \text{a.e.,} \quad \forall n.$$

In addition $z = e^{\lambda f} \in A_p \bigcap S$ implies $f \in BMO$. So we have also $f \in BD$ and hence $f \in BLO$. The proof is finished. □

Remark The condition $z \in S$ could not be taken off in Theorems 6.4.2, 6.4.6. For example when $\{\mathcal{F}_n\}_{n \geq 0}$ (on any space $(\Omega, \mathcal{F}, \mu)$) is taken such that $\mathcal{F}_0$ trivial, and $\mathcal{F}_1 = \mathcal{F}_2 = \cdots = \mathcal{F}$, then $BMO = BLO = L^\infty$, but

$$A_p = \{\text{nonnegative} \quad f : f \in L^1, \ f^{-1} \in L^{\frac{1}{p-1}}\},$$

$$A_1 = \{\text{nonnegative} \quad f : f \in L^1, \ f \geq a > 0\}.$$

This shows that $z \in A_p$ does not imply $\log z \in BMO$ or BLO.

Now for $f = (f_n)_{0 \leq n \leq \infty}$ being a real martingale in L^1, and $\lambda > 0$, we consider another process $\tilde{z} = (\tilde{z}_n)_{0 \leq n \leq \infty}$ with $\tilde{z}_n = e^{\lambda f_n}$. Does the relation between the assertion $f \in BMO$ (or BLO) and the assertion $\tilde{z} \in A_p \bigcap S$ remain being true? The answer is positive. We formulat it as a proposition.

Proposition 6.4.7 *Let $f = (f_n)_{0 \leq n \leq \infty}$ be a real martingale in L^1. If $f \in BMO$ (or BLO), then there exists $\lambda > 0$ such that $\tilde{z} \in A_p \bigcap S$ for some $p > 1$ (or $\tilde{z} \in A_1 \bigcap S$). On the contrary, if $\lambda > 0$ makes $\tilde{z} \in A_p \bigcap S$ (or $A_1 \bigcap S$), then $f \in BMO$ (or BLO).*

Proof. In the arguments of both of sufficience and necessarity, we have

$$1 \leq E(e^{\lambda(f - f_n)}|\mathcal{F}_n) \leq K, \quad \tilde{z}_n \approx z_n = E(e^{\lambda f}|\mathcal{F}_n),$$

which implies the assertions. □

6.5 Factorization of A_p weights

We have shown that the A_p class of weights is increasing when p is so. That is to say the A_1 property is the strongest property. And it is the most convenient object in weight theory indeed. In the classical case, it has been shown that A_p weights can be factorized in terms of A_1 weights. This factorization benefits many problems very much. In this section we will discuss this problem in martingale setting. The preliminaries needed have been done in §4.3, since $z \in A_p \bigcap S$ if and only if $f = \log z \in BMO$, and hence BMO's decomposition to BLO could be used here to factorize A_p as A_1. More precisely, we have

Theorem 6.5.1 *Let $1 \le p \le \infty$. Then weight $z \in A_p \bigcap S$, if and only if there exist $z_1, z_2 \in A_1 \bigcap S$ such that*

$$z = z_1 z_2^{1-p},$$

here z, z_1, z_2 are all special weights.

Proof. Let $z_i \in A_1 \bigcap S, i = 1,2$. Then $z_1 z_2^{1-p} \in A_p \bigcap S$. In fact,

$$\begin{aligned} &E(z_1 z_2^{1-p}|\mathcal{F}_n)E\big(z_2 z_1^{(1-p)^{-1}}|\mathcal{F}_n\big)^{p-1} \\ &\quad = E(z_1 z_{2,n}^{p-1} z_2^{1-p}|\mathcal{F}_n) z_{2,n}^{1-p} E(z_1^{(1-p)^{-1}} z_{1,n}^{(p-1)^{-1}} z_2|\mathcal{F}_n)^{p-1} z_{1,n}^{-1} \\ &\quad \le K_2 E(z_1|\mathcal{F}_n) z_{2,n}^{1-p} K_1 E(z_2|\mathcal{F}_n)^{p-1} z_{1,n}^{-1} \le K_1 K_2. \end{aligned}$$

In additing, denoting $f_i = \log z_i$, then $f_i \in BLO$, and hence $f = \log z = f_1 + (1-p)f_2 \in BMO$. So $z \in S$. This completes the proof of one half of the assertion.

Now return to the main part of the theorem. Assume $z \in A_p \bigcap S$, $p > 1$. Then $z^{1+\varepsilon} \in A_p \bigcap S$ for some $\varepsilon > 0$ (see Corollary 6.3.5.). Thus $f = \log z \in BMO$, and $(1+\varepsilon)f \in \log A_{1,(p-1)^{-1}}$, $f \in \log A_{1+\varepsilon, \frac{1+\varepsilon}{p-1}}$. Now apply f's decomposition established in §4.3. By means of Theorem 4.3.11, we see that there exist $g, h \in BLO$ $\varphi \in L^\infty$ such that

$$g \in \log A_{1+\delta,\tau}, \quad h \in \log A_{\frac{1-\delta}{p-1},\tau}, \quad 0 < \delta < \varepsilon, \quad \tau > 0,$$

$$f = g - h + \varphi = f_1 - f_2, \quad f_1 = g + \varphi, \quad f_2 = h.$$

From Theorem 6.4.6, we see that if $f \in BLO$ and $f \in \log A_{\lambda,\tau}$, then $e^{\lambda f} \in A_1 \bigcap S$. So $e^{(1+\delta)f_1} \in A_1 \bigcap S$, $e^{\frac{1+\delta}{p-1}f_2} \in A_1 \bigcap S$. Of course, we have

$$e^{f_1} \in A_1 \bigcap S, \quad e^{\frac{1}{p-1}f_2} \in A_1 \bigcap S.$$

(That $z \in A_p$ implies $z^\alpha \in A_p, \alpha \le 1$, is a consequence of Hölder's inequality.) Thus, we get the desired factorization

$$z = e^f = e^{f_1} e^{-f_2} = z_1 z_2^{1-p}, \quad z_1 = e^{f_1}, \quad z_2 = e^{\frac{1}{p-1}f_2}.$$

The proof of the theorem is finished. □

6.6 Weighted Φ-inequalities

We want to extend the main results in Chapter 3 to the weighted case. The conditions imposed on weights are mainly A_p and S. The former is the same as in the classical case, but the latter is added. This does not surprise us, since the classical case is regular and hence the condition S holds naturally. But we do not know if this added condition S is superfluous in all the problems we consider. In some of them, for example maximal inequality, the condition S is unuseful indeed.

The first weighted inequality we want to establish is the weakly weighted L^p inequality of the maximal oprator M, which is simple even in the two-weight case. Let (U, V) be a pair of nonnegative measurable functions, $1 \le p < \infty$. In this section, a martingale $f = (f_n)_{n \ge 0} \in L^p(U)$ is meant as $f_n = E(f|\mathcal{F}_n), f \in L^p(U)$. (There is a slight confusion about the concept "$f \in L^p(U)$" with the one in §1.3, there $f \in L^p$ means $\sup_n \|f_n\|_p \le C$.) We want to characterize those (U, V) which make M of weak type $(L^p(U), L^p(V))$, i.e.

$$|\{Mf > \lambda\}|_V \le \Big(\frac{C}{\lambda}\|f\|_{L^p(U)}\Big)^p, \quad \forall\lambda > 0,\ \forall f. \tag{6.6.1}$$

Some prior conditions have to be imposed on (U, V). For example, $U \not\equiv \infty$, $V \not\equiv 0$, otherwise it will be trivial situation. In addition,

$$\int_\Omega V d\mu < \infty, \tag{6.6.2}$$

$$\int_\Omega U^{-\frac{1}{p-1}}\, d\mu < \infty. \tag{6.6.3}$$

In many cases, for example if $\mathcal{F}_0$ is trivial, these two integrabilities are necessary in order to (6.6.1) hold. This can be seen as follows. For N larger enough, we have $|\{U \le N\}| > 0$. By setting $f = \chi_{\{U \le N\}}$, applying (6.6.1) to f with $\lambda = \frac{f_0}{2} = \frac{E(f)}{2}$, we get (6.6.2). Consider $U_\varepsilon = U + \varepsilon$ instead of U, set $f = U_\varepsilon^{-\frac{1}{p-1}}\chi_{\{U \le \frac{1}{\varepsilon}\}}$ (for $p > 1$), assume ε small enough such that $0 < a_\varepsilon = \int_{\{U \le \frac{1}{\varepsilon}\}} U_\varepsilon^{-\frac{1}{p-1}} d\mu < \infty$. Then we have

$$\int_\Omega V dx \le C a_\varepsilon^{-p} \int_{\{U \le \frac{1}{\varepsilon}\}} U_\varepsilon^{-\frac{p}{p-1}} U_\varepsilon d\mu = C a_\varepsilon^{1-p},$$

$$\int_\Omega U^{-\frac{1}{p-1}} d\mu = \lim_{\varepsilon \to 0} a_\varepsilon \le C\Big(\int_\Omega V dx\Big)^{\frac{1}{p-1}} < \infty.$$

A similar argument gives (6.6.3) in the case $p = 1$, by setting $f = \chi_{E_\varepsilon}$, with $E_\varepsilon = \{U \le \operatorname{essinf} U + \varepsilon\}$, and applying (6.6.1).

The two-weights version of A_p is needful.

Definition 6.6.1 *(U, V) is said to satisfy $A_{p,p}$, if*

$$E(V|\mathcal{F}_n)E(U^{-\frac{1}{p-1}}|\mathcal{F}_n)^{p-1} \le C_p, \quad \text{a.e.}, \quad \forall n, \tag{6.6.4}$$

here $E(U^{-\infty}|\mathcal{F}_n)^0$ is meant as U^{-1}.

Theorem 6.6.2 *Let $1 \le p < \infty$. Then following assertions are equivalent.*

(a) *Let τ be a stopping time, E_τ is the conditional expectation operator: $f \to f_\tau$. Then*

$$\sup_\tau \Big(\int_{\{\tau<\infty\}} |f_\tau|^p V d\mu\Big)^{\frac{1}{p}} \le C_p^{(1)}\Big(\int_\Omega |f|^p U d\mu\Big)^{\frac{1}{p}}, \quad \forall f \in L^p(U), \tag{6.6.5}$$

(b) *M is of weak type $(L^p(U), L^p(V))$, i.e.*

$$|\{Mf > \lambda\}|_V^{\frac{1}{p}} \le C_p^{(2)}\lambda^{-1}\Big(\int_\Omega |f|^p U d\mu\Big)^{\frac{1}{p}}, \quad \forall\lambda > 0,\ \forall f \in L^p(U), \tag{6.6.6}$$

(c) *$(U, V) \in A_{p,p}$, i.e.*

$$E(V|\mathcal{F}_n)E(U^{-\frac{1}{p-1}}|\mathcal{F}_n)^{p-1} \le C_p^{(3)p}, \quad \text{a.e.}, \ \forall n.$$

And all $C_p^{(j)}$, $j = 1,2,3$, are equivalent.

Proof. (a) $\Longrightarrow$(b). Let $f = (f_n)_{n\ge 0} \in L^p(U)$. For $\lambda > 0$, define $\tau = \inf\{n : |f_n| > \lambda\}$. Then we have

$$\begin{aligned}|\{Mf > \lambda\}|_V^{\frac{1}{p}} &= \Big(\int_{\{\tau<\infty\}} V d\mu\Big)^{\frac{1}{p}} \le \Big(\lambda^{-p}\int_{\{\tau<\infty\}} |f_\tau|^p V d\mu\Big)^{\frac{1}{p}} \\ &\le C_p^{(1)}\lambda^{-1}\Big(\int_\Omega |f|^p U d\mu\Big)^{\frac{1}{p}}.\end{aligned}$$

This proves (a) $\Longrightarrow$ (b) and $C_p^{(2)} \le C_p^{(1)}$.

(b) $\Longrightarrow$ (a). For n and $B \in \mathcal{F}_n$, and $f \in L^p(U)$ given, set $g = f\chi_B$. Then

$$E(g|\mathcal{F}_n) = E(f|\mathcal{F}_n)\chi_B, \quad |f_n|\chi_B \le Mg.$$

From (b), for any $\lambda > 0$ we get

$$|\{|f_n| > \lambda\}\bigcap B\}|_V \le |\{Mg > \lambda\}|_V \le C_p^{(2)p}\lambda^{-p}\int_B |f|^p U d\mu,$$

$$\lambda^p \int_{B\bigcap\{|f_n|>\lambda\}} V d\mu \le C_p^{(2)p}\int_B |f|^p U d\mu.$$

For all $k \in \mathbf{Z}$ set

$$B_k = \{2^k < |f_n| \le 2^{k+1}\} \subset \{|f_n| > 2^k\}.$$

Then we have

$$\begin{aligned}\int_\Omega |f_n|^p V d\mu &\le 2^p \sum_{-\infty}^{\infty}\int_{B_k\bigcap\{|f_n|>2^k\}} 2^{kp} V d\mu \\ &\le 2^p C_p^{(2)p}\sum_{-\infty}^{\infty}\int_{B_k} |f|^p U d\mu \le (2C_p^{(2)}\|f\|_{L^p(U)})^p.\end{aligned}$$

This proves (b) ⇒ (a), and $C_p^{(1)} \le 2C_p^{(2)}$.

(a) $\Longrightarrow$ (c). For $p > 1$, and any n, any $B \in \mathcal{F}_n$, set $f = U^{-\frac{1}{p-1}}\chi_B$. Then

$$\begin{aligned}\int_B E\left(U^{-\frac{1}{p-1}}|\mathcal{F}_n\right)^p E(V|\mathcal{F}_n)d\mu &= \int_B E\left(U^{-\frac{1}{p-1}}|\mathcal{F}_n\right)^p V d\mu \\ &\le C_p^{(1)p}\int_B U^{-\frac{1}{p-1}}d\mu = C_p^{(1)p}\int_B E\left(U^{-\frac{1}{p-1}}|\mathcal{F}_n\right)d\mu.\end{aligned}$$

Since $E(U^{-\frac{1}{p-1}}|\mathcal{F}_n) < \infty$, a.e. this gives (6.6.4) and $C_p^{(3)} \le C_p^{(1)}$, in the case $p > 1$. For $p = 1$, for all $f \in L^1_+(U)$, we have

$$\int_\Omega f V_n d\mu = \int_\Omega f_n V d\mu \le C_1^{(1)}\int_\Omega f U d\mu.$$

Since f is arbitrary, this gives the answer in the case $p = 1$.

(c) $\Longrightarrow$ (a). Let $f \in L^p(U)$, $1 \le p < \infty$. For any n, we have

$$|E(f|\mathcal{F}_n)|^p \le E(|f|^p U|\mathcal{F}_n)E(U^{-\frac{1}{p-1}}|\mathcal{F}_n)^{p-1}. \tag{6.6.7}$$

Thus we have

$$\begin{aligned}\int_\Omega |f_n|^p V d\mu &= \int_\Omega |f_n|^p E(V|\mathcal{F}_n)d\mu \\ &\le \int_\Omega E(|f|^p U|\mathcal{F}_n)E(V|\mathcal{F}_n)E(U^{-\frac{1}{p-1}}|\mathcal{F}_n)^{p-1}d\mu \\ &\le C_p^{(3)p}\int_\Omega E(|f|^p U|\mathcal{F}_n)d\mu = C_p^{(3)p}\int_\Omega |f|^p U d\mu.\end{aligned}$$

This proves (c) ⇒ (a), and $C_p^{(1)} \le C_p^{(3)}$. In a word, we get

$$C_p^{(2)} \le C_p^{(1)} = C_p^{(3)} \le 2C_p^{(2)}.$$

The proof is finished. □

Now we discuss the weighted maximal inequality in one-weight case. Let z be a measure change (or say z being a special weight).

Theorem 6.6.3 *Let $1 < p < \infty, z$ be a special weight. Then M is $\widehat{L}^p$-bounded if and only if $z \in A_p$.*

Proof. That M's $\widehat{L}^p$-boundedness implies $z \in A_p$ follows from Theorem 6.6.2. Now assume $z \in A_p$. Let $f \in L^p(zd\mu)$. Denote $w = z^{-\frac{1}{p-1}}$. For all $k \in \mathbf{Z}$, define stopping time $\tau_k = \inf\{n : |f_n| > 2^k\}$. Set

$$A_{k,j} = \{x \in \{\tau_k < \infty\} : 2^j < E(w|\mathcal{F}_{\tau_k}) \le 2^{j+1}\}, \quad j \in \mathbf{Z},$$

$$B_{k,j} = \{x \in \{\tau_k < \infty, \tau_{k+1} = \infty\} : 2^j < E(w|\mathcal{F}_{\tau_k}) \le 2^{j+1}\}, \quad j \in \mathbf{Z}.$$

Then $A_{k,j} \in \mathcal{F}_{\tau_k}$, $B_{k,j} \subset A_{k,j}$ and $\{B_{k,j}\}_{k,j}$ is a disjoint family, and

$$\{2^k < Mf \le 2^{k+1}\} = \{\tau_k < \infty, \tau_{k+1} = \infty\} = \bigcup_j B_{k,j}. \tag{6.6.8}$$

We have (for the simplicity, write $E(w)^{-1}w$ as w, a probability measure)

$$f_{\tau_k} = E(f|\mathcal{F}_{\tau_k}) = \widehat{E}_w(fw^{-1}|\mathcal{F}_{\tau_k})E(w|\mathcal{F}_{\tau_k}),$$

and on each $A_{k,j}$, we have

$$\begin{aligned} 2^{kp} &\le \operatorname*{ess\,inf}_{A_{k,j}} |f_{\tau_k}|^p \le \operatorname*{ess\,inf}_{A_{k,j}} |\widehat{E}_w(fw^{-1}|\mathcal{F}_{\tau_k})|^p \operatorname*{ess\,sup}_{A_{k,j}} E(w|\mathcal{F}_{\tau_k})^p \\ &\le 2^p \operatorname*{ess\,inf}_{A_{k,j}} |\widehat{E}_w(fw^{-1}|\mathcal{F}_{\tau_k})|^p |B_{k,j}|_z^{-1} \int_{B_{k,j}} E(w|\mathcal{F}_{\tau_k})^p z\,d\mu. \end{aligned} \tag{6.6.9}$$

Here we have used the fact, on $A_{k,j}, E(w|\mathcal{F}_{\tau_k}) \approx 2^j$. Applying the A_p condition

$$1 \le E(z|\mathcal{F}_\tau)E(w|\mathcal{F}_\tau)^{p-1} \le K, \quad \forall \tau,$$

we have

$$\begin{aligned} \int_{B_{k,j}} E(w|\mathcal{F}_{\tau_k})^p z\,d\mu &\le C \int_{B_{k,j}} E(z|\mathcal{F}_{\tau_k})^{-p'} z\,d\mu \\ &= C \int_{B_{k,j}} \widehat{E}_z(z^{-1}|\mathcal{F}_{\tau_k})^{p'} z\,d\mu. \end{aligned} \tag{6.6.10}$$

Denote

$$\Gamma(\lambda) = \{(k,j) : \operatorname*{ess\,inf}_{A_{k,j}} |\widehat{E}_w(fw^{-1}|\mathcal{F}_{\tau_k})|^p > \lambda\}, \tag{6.6.11}$$

$$G(\lambda) = \bigcup_{(k,j)\in\Gamma_\lambda} A_{k,j}. \tag{6.6.12}$$

By making use of (6.6.9), (6.6.10), we get

$$\begin{aligned} \int_\Omega (Mf)^p z\,d\mu &\le C \sum_{-\infty}^{\infty} 2^{kp} |\{2^k < Mf \le 2^{k+1}\}|_z = C \sum_{k,j} 2^{kp} |B_{k,j}|_z \\ &\le C \sum_{k,j} \operatorname*{ess\,inf}_{A_{k,j}} |\widehat{E}_w(fw^{-1}|\mathcal{F}_{\tau_k})|^p \int_{B_{k,j}} \widehat{E}_z(z^{-1}|\mathcal{F}_{\tau_k})^{p'} z\,d\mu. \end{aligned} \tag{6.6.13}$$

The right-hand side of (6.6.13) is nothing but an integral on a discrete measure space $(\mathbf{Z}^2, d\nu)$ with

$$d\nu : (k,j) \to \int_{B_{k,j}} \widehat{E}_z(z^{-1}|\mathcal{F}_{\tau_k})^{p'} z\,d\mu, \tag{6.6.14}$$

and the integrand

$$F(k,j) = \operatorname*{ess\,inf}_{A_{k,j}} |\widehat{E}_w(fw^{-1}|\mathcal{F}_{\tau_k})|^p. \tag{6.6.15}$$

Now our task is to estimate the distribution function:

$$\sigma(\lambda) = |\Gamma(\lambda)|_\nu = \sum_{(k,j)\in\Gamma_\lambda} \int_{B_{k,j}} \widehat{E}_z(z^{-1}|\mathcal{F}_{\tau_k})^{p'} z d\mu$$

$$\leq \sum_{(k,j)\in\Gamma_\lambda} \int_{B_{k,j}} \widehat{E}_z(z^{-1}\chi_{G(\lambda)}|\mathcal{F}_{\tau_k})^{p'} z d\mu$$

$$\leq \int_{G_\lambda} \widehat{M}_z(z^{-1}\chi_{G(\lambda)})^{p'} z d\mu \leq C \int_{G_\lambda} z^{-p'+1} d\mu = C|G_\lambda|_w. \tag{6.6.16}$$

Here we have used the fact: $\widehat{M}_z$ (with respect to $(\Omega, \mathcal{F}, zd\mu, \{\mathcal{F}_n\})$) is bounded on $\widehat{L}^p(zd\mu)$. Notice again

$$(k,j) \in \Gamma(\lambda) \Rightarrow \operatorname*{ess\,inf}_{A_{k,j}} |\widehat{E}_w(fw^{-1}|\mathcal{F}_{\tau_k})|^p > \lambda \Rightarrow \widehat{M}_w(fw^{-1})^p > \lambda, \text{ on } A_{k,j},$$

this means $G(\lambda) \subset \{\omega \in \Omega : \widehat{M}_w(fw^{-1})^p > \lambda\}$, and hence

$$|G(\lambda)|_w \leq |\{\omega : \widehat{M}_w(fw^{-1})^p > \lambda\}|_w. \tag{6.6.17}$$

Thus, by (6.6.13), (6.6.16), (6.6.17), we get

$$\int_\Omega (Mf)^p z d\mu \leq C \int_0^\infty \sigma(\lambda) d\lambda \leq C \int_0^\infty |G(\lambda)|_w d\lambda$$

$$\leq C \int_0^\infty |\{\widehat{M}_w(fw^{-1})^p > \lambda\}|_w d\lambda$$

$$= C \int_\Omega \widehat{M}_w(fw^{-1})^p w d\mu \leq C \int_\Omega (|f|w^{-1})^p w d\mu$$

$$= C \int_\Omega |f|^p w^{1-p} d\mu = C \int_\Omega |f|^p z d\mu.$$

The proof is finished. □

Now we turn to the weighted Φ-inequality between various operators defined on martingales, such as $Mf, S(f), \sigma(f), f^\#_a$, etc., where $\Phi(u)$ is a continuous and increasing function on $\mathbf{R}^+$ that satisfies $\Phi(0) = 0$, and is either merely moderate (call such Φ "general"), or moderate convex, or concave. Essentially, the weighted results follow immediately by imitating the proofs in the unweighted case by virtue of the condition A_∞, sometimes as well as the condition S. We begin with the general Φ case. For such Φ, in general, the good λ-inequality should be established with respect to new measure $\hat{\mu} = zd\mu$. Under the condition $z \in \tilde{A}_\infty$, say $z^{-1} \in \widehat{A}_p$, we have (see Proposition 6.2.3)

$$\widehat{E}(\chi_F|\mathcal{F}_\tau) \leq CE(\chi_F|\mathcal{F}_\tau)^{\frac{1}{p}}, \quad \forall F \in \mathcal{F}, \quad \forall \tau. \tag{6.6.18}$$

With this fact, we could get good λ-inequalities and rearrangement inequalities by the same way essentially as in Chapter 3.

Lemma 6.6.4 *Let $A = (A_n)_{n\geq 0}, B = (B_n)_{n\geq 0}$ be two nonnegative processes, the former of which be adapted, and the latter be predictable and $B_0 = 0$. Assume that for all stopping times T, τ, we have*

$$E((A_T - A_{T\wedge(\tau-1)})^q|\mathcal{F}_\tau) \leq a^q E(B_T^q|\mathcal{F}_\tau), \tag{6.6.19}$$

where a, q are two positive constants. Assume $z \in \tilde{A}_\infty$ as well. Then the following good λ-inequality and rearrangement inequality hold

$$|\{A_\infty > a\lambda, B_\infty \leq \beta\lambda\}|_{\hat{\mu}} \leq \varepsilon_\beta |\{A_\infty > \lambda\}|_{\hat{\mu}} \quad (\alpha > 1, \varepsilon_\beta \to 0), \tag{6.6.20}$$

$$\widehat{A}_\infty^{\hat{*}}(t) \leq \alpha \widehat{B}_\infty^{\hat{*}}(t/2) + \widehat{A}_\infty^{\hat{*}}(2t), \quad \forall t > 0, \tag{6.6.21}$$

where $\hat{*}$ means the rearrangement with respect to $\hat{\mu}$.

Proof. For $\lambda > 0$, define stopping times

$$T = \inf\{n : B_{n+1} > \beta\lambda\}, \quad \beta > 0 \text{ (notice } B_T \leq \beta\lambda, \text{ even on } \{T = 0\}),$$

$$\tau = \inf\{n : A_n > \lambda\}.$$

Then we have (for $\alpha > 1$)

$$\begin{aligned}\{A_\infty > \alpha\lambda, B_\infty \leq \beta\lambda\} &= \{A_T > \alpha\lambda, T = \infty\} \subset \{A_T > \alpha\lambda\} \\ &\subset \{A_T - A_{T\wedge(\tau-1)} > (\alpha - 1)\lambda\},\end{aligned}$$

$$\begin{aligned}E(\chi_{\{A_\infty > \alpha\lambda, B_\infty \leq \beta\lambda\}}|\mathcal{F}_\tau) &\leq E(\chi_{\{A_T - A_{T\wedge(\tau-1)} > (\alpha-1)\lambda\}}|\mathcal{F}_\tau) \\ &\leq \frac{1}{(\alpha-1)^q} E((A_T - A_{T\wedge(\tau-1)})^q|\mathcal{F}_\tau) \\ &\leq \left(\frac{a}{\alpha-1}\right)^q \frac{1}{\lambda^q} E(B_T^q|\mathcal{F}_\tau) \\ &\leq \left(\frac{\alpha\beta}{\alpha-1}\right)^q.\end{aligned} \tag{6.6.22}$$

By making use of (6.6.18), noticing $z \in \widetilde{A}_\infty$, we get, for some p,

$$\widehat{E}(\chi_{\{A_\infty > \alpha\lambda, B_\infty \leq \beta\lambda\}}|\mathcal{F}_\tau) \leq C\left(\frac{a\beta}{\alpha-1}\right)^{\frac{q}{p}} = \varepsilon_\beta.$$

Since $\{A_\infty > \alpha\lambda, B_\infty \leq \beta\lambda\} \subset \{\tau < \infty\} \in \mathcal{F}_\tau$, we have

$$\begin{aligned}|\{A_\infty > \alpha\lambda, B_\infty \leq \beta\lambda\}|_{\hat{\mu}} &= \int_{\{\tau<\infty\}} \chi_{\{A_\infty > \alpha\lambda, B_\infty \leq \beta\lambda\}} d\hat{\mu} \\ &= \int_{\{\tau<\infty\}} \widehat{E}(\chi_{\{A_\infty > \alpha\lambda, B_\infty \leq \beta\lambda\}}|\mathcal{F}_\tau) d\hat{\mu} \\ &\leq \varepsilon_\beta |\{\tau < \infty\}|_{\hat{\mu}} = \varepsilon_\beta |\{A_\infty > \lambda\}|_{\hat{\mu}}.\end{aligned}$$

This proves (6.6.20).

Now prove (6.6.21). Now define stopping times

$$T = \inf\{n : B_{n+1} > \widehat{B}^{*}_{\infty}(t/2)\}, \quad \tau = \inf\{n : A_n > \widehat{A}^{*}_{\infty}(2t)\}.$$

Then we have (for α determined latter, write $F(t) = \alpha \widehat{B}^{*}_{\infty}(t/2) + \widehat{A}^{*}_{\infty}(2t)$)

$$\{A_\infty > F(t)\} \subset \{T < \infty\} \bigcup \{A_T > F(t)\},$$

$$\{A_T > F(t)\} \subset \{A_T - A_{T\wedge(\tau-1)} > \alpha B^{*}_{\infty}(t/2)\},$$

$$|\{A_\infty > F(t)\}|_{\hat{\mu}} \leq |\{T < \infty\}|_{\hat{\mu}} + |\{A_T - A_{T\wedge(\tau-1)} > \alpha B^{*}_{\infty}(t/2)\}|_{\hat{\mu}}.$$

Since $|\{T < \infty\}|_{\hat{\mu}} = |\{B_\infty > B^{*}_{\infty}(t/2)\}|_{\hat{\mu}} \leq t/2$, it is enough to show

$$|\{A_T - A_{T\wedge\tau-1} > \alpha B^{*}_{\infty}(t/2)\}|_{\hat{\mu}} \leq t/2,$$

provided α suitably large. First we estimate $E(\chi_{\{A_T - A_{T\wedge(\tau-1)} > \alpha B^{*}_{\infty}(t/2)\}}|\mathcal{F}_\tau)$. We have just as in the proof of (6.6.20)

$$E(\chi_{\{A_T - A_{T\wedge(\tau-1)} > \alpha B^{*}_{\infty}(t/2)\}}|\mathcal{F}_\tau) \leq (a/\alpha)^q.$$

Making use of (6.6.18) again, we get, for suitable α,

$$\widehat{E}(\chi_{\{A_T - A_{T\wedge(\tau-1)} > \alpha B^{*}_{\infty}(t/2)\}}|\mathcal{F}_\tau) \leq C(a/\alpha)^{q/p} = 1/4,$$

$$|\{A_T - A_{T\wedge(\tau-1)} > \alpha B^{*}_{\infty}(t/2)\}|_{\hat{\mu}} \leq \frac{1}{4}|\{A_\infty > A^{*}_{\infty}(2t)\}|_{\hat{\mu}} \leq t/2.$$

Thus, $|\{A_\infty > F(t)\}|_{\hat{\mu}} \leq t$, $A^{*}_{\infty}(t) \leq F(t)$. This proves (6.6.22). The proof of the lemma is finished. □

Theorem 6.6.5 *Let $f = (f_n)_{n\geq 0}$ be a martingale such that $|\Delta_n f| \leq D_{n-1}$, for all $n \geq 0$, where $(D_n)_{n\geq 0}$ is a nonnegative, increasing and adapted process, and $z \in \widetilde{A}_\infty$, Φ be general. Then*

$$\widehat{E}(\Phi(S(f))) \leq C\widehat{E}(\Phi(Mf + D_\infty)), \tag{6.6.23}$$

$$\widehat{E}(\Phi(Mf)) \leq C\widehat{E}(\Phi(S(f) + D_\infty)), \tag{6.6.24}$$

$$\widehat{E}(\Phi(\max(Mf, S(f)))) \leq C\widehat{E}(\Phi(\min(Mf, S(f)) + D_\infty)). \tag{6.6.25}$$

Proof. See (6.6.23). According to the condition $|f_0| \leq D_{-1}$, we have $f_0 = 0$. Set $A = (S^2_n(f))_{n\geq 0}$, $B = (B_n)_{n\geq 0}$ with $B_n = (M_{n-1}f + D_{n-1})^2$ ($B_0 = 0$ according to a usual convention). Notice $(M_n f)^2 \leq B_n$, for all $n \geq 0$. Then for any stopping times T, τ, we have

$$\begin{aligned} E(A_T - A_{T\wedge(\tau-1)}|\mathcal{F}_\tau) &= E(S^2_T(f) - S^2_{T\wedge(\tau-1)}(f)|\mathcal{F}_\tau) \\ &= E(S^2(f^{(T)}) - S^2_{\tau-1}(f^{(T)})|\mathcal{F}_\tau) \\ &\leq CE(|f^{(T)} - f^{(T)}_{\tau-1}|^2|\mathcal{F}_\tau) \leq CE(M^2_T(f)|\mathcal{F}_\tau) \\ &\leq CE(B_T|\mathcal{F}_\tau). \end{aligned}$$

The same inequality holds for $A = (M_n^2 f)_{n\geq 0}$, $B = (B_n)$, with $B_n = (S_{n-1}(f) + D_{n-1})^2$, and as well for $A = (\max(M_n f, S_n(f)))_{n\geq 0}$, $B = (B_n)_{n\geq 0}$ with $B_n = \min(M_{n-1} f, S_{n-1}(f)) + D_{n-1}$. So, the desired weighted Φ-inequalities follow from the weighted good λ-inequalities. The proof is finished. □

Some other general Φ-inequalities considered in §3.5 could be extended to the weighted case in a similar way. For example, we have

Lemma 6.6.6 *Let $f \to f_a^\#$, $f \to M_a f$ be the sharp and the maximal operator respectively, where $1 \leq a < \infty$, $f = (f_n)_{n\geq 0}$ be martingales, and $z \in \widetilde{A}_\infty$. Then for some $\alpha > 1$, $\beta > 0$, we have*

$$|\{M_a f > \alpha\lambda\}|_{\hat{\mu}} \leq C|f_a^\# \leq \beta\lambda\}|_{\hat{\mu}} + \varepsilon_\beta |\{M_a f > \lambda\}|_{\hat{\mu}}, \quad \forall \lambda > 0, \tag{6.6.26}$$

$$(M_a f)^*(t) \leq C(f_a^\#)^*(t/4) + (M_a f)^*(5t/4), \quad \forall t > 0. \tag{6.6.27}$$

Proof. For $\lambda > 0$ and $\alpha > 1, \beta > 0$ determined later, define stopping times

$$S = \inf\{n : |f_n| > \lambda\}, \quad T = \inf\{n : |f_n| > \alpha\lambda\},$$

$$R = \inf\{n : \rho_n > \beta\lambda\}, \quad \text{with } \rho_n = E(|f - f_{n-1}|^a | \mathcal{F}_n)^{\frac{1}{a}}.$$

Then we have

$$\begin{aligned} \{T < \infty\} &\subset \{R < \infty\} \bigcup \{T < \infty, S < R\} \\ &\subset \{R < \infty\} \bigcup \{S < R, |f_T - f_{S-1}| > (\alpha - 1)\lambda\} \\ &= \{R < \infty\} \bigcup F. \end{aligned}$$

$$\begin{aligned} E(\chi_F | \mathcal{F}_S) &\leq \frac{1}{(\alpha-1)\lambda} E(\chi_{\{S<R\}} |f_T - f_{S-1}| | \mathcal{F}_S) \\ &= \frac{1}{(\alpha-1)\lambda} E(|E(f - f_{S-1} | \mathcal{F}_T)| | \mathcal{F}_S) \chi_{(S<R)} \\ &\leq \frac{1}{(\alpha-1)\lambda} E(|f - f_{S-1}| | \mathcal{F}_S) \chi_{\{S<R\}} \leq \frac{\beta}{\alpha - 1}, \end{aligned}$$

$$\widehat{E}(\chi_F | \mathcal{F}_S) \leq C\Big(\frac{\beta}{\alpha-1}\Big)^{1/p} = \varepsilon_\beta,$$

$$|F|_{\hat{\mu}} \leq \varepsilon_\beta |\{S < \infty\}|_{\hat{\mu}} = \varepsilon_\beta |\{Mf > \lambda\}|_{\hat{\mu}}.$$

Thus, we get

$$|\{Mf > \alpha\lambda\}|_{\hat{\mu}} \leq |\{f_a^\# > \beta\lambda\}|_{\hat{\mu}} + \varepsilon_\beta |\{Mf > \lambda\}|_{\hat{\mu}}.$$

From $M_a f \leq f_a^\# + Mf$, we get

$$\begin{aligned} |\{M_a f > 2\alpha\lambda\}|_{\hat{\mu}} &\leq |\{f_a^\# > \alpha\lambda\}|_{\hat{\mu}} + |\{Mf > \alpha\lambda\}|_{\hat{\mu}} \\ &\leq C|\{f_a^\# > \beta\lambda\}|_{\hat{\mu}} + \varepsilon_\beta |\{M_a f > \lambda\}|_{\hat{\mu}}. \end{aligned}$$

This proves (6.6.26)

For $t > 0$, define stopping times

$$S = \inf\{n : |f_n| > (Mf)^{\hat{*}}(2t)\},$$

$$T = \inf\{n : |f_n| > Cf_a^{\#\hat{*}}(t/2) + (Mf)^{\hat{*}}(2t)\},$$

$$R = \inf\{n : |\rho_n| > f_a^{\#\hat{*}}(t/2)\}.$$

We get

$$|\{T < \infty\}|_{\hat{\mu}} \leq |\{R < \infty\}|_{\hat{\mu}} + |\{T < \infty, S < R\}|_{\hat{\mu}} \leq \frac{t}{2} + |F|_{\hat{\mu}}.$$

For the estimate of $|F|_{\hat{\mu}}$, we have

$$\widehat{E}(\chi_F|\mathcal{F}_S) \leq CE((\chi_F|\mathcal{F}_S)^{\frac{1}{p}} \leq \varepsilon_\beta,$$

$$|F|_{\hat{\mu}} \leq \varepsilon_\beta |\{S < \infty\}|_{\hat{\mu}} \leq \frac{t}{2}.$$

This proves

$$(Mf)^{\hat{*}}(t) \leq Cf_a^{\#\hat{*}}(t/2) + (Mf)^{\hat{*}}(2t), \quad \forall t > 0. \tag{6.6.28}$$

(6.6.27) follows from (6.6.28) immediately as we have known in §3.6 (Th. 3.6.8). The proof is finished. □

Theorem 6.6.7 *Let Φ be general, $z \in \widetilde{A}_\infty$. Then for all martingales $f = (f_n)_{n \geq 0}$ we have*

$$\widehat{E}(\Phi(M_a f)) \leq C\widehat{E}(\Phi(f_a^{\#})). \tag{6.6.29}$$

Another general Φ-inequality is concerning positive martingales.

Theorem 6.6.8 *Assume that Φ is general one, and $z \in \widetilde{A}_\infty$. Then for all positive martingales $f = (f_n)_{n \geq 0}$, we have*

$$\widehat{E}(\Phi(S(f))) \leq C\widehat{E}(\Phi(Mf)). \tag{6.6.30}$$

Proof. Without loss of generality, assume $f_0 = 0$. As in §3.6, for $\lambda > 0$, define

$$\tau = \inf\{n : f_n > \beta\lambda\}, \quad \beta > 0 \text{ determined later},$$

$$T = \inf\left\{n : \sum_{k=1}^{n} \chi_{\{k \leq \tau\}} |\Delta_k f|^2 > \lambda^2\right\}.$$

Then we have

$$E(\chi_{\{S_{\tau-1}(f) > \alpha\lambda\}}|\mathcal{F}_T) \leq \frac{2\beta^2}{\alpha^2 - 1 - \beta^2},$$

$$\widehat{E}(\chi_{\{S_{\tau-1}(f)>\alpha\lambda\}}|\mathcal{F}_T) \le \varepsilon_\beta,$$

and hence

$$\begin{aligned}|\{S(f) > \alpha\lambda, Mf \le \beta\lambda\}|_{\hat{\mu}} &\le |\{S_{\tau-1}(f) > \lambda\}|_{\hat{\mu}} \\ &\le \varepsilon_\beta|\{T < \infty\}|_{\hat{\mu}} = \varepsilon_\beta|\{S_{\tau-1}(f) > \lambda\}|_{\hat{\mu}} \\ &\le \varepsilon_\beta|\{S(f) > \lambda\}|_{\hat{\mu}}.\end{aligned}$$

This proves the weighted good λ-inequality and hence the weighted general Φ-inequality. The proof is finished. □

Remark Similarly to unweighted case (in §3.6), for positive martingales we have

$$S(f)^*(t) \le \alpha(Mf)^*(t/2) + S(f)^*(2t), \quad \forall t > 0. \tag{6.6.31}$$

In fact, defining

$$\tau = \inf\{n : |f_n| > (Mf)^*(t/2)\}, \quad T = \inf\{n : S_n(f) > S(f)^*(2t)\},$$

we have

$$\begin{aligned}&\{S(f)^2 > \alpha^2(Mf)^{*2}(t/2) + S(f)^{*2}(2t)\} \\ &\qquad \subset \{\tau < \infty\} \bigcup \{S_{\tau-1}(f)^2 > \alpha^2(Mf)^{*2}(t/2) + S(f)^{*2}(2t)\}, \\ &\{S_{\tau-1}(f)^2 > \alpha^2(Mf)^{*2}(t/2) + S(f)^{*2}(2t)\} \\ &\qquad \subset \{T < \tau, S_{\tau-1}(f)^2 - S_T(f)^2 > (\alpha^2 - 1)(Mf)^{*2}(2t)\}.\end{aligned}$$

Denoting $F = \{T < \tau, (S_{\tau-1}f)^2 - (S_Tf)^2 > (\alpha^2-1)(Mf)^{*2}(t/2)\}$, we have

$$\begin{aligned}E(\chi_F|\mathcal{F}_T) &\le \frac{1}{(\alpha^2-1)(Mf)^{*2}(\frac{t}{2})}E(S_{\tau-1}(f)^2 - S_T(f)^2|\mathcal{F}_T)\chi_{\{T<\tau\}} \\ &\le \frac{4}{\alpha^2-1},\end{aligned}$$

$$\widehat{E}(\chi_F|\mathcal{F}_T) \le C\left(\frac{4}{\alpha^2-1}\right)^{\frac{1}{p}} = \varepsilon,$$

provided α is larger enough, and hence

$$|F|_{\hat{\mu}} \le \varepsilon|\{T < \infty\}|_{\hat{\mu}} \le 2\varepsilon t = \frac{t}{2}.$$

This completes the proof of (6.6.31). □

Now we consider the weighted convex or concave inequalities.

Theorem 6.6.9 *Let Φ be moderate convex, and $z \in \widetilde{A}_\infty \bigcap S^-$. Then for all martingales $f = (f_n)_{n\geq 0}$, we have*

$$C\widehat{E}(\Phi(Mf)) \leq \widehat{E}(\Phi(S(f))) \leq C\widehat{E}(\Phi(Mf)), \tag{6.6.32}$$

$$\widehat{E}(\Phi(Mf \vee S(f))) \leq C\widehat{E}(\Phi(Mf \wedge S(f))). \tag{6.6.33}$$

Proof. Without loss of generality, assume $f_0 = 0$. Denote $d_n = \Delta_n f, D_n = M_n d$. Make Davis' decomposition $f = g + h$, with

$$|\Delta_n g| \leq 4D_{n-1}, \quad \forall n,$$

$$\sum_1^\infty |\Delta_n h| \leq \sum_1^\infty \{2(D_n - D_{n-1}) + 2E(D_n - D_{n-1}|\mathcal{F}_{n-1})\}.$$

Since $z \in S^-$, we have

$$\begin{aligned} E(D_n - D_{n-1}|\mathcal{F}_{n-1}) &= z_{n-1}\widehat{E}((D_n - D_{n-1})z_n^{-1}|\mathcal{F}_{n-1}) \\ &\leq K\widehat{E}(D_n - D_{n-1}|\mathcal{F}_{n-1}). \end{aligned}$$

Thus, we get (noting $D_\infty \leq \min(2Mf, S(f))$, $\max(Mh, S(h)) \leq \sum_1^\infty |\Delta_n h|$)

$$\begin{aligned} \widehat{E}(\Phi(Mf)) &\leq C\widehat{E}(\Phi(Mg)) + C\widehat{E}(\Phi(Mh)) \\ &\leq C\widehat{E}(\Phi(S(g) + D_\infty)) + C\widehat{E}\left(\Phi\left(\sum_1^\infty |\Delta_n h|\right)\right) \\ &\leq C\widehat{E}(\Phi(S(f))) + C\widehat{E}\left(\Phi\left(\sum_1^\infty E(D_n - D_{n-1}|\mathcal{F}_{n-1})\right)\right) \\ &\leq C\widehat{E}(\Phi(S(f))) + C\widehat{E}\left(\Phi\left(\sum_1^\infty \widehat{E}(D_n - D_{n-1}|\mathcal{F}_{n-1})\right)\right) \\ &\leq C\widehat{E}(\Phi(S(f))). \end{aligned}$$

The proof of the remainders are similar. The proof is finished. □

Remark As shown in Corollary 6.3.4, the condition $z \in \widetilde{A}_\infty$ together with $z \in S^-$ is slightly stronger than the condition $z \in A_\infty$. However, $z \in A_\infty \bigcap S$ is equivent to $z \in \widetilde{A}_\infty \bigcap S$. We do not know if only the condition $z \in A_\infty$ is sufficient for (6.6.32).

Now consider the moderate convex or concave Φ-inequalities concerning $\sigma(f)$. First we give a lemma.

Lemma 6.6.10 *Assume $z \in S$. Then for all $f = (f_n)_{n\geq 0}$ with $f_0 = 0$, we have*

$$C\widehat{E}(\sigma^2(f)) \leq \widehat{E}(S^2(f)) \leq C\widehat{E}(\sigma^2(f)). \tag{6.6.34}$$

Proof. For the first inequality in (6.6.34), the condition $f_0 = 0$ is not necessary. Under the condition $z_{n-1} \le K z_n$, for all $n \ge 1$, we have

$$\begin{aligned}\widehat{E}(\sigma^2(f)) &= \sum_1^\infty \widehat{E}(E(|\Delta_n f|^2|\mathcal{F}_{n-1})) \\ &= \sum_1^\infty \widehat{E}\left(\widehat{E}\left(\frac{z_{n-1}}{z_n}|\Delta_n f|^2|\mathcal{F}_{n-1}\right)\right) \\ &\le K\widehat{E}(S^2(f)).\end{aligned}$$

Analogously, under the condition $z_n \le C z_{n-1}$, for all $n \ge 1$, we have (since $f_0 = 0$)

$$\widehat{E}(S^2(f)) = \sum_1^\infty \widehat{E}\left(\frac{1}{z_{n-1}}\widehat{E}(|\Delta_n f|^2 z_n|\mathcal{F}_{n-1})\right) \le K\widehat{E}(\sigma^2(f)).$$

The proof is finished. □

Theorem 6.6.11 *Assume $z \in S^-$, and Φ be moderate convex. Then*

$$\widehat{E}(\Phi(\sigma^2(f))) \le C\widehat{E}(\Phi(S^2(f))). \tag{6.6.35}$$

Proof. For $\lambda > 0$, define $\tau = \inf\{n : \sigma^2_{n+1}(f) > \lambda\}$. Then

$$\begin{aligned}\widehat{E}(\sigma^2(f) - \lambda|\mathcal{F}_\tau) &\le \widehat{E}\left(\sum_{k=\tau+1}^\infty E(|\Delta_k f|^2|\mathcal{F}_{k-1})\Big|\mathcal{F}_\tau\right) \\ &\le K\widehat{E}\left(\sum_{k=\tau+1}^\infty \widehat{E}(|\Delta_k f|^2|\mathcal{F}_{k-1})\Big|\mathcal{F}_\tau\right) \\ &= K\sum_{k=1}^\infty \widehat{E}(\widehat{E}(\chi_{\{\tau\le k-1\}}|\Delta_k f|^2|\mathcal{F}_{k-1})|\mathcal{F}_\tau) \\ &= K\widehat{E}\left(\sum_{\tau+1}^\infty |\Delta_k f|^2\Big|\mathcal{F}_\tau\right) \\ &\le K\widehat{E}(S^2(f)|\mathcal{F}_\tau).\end{aligned}$$

Thus, we get

$$\int_{\{\sigma^2(f)>\lambda\}} (\sigma^2(f) - \lambda)d\widehat{\mu} \le \int_{\{\sigma^2(f)>\lambda\}} S^2(f)d\widehat{\mu}.$$

From the convexity lemma, we get (6.6.35). The proof is finished. □

Theorem 6.6.11′ *Assume that $z \in \widetilde{A}_\infty \bigcap S^-$, Φ is moderate convex. Then*

$$\widehat{E}(\Phi(\sigma^2(f))) \le C\widehat{E}(\Phi((Mf)^2)). \tag{6.6.36}$$

It follows by combining Theorems 6.6.9 and 6.6.10.

Remark In particular, $\Phi(u) = u^{\frac{p}{2}}, p \geq 2$, and $z \in \widetilde{A}_p \bigcap S^-$, then we have

$$\widehat{E}(\sigma^p(f)) \leq C\widehat{E}(|f|^p).$$

Theorem 6.6.12 *Assume that $z \in S^+$, and Φ is concave. Then for $f = (f_n)_{n\geq 0}$, $f_0 = 0$,*

$$\widehat{E}(\Phi(S^2(f))) \leq C\widehat{E}(\Phi(\sigma^2(f))). \tag{6.6.37}$$

Proof. Denote

$$W = \sigma^2(f) = \sum_1^\infty w_k = \sum_1^\infty E(|\Delta_k f|^2 | \mathcal{F}_{k-1}),$$

$$W_n = \sum_1^n w_k, \quad W_0 = 0.$$

$$Y = S^2(f) = \sum_1^\infty |\Delta_k f|^2, \quad Y_n = \sum_1^n |\Delta_k f|^2, \quad Y_0 = 0.$$

Define the stopping time $\tau = \inf\{n : W_{n+1} > \lambda\}$. Since

$$Y_\tau = S^2(f^{(\tau)}), \quad W_\tau = \sigma^2(f^{(\tau)}),$$

and

$$\{\tau < \infty\} = \{W > \lambda\}, \quad W_\tau \leq W \wedge \lambda, \quad Y \wedge \lambda \leq Y_\tau + \lambda\chi_{\{\tau<\infty\}},$$

by means of the second inequality in (6.6.34), we get

$$\widehat{E}(Y_\tau) = \widehat{E}(S^2(f^{(\tau)})) \leq C\widehat{E}(\sigma^2(f^{(\tau)})) = C\widehat{E}(W_\tau) \leq C\widehat{E}(W \wedge \lambda).$$

Meanwhile, we have

$$\widehat{E}(\lambda\chi_{\{\tau<\infty\}}) = \lambda\hat{\mu}(\{\tau < \infty\}) \leq \widehat{E}(W \wedge \lambda).$$

Thus, we get

$$\widehat{E}(Y \wedge \lambda) \leq C\widehat{E}(W \wedge \lambda), \quad \forall\lambda \geq 0.$$

By means of Lemma 3.4.1, we get (6.6.37). The proof is finished. □

Theorem 6.6.12′ *Assume that $z \in A_\infty \bigcap S$, Φ is concave. Then*

$$\widehat{E}(\Phi((Mf)^2)) \leq C\widehat{E}(\Phi(\sigma^2(f))). \tag{6.6.38}$$

Proof. Since $z \in A_\infty \bigcap S = \widetilde{A}_\infty \bigcap S$, by combining Theorems 6.6.9 and 6.6.12, we get (6.6.38). □

Remark In particular, if $\Phi(u) = u^{\frac{p}{2}}, 0 < p \leq 2$, and $z \in A_\infty \bigcap S$, then we have

$$\widehat{E}((Mf)^p) \leq C\widehat{E}(\sigma^p(f)). \tag{6.6.39}$$

By the way, we want to show that the proof of (6.6.39) could be more immediate. Define a stopping time $\tau = \inf\{n : \sigma_{n+1}(f) > \lambda\}$. We have (noticing $z \in A_\infty \bigcap S$)

$$\hat{\mu}(\{Mf > \lambda\}) \leq \hat{\mu}(\{Mf^{(\tau)} > \lambda\}) + \hat{\mu}(\{\tau < \infty\}),$$

$$\begin{aligned}\hat{\mu}(\{Mf^{(\tau)} > \lambda\}) &\leq \frac{C}{\lambda^2}\int_\Omega \sigma^2(f^{(\tau)})d\hat{\mu} \\ &\leq \frac{C}{\lambda^2}\int_{\{\sigma(f)\leq\lambda\}} \sigma^2(f)d\hat{\mu} + \frac{C}{\lambda^2}\int_{\{\tau<\infty\}} \sigma_\tau^2(f)d\hat{\mu} \\ &\leq \frac{C}{\lambda^2}\int_{\{\sigma(f)\leq\lambda\}} \sigma^2(f)d\hat{\mu} + C\hat{\mu}(\{\tau < \infty\}).\end{aligned}$$

Thus (for $0 < p < 2$)

$$\lambda^{p-1}\hat{\mu}(\{Mf > \lambda\}) \leq C\lambda^{p-3}\int_{\{\sigma(f)\leq\lambda\}} \sigma^2(f)d\hat{\mu} + C\lambda^{p-1}\hat{\mu}(\{\sigma(f) > \lambda\}).$$

Integrating two sides with respect to $\lambda \in (0, \infty)$, we get (6.6.39).

Notes to Chapter 6

§6.1–6.3. The main results are due to Doléans-Dade & Meyer [2], but the martingale version of the A_p conditions was introduced first by Japanese mathematicians, see for example Izumisawa-Kazamaki [1]. The $\varepsilon = 1$ case in Lemma 6.3.1, is due to Coifman-Fefferman [1], in the classical case. It was R. Coifman and C. Fefferman, who opened an approach to the weighted inequality theory via inverse Hölder's inequality which we will say more about, later in notes to §6.6. Several corollaries and propositions in §6.3, have not been stated explicitly before. The last example in §6.3 is due to Bonami-Lepingle [1].

§6.4. The relation between A_p weights and BMO martingales is well known.

§6.5. The factorization of A_p weights was discovered by P. Jones [1] in the classical case. Its martingale version but with "continuous path hypothesis" is due to Varopoulos [1]. The result in general case is due to Long-Peng [1].

§6.6. The characterization of the weak type $(L^p(z), L^p(z))$ of M is due to A. Uchiyama. Theorem 6.6.2 is due to Long-Peng [2]. As for the type $(L^p(z), L^p(z))$ of M, Izumisawa-Kazamaki [1] obtained that $z \in A_p \bigcap S^-$ implies M's $L^p(z)$-boundedness. They went along the classical approach opened by Coifman-Fefferman. We see that under the A_p condition, we have

$$|E(f|\mathcal{F}_n)|^p \leq C\widehat{E}(|f|^p|\mathcal{F}_n), \quad Mf \leq \widehat{M}(|f|^p)^{\frac{1}{p}}.$$

So, $z \in A_p$ implies $\widehat{E}((Mf)^{p+\varepsilon}) \leq C_\varepsilon \widehat{E}(|f|^{p+\varepsilon})$, for all $\varepsilon > 0$. Thus, the problem is reduced to find out under what condition, $z \in A_{p-\varepsilon}$ for some $\varepsilon > 0$. The inverse Hölder's inequality was discovered for this purpose. But in martingale setting this approach needs some added condition (S^- for example). The new approach we adopted in the book was discovered by Jawerth [1]. This nice idea has been influenced very much by E. Sawyer's works on classical two-weights maximal inequality (Sawyer [1]). For related topic, see Long-Peng [2] where two-weights maximal (p, q) inequality was discussed. Bonami-Lepingle [1] obtained the good λ-inequality between Mf and $S(f) + D_\infty$, or $S(f)$ and $Mf + D_\infty$, the remained parts of Lemma 6.6.4 have not seen elsewhere. Theorem 6.6.5 was well known by specialists. Lemma 6.6.6 and Theorem 6.6.7 have not been seen before. Theorem 6.6.8 in unweighted case is due to Burkholder [2] as we have shown in Notes to Chapter 3. Theorem 6.6.9 was known, too. The results concerning $\sigma(f)$ are due to Long [4], the particular case $\Phi(u) = u^{\frac{p}{2}}$ is due to Kazamaki [2].

7 Regular Martingales

In the preceding chapters, we have met an extremely irregular $\{\mathcal{F}_n\}_{n\geq 0}$, i.e. the one with $\mathcal{F}_0$ = trivial, $\mathcal{F}_1 = \mathcal{F}_2 = \cdots = \mathcal{F}$. In this case all martingales are in L^1_u, no $H_p(p<1)$ martingale theory could be developed. In addition, in this case many counterexamples could be constructed easily. For example, as we have shown that in this case, we have $bmo_a = {}_ak_p = L^a$(modulo constants), for all $p \geq a$, and hence $bmo_b \subsetneq bmo_a$ in general when $a < b$; and we have also $h_p = L^2$(modulo constants), $0 < p < \infty$, etc. So, in order to obtain reasonable results in various problems, some regularity assumption were added. In this chapter we will do some investigations about $H_p(0 < p \leq 1)$ under a kind of regularity condition. Such regularity has occurred ever in preceding chapters, for example in §1.3 characterizing the uniform integrability; in Theorem 4.6.7 characterizing the distance from $f \in BMO$ to L^∞; and in §6.3 obtaining some further properties of A_p weights, etc. In §7.1, we will introduce and study in detail such regularity. §7.2 will be devoted to the regular $H_p(0 < p \leq 1)$ theorey such as atomic decomposition, duality and interpolation etc. Besides these, some results (for example, the singular integral characterization of H_p and Fefferman-Stein's decomposition of BMO) concerning a particular example of regular martingales, i.e. so-called q-regular ones, will be given. In §7.3, we will study the weighted Φ-inequalities among Mf, $S(f)$, $\sigma(f)$ etc. under such regularity condition. In order to show that not only the q-regular martingales are the examples of such regular martingales, we will give a new example occurred in Harmonic Analysis in the next chapter. The assumptions imposed on $(\Omega, \mathcal{F}, \mu, \{\mathcal{F}_n\}_{n\geq 0})$ are unchanged.

7.1 A kind of regularity

Definition 7.1.1 *Let* $(\Omega, \mathcal{F}, \mu, \{\mathcal{F}_n\}_{n\geq 0})$ *be as usual.*

(a) It is said to satisfy the condition "R", if

$$\chi_F \leq dE(\chi_F|\mathcal{F}_{n-1}), \quad \forall F \in \mathcal{F}_n, n \geq 1; \tag{7.1.1}$$

(b) It is said to satisfy the condition "R_ω", if for all $n \geq 1$, $F_n \in \mathcal{F}_n$, there exists at least a $G_n \in \mathcal{F}_{n-1}$ such that

$$F_n \subset G_n, \quad |G_n| \leq d|F_n|; \tag{7.1.2}$$

(c) It is said to satisfy "stopping time predictability", if for any nonvanishing stopping time T, there exists a stopping time τ such that

$$0 \leq \tau < T \text{ on the set } \{T < \infty\}, \quad \text{and } |\{\tau < \infty\}| \leq d|\{T < \infty\}|; \tag{7.1.3}$$

(d) It is said to satisfy "weak good stopping time" property, if for all nonnegative adapted processes $\gamma = (\gamma_n)_{n \geq 0}$, for all $\lambda \geq \|\gamma_0\|_\infty$, there exists a stopping time τ_λ such that

$$\{M\gamma > \lambda\} \subset \{\tau_\lambda < \infty\}, \tag{7.1.4.i}$$

$$|\{\tau_\lambda < \infty\}| \leq d|\{M\gamma > \lambda\}|, \tag{7.1.4.ii}$$

$$\sup_{n \leq \tau_\lambda} \gamma_n = M_{\tau_\lambda}\gamma \leq \lambda; \tag{7.1.4.iii}$$

(e) It is said to satisfy the "inverse maximal inequality", if for all $f \in L^1_+$, for all $\lambda \geq \|f_0\|_\infty$, we have

$$\int_{\{Mf > \lambda\}} f d\mu \leq d\lambda|\{Mf > \lambda\}|; \tag{7.1.5}$$

(f) It is said to satisfy the "strong good stopping time" property, if the conditions in (d) and an added condition imposed on τ_λ: $\lambda_1 \leq \lambda_2$ implies $\tau_{\lambda_1} \leq \tau_{\lambda_2}$, hold.

Remark d implied in preceding expressions is a constant not less than 1. Here the set inclusion is meant modulo sets of measure 0. Notice that (7.1.1) could be formulated as

$$E(f|\mathcal{F}_n) \leq dE(f|\mathcal{F}_{n-1}), \quad \forall f \in L_1^+, \ n \geq 1. \tag{7.1.6}$$

This can be seen as follows. Assume that (7.1.1) hold and $f \in L_1^+(\mathcal{F}_n)$. Let $\{f^{(m)}\}$ be a sequence of finite sums each of which is like $\sum c_i \chi_{F_i}$ with $c_i \geq 0$, $F_i \in \mathcal{F}_n$. Say $f^{(m)} = \sum c_i^{(m)} \chi_{F_i^{(m)}}$. Then

$$\begin{aligned} f^{(m)} &= \sum c_i^{(m)} \chi_{F_i^{(m)}} \leq d \sum c_i^{(m)} E\left(\chi_{F_i^{(m)}}|\mathcal{F}_{n-1}\right) \\ &= dE\left(\sum c_i^{(m)} \chi_{F_i^{(m)}}|\mathcal{F}_{n-1}\right) = dE(f^{(m)}|\mathcal{F}_n). \end{aligned}$$

Assuming that $\{f^{(m)}\}$ is increasing and convergent to f a.e., then (7.1.6) holds for $f \in L_1^+(\mathcal{F}_n)$. Now for $f \in L^1_+$ given arbitarily, we have

$$E(f|\mathcal{F}_n) = f_n \leq dE(f_n|\mathcal{F}_{n-1}) = dE(f|\mathcal{F}_{n-1}).$$

The assertion is proved.

Examples

1. Let $(\Omega, \mathcal{F}, \mu, \{\mathcal{F}_n\}_{n\geq 0})$ be such that each $\mathcal{F}_n$ is purely atomic, and $|I_{i,j}^{(n+1)}| \geq \alpha |I_i^{(n)}|$, $n \geq 0$, where $I_i^{(n)}$ is a $\mathcal{F}_n$-atom, and $I_{i,j}^{(n+1)}$ is any $\mathcal{F}_{n+1}$-atom with $I_{i,j}^{(n+1)} \subset I_i^{(n)}$, and $\alpha > 0$ is a constant. Then the condition "R" holds. In fact, for all $f \in L_+^1$, on $I_{i,j}^{(n+1)} \subset I_i^{(n)}$, we have

$$\begin{aligned} E(f|\mathcal{F}_{n+1}) &= \frac{1}{|I_{i,j}^{(n+1)}|} \int_{I_{i,j}^{(n+1)}} f d\mu \\ &\leq \frac{1}{\alpha} \frac{1}{|I_i^{(n)}|} \int_{I_i^{(n)}} f d\mu = \frac{1}{\alpha} E(f|\mathcal{F}_n). \end{aligned}$$

2. Let $(\Omega, \mathcal{F}, \mu, \{\mathcal{F}_n\}_{n\geq 0})$ be such that for all nonnegative martingales $f = (f_n)_{n\geq 0}$, $\Delta_n f$ can be written as

$$\Delta_n f = \sum_{j=1}^{N} V_n^{(j)} \gamma_n^{(j)}, \quad n \geq 1,$$

with $V_n^{(j)}$ being $\mathcal{F}_{n-1}$-measurable and $\gamma_n^{(j)}$ satisfying

$$\|\gamma_n^{(j)}\|_\infty \leq B, \quad E(\gamma_n^{(j)}|\mathcal{F}_{n-1}) = 0, \quad E(\gamma_n^{(j)}\bar{\gamma}_n^{(l)}|\mathcal{F}_{n-1}) = \delta_{j,l}.$$

$(\{V_n^{(j)}\}, \{\gamma_n^{(j)}\})$ may depend on f, but N, B are independent of f. Then the condition "R" holds in this case. In fact, $W = \left(\sum_{j=1}^N |V_n^{(j)}|^2\right)^{\frac{1}{2}}$ is $\mathcal{F}_{n-1}$-measurable, and

$$|\Delta_n f| \leq \left(\sum_{j=1}^{N} |V_n^{(j)}|^2\right)^{\frac{1}{2}} \left(\sum_{j=1}^{N} |\gamma_n^{(j)}|^2\right)^{\frac{1}{2}} \leq N^{\frac{1}{2}} B W,$$

$$E(|\Delta_n f|^2|\mathcal{F}_{n-1}) = W^2.$$

Thus we have

$$|\Delta_n f| \leq N^{\frac{1}{2}} B W^{-1} E(|\Delta_n f|^2|\mathcal{F}_{n-1}) \leq N B^2 E(|\Delta_n f||\mathcal{F}_{n-1}),$$

$$f_n \leq f_{n-1} + |\Delta_n f| \leq f_{n-1} + N B^2 E(|\Delta_n f||\mathcal{F}_{n-1}) \leq (1 + 2NB^2) f_{n-1}.$$

This proves the assertion.

Our main result concerning these regularity conditions is as follows.

Theorem 7.1.2 *All of the regularity conditions in Definition* 7.1.1 *are equivalent.*

Proof. (a) $\Rightarrow$ (b). Let $n \geq 1$, and $F_n \in \mathcal{F}_n$. Denote

$$G_n = \left\{E(\chi_{F_n}|\mathcal{F}_{n-1}) \geq \frac{1}{d}\right\}.$$

Then $G_n \in \mathcal{F}_{n-1}$, and $F_n \subset G_n$ owing to (7.1.1), and

$$|G_n| \leq dE(E(\chi_{F_n}|\mathcal{F}_{n-1})) = d|F_n|.$$

(b) $\Rightarrow$ (c). Let T be a stopping time not taking the value 0. For each $n \geq 1$, for $F_n = \{T = n\}$, take $G_n \in \mathcal{F}_{n-1}$ as in (7.1.2), and define $\tau = \inf\{n : \omega \in G_{n+1}\}$, then τ is what we want. In fact,

$$T(\omega) = n \Rightarrow \omega \in G_n \Rightarrow \tau(\omega) \leq n - 1 = T(\omega) - 1,$$

that is $\tau < T$ on the set $\{T < \infty\}$. In addition,

$$|\{\tau < \infty\}| \leq \sum_{n=0}^{\infty} |G_{n+1}| \leq d \sum_{n=1}^{\infty} |\{T = n\}| = d|\{T < \infty\}|.$$

(c) $\Rightarrow$ (d). Let $(\gamma_n)_{n\geq 0}$ be any nonnegative adapted process, and $\lambda \geq \|\gamma_0\|_\infty$. Define the stopping time $T = \inf\{n : \gamma_n > \lambda\}$. Then T does not take the value 0. So there exists a stopping time τ_λ satisfying (7.1.3). Thus we have

$$\{M\gamma > \lambda\} = \{T < \infty\} \subset \{\tau_\lambda < \infty\},$$

$$|\{\tau_\lambda < \infty\}| \leq d|\{T < \infty\}| = d|\{M\gamma > \lambda\}|.$$

It remains to check (7.1.4.iii). We have

$$M_{\tau_\lambda}\gamma\chi_{(T<\infty)} \leq M_{T-1}\gamma\chi_{(T<\infty)} \leq \lambda,$$

$$M_{\tau_\lambda}\gamma\chi_{(T=\infty)} \leq M\gamma\chi_{(T=\infty)} \leq \lambda.$$

(d) $\Rightarrow$ (e). Let $f \in L^1_+$. Consider the martingale $f = (f_n)_{n\geq 0}$. For this process and $\lambda \geq \|f_0\|_\infty$, we have a stopping time τ_λ such that (7.1.4.i)–(7.1.4.iii) hold. So we get

$$\begin{aligned}\int_{\{Mf>\lambda\}} f d\mu \leq \int_{\{\tau_\lambda<\infty\}} f d\mu &= \int_{\{\tau_\lambda<\infty\}} f_{\tau_\lambda} d\mu \\ &\leq \lambda|\{\tau_\lambda < \infty\}| \leq d\lambda|\{Mf > \lambda\}|.\end{aligned}$$

(e) $\Rightarrow$ (a). Suppose that (a) was not true but (e) holds. For any $\varepsilon > 0$, there exist n and $B \in \mathcal{F}_n$ such that

$$|\{\omega \in B : E(\chi_B|\mathcal{F}_{n-1}) \leq \varepsilon\chi_B\}| > 0.$$

It says nothing but that for any $\varepsilon > 0$, there exist n and $A \in \mathcal{F}_n$ such that

$$|A| > 0, \quad E(\chi_A|\mathcal{F}_{n-1}) \leq \varepsilon.$$

This equivalence can be seen as follows. By taking $B = A$, the former follows from the latter. On the contrary, by taking

$$A = \{E(\chi_B|\mathcal{F}_{n-1}) \leq \varepsilon\} \bigcap B,$$

the latter follows from the former. In fact, $|A| > 0$ and

$$E(\chi_A|\mathcal{F}_{n-1}) = \chi_{\{E(\chi_B|\mathcal{F}_{n-1}) \leq \varepsilon\}} E(\chi_B|\mathcal{F}_{n-1}) \leq \varepsilon.$$

Now let ε and λ be given such that $0 < \varepsilon < \lambda < 1$. For such ε, take $A \in \mathcal{F}_n$ be as above. Then owing to $\{M\chi_A > \lambda\} = A$, which follows from the facts

$$M\chi_A = \max(M_{n-1}\chi_A, \chi_A),$$

$$E(\chi_A|\mathcal{F}_k) = E(E(\chi_A|\mathcal{F}_{n-1})|\mathcal{F}_k) \leq \varepsilon < \lambda, \quad \forall k \leq n-1,$$

we get

$$\int_{\{M\chi_A > \lambda\}} \chi_A d\mu = |A| = \frac{1}{\lambda}\lambda|\{M\chi_A > \lambda\}.$$

Since $\frac{1}{\lambda}$ could be big arbitrarily, we get a contradiction to (7.1.5). The implication from (e) to (a) is proved.

(f) $\Rightarrow$ (d). Obvious.

(a) $\Rightarrow$ (f). Let $(\gamma_n)_{n \geq 0}$ be any nonnegative adapted process, and $\lambda \geq \|\gamma_0\|_\infty$. Define the stopping time

$$\tau_\lambda = \inf\left\{n : \omega \in \left\{E(\chi_{\{\gamma_{n+1} > \lambda\}}|\mathcal{F}_n) \geq \frac{1}{d}\right\}\right\} = \inf\{n : \omega \in G_{n+1}\}. \tag{7.1.7}$$

We claim $\{M\gamma > \lambda\} \subset \{\tau_\lambda < \infty\}$. In fact if $\omega \in \{M\gamma > \lambda\}$, then there exists an $n \geq 0$ such that $\omega \in \{\gamma_{n+1} > \lambda\}$ (since $\lambda \geq \|\gamma_0\|_\infty$), We have

$$\chi_{\{\gamma_{n+1} > \lambda\}} = E(\chi_{\{\gamma_{n+1} > \lambda\}}|\mathcal{F}_{n+1}) \leq dE(\chi_{\{\gamma_{n+1} > \lambda\}}|\mathcal{F}_n).$$

This means $\{\gamma_{n+1} > \lambda\} \subset G_{n+1}$. So $\tau_\lambda(\omega) \leq n < \infty$. Now prove $|\{\tau_\lambda < \infty\}| \leq d|\{M\gamma > \lambda\}|$. In fact, we have

$$\begin{aligned}
\{\tau_\lambda = n\} &= G_{n+1} \bigcap \{\tau_\lambda = n\} \\
&= \left\{E(\chi_{\{\gamma_{n+1} > \lambda\}}|\mathcal{F}_n)\chi_{\{\tau_\lambda = n\}} \geq \frac{1}{d}\right\} \\
&= \left\{E(\chi_{\{\gamma_{n+1} > \lambda\} \cap \{\tau_\lambda = n\}}|\mathcal{F}_n) \geq \frac{1}{d}\right\},
\end{aligned}$$

and hence

$$\begin{aligned}
|\{\tau_\lambda < \infty\}| &= \sum_0^\infty |\{\tau_\lambda = n\}| \leq d\sum_0^\infty E(E(\chi_{\{\gamma_{n+1} > \lambda\} \cap \{\tau_\lambda = n\}}|\mathcal{F}_n)) \\
&= d\sum_0^\infty |\{\gamma_{n+1} > \lambda\} \bigcap \{\tau_\lambda = n\}| \\
&\leq d\sum_0^\infty |\{M\gamma > \lambda\} \bigcap \{\tau_\lambda = n\}| = d|\{M\gamma > \lambda\}|.
\end{aligned}$$

We now prove $M_{\tau_\lambda}\gamma \le \lambda$. When $\tau_\lambda = 0$, it follows from the assumption $\|\gamma_0\|_\infty \le \lambda$. When $\tau_\lambda = \infty$, if follows from

$$\{\tau_\lambda = \infty\} \subset \{M_\gamma \le \lambda\}, \quad \text{since} \quad \{M\gamma > \lambda\} \subset \{\tau_\lambda < \infty\}.$$

In the remaining cases, we have

$$\tau_\lambda(\omega) = n \ge 1 \Rightarrow \omega \notin \bigcup_{1\le j\le n} G_j \Rightarrow \omega \notin \bigcup_{1\le j\le n} \{\gamma_j > \lambda\},$$

and hence $M_{\tau_\lambda}\gamma = M_n\gamma \le \lambda$. Thus we get $M_{\tau_\lambda}\gamma\chi_{\{0<\tau_\lambda<\infty\}} \le \lambda$, and hence $M_{\tau_\lambda}\gamma \le \lambda$ is proved. Finally, assume $\lambda_1 \le \lambda_2$. We have

$$E(\chi_{\{\gamma_{n+1}>\lambda_1\}}|\mathcal{F}_n) = E(\chi_{\{\gamma_{n+1}>\lambda_2\}}|\mathcal{F}_n) + E(\chi_{\{\lambda_2\ge\gamma_{n+1}>\lambda_1\}}|\mathcal{F}_n).$$

When $\tau_{\lambda_2}(\omega) = n$, we see

$$E(\chi_{\{\gamma_{n+1}>\lambda_1\}}|\mathcal{F}_n) \ge E(\chi_{\{\gamma_{n+1}>\lambda_2\}}|\mathcal{F}_n) \ge \frac{1}{d}.$$

So $\tau_{\lambda_1}(\omega) \le n = \tau_{\lambda_2}(\omega)$. The proof of the theorem is finished. □

Remark We want to give a direct proof of (b) ⇒ (a), from which we could get some information about the construction of $\{\mathcal{F}_n\}_{n\ge 0}$ having such regularity.

Proof of (b) ⇒ (a). Assume "R_ω" holds. We want to show that for $n > 1$, and $F_n \in \mathcal{F}_n$, $\{E(\chi_{F_n}|\mathcal{F}_{n-1}) > 0\}$ is the smallest set which is in $\mathcal{F}_{n-1}$ and contains F_n. That is to say, for all $F_n \in \mathcal{F}_n$ and $G_n \in \mathcal{F}_{n-1}$ such that $F_n \subset G_n$, We have

$$F_n \subset \{E(\chi_{F_n}|\mathcal{F}_{n-1}) > 0\} \subset G_n \quad \text{(modulo sets of measure 0)}.$$

In fact, denoting $F_0 = \{E(\chi_{F_n}|\mathcal{F}_{n-1}) = 0\}$, then

$$|F_n \textstyle\bigcap F_0| = \displaystyle\int_{F_0} \chi_{F_n} d\mu = \int_{F_0} E(\chi_{F_n}|\mathcal{F}_{n-1}) d\mu = 0.$$

That means $F_n \subset F_0^c$. Meanwhile, since

$$E(\chi_{F_n}|\mathcal{F}_{n-1})\chi_{G_n^c} = E(\chi_{F_n\cap G_n^c}|\mathcal{F}_{n-1}) = 0,$$

we have $G_n^c \subset F_0$, and hence $F_0^c \subset G_n$. This proves the assertion. Now the condition "R_ω" implies

$$|\{E(\chi_{F_n}|\mathcal{F}_{n-1}) > 0\} \le d|F_n|, \quad \forall n \ge 1, \quad \forall F_n \in \mathcal{F}_n. \tag{7.1.8}$$

Let $F_n \in \mathcal{F}_n$ be given. Set

$$H_n = \left\{E(\chi_{F_n}|\mathcal{F}_{n-1}) < \frac{1}{d}\right\} \bigcap F_n.$$

Since

$$E(\chi_{H_n}|\mathcal{F}_{n-1})\chi_{H_n} \le E(\chi_{F_n}|\mathcal{F}_{n-1})\chi_{H_n} < \frac{1}{d},$$

we have

$$H_n \subset \{0 < E(\chi_{H_n}|\mathcal{F}_{n-1})\} \bigcap \left\{E(\chi_{H_n}|\mathcal{F}_{n-1}) < \frac{1}{d}\right\} = K_n \in \mathcal{F}_{n-1},$$

and

$$|K_n| \leq |\{E(\chi_{H_n}|\mathcal{F}_{n-1}) > 0\}| \leq d|H_n|.$$

If $|K_n| > 0$, we would have a contradiction

$$|H_n| = \int_{K_n} \chi_{H_n} d\mu = \int_{K_n} E(\chi_{H_n}|\mathcal{F}_{n-1}) d\mu < \frac{1}{d}|K_n| \leq |H_n|.$$

This means $|K_n|$=0, and hence $|H_n| = 0$, that is

$$F_n \subset \left\{E(\chi_{F_n}|\mathcal{F}_{n-1}) \geq \frac{1}{d}\right\}.$$

The proof is finished. □

Now we give a further discussion for such regularity. Consider the situation where d in (7.1.5) is depending on f.

(g) $(\Omega, \mathcal{F}, \mu, \{\mathcal{F}_n\}_{n\geq 0})$ is said to satisfy the "weak inverse maximal inequality", if for all $f \in L^1_+$, there exists constant d_f, such that for all $\lambda \geq \lambda_f$, we have

$$\int_{\{Mf>\lambda\}} f d\mu \leq d_f \lambda |\{Mf > \lambda\}|. \tag{7.1.9}$$

Theorem 7.1.2′ (g) *is equivalent to the others listed in Definition* 7.1.1.

Proof. Only (g) ⇒ (a) is to be proved. Construct the sequences $\{n_i\}$ (of positive integers) and $\{\lambda_i\}$, $\{\varepsilon_i\}$ (of positive nombers) satisfying

$$1 + n_1 + \cdots + n_{i-1} < \lambda_i < n_i \nearrow \infty, \quad \lambda_i = o(n_i), \quad \sum n_i \varepsilon_i \leq 1.$$

Suppose (7.1.1) were not true. There would exist sequences $\{k_i\}$ (of positive integers) and $\{A_i\}$ with $A_i \in \mathcal{F}_{k_i}$, such that

$$|A_i| > 0, \quad E(\chi_{A_i}|\mathcal{F}_{k_i-1}) \leq \varepsilon_i.$$

Without loss of generality we can assume

$$\sum_{j\geq i} |A_j| \leq 2|A_i|, \quad \forall i.$$

(Otherwise we take a subsequence, for example take $A_{i_1} = A_1$, and leave out a finite number of the sets after A_1 such that

$$\sum{}' |A_i| \leq |A_1|, \quad \text{here } \textstyle\sum' \text{ means only over remained } i,$$

owing to the fact $\sum |A_i| \leq \sum \varepsilon_i < \infty$, then call the first one among the remaining sets A_{i_2}, go on like this, $\cdots$.) Now set

$$g = \sum_1^\infty n_i \chi_{A_i}.$$

Then $g \in L^1$, and

$$Mg \leq \sum_1^\infty n_i M\chi_{A_i} \leq \sum_1^\infty n_i(\varepsilon_i + \chi_{A_i}) \leq g + 1.$$

Notice that $\{g > \lambda_i - 1\} \subset \bigcup_{j \geq i} A_j$ because of

$$g \leq \sum_1^{i-1} n_j \leq \lambda_i - 1 \quad \text{on the set} \quad \bigcap_{j \geq i} A_j^c,$$

we get

$$|\{Mg > \lambda_i\}| \leq |\{g > \lambda_i - 1\}| \leq \sum_{j \geq i} |A_j| \leq 2|A_i|.$$

But we have also (noticing $A_i = \{n_i \chi_{A_i} > \lambda_i\}$)

$$n_i |A_i| = \int_{\{n_i \chi_{A_i} > \lambda_i\}} n_i d\mu \leq \int_{\{g > \lambda_i\}} g d\mu \leq d_g \lambda_i |\{Mg > \lambda_i\}| \leq 2 d_g \lambda_i |A_i|.$$

This would imply

$$d_g \geq \frac{n_i}{2\lambda_i} \longrightarrow \infty.$$

This contradiction proves the theorem. □

Henceforth, such regularity will be called "R" condition.

7.2 Regular $H_p(0 < p \leq 1)$ martingales

Let $(\Omega, \mathcal{F}, \mu, \{\mathcal{F}_n\}_{n \geq 0})$ be the usual one. Assume the condition "R" holds in this section. We have defined the spaces H_p^* and H_p^S. We will see in §7.3 that in regular case, these two definitions are equivalent. We denote it as H_p, and adopt any one of its two definitions. In addition, in this section we will only consider those martingales $f = (f_n)_{n \geq 0}$ with $f_0 = 0$. But it is not an essential restriction which we will explain in remarks.

7.2.1 Atomic decomposition of $H_p(0 < p \leq 1)$

We have defined the concept of atoms and made the atomic decomposition of $\mathcal{P}_1$ in §2.7. The method and the result remain valid for $H_p(0 < p \leq 1)$, in the regular case. We can make the atomic decomposition for each f in a dense subspace of H_p, with a series converging both in H_p and pointwise.

Theorem 7.2.1.1 *Let $f \in L^1 \subset H_p(p < 1)$ or $f \in H_1$, $f_0 = 0$. Then there exist sequences $\{\lambda_j\}$ (of positive numbers) and $\{a_j\}$ (of p-atoms), such that*

$$f = \sum_{1}^{\infty} \lambda_j a_j, \quad \text{in } H_p \text{ and pointwise,} \tag{7.2.1.1}$$

and

$$\|f\|_{H_p}^p \leq \sum_{1}^{\infty} \lambda_j^p \leq C_p d \|f\|_{H_p}^p. \tag{7.2.1.2}$$

Proof. Let $f = (f_n)_{n\geq 0}$ be as above (noticing $f_0 = 0$), $\lambda = 2^k$, $k \in \mathbf{Z}$. For the process $(|f_n|)_{n\geq 0}$ and $\lambda = 2^k$, define the stopping time τ_k satisfying the properties as shown in (f) of Definition 7.1.1. Since $\{\tau_k\}$ is increasing and $|\{\tau_k < \infty\}| \to 0$, we see $\lim_{k\to\infty} \tau_k = \infty$, a.e, and $\lim_{k\to\infty} f_{\tau_k} = f$, a.e. In addition, we have also $\lim_{k\to-\infty} |f_{\tau_k}| \leq \lim_{k\to-\infty} 2^k = 0$, a.e. Thus we get the following decomposition which converges pointwise

$$f = \sum_{k=-\infty}^{\infty} (f_{\tau_k} - f_{\tau_{k-1}}). \tag{7.2.1.3}$$

Denote

$$\lambda_k = 2^{k+1} |\{\tau_{k-1} < \infty\}|^{\frac{1}{p}}, \quad a_k = \lambda_k^{-1}(f_{\tau_k} - f_{\tau_{k-1}})\chi_{\{\tau_{k-1}<\infty\}}.$$

Then a_k are all p-atoms. In fact, we have

$$\|a_k\|_\infty \leq |\{\tau_{k-1} < \infty\}|^{-\frac{1}{p}},$$

$$(a_k)_n \chi_{\{n\leq\tau_{k-1}\}} = \lambda_k^{-1}(f_{\tau_k\wedge n} - f_{\tau_{k-1}\wedge n})\chi_{\{n\leq\tau_{k-1}\}} = 0, \quad \forall n.$$

Denote $\sigma(\lambda) = |\{Mf > \lambda\}|$. We have

$$\sum_{-\infty}^{\infty} \lambda_k^p \leq d \sum_{-\infty}^{\infty} 2^{(k+1)p} |\{Mf > 2^{k-1}\}|$$

$$\leq C_p d \sum_{-\infty}^{\infty} \int_{2^{k-1}}^{2^k} \lambda^{p-1}\sigma(\lambda)d\lambda = C_p d \|f\|_{H_p}^p.$$

It remains to prove the first inequality in (7.2.1.2), and the convergence in H_p of (7.2.1.1). First we show that all p-atoms are in the unit ball of H_p. In fact, let a be a p-atom accociated with stopping time T. Then $Ma\chi_{(T=\infty)} = 0$, and hence

$$\int_\Omega (Ma)^p d\mu \leq |\{T < \infty\}|^{-1}|\{T < \infty\}| = 1.$$

In addition, we show that we have not only (7.2.1.3), but also

$$f_n = \sum_{-\infty}^{\infty}(f_{\tau_k \wedge n} - f_{\tau_{k-1}\wedge n}) = \sum_{-\infty}^{\infty} \lambda_k (a_k)_n, \quad \forall n, \tag{7.2.1.4}$$

where the series converges pointwise. The reason is the same, since

$$\lim_{k\to\infty} f_{\tau_k \wedge n} = f_n, \quad \text{a.e.}, \quad \lim_{k\to-\infty} |f_{\tau_k \wedge n}| \leq \lim_{k\to-\infty} 2^k = 0, \quad \text{a.e.}$$

Thus we get

$$Mf = \sup_n |f_n| \leq \sum_{-\infty}^{\infty} \lambda_k Ma_k, \quad \|f\|_{H_p}^p \leq \sum_{-\infty}^{\infty} \lambda_k^p \int_\Omega (Ma_k)^p d\mu \leq \sum_{-\infty}^{\infty} \lambda_k^p.$$

Likewise, we have also

$$(f - f_{\tau_{N-1}})_n = \sum_{k=N}^{\infty} \lambda_k (a_k)_n, \quad \forall n; \quad (f_{\tau_M})_n = \sum_{k=-\infty}^{M} \lambda_k (a_k)_n, \quad \forall n.$$

Thus we get

$$\lim_{N\to\infty} \|f - f_{\tau_{N-1}}\|_{H_p}^p \leq \lim_{N\to\infty} \sum_{N}^{\infty} \lambda_k^p = 0,$$

$$\lim_{M\to-\infty} \|f_{\tau_M}\|_{H_p}^p \leq \lim_{M\to-\infty} \sum_{-\infty}^{M} \lambda_k^p = 0.$$

This proves the assertions. Finally rewrite $\sum_{-\infty}^{\infty} \lambda_k(a_k)$ as $\sum_1^\infty \mu_j b_j$, which completes the proof of the theorem. □

Remarks

1. For $f = (f_n)_{n\geq 0} \in H_p \bigcap L^1$ without the assumption $f_0 = 0$, we have the atomic decomposition $f = f_0 + \sum_1^\infty \lambda_j a_j = \sum_0^\infty \lambda_j a_j$ with the convention: all $\mathcal{F}_0$-measurable a with $\|a\|_p = 1$ are meant as special atoms.

2. When all $\mathcal{F}_n$ are atomic, we could get atomic decompositions with the meaning of atom more coincident with the classical one. We omit the details here.

7.2.2 Dual spaces of $H_p(0 < p \leq 1)$

This is a solved problem. As we have known in §2.8(see the remark after Theorem 2.8.3) that we have $(H_p^S)' = {}_2\Lambda_\alpha$, $\alpha = \frac{1}{p} - 1$. In this subsection, we reformulate this problem by making use of atomic decomposition, and consider only those martingales $f = (f_n)_{n\geq 0}$ with $f_0 = 0$.

We need two facts ablut ${}_1\lambda_\alpha$.

Lemma 7.2.2.1 *Let $\alpha \geq 0$. Then we have*

$$\|\varphi\|_{{}_1\lambda_\alpha} = \sup_T |\{T < \infty\}|^{-1-\alpha}\|\varphi - \varphi_T\|_1, \tag{7.2.2.1}$$

where "sup" is taken over all stopping times.

Proof. Denote $\beta = \sup_T |\{T < \infty\}|^{-1-\alpha}\|\varphi - \varphi_T\|_1$. For any n and $F \in \mathcal{F}_n$, defining

$$T_F = n, \quad \omega \in F; \quad T_F = \infty, \quad \omega \notin F,$$

we get

$$|F|^{-1-\alpha}\int_F |\varphi - \varphi_n| d\mu \leq \beta.$$

This implies $E(|\varphi - \varphi_n||\mathcal{F}_n) \leq \beta\omega_n^\alpha$, $\|\varphi\|_{{}_1\lambda_\alpha} \leq \beta$, as shown in the proof of Lemma 2.8.1. On the contrary, from $\varphi \in {}_1\lambda_\alpha$, we have

$$E(|\varphi - \varphi_T||\mathcal{F}_T) = \sum_0^\infty E(|\varphi - \varphi_n||\mathcal{F}_n)\chi_{(T=n)} \leq \|\varphi\|_{{}_1\lambda_\alpha}\sum_0^\infty \omega_n^\alpha\chi_{(T=n)}.$$

Since for all n,

$$\omega_n\chi_{(T=n)} = \begin{cases} |I|, & \text{when } I \subset \{T = n\},\ I \text{ is } \mathcal{F}_n\text{-atom}, \\ 0, & \text{otherwise}, \end{cases}$$

we get, for all stopping time T,

$$E(|\varphi - \varphi_T||\mathcal{F}_T) \leq \|\varphi\|_{{}_1\lambda_\alpha}|\{T < \infty\}|^\alpha,$$

$$\begin{aligned} &|\{T < \infty\}|^{-1-\alpha}\|\varphi - \varphi_T\|_1 \\ &\quad = |\{T < \infty\}|^{-1-\alpha}\int_{\{T<\infty\}} E(|\varphi - \varphi_T||\mathcal{F}_T)d\mu \leq \|\varphi\|_{{}_1\lambda_\alpha}. \end{aligned}$$

This proves $\beta \leq \|\varphi\|_{{}_1\lambda_\alpha}$. The proof of the lemma is finished. □

Lemma 7.2.2.2 *Let $0 < p \leq 1, \alpha = \frac{1}{p} - 1$. Then*

$$\frac{1}{2}\|\varphi\|_{{}_1\lambda_\alpha} \leq \sup_{a:p\text{-atom}} |E(a\varphi)| \leq \|\varphi\|_{{}_1\lambda_\alpha}. \tag{7.2.2.2}$$

The proof is similar to that of Lemma 2.5.7.

Theorem 7.2.2.3 *Assume the condition "R" holds. Let* $0 < p \leq 1, \alpha = \frac{1}{p} - 1$. *Then* $(H_p)'_0 = {}_1\lambda_\alpha$. *More precisely, the mapping* $\varphi \to l_\varphi$ *from* ${}_1\lambda_\alpha$ *to* $(H_p)'_0$ *satisfies*

$$\frac{1}{2}||\varphi||_{{}_1\lambda_\alpha} \leq ||l|| \leq C_p||\varphi||_{{}_1\lambda_\alpha}. \tag{7.2.2.3}$$

Proof. First we prove that each $\varphi \in {}_1\lambda_\alpha$ yields a bounded linear functional l_φ on $(H_p)_0$. Let $\varphi \in {}_1\lambda_\alpha$ be given, and $f \in (H_1)_0 \subset (H_p)_0$ be any one which has an atomic decomposition $f = \sum_1^\infty \lambda_j a_j$ converging both in H_1 and H_p (This is possible, since both of the decompositions in H_1 and H_p are same essentially in the sense they come from the same decomposition $f = \sum_{-\infty}^{\infty}(f_{\tau_k} - f_{\tau_k - 1})$.) and satisfying $\sum_1^\infty \lambda_j^p \leq C_p d||f||_{H_p}^p$. Then we have

$$\begin{aligned} \Big|E\Big(\varphi\sum_1^n \lambda_j a_j\Big)\Big| &= \Big|\sum_1^n \lambda_j E(\varphi a_j)\Big| \leq ||\varphi||_{{}_1\lambda_\alpha}\sum_1^n \lambda_j \\ &\leq ||\varphi||_{{}_1\lambda_\alpha}\Big(\sum_1^n \lambda_j^p\Big)^{\frac{1}{p}} \leq C_p d||\varphi||_{{}_1\lambda_\alpha}||f||_{H_p}. \end{aligned} \tag{7.2.2.4}$$

Since we have

$$E(|\varphi - \varphi_n||\mathcal{F}_{n+1}) \leq dE(|\varphi - \varphi_n||\mathcal{F}_n) \leq d||\varphi||_{{}_1\lambda_\alpha}\omega_n^\alpha \leq d||\varphi||_{{}_1\lambda_\alpha},$$

this means $\varphi \in BMO$. Since $\sum_1^\infty \lambda_j a_j \to f$ in H_1, the series in (7.2.2.4) converges to a value not depending on the decomposition of f. Thus we can define $l_\varphi(f)$ for $f \in (H_1)_0 \subset (H_p)_0$ as

$$l_\varphi(f) = \lim_n E\Big(\varphi\sum_1^n \lambda_j a_j\Big)$$

which satisfies still

$$|l_\varphi(f)| \leq C_p d||\varphi||_{{}_1\lambda_\alpha}||f||_{H_p}.$$

This proves ${}_1\lambda_\alpha \subset (H_p)'_0$.

Now let $l \in (H_p)'_0$. Since $H_2 \subset H_p$, $l \in (H_2)'_0$. Thus there exists $\varphi \in (H_2)_0$ such that

$$l_\varphi(f) = E(f\varphi), \quad \forall f \in (H_2)_0.$$

In particular, for all p-atoms a we have

$$|E(\varphi a)| = |l(a)| \leq ||l||\ ||a||_{H_p} \leq ||l||.$$

From (7.2.2.2), we get $||\varphi||_{{}_1\lambda_\alpha} \leq 2||l||$. This proves $(H_p)'_0 \subset {}_1\lambda_\alpha$. The proof of the theorem is finished. □

Remarks

1. We know that $h_p' = {}_2\lambda_\alpha$, $0 < p \leq 1$, and we will see in §7.3 that $(H_p)_0 = h_p$(modulo $\mathcal{F}_0$-martingales). So we have also $h_p' = {}_1\lambda_\alpha$, $0 < p \leq 1$. This means that in the regular case, ${}_1\lambda_\alpha = {}_2\lambda_\alpha$, $\alpha \geq 0$.

2. How could we get H_p' from $(H_p)_0'$? Obviously we have $(L^p(\mathcal{F}_0))' = {}_1\lambda_\alpha(\mathcal{F}_0) = \{\varphi_0 : \mathcal{F}_0\text{-measurable}, \|\varphi_0\|_{{}_1\lambda_\alpha(\mathcal{F}_0)} = \|\varphi_0\omega_0^{-\alpha}\|_\infty < \infty\}$. Thus, we have

$$H_p' = (L^p(\mathcal{F}_0) \oplus (H_p)_0)' = {}_1\lambda_\alpha(\mathcal{F}_0) \oplus {}_1\lambda_\alpha,$$

here the third space is endowed with the norm $\|\varphi_0\|_{{}_1\lambda_\alpha(\mathcal{F}_0)} + \|\varphi\|_{{}_1\lambda_\alpha}$. This space is nothing but ${}_1\Lambda_\alpha$. In fact

$$\|E(|\varphi - \varphi_n||\mathcal{F}_{n-1})\|_\infty \approx \|E(|\varphi - \varphi_n||\mathcal{F}_n)\|_\infty, \quad n \geq 1,$$

$$E(|\varphi||\mathcal{F}_0) \approx E(|\varphi_0||\mathcal{F}_0) + E(|\varphi - \varphi_0||\mathcal{F}_0).$$

3. Let $0 < p < 1$. Define

$$B_p = \Big\{f = \sum_1^\infty \lambda_j a_j : a_j \ p\text{-atoms}, \sum_1^\infty \lambda_j < \infty\Big\}, \tag{7.2.2.5}$$

$$\|f\|_{B_p} = \inf\Big\{\sum_1^\infty \lambda_j : \text{over all possible expressions}\Big\}, \tag{7.2.2.6}$$

here the series in (7.2.2.5) converges in $({}_1\lambda_\alpha)'$. That $\|\cdot\|_{B_p}$ is a norm is easy to prove (the positive homogeneity and subadditivity are obvious, and $\|f\|_{B_p} = 0 \Rightarrow f = 0$ follows from $\|f\|_{({}_1\lambda_\alpha)'} \leq c\|f\|_{B_p}$.) That B_p is complete could be seen as follows. Let $\{f^{(n)}\}$ be a Cauchy sequence. Let $\{f^{(n_k)}\}$ be such that

$$f^{(n_k)} - f^{(n_{k-1})} = \sum_1^\infty \lambda_j^{(k)} a_j^{(k)}, \quad \sum_1^\infty \lambda_j^{(k)} \leq 2^{-k}.$$

Then $\lim_{k\to\infty} f^{(n_k)} = f \in B_p$ which comes from $f - f^{(n_0)} = \sum_{k=1}^\infty \sum_{j=1}^\infty \lambda_j^{(k)} a_j^{(k)}$. f must be the limit in B_p of $\{f^{(n)}\}$. The proof of the theorem gives really $B_p' = {}_1\lambda_\alpha$. So we get the containing Banach space of H_p, $0 < p < 1$.

7.2.3 Interpolation theory of H_p

Let $0 < p_1 < p_2 < \infty$, $0 < q \leq \infty$, $0 < \theta < 1$. We want to consider the real interpolation spaces $(H_{p_0}, H_{p_1})_{\theta,q}$, and something more about the operator interpolation. We need a kind of martingale spaces, so-called Hardy-Lorentz spaces.Remember the definition of the classical Lorentz spaces

$$L^{p,q} = \Big\{f : \|f\|_{p,q} = \Big(\frac{q}{p}\int_0^\infty \big(t^{\frac{1}{p}} f^*(t)\big)^q \frac{dt}{t}\Big)^{\frac{1}{q}} < \infty\Big\}, \ 0 < p, q < \infty, \tag{7.2.3.1.i}$$

$$L^{p,\infty} = \Big\{f : \|f\|_{p,\infty} = \sup_{t>0} t^{\frac{1}{p}} f^*(t) < \infty\Big\}, \quad 0 < p < \infty, \tag{7.2.3.1.ii}$$

where f is a measurable function on any measure space, f^* is the nonincreasing rearrangement function defined as

$$f^*(t) = \inf\{\lambda : \sigma_f(\lambda) = |\{|f| > \lambda\}| \leq t\}, \quad 0 < t < \infty. \tag{7.2.3.2}$$

The Hardy-Lorentz spaces of martingales are defined as

$$H_{p,q} = \{f = (f_n)_{n\geq 0} : \quad Mf \in L^{p,q}\}, \quad 0 < p \leq \infty, \quad 1 < q \leq \infty. \tag{7.2.3.3}$$

We will show that $H_{p,q}$ is colsed under the K-method interpolation. For the Interpolation Theory, see Bergh-Löfström [1]. By the stability theorem, it is enough to consider $(H_p, L^\infty)_{\theta,q}$.

Lemma 7.2.3.1 *Assume the condition* "R". *Let* $0 < p < \infty$. *Then for all* $f \in H_p + L^\infty$,

$$C^p \int_0^{dt^p} (Mf)^{*p}(\tau)d\tau \leq K(t, f, H_p, L^\infty)^p$$
$$\leq C_p \int_0^{dt^p} (Mf)^{*p}(\tau)d\tau, \quad \forall t > 0, \tag{7.2.3.4}$$

where $K(t, f, H_p, L^\infty)$ *(simplifird as* $K(t, f)$*) is defined as*

$$K(t, f) = K(t, f, H_p, L^\infty)$$
$$= \inf_{f=f_0+f_1} (\|f_0\|_{H_p} + t\|f_1\|_\infty), \quad \forall f \in H_p + L^\infty. \tag{7.2.3.5}$$

Proof. Let $f = (f_n)_{n\geq 0} \in H_p + L^\infty$, and $f = g + h$ be its decomposition in $H_p + L^\infty$. We have

$$Mf \leq Mg + Mh, \quad (Mf)^* \leq (Mg)^* + \|Mh\|_\infty.$$

And so

$$\int_0^{dt^p} (Mf)^{*p}(\tau)d\tau \leq C_p\Big(\int_0^{dt^p} (Mg)^{*p}(\tau)d\tau + t^p\|Mh\|_\infty^p\Big)$$
$$\leq C_p\Big(\|g\|_{H_p}^p + t^p\|h\|_\infty^p\Big).$$

Taking "inf" over all possible decompositions, we get the first inequality in (7.2.3.4).

Now let $f \in H_p + L^\infty$ and such that $f_0 = 0$. For $t > 0$, set $\lambda = (Mf)^*(t^p)$. If $\lambda = 0$, then the second inequality in (7.2.3.4) holds naturally, owing to $\int_0^{dt^p} (Mg)^{*p}(\tau)d\tau = \|f\|_{H_p}^p$. Now for the process $(|f_n|)_{n\geq 0}$ (noting $|f_0| = 0$) and $\lambda = (Mf)^*(t^p)$, take the stopping time T as shown in Theorem 7.1.2. We have

$$M_T f \leq \lambda, \quad |\{T < \infty\}| \leq d|\{Mf > \lambda\}|.$$

Now make the decomposition

$$f = f - f^{(T)} + f^{(T)} = g + h,$$

with

$$g = (g_n)_{n\geq 0} = (f_n - f_{n\wedge T})_{n\geq 0}, \quad h = (h_n)_{n\geq 0} = (f_{n\wedge T})_{n\geq 0}.$$

Then $h \in L^\infty$, $\|h\|_\infty \leq \lambda = (Mf)^*(t^p)$. And since

$$|g_n|\chi_{\{T=\infty)\}} = |f_n - f_{n\wedge T}|\chi_{\{T=\infty\}} = 0,$$

we have $Mg\chi_{\{(T=\infty)\}} = 0$. Notice that

$$Mg \leq Mf + \|Mh\|_\infty, \quad |\{T < \infty\}| \leq d|\{Mf > (Mf)^*(t^p)\}| \leq dt^p.$$

Thus we get

$$\begin{aligned} K(t,f)^p &\leq C_p(\|g\|_{H_p}^p + t^p\|h\|_\infty^p) \\ &\leq C_p\Big\{\int_{\{T<\infty\}} (Mg)^p d\mu + t^p(Mf)^{*p}(t^p)\Big\} \\ &\leq C_p \int_0^{dt^p} (Mf)^{*p}(\tau)d\tau. \end{aligned}$$

For general $f = (f_n)_{n\geq 0} \in H_p + L^\infty$. We have $f = f_0 + f - f_0$, and

$$\begin{aligned} K(t,f,H_p,L^\infty) &\leq K(t,f_0,H_p,L^\infty) + K(t,f-f_0,H_p,L^\infty) \\ &\leq C_p \int_0^{t^p} f_0^{*p}(\tau)d\tau + C_p\int_0^{dt^p} M(f-f_0)^{*p}(\tau)d\tau \\ &\leq C_p \int_0^{dt^p} M(f)^{*p}(\tau)d\tau. \end{aligned}$$

The proof is finshed. □

Theorem 7.2.3.2 *Let* $0 < p < \infty$, $0 < q \leq \infty$, $0 < \theta < 1$. *Denote* $r = \frac{p}{1-\theta}$. *Then*

$$(H_p, L^\infty)_{\theta,q} = H_{r,q}. \tag{7.2.3.6}$$

Proof. Let $f \in H_p + L^\infty$. Then

$$(Mf)^*(t^p) \leq \Big(t^{-p}\int_0^{dt^p} M(f)^{*p}(\tau)d\tau\Big)^{\frac{1}{p}} \leq Ct^{-1}K(t,f).$$

And hence

$$\begin{aligned} \|Mf\|_{r,q} &= C\Big(\int_0^\infty \Big(t^{\frac{1-\theta}{p}}(Mf)^*(t)\Big)^q\frac{dt}{t}\Big)^{\frac{1}{q}} \\ &= C\Big(\int_0^\infty \Big(t^{1-\theta}(Mf)^*(t^p)\Big)^q\frac{dt}{t}\Big)^{\frac{1}{q}} \\ &\leq C\Big(\int_0^\infty \Big(t^{-\theta}K(t,f)\Big)^q\frac{dt}{t}\Big)^{\frac{1}{q}} = C\|f\|_{(H_p,L^\infty)_{\theta,q}}. \end{aligned}$$

This proves $(H_p, L^\infty)_{\theta,q} \subset H_{r,q}$.

Now prove $\|f\|_{(H_p,L^\infty)_{\theta,q}} \le C\|Mf\|_{r,q}$. When $q=\infty$, we have

$$\begin{aligned} t^{-\theta}K(t,f) &\le Ct^{-\theta}\Big(\int_0^{dt^p} \tau^{\theta-1}\Big(\tau^{\frac{1-\theta}{p}}(Mf)^*(\tau)\Big)^p d\tau\Big)^{\frac{1}{p}} \\ &\le \sup_{\tau>0}\Big\{\tau^{\frac{1-\theta}{p}}(Mf)^*(\tau)\Big\}Ct^{-\theta}\Big(\int_0^{dt^p}\tau^{\theta-1}d\tau\Big)^{\frac{1}{p}} \\ &\le C\|Mf\|_{r,\infty}. \end{aligned}$$

When $q<\infty$, by means of the following Hardy's inequality (see Lemma 7.2.3.3), we get

$$\begin{aligned} \Big(\int_0^\infty \big(t^{-\theta}K(t,f)\big)^q\frac{dt}{t}\Big)^{\frac{1}{q}} &\le C\Big(\int_0^\infty t^{-\theta q}\Big(\int_0^{dt^p}(Mf)^{*p}(\tau)d\tau\Big)^{\frac{q}{p}}\frac{dt}{t}\Big)^{\frac{1}{q}} \\ &\le C\Big(\int_0^\infty t^{(1-\theta)\frac{q}{p}}(Mf)^{*p}(t)\frac{dt}{t}\Big)^{\frac{1}{q}} \\ &= C\|Mf\|_{r,q}. \end{aligned}$$

The proof is finished. □

Lemma 7.2.3.3 *Let* $0<r<q\le\infty$, f *be nonnegative and nonincreasing on* $(0,\infty)$. *Then (denoting* $q_0=\min(1,q)$)

$$\Big(\int_0^\infty\Big(\frac{1}{t}\int_0^t f(\tau)d\tau\Big)^q t^r\frac{dt}{t}\Big)^{\frac{1}{q}} \le \Big(\frac{q}{q-r}\Big)^{\frac{1}{q_0}}\Big(\int_0^\infty f(t)^q t^r\frac{dt}{t}\Big)^{\frac{1}{q}}. \tag{7.2.3.7}$$

Proof. When $q\ge1$, without the monotonic property of f, it is well known as Hardy's inequality. Now assume $q<1$ and $\int_0^\infty f(t)^q t^{r-1}dt<\infty$. Owing to the monotonic property of f, we have

$$f(t)^q t^r \le 2\int_{\frac{t}{2}}^t f(\tau)^q\tau^{r-1}d\tau = o(1), \quad t\to0.$$

Denote $F(t)=\int_0^t f(\tau)d\tau$. We have $F(t)^q=o(t^{q-r})$, $t\to0$. Thus we get

$$\begin{aligned} \int_0^\infty F(t)^q t^{r-q-1}dt &= \frac{t^{r-q}}{r-q}F(t)^q\Big|_0^\infty + \frac{q}{q-r}\int_0^\infty F(t)^{q-1}f(t)t^{r-q}dt \\ &\le \frac{q}{q-r}\int_0^\infty\Big(\frac{tf(t)}{F(t)}\Big)^{1-q} f(t)^q t^{r-1}dt \\ &\le \frac{q}{q-r}\int_0^\infty f(t)^q t^{r-1}dt. \end{aligned}$$

The proof is finished. □

Now return to the interpolation.

Theorem 7.2.3.4 *Let $0 < p_i < \infty$, $0 < q_i, q \leq \infty$, $p_0 \neq p_1$, $i = 0, 1$, $0 < \theta < 1$. Let p be such that $\frac{1}{p} = \frac{1-\theta}{p_0} + \frac{\theta}{p_1}$. Then*

$$(H_{p_0}, H_{p_1})_{\theta,q} = H_{p,q}. \tag{7.2.3.8}$$

Proof. Choose $r < \min(p_0, p_1)$. Then there exist $\theta_i \in (0,1)$, $i = 0,1$, such that $p_i = \frac{r}{1-\theta_i}$, $i = 0, 1$. From Theorem 7.2.3.2, we have

$$H_{p_i,q_i} = (H_r, L^\infty)_{\theta_i,q_i}, \quad i = 0, 1.$$

By making use of the stability theorem, we see that $\delta = (1-\theta)\theta_0 + \theta\theta_1$ makes

$$\begin{aligned}(H_{p_0,q_0}, H_{p_1,q_1})_{\theta,q} &= ((H_r, L^\infty)_{\theta_0,q_0}, (H_r, L^\infty)_{\theta_1,q_1})_{\theta,q} \\ &= (H_r, L^\infty)_{\delta,q} = H_{\frac{r}{1-\delta},q} = H_{p,q}.\end{aligned}$$

Here the identity $p = \frac{r}{1-\delta}$ follows from the fact

$$\frac{r}{p_i} = 1 - \theta_i, \quad i = 0, 1; \quad \delta = (1-\theta)\theta_0 + \theta\theta_1.$$

In particular, $q_i = p_i$, $i = 0, 1$, gives (7.2.3.8). The proof is finished. □

Theorem 7.2.3.5 *Let $0 < p_i, \tilde{p}_i < \infty$, $0 < q_i, \tilde{q}_i, q \leq \infty$, $i = 0, 1$, $p_0 \neq p_1$, $\tilde{p}_0 \neq \tilde{p}_1$ and T be a quasilinear operator from H_{p_i,q_i} to $H_{\tilde{p}_i,\tilde{q}_i}$ (or to $L^{\tilde{p}_i,\tilde{q}_i}$), $i = 0, 1$. Then T is also bounded from $H_{p,q}$ to $H_{\tilde{p},q}$ (or to $L^{\tilde{p},q}$), where*

$$\frac{1}{p} = \frac{1-\theta}{p_0} + \frac{\theta}{p_1}, \quad \frac{1}{\tilde{p}} = \frac{1-\theta}{\tilde{p}_0} + \frac{\theta}{\tilde{p}_1}.$$

This folllows from the interpolation theorem.

When we consider only the operator interpolation but not the space interpolation, then it will be simpler by making use of atomic decomposition. We need an addition principle about weak estimates.

Lemma 7.2.3.6 *Let $\{f_k\}$ be a sequence of measurable functions (on any σ-finite measure space) satisfying*

$$|\{|f_k| > \lambda\}| \leq \lambda^{-p}, \quad \forall \lambda > 0, 0 < p < \infty, \tag{7.2.3.9}$$

and $\{m_k\}$ be a sequence of positive numbers. Then

$$\Big|\Big\{\sum_k m_k|f_k| > \lambda\Big\}\Big| \leq \begin{cases} \dfrac{2-p}{1-p}\sum_k m_k^p \lambda^{-p}, & 0 < p < 1, \\ 4\sum_k m_k\Big(1 + \log \sum_l \dfrac{m_l}{m_k}\Big)\lambda^{-1}, & p = 1, \\ p'2^p\Big(\sum_k m_k\Big)^p \lambda^{-p}, & p > 1. \end{cases} \tag{7.2.3.10}$$

Proof. Consider the case $0 < p < 1$. For $\lambda > 0$, set

$$g_k = |f_k|\chi_{\{|f_k| \le \frac{\lambda}{m_k}\}}, \quad h_k = |f_k| - g_k, \quad E_\lambda = \bigcup_k \{h_k \ne 0\}.$$

Then we have

$$|\{h_k \ne 0\}| = \left|\left\{|f_k| > \frac{\lambda}{m_k}\right\}\right| \le \frac{m_k^p}{\lambda^p}, \quad |E_\lambda| \le \sum_k m_k^p \lambda^{-p}.$$

Meanwhile, denote $\sigma_k(t) = |\{|f_k| > t\}|$, we have

$$E(g_k) = \int_{\left\{|f_k| \le \frac{\lambda}{m_k}\right\}} |f_k| d\mu \le \int_0^{\frac{\lambda}{m_k}} \sigma_k(t) dt \le \frac{1}{1-p}\left(\frac{\lambda}{m_k}\right)^{1-p}.$$

Thus we get

$$\begin{aligned}\left|\left\{\sum_k m_k|f_k| > \lambda\right\}\right| &\le |E_\lambda| + \left|\left\{x \notin E_\lambda : \sum_k m_k|f_k| > \lambda\right\}\right| \\ &\le |E_\lambda| + \left|\left\{\sum_k m_k g_k > \lambda\right\}\right| \\ &\le \frac{2-p}{1-p}\sum_k m_k^p \lambda^{-p}.\end{aligned}$$

Now assume $p = 1$. Without loss of generality, assume $\sum_l m_l = 1$. (Otherwise, by homogeneity, consider $\left\{\left(\sum_l m_l\right)^{-1} m_k\right\}$ instead.) So, $m_k \le 1$. For $\lambda > 0$, set

$$|f_k| = |f_k|\left\{\chi_{\{|f_k| \le \lambda\}} + \chi_{\{\lambda < |f_k| \le \frac{\lambda}{m_k}\}} + \chi_{\{|f_k| > \frac{\lambda}{m_k}\}}\right\} = \theta_k + g_k + h_k,$$

$$\sum_k m_k|f_k| = \theta + g + h.$$

Since $\theta \le \lambda \sum_l m_l = \lambda$, we have

$$\left\{\sum_k m_k|f_k| > 2\lambda\right\} \subset \left\{h = 0, \sum_k m_k g_k > \lambda\right\} \bigcup \{h \ne 0\}.$$

Thus we get

$$|\{h \ne 0\}| = \left|\bigcup \{h_k \ne 0\}\right| \le \sum_k \left|\left\{|f_k| > \frac{\lambda}{m_k}\right\}\right| \le \frac{1}{\lambda}\sum_k m_k = \frac{1}{\lambda},$$

$$\begin{aligned}\left|\left\{h = 0, \sum_k m_k g_k > \lambda\right\}\right| &\le \left|\left\{\sum_k m_k g_k > \lambda\right\}\right| \\ &= \left|\left\{\sum_k m_k|f_k|\chi_{\{\lambda < |f_k| \le \frac{\lambda}{m_k}\}} > \lambda\right\}\right| \\ &\le \frac{1}{\lambda}\sum_k m_k \int_{\{\lambda < |f_k| \le \frac{\lambda}{m_k}\}} |f_k| d\mu \\ &\le \frac{1}{\lambda} + \frac{1}{\lambda}\sum_k m_k \log \frac{1}{m_k},\end{aligned}$$

$$\Big|\Big\{\sum_k m_k|f_k| > 2\lambda\Big\}\Big| \leq \frac{2}{\lambda}\sum_k m_k \log\frac{1}{m_k}.$$

Finally, consider the case $p > 1$. Assume $\sum_l m_l = 1$ again. For $\lambda > 0$, set

$$g_k = |f_k|\chi_{\{|f_k|\leq\lambda\}}, \quad h_k = |f_k| - g_k.$$

Then $g \leq \lambda$, and hence

$$\begin{aligned}\Big|\Big\{\sum_k m_k|f_k| > 2\lambda\Big\}\Big| &\leq \Big|\Big\{\sum_k m_k h_k > \lambda\Big\}\Big| \leq \frac{1}{\lambda}\sum_k m_k E(h_k)\\ &= \frac{1}{\lambda}\sum_k m_k\Big[-t\sigma_k(t)|_\lambda^\infty + \int_\lambda^\infty \sigma_k(t)dt\Big]\\ &\leq \frac{1}{\lambda}\sum_k m_k\Big[\frac{1}{\lambda^{p-1}} + \int_\lambda^\infty \frac{dt}{t^p}\Big] = p'\lambda^{-p}.\end{aligned}$$

The proof of the lemma is finished. □

Now consider a so-called countably sublinear operator T (It means that $|Tf| \leq \sum_j \lambda_j|Ta_j|$, for all $f \in H_1 \subset H_p$, for all decompositions of $f, 0 < p \leq 1$.) and its weak boundedness on H_p. By weak type (H_p, L^q) we mean the following

$$\Big|\Big\{|Tf| > \lambda\Big\}\Big|_\nu \leq \Big(\frac{C}{\lambda}\|f\|_{H_p}\Big)^q, \quad \forall\lambda > 0, \quad 0 < p, \quad q < \infty. \tag{7.2.3.11}$$

Theorem 7.2.3.7 *Let T be a countably sublinear operator, $0 < p \leq 1$, $0 < q < \infty$, $q \neq 1$, $p \leq q$. Then T is of weak type (H_p, L^q), if and only if*

$$|\{|Ta| > \lambda\}| \leq C\lambda^{-q}, \quad \forall p\text{-atoms } a. \tag{7.2.3.12}$$

Proof. Necessarity is obvious. Assume (7.2.3.12) holds. For $f \in H_1 \subset H_p$, make the atomic decomposition $\sum_j \lambda_j a_j$ of f. Then we have

$$|Tf| \leq \sum_j \lambda_j|Ta_j|,$$

$$|\{|Tf| > \lambda\}| \leq \begin{cases} C(\sum_j \lambda_j)^q\lambda^{-q}, & q > 1,\\ C(\sum_j \lambda_j^q)\lambda^{-q}, & q < 1.\end{cases}$$

Since $p \leq q$, in both of cases, we have

$$|\{|Tf| > \lambda\}| \leq C\Big(\sum_j \lambda_j^p\Big)^{\frac{q}{p}}\lambda^{-q} \leq C(\|f\|_{H_p}\lambda^{-1})^q.$$

The proof of the theorem is finished. □

We could interpolate the type (H_p, L^q) of T from its weak type (H_{p_i}, L^{q_i}), $i = 0, 1$.

Theorem 7.2.3.8 *Let $0 < p_1 < \min(1, p_2) \le p_2 \le \infty$, $q_1 \neq q_2$, $p_i \le q_i$, $i = 0, 1$. Let T be a countably sublinear operator of weak type (H_{p_i}, L^{q_i}) simultaneously. Then for p, q such that*

$$\frac{1}{p} = \frac{1-t}{p_0} + \frac{t}{p_1} \ge 1, \quad \frac{1}{q} = \frac{1-t}{q_0} + \frac{t}{q_1}, \quad 0 < t < 1, \tag{7.2.3.13}$$

T is of type (H_p, L^q).

Proof. It is enough to prove $||Ta||_q \le C$, for all p-atoms. Once it is proved, noting $p \le q$, we would have

$$||Tf||_q^q \le \Sigma\lambda_j^q E(|Ta_j|^q) \le C\Sigma\lambda_j^q \le C\left(\Sigma\lambda_j^p\right)^{\frac{q}{p}} \le C||f||_{H_p}^q, \quad q \le 1,$$

$$||Tf||_q \le \Sigma\lambda_j(||Ta_j||_q) \le C\Sigma\lambda_j \le C\left(\Sigma\lambda_j^p\right)^{\frac{1}{p}} \le C||f||_{H_p}, \quad q > 1.$$

Now prove $||Ta||_q \le C$. Consider the case $q_1, q_2 < \infty$. Let a be any p-atom associated with a stopping time τ. We have

$$||a||_{H_{p_i}}^{p_i} = E((Ma)^{p_i}) \le |\{\tau < \infty\}|^{1-\frac{p_i}{p}}, \quad i = 1, 2.$$

Thus, say $q_1 < q < q_2$, we get

$$\begin{aligned}\frac{1}{q}||Ta||_q^q &= \int_0^\infty \lambda^{q-1}|\{|Ta| > \lambda\}|d\lambda \\ &\le \int_0^\delta \lambda^{q-1}\Big(\frac{C}{\lambda}||a||_{H_{p_1}}\Big)^{q_1} d\lambda + \int_\delta^\infty \lambda^{q-1}\Big(\frac{C}{\lambda}||a||_{H_{p_2}}\Big)^{q_2} d\lambda \\ &\le C\delta^{q-q_1}|\{\tau < \infty\}|^{\left(\frac{1}{p_1}-\frac{1}{p}\right)q_1} + C\delta^{q-q_2}|\{\tau < \infty\}|^{\left(\frac{1}{p_2}-\frac{1}{p}\right)q_2}.\end{aligned}$$

Taking $\delta = |\{\tau < \infty\}|^\alpha$, with α satisfying

$$q\alpha = \left(\frac{1}{p_i} - \frac{1}{p}\right) \Big/ \left(\frac{1}{q_i} - \frac{1}{q}\right), \quad i = 1, 2,$$

then $||Ta|| \le C$. When one of q_i is ∞, say $q_2 = \infty$, the proof is unchanged. More precisely, we have

$$||Ta||_\infty \le C||a||_{H_{p_2}} \le C|\{\tau < \infty\}|^{\left(\frac{1}{p_2}-\frac{1}{p}\right)},$$

$$\begin{aligned}\frac{1}{q}||Ta||_q^q &= \int_0^{||Ta||_\infty} \lambda^{q-1}\Big(\frac{C}{\lambda}||a||_{H_{p_1}}\Big)^{q_1} d\lambda \\ &\le C|\{\tau < \infty\}|^{(\frac{1}{p_1}-\frac{1}{p})q_1}|\{\tau < \infty\}|^{(\frac{1}{p_2}-\frac{1}{p})(q-q_1)} = C.\end{aligned}$$

The proof is finished. □

By the way we give a kind of decomposition of martingales which could be used in operator interpolation. This decomposition is parallel to the trivial decomposition of L^p. Let $0 < p_0 \leq p \leq p_1 < \infty$, and $\lambda > 0$, then there exists following decomposition for $f \in L^p$,

$$f = g + h \text{ with } \|h\|_{p_0}^{p_0} \leq \lambda^{p_0-p}\|f\|_p^p, \quad \|g\|_{p_1}^{p_1} \leq \lambda^{p_1-p}\|f\|_p^p,$$

by setting

$$h = f\chi_{\{|f|>\lambda\}}, \quad g = f\chi_{\{|f|\leq\lambda\}}.$$

Theorem 7.2.3.9 *Assume the condition "R" holds again. Let $0 < p_0 \leq 1 < p \leq p_1 < \infty$. Then for all $f \in L^p$ and $\lambda > 0$, we have the decomposition $f = g + h$ of $f \in L^p$ such that*

$$\|h\|_{H_{p_0}}^{p_0} \leq C\lambda^{p_0-p}\|f\|_p^p, \quad \|g\|_{p_1}^{p_1} \leq C\lambda^{p_1-p}\|f\|_p^p. \tag{7.2.3.14}$$

Proof. First consider the case $f_0 = 0$. For $(|f_n|)_{n\geq 0}$ and $\lambda > 0$, define a stopping time τ such that

$$M_\tau f \leq \lambda, \quad |\{\tau < \infty\}| \leq d|\{Mf > \lambda\}|. \tag{7.2.3.15}$$

Then $f = f^{(\tau)} + (f - f^{(\tau)}) = g + h$ is just what we want. In fact

$$\|g\|_{p_1}^{p_1} = E(|f_\tau|^{p_1}) \leq \lambda^{p_1-p}E(|f_\tau|^p) \leq \lambda^{p_1-p}E(|f|^p).$$

Meanwhile, noticing $Mf\chi_{(\tau=\infty)} = 0$, we have

$$\begin{aligned}
\|Mh\|_{p_0}^{p_0} &\leq \left(\int_{\{\tau<\infty\}} (Mh)^p d\mu\right)^{\frac{p_0}{p}} |\{\tau < \infty\}|^{1-\frac{p_0}{p}} \\
&\leq d^{1-\frac{p_0}{p}}\left(\frac{p}{p-1}\right)^{p_0} E(|h|^p)^{\frac{p_0}{p}} |\{Mf > \lambda\}|^{1-\frac{p_0}{p}} \\
&\leq C\|f\|_p^{p_0}(\lambda^{-p}\|f\|_p^p)^{1-\frac{p_0}{p}} = C\lambda^{p_0-p}\|f\|_p^p.
\end{aligned}$$

In the case $f_0 \neq 0$, we make the decomposition for f_0 and $f - f_0$ separately, the former of which is just the L^p-case. The proof of the theorem is finished. □

Remark For $p > 1$ and $\lambda > 0$ given, the preceding decomposition is valid for all $p_0, p_1, 0 < p_0 \leq 1 < p \leq p_1 < \infty$.

We do not know if the condition "R" could be got rid of in the preceding arguments. But sometimes it is really so as shown in following theorem.

Theorem 7.2.3.10 *Let $1 \leq p_0 < p_1 \leq \infty$, $1 < q_1 < \infty$. Assume that T is a linear operator being of type $(H_1, L^{p_0'})$ and $(L^{q_1'}, L^{p_1'})$ simultaneously; or Q is a linear operator being of type (L^{p_0}, BOM), and (L^{p_1}, L^{q_1}) simultaneously. Then T is of type $(L^{q'}, L^{p'})$; or Q is of type (L^p, L^q), where*

$$\frac{1}{p} = \frac{1-t}{p_0} + \frac{t}{p_1}, \quad \frac{1}{q} = \frac{t}{q_1}, \quad 0 \leq t \leq 1.$$

Proof. The assertion concerning T follows from the one concerning Q by a duality argument. Consider the Q case. Define the sublinear operator $\widetilde{Q}$ as

$$\widetilde{Q}f = (Qf)^{\#} = \sup_n E(|Qf - (Qf)_{n-1}|\mathcal{F}_n). \tag{7.2.3.16}$$

Then $\widetilde{Q}$ is of type (L^{p_0}, L^∞). and (L^{p_1}, L^{q_1}). In fact, we have

$$\|\widetilde{Q}f\|_\infty = \|Qf\|_* \leq C_1\|f\|_{p_0}, \quad \|\widetilde{Q}f\|_{q_1} \leq C\|Qf\|_{q_1} \leq C_2\|f\|_{p_1}.$$

From Marcinkiewicz' interpolation theorem (which is only related to Lebesgue spaces, see for example, Zygmund [1]), we see that $\widetilde{Q}$ is of type (L^p, L^q). By making use of Theorem 3.5.7 we get

$$\|Qf\|_q \leq C\|\widetilde{Q}f\|_q \leq \|f\|_p.$$

The proof is finished. □

7.2.4 H_1 and $L\log^+ L$

In classical H_1 theorey, one of Zygmund's theorems tells us the relation between H_1 and $L\log^+ L$. We have its martingale version as follows.

Theorem 7.2.4.1 *Let $f = (f_n)_{n\geq 0} \in L\log^+ L$. Then $f \in H_1$. On the contrary, under the condition "R", each positive martingle $f = (f_n)_{n\geq 0}$ which is in H_1 and such that $f_0 \in L\log^+ L$, must be in $L\log^+ L$.*

Proof. The first part of the assertions follows from

$$|\{Mf > \lambda\}| \leq \frac{1}{\lambda}\int_{\{Mf>\lambda\}} |f|d\mu, \quad \forall\lambda > 0.$$

In fact, integrating both sides of this inequality on $\lambda \in [1, \infty)$, we get

$$E((Mf-1)^+) \leq \int_{\{Mf>1\}} |f|\log Mf d\mu = E(|f|\log^+ Mf).$$

But for all $b > a > 0$, we have

$$a\log^+ b \leq a\log^+ a + a\log^+\frac{b}{a} \leq a\log^+ a + \frac{b}{e},$$

we deduce

$$E((Mf-1)^+) \leq E(|f|\log^+|f|) + \frac{1}{e}E(Mf),$$

$$E(Mf) \leq \frac{e}{e-1}E(|f|\log^+|f|) + \frac{e}{e-1}.$$

On the contrary, let $f = (f_n)_{n\ge 0} \in H_1$, $f \ge 0$, and $f_0 \in L\log^+ L$, and $\lambda > 0$. Define the stopping time $\tau = \inf\{n : f_n > \lambda\}$. Then $f_\tau \chi_{\{\tau > 0\}} \le d\lambda$. And we have

$$\begin{aligned}\int_{\{f>\lambda\}} f d\mu &\le \int_{\{\tau<\infty\}} f d\mu = \int_{\{\tau<\infty\}} f_\tau d\mu \\ &\le \int_{\{\tau=0\}} f_0 d\mu + d\lambda|\{0<\tau<\infty\}| \\ &\le \int_{\{f_0>\lambda\}} f_0 d\mu + d\lambda\{Mf > \lambda\}|.\end{aligned}$$

For λ_0 taken arbitrarily, we get

$$\int_{\{f>\lambda_0\}} f \log \frac{f}{\lambda_0} d\mu = \int_{\lambda_0}^{\infty} \frac{1}{\lambda} \int_{\{f>\lambda\}} f d\mu \le d\|f\|_{H_1} + \int_{\{f_0>\lambda_0\}} f_0 \log \frac{f_0}{\lambda_0} d\mu,$$

and hence by taking $\lambda_0 = 1$,

$$E(f\log^+ f) \le d\|f\|_{H_1} + E(f_0 \log^+ f_0).$$

The proof is finished. □

Remark The proof for the first part could be formulated being valid for any quasilinear operator which is both of type (L^∞, L^∞) and weak type (L^1, L^1). In fact, for any $\lambda > 0$, set

$$f = f\chi_{\{|f|\le\lambda\}} + \chi_{\{|f|>\lambda\}} = f_1 + f_2.$$

Then owing to

$$\begin{aligned}&|Tf| \le k(|Tf_1| + |Tf_2|) \le C(\lambda + k|Tf_2|) \\ &|\{Tf| > 2C\lambda\}| \subset \{|Tf_2| > k^{-1}C\lambda\}, \\ &|\{|Tf| > \lambda\}| \le \frac{C}{\lambda}\int_{\{|f|\le\lambda\}} |f| d\mu,\end{aligned}$$

we get

$$E((|Tf| - 1)^+) \le CE(|f|\log^+|f|) + C.$$

7.2.5 Rearrangement of functions in $H_p \bigcap \operatorname{Re} L^1 (0 < p \le 1)$

As an extension of the result in §7.2.4, a property of the rearrangement of functions in $H_p \bigcap \operatorname{Re} L^1 (0 < p \le 1)$ will be stated as a theorem. The condition "R" is still assumed. Let f be a real measurable function on the probability space $(\Omega, \mathcal{F}, \mu)$, and f_d be f's rearrangement on $[-1, 1]$ satisfying

$$f_d \ge 0, \quad \text{and decreasing on } [0,\, 1], \tag{7.2.5.1.i}$$

$$f_d < 0, \quad \text{and decreasing on } [-1, 0], \tag{7.2.5.1.ii}$$

$$|\{\omega :\ f \in B\}| = |\{x :\ f_d \in B\}|, \quad \forall \text{ Borel set } B \subset \mathbf{R}. \tag{7.2.5.1.iii}$$

Theorem 7.2.5.1 *Assume "R" holds. Let $f \in H_p \bigcap \operatorname{Re} L^1$, $0 < p \leq 1$, $f_0 = 0$, then we have*

$$\int_0^1 \left|\frac{1}{x}\int_{-x}^{x} f_d(t)dt\right|^p dx \leq C_{p,d}\|f\|_{H_p}^p. \tag{7.2.5.2}$$

Proof. Denote

$$f_n = E(f|\mathcal{F}_n), \quad Mf = \sup_n |f_n|,$$

$$g_n = (f_n)^+, \quad Mg = \sup_n |g_n| = (\sup_n f_n)^+,$$

$$h_n = (f_n)^-, \quad Mh = \sup_n |h_n| = (-\inf_n f_n)^+.$$

For nonnegative adapted process $(g_n)_{n \geq 0}$ and $\lambda > 0$, define a stopping time τ_λ satisfying

$$|\{\tau_\lambda < \infty\}| \leq d|\{Mg > \lambda\}|,$$

$$\{f > \lambda\} \subset \{\sup_n f_n > \lambda\} = \{Mg > \lambda\} \subset \{\tau_\lambda < \infty\},$$

$$f_{\tau_\lambda}\chi_{\{\tau_\lambda < \infty\}} \leq g_{\tau_\lambda}\chi_{\{\tau_\lambda < \infty\}} \leq \lambda.$$

Then we have

$$\begin{aligned}\int_{\{\tau_\lambda<\infty\}} f d\mu &= \int_{\{\tau_\lambda<\infty\}} f_{\tau_\lambda} d\mu \leq \lambda|\{\tau_\lambda < \infty\}| \\ &\leq d\lambda|\{Mg > \lambda\}| = d\lambda\sigma(\lambda).\end{aligned} \tag{7.2.5.3}$$

Denote

$$\theta_\lambda = \sup\{x : \ f_d(x) > \lambda\}.$$

Notice the following facts: both functions $\int_\varphi^{\theta_\lambda} f_d(x)dx$ and $\lambda\ (\theta_\lambda - \varphi)$ are continuous with respect to $\varphi \in [-1, 0]$, and when $|\varphi|(\varphi \leq 0)$ is increasing, the former is decreasing, the latter is increasing, and

$$\int_0^{\theta_\lambda} f_d(x)dx > \lambda\theta_\lambda,$$

and

$$E(f) = 0, \quad -\int_{-1}^{0} f_d(x)dx = \int_0^1 f_d(x)dx \geq \int_0^{\theta_\lambda} f_d(x)dx.$$

From these facts, we see that there exists a unique $\varphi_\lambda < 0$ such that

$$\int_{\varphi_\lambda}^{\theta_\lambda} f_d(x)dx = \lambda(\theta_\lambda - \varphi_\lambda). \tag{7.2.5.4}$$

Denote $\psi_\lambda = \theta_\lambda - \varphi_\lambda$, then ψ_λ is decreasing with respect to $\lambda \in (0, \infty)$. In fact, for $\lambda_1 > \lambda_2$, we have $\theta_{\lambda_1} \leq \theta_{\lambda_2}$, and $\varphi_{\lambda_2} \leq \varphi_{\lambda_1}$ the latter of which follows from the fact that for all $\varphi \in [\varphi_{\lambda_1}, 0)$, we have

$$\int_{\varphi}^{\theta_{\lambda_2}} f_d(x)dx = -\int_{\varphi_{\lambda_1}}^{\varphi} + \int_{\varphi_{\lambda_1}}^{\theta_{\lambda_1}} + \int_{\theta_{\lambda_1}}^{\theta_{\lambda_2}}$$
$$> \lambda_1(\theta_{\lambda_1} - \varphi_{\lambda_1}) + \lambda_2(\theta_{\lambda_2} - \theta_{\lambda_1}) > \lambda_2(\theta_{\lambda_2} - \varphi).$$

Now from (7.2.5.4), we see that there exists a set E with

$$E = \{f > \lambda\} \bigcup \{f < \beta\} \bigcup F, \quad F \subset \{f = \beta\}, \quad \beta \leq 0,$$

such that $|\{f < \beta\} \bigcup F| = -\varphi_\lambda$. For such E we have the following two facts

$$\int_E f d\mu = \int_{\varphi_\lambda}^{\theta_\lambda} f_d(x)dx = \lambda\psi_\lambda = \lambda|E|, \tag{7.2.5.5}$$

$$\int_E f d\mu = \inf\Big\{\int_F f d\mu : \; F \supset \{f > \lambda\}, |F| \leq |E|\Big\}. \tag{7.2.5.6}$$

We claim that

$$\psi_\lambda \leq d|\{Mg > \lambda\}| = d\sigma(\lambda). \tag{7.2.5.7}$$

In fact, when $\int_E f d\mu \leq \int_{\{\tau_\lambda < \infty\}} f d\mu$, (7.2.5.7) follows from (7.2.5.5) and (7.2.5.3). When $\int_E f d\mu > \int_{\{\tau_\lambda < \infty\}} f d\mu$, (7.2.5.7) follows from (7.2.5.6): since $\{f > \lambda\} \subset \{\tau_\lambda < \infty\}$, $F = \{\tau_\lambda < \infty\}$ could not make $|F| \leq |E|$, this means $\psi_\lambda = |E| < |\{\tau_\lambda < \infty\}| \leq d\sigma(\lambda)$. The claim is proved. Now denote

$$\alpha_n = \Big(2^{n-1}\int_{-2^{-n}}^{2^{-n}} f_d(x)dx\Big)^+.$$

We claim that when $\alpha_n > 0$, we have $\psi_{\alpha_n} \geq 2^{-n}$. In fact, when $\theta_{\alpha_n} \geq 2^{-n}$, it is obvious. When $0 < \theta_{\alpha_n} < 2^{-n}$, we have

$$\int_{-2^{-n}}^{\theta_{\alpha_n}} f_d(x)dx = \int_{-2^{-n}}^{2^{-n}} - \int_{\theta_{\alpha_n}}^{2^{-n}} \geq \alpha_n(2^{1-n} - 2^{-n} + \theta_{\alpha_n}) = \alpha_n(2^{-n} + \theta_{\alpha_n}),$$

which means $\varphi_{\alpha_n} < -2^{-n}$. Thus we get

$$\psi_{\alpha_n} = \theta_{\alpha_n} - \varphi_{\alpha_n} \geq 2^{-n}.$$

Now set $j_0 = 0$, $j_1 (> j_0)$ being the least integer such that $\alpha_{j_1} > 2\alpha_{j_0}, \cdots, j_i(> j_{i-1})$ being the least integer such that $\alpha_{j_i} > 2\alpha_{j_{i-1}}, \cdots$. Thus we have

$$\sum_0^\infty 2^{-n}\alpha_n^p \leq (1+2^p)\sum_{i=0}^\infty 2^{-j_i}\alpha_{j_i}^p \leq (1+2^p)\sum_{i=0}^\infty \psi_{\alpha_{j_i}}\alpha_{j_i}^p$$
$$\leq C_p \sum_{i=0}^\infty \int_{\frac{\alpha_{j_i}}{2}}^{\alpha_{j_i}} \psi_\lambda \lambda^{p-1} d\lambda \leq C_{p,d}\|Mg\|_p^p,$$

i.e.

$$\sum_0^\infty 2^{-n}\left(\left(2^{n-1}\int_{-2^{-n}}^{2^{-n}} f_d(x)dx\right)^+\right)^p \le C_{p,d}\|f\|_{H_p}^p.$$

Similarly, we also have

$$\sum_0^\infty 2^{-n}\left(\left(2^{n-1}\int_{-2^{-n}}^{2^{-n}} f_d(x)dx\right)^-\right)^p \le C_{p,d}\|f\|_{H_p}^p.$$

Appealling to an equaivalence given in Davis [3]

$$\int_0^1\left|\frac{1}{x}\int_{-x}^{x} f_d(t)dt\right|^p dx \approx \sum_0^\infty 2^{n(p-1)}\left|\int_{-2^{-n}}^{2^{-n}} f_d(t)dt\right|^p (\text{modulo } \|f\|_p^p), \quad (7.2.5.8)$$

we get (7.2.5.2). The proof is finished. □

7.2.6 Operator characterization of H_p in the q-martingale case

The classical H_1 could be defined as

$$H_1(\mathbf{T}) = \{f \in L^1 : \tilde{f} \in L^1, \text{ with } \tilde{f} \text{ conjugate function of } f\},$$

$$H_1(\mathbf{R}^n) = \{f \in L^1 : R_j f \in L^1, j = 1, \cdots, n, \text{ with } R_j \text{ Riesz transforms}\}.$$

What about the problem in martingale setting? In the so-called q-martingale case, there is a nice version of the problem. Let $q \in \mathbf{Z}^+$, $q \ge 3$. Let $(\Omega, \mathcal{F}, \mu, \{\mathcal{F}_n\}_{n\ge 0})$ be such that $\mathcal{F}_n$ is generated by q^n atoms, each of them is of measure $q^{-n}, n \ge 0$, and $\{\mathcal{F}_n\}_{n\ge 0}$ is increasing. We could index all atoms as

$$I_0, \quad I_{i_1,\cdots,i_n}, \quad 1 \le i_1, \cdots, i_n \le q, \quad n = 1, 2, \cdots. \quad (7.2.6.1)$$

Our task is to introduce a singular integral operator defined on martingales, and to characterize H_p (mainly H_1), and give the Fefferman-Stein decomposition of BMO constructively, in this case.

Let $V \subset \mathbf{C}^q$ be the superplane defined by

$$V = \left\{x \in \mathbf{C}^q : \sum_{i=1}^q x_i = 0\right\}. \quad (7.2.6.2)$$

Assume that A is a linear transform in V, and its extension in $\mathbf{C}^q$ has the matrix $A = (a_{ij})$. Let $f = (f_n)_{n\ge 0}$ with $f_0 = 0$ be a q-martingale. Then for all $n \ge 1$, and for all $(n-1)$-tuples $(i_1, \cdots, i_{n-1})$,

$$(\Delta_n f|_{I_{i_1,\cdots,i_{n-1},j}})_{j=1}^q \in V.$$

And so $g = (g_n)_{n \ge 0}$ defined as follows

$$g_n = \sum_1^n \Delta_k g, \quad g_0 = 0, \tag{7.2.6.3}$$

with

$$\begin{aligned}(\Delta_n g|_{I_{i_1,\cdots,i_{n-1},j}})_{j=1}^q &= \Big(\sum_{k=1}^q a_{j,k}\Delta_n f|_{I_{i_1,\cdots,i_{n-1},k}}\Big)_{j=1}^q \\ &= A\Big((\Delta_n f|_{I_{i_1,\cdots,i_{n-1},k}})_{k=1}^q\Big),\end{aligned}$$

is a q-martingale. Donote it by $g = Tf$. Thus, each linear transform A in V naturally yields a linear operator T on q-martingales. Notice that for all q-martingales $f = (f_n)_{n \ge 0}$, and its stopped martingales $f^{(n)} = f_n$, we have

$$Tf_n = (Tf)_n, \quad n \ge 0.$$

And hence we could write Tf as $(Tf_n)_{n \ge 0}$. We need some lemmas.

Lemma 7.2.6.1 *Let $x = (x_1, \cdots, x_q)$ be a real vector in V and such that* $\min x_i = -1$. *Then there exists a q-martingale $f \in L^1 - H_1$ such that*

$$\Delta_n f|_{I_{i_1,\cdots,i_{n-1},j}} = \lambda_{i_1,\cdots,i_{n-1}} x_j, \quad j = 1, \cdots, q. \tag{7.2.6.4}$$

Proof. Define $g = (g_n)_{n \ge 0}$ as follows

$$g_n|_{I_{i_1,\cdots,i_n}} = \prod_{k=1}^n (1 + x_{i_k}), \quad n \ge 1, \ g_0 = 1.$$

It is a q-martingale. In fact $g_n|_{I_{i_1,\cdots,i_n}} = (1 + x_{i_n})g_{n-1}|_{I_{i_1,\cdots,i_{n-1}}}$, and so

$$\begin{aligned}&|I_{i_1,\cdots,i_{n-1}}|^{-1}\int_{I_{i_1,\cdots,i_{n-1}}} g_n d\mu \\ &= |I_{i_1,\cdots,i_{n-1}}|^{-1}\sum_{i_n=1}^q \int_{I_{i_1,\cdots,i_{n-1},i_n}} (1 + x_{i_n})g_{n-1}d\mu \\ &= \Big(|I_{i_1,\cdots,i_{n-1}}|^{-1}\sum_{i_n=1}^q |I_{i_1,\cdots,i_{n-1},i_n}|\Big)g_{n-1}|_{i_1,\cdots,i_{n-1}} = g_{n-1}|_{i_1,\cdots,i_{n-1}}.\end{aligned}$$

In addition g is nonnegative and $g \in L^1, E(g_n) = E(g_0) = 1$. Now set $f = \sum_1^\infty \frac{g_k}{k^2}$, and consider the martingale $f = (f_n)_{n \ge 0}$ generated by f. Then

$$f_n = \sum_{k=1}^{n-1}\frac{g_k}{k^2} + \sum_{k \ge n}\frac{g_n}{k^2} \ge \frac{g_n}{n}, \ n \ge 1, \ f_0 = \sum_1^\infty \frac{1}{k^2},$$

and

$$\Delta_n f = f_n - f_{n-1} = \sum_{k \ge n}\frac{g_n - g_{n-1}}{k^2}, \quad n \ge 1,$$

$$\begin{aligned}\Delta_n f|_{I_{i_1,\cdots,i_n}} &= C(g_n - g_{n-1})|_{I_{i_1,\cdots,i_n}}\\ &= (Cg_{n-1}|_{I_{i_1,\cdots,i_{n-1}}})x_{i_n} = \lambda_{i_1,\cdots,i_{n-1}}x_{i_n}, \quad n \geq 1.\end{aligned}$$

It remains to prove $f \notin H_1$. Assume $x_1 = -1$. Notice that

$$g_n|_{I_{i_1,\cdots,i_n}} = 0 \text{ if at least one of } i_1, i_2, \cdots, i_n \text{ is } 1, \ \forall n \geq 1.$$

Denote

$$F_n = \bigcup_{i_1,\cdots,i_{n-1}\neq 1} I_{i_1,\cdots,i_{n-1},1}, \quad n \geq 1,$$

then $\{F_n\}$ is disjoint, and

$$\int_{F_n} g_{n-1}d\mu = \frac{1}{q}\int_{\bigcup_{i_1,\cdots,i_{n-1}\neq 1} I_{i_1,\cdots,i_{n-1}}} g_{n-1}d\mu = \frac{1}{q}E(g_{n-1}) = \frac{1}{q}, \quad n \geq 1.$$

Thus we get

$$\int_\Omega Mfd\mu \geq \sum_n \int_{F_n} Mfd\mu \geq \sum_{n\geq 2}\int_{F_n}\frac{g_{n-1}}{n-1} = \frac{1}{q}\sum_{n\geq 2}\frac{1}{n-1} = \infty.$$

The proof is finished. □

Here is the main lemma in this topic.

Lemma 7.2.6.2 *Let $A_1, \cdots, A_m$ be linear transforms in V which have no nonzero common eigenvector in $\mathbf{R}^q \bigcap V$. Then there exist $p_0 < 1$, such that for all $p > p_0$, and for all $a = (a_i)_0^m \in \mathbf{C}^{m+1}$, for all $x_0 \in V$, $x_i = A_i x_0, i = 1, \cdots, m$, $(x_i = (x_{i,k})_{k=1}^q)$, we have*

$$\|a\|^p \leq \frac{1}{q}\sum_{k=1}^q \|(a_i + x_{i,k})_{i=0}^m\|^p, \tag{7.2.6.5}$$

where the norm $\|\cdot\|$ is Euclidean one.

Proof. Since $x_i \in V$, we have

$$a_i = \frac{1}{q}\sum_{k=1}^q (a_i + x_{i,k}), \quad \forall i. \tag{7.2.6.6}$$

So when $p \geq 1$, we have

$$\|a\| \leq \frac{1}{q}\sum_{k=1}^q \|(a_i + x_{i,k})_{i=0}^m\| \leq \Big(\frac{1}{q}\sum_{k=1}^q \|(a_i + x_{i,k})_{i=0}^n\|^p\Big)^{\frac{1}{p}}.$$

Now consider the case $p < 1$.

First, let $\varepsilon > 0$ be small enough and determined later. Assume $\frac{\|x_0\|}{\|a\|} \leq \varepsilon$, where $\|\cdot\|$ are norms in $\mathbf{C}^q$ and $\mathbf{C}^{m+1}$ respectively. Denote A_0 as the identity operator, and

$$K_1 = \left\{(a, x_0, \cdots, x_m) : a \in \mathbf{C}^{m+1}, x_i = A_i x_0,\ x_0 \in V, \|a\| = \sum_{i=0}^{m} \|x_i\|^2 = 1\right\},$$

and

$$\alpha = \max\left(\sum_k \left(\operatorname{Re} \sum_i \bar{a}_i x_{i,k}\right)^2\right) \text{ on this compact set } K_1.$$

Since on the set K_1, we have

$$\sum_k \left(\operatorname{Re} \sum_i \bar{a}_i x_{i,k}\right)^2 \leq \sum_k \sum_i |a_i|^2 \sum_i |x_{i,k}|^2 \leq \|a\|^2 \sum_i \|x_i\|^2 = 1,$$

we get $\alpha \leq 1$. If $\alpha = 1$, there would be $(a, x_0, \cdots, x_m) \in K_1$ such that

$$\sum_k \left(\operatorname{Re} \sum_i \bar{a}_i x_{i,k}\right)^2 = 1.$$

This would imply that all $\sum_i \bar{a}_i x_{i,k}$ are real, and

$$x_{i,j} = \lambda_j a_i, \quad \forall i, j, \tag{7.2.6.7}$$

and so

$$\lambda_j a_i x_{k,l} = \lambda_l a_k x_{i,j}, \quad \forall i, j, k, l.$$

In particular, taking $l = j$, then when $\lambda_j \neq 0$, for all j, we would have

$$a_i x_{k,j} = a_k x_{i,j}, \quad \forall i, j, k. \tag{7.2.6.8}$$

If $\lambda_j = 0$ for some j, then $x_{i,j} = 0$, for all i owing to (7.2.6.7). This means that we could assume $\lambda_j \neq 0$, for all j, by taking off those j with $\lambda_j = 0$. In a word, (7.2.6.8) would hold. Taking $i = 0$ in (7.2.6.8), and noticing $a_0 \neq 0$ (otherwise, we would have $x_0 = 0$ owing to (7.2.6.7) again), we have

$$x_{k,j} = \frac{a_k}{a_0} x_{0,j}, \quad \forall k, j,$$

which means that $\frac{x_0}{a_0}$ is the common eigenvector of all A_k, $k = 1, \cdots, m$. Since $\frac{x_0}{a_0} \in \mathbf{R}^q$ which can be seen from

$$\bar{a}_k a_0 x_{k,j} = |a_k|^2 x_{0,j}, \quad a_0^{-1} x_{0,j} = \|a\|^{-2} \sum_{k=0}^{m} \bar{a}_k x_{k,j},$$

we would get a contradiction. So we get, for some $\alpha < 1$,

$$\sum_k \left(\operatorname{Re} \sum_i \bar{a}_i x_{i,k}\right)^2 \leq \alpha \|a\|^2 \sum_i \|x_i\|^2, \quad \forall a \in \mathbf{C}^{m+1},$$
$$\forall x_0 \in V,\ x_i = A_i x_0,\ i = 0, \cdots, m.$$

By the binomial expansion, we get

$$\begin{aligned}
\sum_k \|(a_i + x_{i,k})_{i=0}^m\|^p &= \sum_k \Big(\sum_i |a_i + x_{i,k}|^2\Big)^{\frac{p}{2}} \\
&= \|a\|^p \sum_k \Big(1 + \|a\|^{-2}\Big(2\,\mathrm{Re} \sum_i \bar{a}_i x_{i,k} + \sum_i |x_{i,k}|^2\Big)\Big)^{\frac{p}{2}} \\
&= \|a\|^p \sum_k \Big\{1 + \frac{p}{2}\|a\|^{-2}\Big(2\,\mathrm{Re} \sum_i \bar{a}_i x_{i,k} + \sum_i |x_{i,k}|^2\Big) \\
&\quad + \frac{p}{4}\Big(\frac{p}{2} - 1\Big)\|a\|^{-4}\Big(2\,\mathrm{Re} \sum_i \bar{a}_i x_{i,k}\Big)^2 + O(\|a\|^{-3}\|x_0\|^3)\Big\} \\
&= \|a\|^p \sum_k \Big\{1 + \frac{p}{2}\|a\|^{-2} \sum_i |x_{i,k}|^2 \\
&\quad + \frac{p}{2}(p-2)\|a\|^{-4}\Big(\mathrm{Re} \sum_i \bar{a}_i x_{i,k}\Big)^2 + O(\|a\|^{-3}\|x_0\|^3)\Big\}.
\end{aligned}$$

Thus, we get

$$\begin{aligned}
&\|a\|^{-p} \sum_k \|(a_i + x_{i,k})_{i=0}^m\|^p \\
&\geq q + \frac{p}{2}\|a\|^{-2} \sum_i \|x_i\|^2 + \frac{p}{2}(p-2)\alpha\|a\|^{-2} \sum_i \|x_i\|^2 + O(\|a\|^{-3}\|x_0\|^3).
\end{aligned}$$

Since $\alpha < 1$, when p near 1, say $p > p_0$, and ε small enough, we have

$$\frac{p}{2}(1 + (p-2)\alpha)\|a\|^{-2} \sum_i \|x_i\|^2 + O(\|a\|^{-3}\|x_0\|^3) \geq 0,$$

and hence (7.2.6.5) in the case $\frac{\|x_0\|}{\|a\|} \leq \varepsilon$.

Now consider the case $\frac{\|x_0\|}{\|a\|} \geq \varepsilon$. Set

$$\begin{aligned}
K_2 = \Big\{(a, x_0, \cdots, x_m) : \ & a \in \mathbf{C}^{m+1},\ x_0 \in V,\ x_i = A_i x_0,\ i = 0, \cdots, m, \\
& \frac{\|x_0\|}{\|a\|} \geq \varepsilon,\ \frac{1}{q} \sum_k \|(a_i + x_{i,k})_{i=0}^m\| = 1\Big\}.
\end{aligned}$$

It is a compact set in $\mathbf{C}^{m+1} \times V^{m+1}$, since it is closed and bounded owing to

$$\|a\| \leq \frac{1}{q} \sum_k \|(a_i + x_{i,k})_{i=0}^m\| = 1,$$

$$\Big(\sum_{i,k} |x_{i,k}|^2\Big)^{\frac{1}{2}} \leq \Big(\sum_{i,k} |a_i + x_{i,k}|^2\Big)^{\frac{1}{2}} + \Big(\sum_{i,k} |a_i|^2\Big)^{\frac{1}{2}} \leq C.$$

Denote

$$\beta = \max\{\|a\| : (a, x_0, \cdots, x_m) \in K_2\}.$$

Then $\beta \leq 1$. If $\beta = 1$, then there would be (owing to the compactness) $(a, x_0, \cdots, x_m) \in K_2$ such that

$$||a|| = \frac{1}{q}\sum_k ||(a_i + x_{i,k})_{i=0}^m||. \tag{7.2.6.9}$$

This implies (combining (7.2.6.6) and (7.2.6.9))

$$a_i + x_{i,k} = \mu_k a_i, \mu_k \geq 0,\ \forall i, k, \quad x_{i,k} = (\mu_k - 1)a_i = \frac{a_i}{a_0}x_{0,k},\ \forall i, k. \tag{7.2.6.10}$$

Since $||x_0|| \geq \varepsilon ||a|| = \varepsilon$, $a_0 \neq 0$(owing to $x_{0,k} = (\mu_k - 1)a_0$), $\frac{x_0}{a_0} = (\mu_1 - 1, \cdots, \mu_q - 1)$ is a nonzero real vector in $\mathbf{R}^q$ which is an eigenvector of A_i with the eigenvalue a_i. This would be a contradiction. So $\beta < 1$. Thus we get with $\beta < 1$,

$$||a|| \leq \frac{\beta}{q}\sum_k ||(a_i + x_{i,k})_{i=0}^m||,\ \forall a \in \mathbf{C}^{m+1},\ \forall x_0 \in V, ||x_0|| \geq \varepsilon||a||, x_i = A_i x_0.$$

and hence for all $p < 1$,

$$||a||^p \leq \left(\frac{\beta}{q}\right)^p \sum_k ||(a_i + x_{i,k})_{i=0}^m||^p.$$

For p near 1, say $1 > p > \log q / \log \frac{q}{\beta}$, then

$$||a||^p \leq \frac{1}{q}\sum_k ||(a_i + x_{i,k})_{i=0}^m||^p.$$

The proof of the lemma is finished. □

The next lemma is essentially of the same type as the extrapolation lemmas in §5.2, §5.4, which is valid in the regularity case shown as in §7.1.

Lemma 7.2.6.3 *Assume the condition "R" holds. Let T be a quasilinear operator defined on finite martingales which is bounded from H_2 to H_2 and from $L^p(\mathcal{F}_0)$ to $L^p(\mathcal{F}_0)$ for $p \in (0, 2]$, and such that*

$$Tf^{(\tau)} = (Tf)^{(\tau)}, \quad \text{for all stopping times } \tau \text{ and } f. \tag{7.2.6.11}$$

Then T is bounded from H_p to $H_p, p \in (0, 2]$.

Proof. According to (7.2.6.11), we have $Tf_0 = (Tf)_0$, and

$$||Tf_0||_p \leq C||f_0||_p \leq C||f||_{H_p}.$$

This means that we can assume $f_0 = 0$ without loss of generality. Now let $f = (f_n)_{n\geq 0}, f_0 = 0$, be a finite martingale. For $\lambda > 0$ and the process $(|f_n|)_{n\geq 0}$ define stopping time τ_λ satisfying

$$|\{\tau_\lambda < \infty\}| \leq d|\{Mf > \lambda\}|, \quad \{Mf > \lambda\} \subset \{\tau_\lambda < \infty\}, \quad M_{\tau_\lambda} f \leq \lambda.$$

Since $Tf^{(\tau_\lambda)} = (Tf)^{(\tau_\lambda)}$, we have

$$M\left(Tf^{(\tau_\lambda)}\right) = \sup_n |(Tf)_{\tau_\lambda \wedge n}| = M_{\tau_\lambda}(Tf).$$

And hence

$$|\{M(Tf) > \lambda\}| \le |\{M_{\tau_\lambda}(Tf) > \lambda\}| + |\{\tau_\lambda < \infty\}|,$$

$$\begin{aligned} |\{M_{\tau_\lambda}(Tf) > \lambda\}| &= |\{M(Tf^{(\tau_\lambda)}) > \lambda\}| \le \frac{C}{\lambda^2}\int_\Omega |f_{\tau_\lambda}|^2 d\mu \\ &= \frac{C}{\lambda^2}\left(\int_{\{\tau_\lambda=\infty\}} + \int_{\{\tau_\lambda<\infty\}}\right) \\ &\le \frac{C}{\lambda^2}\int_{\{Mf\le\lambda\}} |f|^2 d\mu + C|\{\tau_\lambda < \infty\}|, \end{aligned}$$

$$\lambda^{p-1}|\{M(Tf) > \lambda\}| \le C\lambda^{p-3}\int_{\{Mf\le\lambda\}} |f|^2 d\mu + C\lambda^{p-1}|\{Mf > \lambda\}|.$$

From this $\|Tf\|_{H_p} \le C\|f\|_{H_p}$ follows immediately. The proof is finished. □

As a result, we get the H_p- and BMO-boundedness of T defined on q-martingales as above.

Corollary 7.2.6.4 *Let T be the operator defined by the linear transform A in V with the associated matrix $A = (a_{k,j})$ and $\|A\| = 1$. Then T is bounded on H_p, $0 < p < \infty$, and BMO.*

Proof. Assume $f_0 = 0$ without loss of generality. Notice

$$(Tf)_n = Tf_n, \quad \text{and} \quad (Tf)_n - (Tf)_{n-1} = A(\Delta_n f), \quad n \ge 1.$$

We have

$$\begin{aligned} E(|\Delta_n(Tf)|^2|\mathcal{F}_{n-1})|_{I_{i_1,\dots,i_{n-1}}} &= \frac{1}{q}\sum_{k=1}^q \left|\sum_{j=1}^q a_{k,j}\Delta_n f|_{I_{i_1,\dots,i_{n-1},j}}\right|^2 \\ &\le \frac{1}{q}\|A\|^2 \sum_{j=1}^q \left|\Delta_n f|_{I_{i_1,\dots,i_{n-1},j}}\right|^2 = E(|\Delta_n f|^2|\mathcal{F}_{n-1})|_{I_{i_1,\dots,i_{n-1}}}, \end{aligned}$$

and hence, for all $n > m \ge 0$,

$$E(|\Delta_n(Tf)|^2|\mathcal{F}_m) \le E(|\Delta_n f|^2|\mathcal{F}_m). \tag{7.2.6.12}$$

Thus,

$$\begin{aligned} E(|Tf - Tf_n|^2|\mathcal{F}_n) &= \sum_{n+1}^\infty E(|\Delta_k(Tf)|^2|\mathcal{F}_n) \le \sum_{n+1}^\infty E(|\Delta_k f|^2|\mathcal{F}_n) \\ &= E(|f - f_n|^2|\mathcal{F}_n) \le \|f\|_{BMO}^2, \end{aligned}$$

$$||Tf||^2_{BMO_2} \leq q \sup_n E(|Tf - Tf_n|^2|\mathcal{F}_n) \leq q||f||^2_{BMO_2}.$$

The assertion concerning BMO is proved.

(7.2.6.12) implies the H_2-boundedness of T. Now we verify (7.2.6.11). Notice

$$f^{(\tau)} = (f_{\tau\wedge n})_{n\geq 0}, \; f_{\tau\wedge n} = \sum_{k=0}^{n} \Delta_k f \chi_{\{k\leq\tau\}}, \quad (Tf^{(\tau)})_n = \sum_{k=0}^{n} A(\Delta_k f \chi_{\{k\leq\tau\}}).$$

Since $\chi_{\{k\leq\tau\}}$ is $\mathcal{F}_{k-1}$-measurable, and hence is constant on each $\mathcal{F}_{k-1}$-atom, we have

$$(Tf^{(\tau)})_n = \sum_{k=0}^{n} A(\Delta_k f)\chi_{\{k\leq\tau\}} = (Tf)_{\tau\wedge n} = ((Tf)^{(\tau)})_n.$$

Thus, Lemma 7.2.6.3 gives the H_p-boundedness of T, for $0 < p \leq 2$. Since T^*, the adjoint operator of T, is defined by A^*, the same assertion hlds for T^*, too. So T is H_p-bounded for $0 < p < \infty$. The proof is finished. □

Now we turn to the main theorem on the operator characterization of H_p (mainly H_1).

Theorem 7.2.6.5 *Let $\{A_i\}_1^m$ be a family of linear transforms in V, and $\{T_i\}_1^m$ be the corresponding operator family as above. Then*

$$H_1 = \{f \in L^1 : T_i f \in L^1, i = 1, \cdots, m\}(= K), \tag{7.2.6.13}$$

if and only if $\{A_i\}_1^m$ has no nonzero common eigenvector in $\mathbf{R}^q \bigcap V$.

Proof. Assume that $\{A_i\}_1^m$ has a nonzero common eigenvector $x = (x_k)_1^q$ in $\mathbf{R}^q \bigcap V$, $A_i x = \lambda_i x, i = 1, \cdots, m$. Without loss of generality, assume $\min x_k = -1$. From Lemma 7.2.6.1, we see that there exists a $f \in L_u^1 - H_1$ such that

$$\Delta_n f|_{I_{i_1,\cdots,i_{n-1},k}} = \lambda_{i_1,\cdots,i_{n-1}} x_k, \quad k = 1, \cdots, q.$$

Since

$$\begin{aligned}\left(\Delta_n(T_i f)|_{I_{i_1,\cdots,i_{n-1},k}}\right)_{k=1}^q &= A_i\left((\Delta_n f|_{I_{i_1,\cdots,i_{n-1},k}})_{k=1}^q\right)\\ &= \lambda_i \lambda_{i_1,\cdots,i_{n-1}} x = \lambda_i(\Delta_n f|_{I_{i_1,\cdots,i_{n-1},k}})_{k=1}^q,\end{aligned}$$

$T_i f = \lambda_i f$, $i = 1, \cdots, m$. Thus, $f \in K$, $f \notin H_1$. So $H_1 \neq K$.

Assume that $\{A_i\}_1^m$ has no nonzero common eigenvector in $\mathbf{R}^q \bigcap V$. It is enough to prove $K \subset H_1$, since $H_1 \subset K$ owing to Corollary 7.2.6.4. Assume that $f = (f_n)_{n\geq 0}$ is such that

$$T_i f = (T_i f_n)_{n\geq 0} \in L^1, \quad i = 0, 1, \cdots, n.$$

Denote

$$g_n(\omega) = ||T_i f_n(\omega))_{i=0}^m|| = \left(\sum_{i=0}^{m} |T_i f_n(\omega)|^2\right)^{\frac{1}{2}}. \tag{7.2.6.14}$$

For any $n \geq 1$, and any n-tuples $(i_1, \cdots, i_n)$, denote

$$a_i = T_i f_n|_{I_{i_1,\cdots,i_n}}, \quad x_{i,k} = \Delta_{n+1}(T_i f)|_{I_{i_1,\cdots,i_n,k}}.$$

Applying Lemma 7.2.6.2, we get for $p_0 < p < 1$,

$$g_n^p|_{I_{i_1,\cdots,i_n}} = ||\tilde{a}||^p \leq \frac{1}{q}\sum_{k=1}^{q} ||(a_i + x_{i,k})_{i=0}^m||^p$$

$$= \frac{1}{q}\sum_{k=1}^{q} ||(T_i f_{n+1}|_{I_{i_1,\cdots,i_n,k}})_{i=0}^m||^p = E(g_{n+1}^p|\mathcal{F}_n)|_{I_{i_1,\cdots,i_n}}.$$

This means that $(g_n^p)_{n\geq 0}$ is a submartingale which is in $L^{\frac{1}{p}}$, since

$$||g_n^p||_{\frac{1}{p}} = ||g_n||_1^p = E\Big(\Big(\sum_i |T_i f_n|^2\Big)^{\frac{1}{2}}\Big)^p \leq \Big(\sum_i ||T_i f_n||_1\Big)^p.$$

Thus, we get

$$||Mg||_1^p \leq \Big\|\sup_n g_n^p\Big\|_{\frac{1}{p}} \leq C\sup_n ||g_n^p||_{\frac{1}{p}} \leq C\sup_n \Big(\sum_i ||T_i f_n||_1\Big)^p.$$

This proves $K \subset H_1$. The proof is finished. □

Remark (7.2.6.13) could be rewritten as $H_1 = \{f \in L_u^1 : T_i f \in L_u^1\}$. In addition, the same argument gives the operator characterzation of H_p for $p_0 < p < 1$. The restriction $p_0 < p$ comes from Lemma 7.2.6.2. Uchiyama [2] showed that in martingale setting this restriction could not be got rid of. This is different from the classical case. We omit the detail.

As a consequence, we get Fefferman-Stein's decomposition of BMO.

Theorem 7.2.6.5′ *Let $\{A_i\}_1^m$ and $\{T_i\}_1^m$ be as in Theorem 7.2.6.5, and $*$ denotes the adjoint operator. Then*

$$BMO = \Big\{\sum_0^m T_i g_i : g_i \in L^\infty\Big\}(= L), \tag{7.2.6.15}$$

if and only if $\{A_i^*\}_1^m$ has no nonzero common eigenvector in $\mathbf{R}^q \bigcap V$.

Proof. It is enough to prove that the assertion $BMO = L$ is equivalent to the assertion $H_1 = \{f \in L_u^1 : T_i^* f \in L_u^1\}(= B)$. Assume $H_1 = \{f \in L_u^1 : T_i^* f \in L_u^1\}$. Obviously $L \subset BMO$ owing to $T_i(BMO) \subset BMO$. Now let $\varphi \in BMO$. φ yields a bounded linear functional l_φ on $H_1 = \{f \in L_u^1 : T_i^* f \in L_u^1\} \subset \bigoplus_{i=0}^m L^1$. l_φ could be extended to whole $\bigoplus_{i=0}^m L^1$. So these exist $g_i \in L^\infty, 1 = 0, \cdots, m$, such that

$$E(f\bar{\varphi}) = l_\varphi(f) = \sum_0^m E(T_i^* f\bar{g}_i) = \sum_0^m E(fT_i\bar{g}_i), \quad \forall f \in L^\infty \subset H_1.$$

Since L^∞ is dense in H_1, we get

$$\varphi = \sum_0^m T_i g_i, \quad g_i \in L^\infty.$$

This proves $BMO \subset L$. Notice that we have already obtained the equivalence between the two norms. In fact, for any one of such expressions of φ, we have

$$\|\varphi\|_* \leq C \sum_0^m \|g_i\|_\infty;$$

and for any $\varphi \in BMO$, there exists the desired expressions of φ such that

$$\sum_0^m \|g_i\|_\infty \leq C\|l_\varphi\| \leq C\|\varphi\|_*.$$

This completes the proof of the reasoning from "$H_1 = B$" to "$BMO = L$".

Now assume $BMO = L$. Notice that $H_1 \subset \{f \in L_u^1 : T_i f \in L_u^1\} = B$ and $H_1' = BMO$. For any $f \in L^\infty \subset H_1$, we could close $\|f\|_{H_1}$ by $E(f\bar{\varphi})$ with $\varphi \in BMO$, $\varphi = \sum_0^m T_i g_i$, $g_i \in L^\infty$, satisfying

$$\|l_\varphi\| = 1, \quad \sum_0^m \|g_i\|_\infty \leq C\|l_\varphi\| = C. \tag{7.2.6.16}$$

Since

$$\begin{aligned} |E(f\bar{\varphi})| &= \Big|E\Big(f\sum_0^m \overline{T_i g_i}\Big)\Big| = \Big|E\Big(\sum_0^m T_i^* f \bar{g}_i\Big)\Big| \\ &\leq \sup_{\{g_i\}:\sum_0^m \|g_i\|_\infty \leq C} \Big|E\Big(\sum_0^m T_i^* f \bar{g}_i\Big)\Big| \leq C\sum_0^m \|T_i^* f\|_1, \end{aligned}$$

we get, for all $f \in L^\infty \subset H_1$,

$$\|f\|_{H_1} \leq C\sum_0^m \|T_i^* f\|_1.$$

But the opposite inequality is shown earlier

$$\sum_0^m \|T_i^* f\|_1 \leq C\|f\|_{H_1}.$$

Thus we have already prove that the H_1-norm and the B-norm are equivalent on L^∞, this proves that H_1 is a closed subspace. If H_1 were a proper one, then there would be $l \in B'$, $l \neq 0$, but $l|_{H_1} = 0$. That is to say some $\varphi \in \Big\{\sum_0^m T_i g_i : g_i \in L^\infty\Big\} = BMO$ would make $l_\varphi|_{H_1} = 0$. This would be a contradiction. This proves $H_1 = B$. The proof is finished. □

How could we get the Fefferman-Stein's decomposition constructively first, and then get the operator characterization of H_1? This was a question proposed by L. Carleson in the classical case. And A.Uchiyama gave an affirmative answer to it first in q-martingale case and then in the classical case. Now we introduce it. The construction is done by induction. The crucial point is to investigate the solution

$$\sum_{j=0}^{m} A_j \mathcal{X}^{(j)} = \mathcal{Y}, \quad \forall \mathcal{Y} \in V \text{ given}, \tag{7.2.6.17}$$

where $\{A_j\}_0^m$ are given family of linear transforms in V, and $\{\mathcal{X}^{(j)}\}_{j=0}^m$ is the solution vector with $\mathcal{X}^{(j)} \in V$, $j = 0, \cdots, m$. This comes from the induction: for the construction of $\{g_j\}_{j=0}^m$ with $g_j \in L^\infty$, such that $f = \sum_0^m A_j g_j$ for $f \in BMO$ given arbitrarily, it is enough to construct $g_{j,n} \in L^\infty(\mathcal{F}_n)$, such that $f_n = \sum_0^m A_j g_{j,n}$, for all n; meanwhile, by induction, we get $\{g_{j,n+1}\}_0^m$ owing to the identity

$$f_{n+1} = f_n + \Delta_n f = \sum_0^m A_j g_{j,n} + \sum_0^m A_j \Delta_n g_j = \sum_0^m A_j g_{j,n+1},$$

here $\{\Delta_n g_j\}_0^m$ comes just from solving the equation (7.2.6.17).

Now we investigate equation (7.2.6.17), and give the following two lemmas. Let $q \geq 3$. Consider a complex $q \times (m+1)$-matrix. Denote its row vectors by $X = (X_k)_0^m$, $Y = (Y_k)_0^m, \cdots$, and its column vectors by $\mathcal{X} = (\mathcal{X}_j)_1^q$, $\mathcal{Y} = (\mathcal{Y}_j)_1^q, \cdots$. In addition, $a = (a_k)_0^m$ denotes always a vector in $\mathbf{C}^{m+1}$.

Lemma 7.2.6.6 *The following two assertions are equivalent.*

(a) $A_1^*, \cdots, A_m^*$ *have no nonzero common eigenvector in* $\mathbf{R}^q \bigcap V$.

(b) *For any* $a = (a_k)_0^m$ *and* $\mathcal{Y} \in V$, *there exist* $\mathcal{X}^{(k)} \in V$, $k = 0, \cdots, m$, *such that*

$$\sum_0^m A_k \mathcal{X}^{(k)} = \mathcal{Y}, \tag{7.2.6.18}$$

$$\sum_{k=0}^m \{ \operatorname{Re} a_k \operatorname{Re} \mathcal{X}_j^{(k)} + \operatorname{Im} a_k \operatorname{Im} \mathcal{X}_j^{(k)} \} = 0, \quad \forall j = 1, \cdots, q, \tag{7.2.6.19}$$

$$\sum_0^m \|\mathcal{X}^{(k)}\| \leq C\|\mathcal{Y}\|, \tag{7.2.6.20}$$

with C *depending only on* q *and* $(A_k)_{k=0}^m$, *but independent of* $\mathcal{Y}$ *and* a.

Proof. (b) $\Rightarrow$ (a). Suppose that $A_1^*, \cdots, A_m^*$ have a nonzero common eigenvector $\mathcal{Y}_0$ in $\mathbf{R}^q \bigcap V$, with eigenvalues $a_1, \cdots, a_m$. For $a = (1, a_1, \cdots, a_m)$ and $\mathcal{Y}_0$ solve (7.2.6.18) under the restriction (7.2.6.19). Suppose $\{\mathcal{X}^{(k)}\}_{k=0}^m$ were the solution. Then we would have

$$\begin{aligned} \|\mathcal{Y}_0\|^2 &= \operatorname{Re} \left\langle \sum_{k=0}^m A_k \mathcal{X}^{(k)}, \mathcal{Y}_0 \right\rangle = \operatorname{Re} \sum_{k=0}^m \langle \mathcal{X}^{(k)}, A_k^* \mathcal{Y}_0 \rangle \\ &= \operatorname{Re} \left\langle \sum_{k=0}^m \bar{a_k} \mathcal{X}^{(k)}, \mathcal{Y}_0 \right\rangle = \left\langle \operatorname{Re} \sum_{k=0}^m \bar{a_k} \mathcal{X}^{(k)}, \mathcal{Y}_0 \right\rangle = 0. \end{aligned}$$

This means no (b) holds. The proof of (b) $\Rightarrow$ (a) is finished.

(a) $\Rightarrow$ (b). The complex inner product space $(\mathbf{C}^k, \langle , \rangle)$ becomes a real inner product space $(\mathbf{C}^k, (,))$ with

$$(a, b) = \operatorname{Re}\langle a, b\rangle = \sum_j (\operatorname{Re} a_j \operatorname{Re} b_j + \operatorname{Im} a_j \operatorname{Im} b_j).$$

So V is a real inner product space of dimension $2(q-1)$. For $a = (a_0, \cdots, a_m) \in \mathbf{C}^{m+1}$ given, denote

$$D_a = \left\{ \{\mathcal{X}^{(k)}\}_{k=0}^m \in V^{m+1} : (7.2.6.19) \text{ holds} \right\} \tag{7.2.6.21}$$

Then D_a is closed in V^{m+1}, and hence

$$L = \left\{ \sum_{k=0}^m A_k \mathcal{X}^{(k)} : \ \{\mathcal{X}^{(k)}\}_{k=0}^m \in D_a \right\}, \tag{7.2.6.22}$$

is also closed in V. So for proving $L = V$, it is enough to prove $L^\perp = \{0\}$, that is to say to prove that for all $\mathcal{Y} \in V$, there exist $\{\mathcal{X}^{(k)}\}_{k=0}^m \in D_a$, such that

$$\left(\sum_0^m A_k \mathcal{X}^{(k)}, \mathcal{Y} \right) = \operatorname{Re} \left\langle \sum_0^m A_k \mathcal{X}^{(k)}, \mathcal{Y} \right\rangle \neq 0.$$

When $a_0 = 0$, it is obvious since any $\{\mathcal{X}^{(k)}\}_{k=0}^m$ with $\mathcal{X}^{(1)} = \cdots = \mathcal{X}^{(m)} = 0$ satisfies (7.2.6.19), and hence $L = V$. When $a_0 \neq 0$, consider two cases.

First case. There exists no real vector $\mathcal{Y}_0$, no complex λ such that $\mathcal{Y} = \lambda \mathcal{Y}_0$. In this case the components $\mathcal{Y}_j, j = 1, \cdots, q$ of $\mathcal{Y}$ are vectors in $\mathbf{C}$ which do not lie in a line. So $\operatorname{Re}\langle i a_0, \mathcal{Y}_j\rangle \neq 0$ for at least one j. But $\operatorname{Re}\langle i a_0, \sum_1^q \mathcal{Y}_j\rangle = 0$. So at least two of $\operatorname{Re}\langle i a_0, \mathcal{Y}_j\rangle$ are of opposite sign, say $\operatorname{Re}\langle i a_0, \mathcal{Y}_1\rangle > 0$, $\operatorname{Re}\langle i a_0, \mathcal{Y}_2\rangle < 0$. Set $\mathcal{X}^{(1)} = \cdots = \mathcal{X}^{(m)} = 0$, and $\mathcal{X}^{(0)} = (i a_0, -i a_0, 0, \cdots, 0)$, then we have

$$\sum_{k=0}^m \left(\operatorname{Re} a_k \operatorname{Re} \mathcal{X}_j^{(k)} + \operatorname{Im} a_k \operatorname{Im} \mathcal{X}_j^{(k)} \right)$$

$$= \operatorname{Re} \sum_{k=0}^m a_k \overline{\mathcal{X}}_j^{(k)} = \operatorname{Re}(a_0 \overline{\mathcal{X}}_j^{(0)}) = 0, \quad \forall j$$

(since $a_0 \overline{\mathcal{X}}_1^{(0)} = -|a_0|^2 i$, $a_0 \overline{\mathcal{X}}_2^{(0)} = |a_0|^2 i$, $a_0 \overline{\mathcal{X}}_j^{(0)} = 0$, $j \geq 3$) and

$$\operatorname{Re} \left\langle \sum_{k=0}^m A_k \mathcal{X}^{(k)}, \mathcal{Y} \right\rangle = \operatorname{Re}\langle \mathcal{X}^{(0)}, \mathcal{Y}\rangle = \operatorname{Re}\langle i a_0, \mathcal{Y}_1\rangle + \operatorname{Re}\langle -i a_0, \mathcal{Y}_2\rangle > 0.$$

Second case. Assume $\mathcal{Y} = \lambda \mathcal{Y}_0$ with $\lambda \in \mathbf{C}$, $\mathcal{Y}_0$ a real vector. Then there exists at least one, say A_1^*, such that $A_1^* \mathcal{Y}$ and $\mathcal{Y}$ are linearly independent. So there exists $\mathcal{Z} \in V$ such that

$$\langle \mathcal{Z}, A_1^* \mathcal{Y}\rangle = a_0, \quad \langle \mathcal{Z}, \mathcal{Y}\rangle = 0.$$

Set $\mathcal{X}^{(0)} = \bar{a}_0 \mathcal{Z}, \mathcal{X}^{(1)} = -\bar{a}_0 \mathcal{Z}, \mathcal{X}^{(k)} = 0,\ k \geq 2$. Then

$$\mathrm{Re} \sum_{k=0}^{m} a_k \overline{\mathcal{X}}_j^{(k)} = \mathrm{Re}\,(a_0 a_1 \bar{\mathcal{Z}}_j - a_1 a_0 \bar{\mathcal{Z}}_j) = 0, \quad \forall j,$$

and

$$\mathrm{Re} \left\langle \sum_{k=0}^{m} A_k \mathcal{X}^{(k)}, \mathcal{Y} \right\rangle = \mathrm{Re}\,\langle \mathcal{X}^{(0)}, \mathcal{Y} \rangle + \mathrm{Re}\,\langle A_1 \mathcal{X}^{(1)}, \mathcal{Y} \rangle = -|a_0|^2 \neq 0.$$

This completes the proof of the assertion $L = V$, and hence (7.2.6.18), (7.2.6.19). Now we prove (7.2.6.20). It is enough to prove that for all $a = (a_0, \cdots, a_m) \in \sum$ (unit ball of $\mathbf{C}^{m+1}$), there exists a uniform C such that (7.2.6.20) is true. For $a \in \sum$, there exist vectors $\{\mathcal{X}^{(k,j)}\}_{k=0}^{m} (j = 1, \cdots, 2(q-1))$ in D_a such that $\left\{ \sum_{k=0}^{m} A_k \mathcal{X}^{(k,j)} \right\}_{j=1}^{2(q-1)}$ is a basis of V owing to (7.2.6.18). We claim that for a compact neighborhood U_a being small enough, for all $a^* \in U_a$, we can find $\{\mathcal{X}^{(k,j)*}\}_{k=0}^{m} (j = 1, \cdots, 2(q-1))$ in D_{a^*} such that $\left\{ \sum_{k=0}^{m} A_k \mathcal{X}^{(k,j)*} \right\}_{j=1}^{2(q-1)}$ is also a basis of V. In fact, since $a \neq 0$, at least one component of a is not zero, say $\mathrm{Im}\, a_m \neq 0$. Set

$$\begin{aligned}
\mathrm{Re}\, \mathcal{X}^{(k,j)*} &= \mathrm{Re}\, \mathcal{X}^{(k,j)}, \quad k = 0, \cdots, m,\ j = 1, \cdots, 2(q-1), \\
\mathrm{Im}\, \mathcal{X}^{(k,j)*} &= \mathrm{Im}\, \mathcal{X}^{(k,j)}, \quad k = 0, \cdots, m-1,\ j = 1, \cdots, 2(q-1), \\
\mathrm{Im}\, \mathcal{X}^{(m,j)*} &= -\frac{1}{\mathrm{Im}\, a_m^*} \Big\{ \sum_{k=0}^{m} \mathrm{Re}\, a_k^* \,\mathrm{Re}\, \mathcal{X}^{(k,j)} + \sum_{k=0}^{m-1} \mathrm{Im}\, a_k^* \,\mathrm{Im}\, \mathcal{X}^{(k,j)} \Big\}, \\
& \qquad j = 1, \cdots, 2(q-1).
\end{aligned}$$

Then $\{\mathcal{X}^{(k,j)*}\}_{k=0}^{m} \in D_{a^*}$. In addition, when U_a is small enough, $\left\{ \sum_{k=0}^{m} A_k \mathcal{X}^{(k,j)*} \right\}_{j=1}^{2(q-1)}$ is also a basis of V, since $\left\{ \sum_{k=0}^{m} A_k \mathcal{X}^{(k,j)*} \right\}_{j=1}^{2(q-1)}$ is near to $\left\{ \sum_{k=0}^{m} A_k \mathcal{X}^{(k,j)} \right\}_{j=1}^{2(q-1)}$ when a^* is near to a. Now let B be the unit sphere in $\mathbf{R}^{2(q-1)}$, and consider the function

$$f = \sum_{k=0}^{m} \Big\| \sum_{j=1}^{2(q-1)} r_j \mathcal{X}^{(k,j)*} \Big\| \Big/ \Big\| \sum_{j=1}^{2(q-1)} r_j \sum_{k=0}^{m} A_k \mathcal{X}^{(k,j)*} \Big\|,$$

on $U_a \times B$, where $\{\mathcal{X}^{(k,j)*}\}_{k=0}^{m},\ j = 1, \cdots, 2(q-1)$ is defined as above for $a^* \in U_a, a \in \sum$. Then f is continuous on $U_a \times B$, and hence is bounded. Thus we get

$$\sum_{k=0}^{m} \Big\| \sum_{j=1}^{2(q-1)} r_j \mathcal{X}^{(k,j)*} \Big\| \leq C_a \Big\| \sum_{j=1}^{2(q-1)} r_j \sum_{k=0}^{m} A_k \mathcal{X}^{(k,j)*} \Big\|, \quad \forall \{r_j\} \in B,\ \forall a^* \in U_a.$$

Since $\sum$ is compact, only finite C_a's are concerned, it can be replaced by a uniform constant C. In addition, by homogeneity, B can be replaced by $\mathbf{R}^{2(q-1)}$. Thus we get

$$\sum_{k=0}^{m} \Big\| \sum_{j=1}^{2(q-1)} r_j \mathcal{X}^{(k,j)} \Big\| \leq C \Big\| \sum_{j=1}^{2(q-1)} r_j \sum_{k=0}^{m} A_k \mathcal{X}^{(k,j)} \Big\|,$$

$$\forall \{r_j\} \in \mathbf{R}^{2(q-1)}, \ \forall a \in \sum. \tag{7.2.6.23}$$

Now for any $\mathcal{Y} \in V$, there exists $\{r_j\} \in \mathbf{R}^{2(q-1)}$ such that

$$\mathcal{Y} = \sum_{j=1}^{2(q-1)} r_j \sum_{k=0}^{m} A_k \mathcal{X}^{(k,j)} = \sum_{k=0}^{m} A_k \sum_{j=1}^{2(q-1)} r_j \mathcal{X}^{(k,j)} = \sum_{k=0}^{m} A_k \mathcal{X}^{(k)}.$$

With $\mathcal{X}^{(k)} = \sum_{j=1}^{2(q-1)} r_j \mathcal{X}^{(k,j)}$, (7.2.6.23) gives (7.2.6.20). The proof of the lemma is finished. □

Lemma 7.2.6.7 *Let $\{A_k\}_0^m$ be such that $\{A_k^*\}_0^m$ has no nonzero common eigenvector in $\mathbf{R}^q \bigcap V$, $R > 0$ be large enough and dependent on q and $\{A_k\}_0^m$. Let $\mathcal{Y} \in V$, $a = (a_k)_0^m$, and $b_1, \cdots, b_q \geq 0$ be such that*

$$\min(R - \|a\|, 1) \geq \Big\{ \frac{1}{q} \Big(\|\mathcal{Y}\|^2 + \sum_1^q b_j^2 \Big) \Big\}^{\frac{1}{2}}. \tag{7.2.6.24}$$

Then there exists $\{\mathcal{X}^{(k)}\}_0^m \subset V$ such that

$$\sum_0^m A_k \mathcal{X}^{(k)} = \mathcal{Y}, \tag{7.2.6.25}$$

$$R - \|a + X^{(j)}\| \geq b_j, \quad 1 \leq j \leq q, \tag{7.2.6.26}$$

where $X^{(j)} = (\mathcal{X}_j^{(0)}, \mathcal{X}_j^{(1)}, \cdots, \mathcal{X}_j^{(m)})$, $j = 1, \cdots, q$.

Proof. When $R - \|a\| \geq 2q^{\frac{1}{2}}$, by denoting $\mathcal{X}^{(0)} = \mathcal{Y}$, $\mathcal{X}^{(k)} = 0$, $k \geq 1$, we get $\sum_{k=0}^m A_k \mathcal{X}^{(k)} = \mathcal{Y}$, and

$$R - \|a + X^{(j)}\| \geq R - \|a\| - \|X^{(j)}\| \geq 2q^{\frac{1}{2}} - \|\mathcal{Y}\| \geq q^{\frac{1}{2}} \geq b_j.$$

Now consider the case $2q^{\frac{1}{2}} \geq R - \|a\| \geq 1 = \Big\{ \frac{1}{q} \Big(\|\mathcal{Y}\|^2 + \sum_1^q b_j^2 \Big) \Big\}^{\frac{1}{2}}$. When $b_j < \frac{1}{2}$, for all j, we see by making use of Lemma 7.2.6.6, there exists $\{\mathcal{X}^{(k)}\}_0^m \subset V$ such that (7.2.6.18)–(7.2.6.20) hold. Notice that (7.2.6.19) means that the vectors $X^{(j)}$'s are all perpendicular to a in $\mathbf{R}^{2(m+1)}$, and $\|X^{(j)}\| \leq \sum_{k=0}^m \|\mathcal{X}^{(k)}\| \leq C\|\mathcal{Y}\| \leq C$, we get

$$\|a + X^{(j)}\| = (\|a\|^2 + \|X^{(j)}\|^2)^{\frac{1}{2}} \leq \|a\| \Big(1 + \frac{1}{2} \frac{\|X^{(j)}\|^2}{\|a\|^2} \Big)$$
$$\leq (R-1) + \frac{C}{R} \leq R - b_j.$$

If $\max_j b_j \geq \frac{1}{2}$, the idea is to write $a + X^{(j)}$ as a direct sum $(a + X^{(j)\prime}) + X^{(j)\prime\prime}$ and expect $R - \|a + X^{(j)\prime}\|$ might control b_j a little more. More precisely, we have (assuming $b_1 \geq b_2 \geq \cdots \geq b_q$)

$$\frac{1}{q^2}\Big(\sum_1^q b_j\Big)^2 \leq \frac{1}{q}(b_1^2 + \cdots + b_q^2) - \frac{1}{q^2}(b_1 - b_q)^2,$$

$$\begin{aligned}\bar{b} = \frac{1}{q}\sum_1^q b_j &\leq \Big(1 - \frac{\|\mathcal{Y}\|^2}{q} - \Big(\frac{b_1 - b_q}{q}\Big)^2\Big)^{\frac{1}{2}} \\ &\leq 1 - \frac{1}{2}\frac{1}{q^2}(q\|\mathcal{Y}\|^2 + (b_1 - b_q)^2) = 1 - C(\|\mathcal{Y}\|^2 + (b_1 - b_q)^2).\end{aligned}$$

Thus we can find $\mathcal{Z} = (\mathcal{Z}_j)_{j=1}^q \in \mathbf{R}^q \bigcap V$ such that (for example, $\mathcal{Z}_j = \bar{b} - b_j$, $1 \leq j \leq q$, will do)

$$\|\mathcal{Z}\| \leq C(b_1 - b_q), \tag{7.2.6.27}$$

$$-b_j \leq \mathcal{Z}_j \leq 1 - b_j - C(\|\mathcal{Y}\|^2 + (b_1 - b_q)^2), \quad 1 \leq j \leq q. \tag{7.2.6.28}$$

Now set

$$\mathcal{X}^{(k)\prime} = \frac{a_k}{\|a\|}\mathcal{Z}, \quad k = 0, \cdots, m, \quad \mathcal{Y}' = \mathcal{Y} - \sum_{k=0}^m A_k \mathcal{X}^{(k)\prime}. \tag{7.2.6.29}$$

Then $\|\mathcal{Y}'\| \leq C(\|\mathcal{Y}\| + (b_1 - b_q))$, and hence by Lemma 7.2.6.6, there exists $\{\mathcal{X}^{(k)\prime\prime}\}_{k=0}^m \in D_a$ such that

$$\sum_{k=0}^m A_k \mathcal{X}^{(k)\prime\prime} = \mathcal{Y}', \quad \sum_{k=0}^m \|\mathcal{X}^{(k)\prime\prime}\| \leq C(\|\mathcal{Y}\| + (b_1 - b_q)).$$

Denote $\mathcal{X}^{(k)} = \mathcal{X}^{(k)\prime} + \mathcal{X}^{(k)\prime\prime}$, we have

$$\sum_{k=0}^m A_k \mathcal{X}^{(k)} = \sum_{k=0}^m A_k \mathcal{X}^{(k)\prime} + \sum_{k=0}^m A_k \mathcal{X}^{(k)\prime\prime} = \mathcal{Y}.$$

Noticing that $\|a\| \geq R - 2q^{\frac{1}{2}}$, and $X^{(j)\prime} = \frac{\mathcal{Z}_j}{\|a\|}a$ owing to (7.2.6.29), so $X^{(j)\prime\prime}$ is orthogonal to $a + X^{(j)\prime}$ for each j, and

$$CR \leq \|a + X^{(j)\prime}\| = \|a\| + \mathcal{Z}_j \leq R - 1 + \mathcal{Z}_j \leq R - b_j - C(\|\mathcal{Y}\|^2 + (b_1 - b_q)^2).$$

In addition, noticing $\|X^{(j)\prime}\| + \|X^{(j)\prime\prime}\| \leq C(\|\mathcal{Y}\| + (b_1 - b_q))$, we get

$$\begin{aligned}\|a + X^{(j)}\| &= (\|a + X^{(j)\prime}\|^2 + \|X^{(j)\prime\prime}\|^2)^{\frac{1}{2}} \\ &\leq \|a + X^{(j)\prime}\|\Big(1 + \frac{C\|X^{(j)\prime\prime}\|^2}{\|a + X^{(j)\prime}\|^2}\Big)^{\frac{1}{2}} \\ &\leq \|a + X^{(j)\prime}\| + \frac{C}{R}(\|\mathcal{Y}\|^2 + (b_1 - b_q)^2) \\ &\leq R - b_j + \Big(\frac{C}{R} - C\Big)(\|\mathcal{Y}\|^2 + (b_1 - b_q)^2) \leq R - b_j, \quad \forall j.\end{aligned}$$

Now the case $R - \|a\| \geq \alpha = \left(\frac{1}{q}\left(\|\mathcal{Y}\|^2 + \sum_1^q b_j^2\right)\right)^{\frac{1}{2}}$, $\alpha \leq 1$, remains to be considered. We have

$$\frac{R}{\alpha} - \left\|\frac{a}{\alpha}\right\| \geq \left(\frac{1}{q}\left(\left\|\frac{\mathcal{Y}}{\alpha}\right\|^2 + \sum_1^q \left(\frac{b_j}{\alpha}\right)^2\right)\right)^{\frac{1}{2}} = 1.$$

Applying the preceding argument to $\frac{R}{\alpha}$, $\frac{a}{\alpha}$, $\frac{\mathcal{Y}}{\alpha}$ and $\left\{\frac{b_j}{\alpha}\right\}_1^q$, we get $\{\widetilde{\mathcal{X}}^{(k)}\}_{k=0}^m \subset D_{\frac{a}{\alpha}}$, such that $\sum_0^m A_k \widetilde{\mathcal{X}}^{(k)} = \frac{\mathcal{Y}}{\alpha}$ and $\frac{R}{\alpha} - \|\frac{a}{\alpha} + \widetilde{X}^{(j)}\| \geq \frac{b_j}{\alpha}$. This means $\{\mathcal{X}^{(k)}\} = \{\alpha\widetilde{\mathcal{X}}^{(k)}\}$ will fit the requiment. The proof of the lemma is finished. □

Now we are in the positition to get the main result: the constructive proof of Theorem 7.2.6.5'.

Constructive proof of **Theorem 7.2.6.5'**. Assume that $\{A_k^*\}_{k=0}^m$ has no nonzero common eigenvector in $\mathbf{R}^q \bigcap V$. Let $f \in BMO$, $f_0 = 0$, $\|f\|_* = 1$, $R > 1$ be determined in Lemma 7.2.6.7. Set $\mathcal{G}_0 = 0$ as a vector in $\mathbf{C}^{m+1}$. Assume that we have got $\mathcal{G}_n = (g_{k,n})_{k=0}^m$ with $g_{k,n} \in L^\infty(\mathcal{F}_n)$ satisfying

$$\sum_{k=0}^m T_k g_{k,n} = f_n, \tag{7.2.6.30}$$

$$R - \|\mathcal{G}_n\| \geq E\Big(\sum_{l=n+1}^\infty |\Delta_l f|^2 | \mathcal{F}_n\Big)^{\frac{1}{2}}. \tag{7.2.6.31}$$

For any $I = I_{i_1,\cdots,i_n} \in \mathcal{F}_n$, denote $I_j = I_{i_1,\cdots,i_n,j}$, and

$$\mathcal{Y} = (\Delta_{n+1} f\,|_{I_1}, \cdots, \Delta_{n+1} f\,|_{I_q}) \in V, \quad a = (g_{k,n}\,|_I)_{k=0}^m,$$

$$b_j = \Big\{E\Big(\sum_{l=n+2}^\infty |\Delta_l f|^2 | \mathcal{F}_{n+1}\Big)\Big|_{I_j}\Big\}^{\frac{1}{2}}, \quad j = 1, \cdots, q.$$

Then

$$\begin{aligned} E\Big(\sum_{l=n+1}^\infty |\Delta_l f|^2 | \mathcal{F}_n\Big)\Big|_I &= \frac{1}{q}\|\mathcal{Y}\|^2 + E\Big(\Big(\sum_{l=n+2}^\infty |\Delta_l f|^2 | \mathcal{F}_{n+1}\Big)\Big| \mathcal{F}_n\Big)\Big|_I \\ &= \frac{1}{q}\|\mathcal{Y}\|^2 + \frac{1}{q}\sum_{j=1}^q b_j^2 \leq \min(R - \|a\|, 1). \end{aligned}$$

Applying Lemma 7.2.6.7 to such $a, \mathcal{Y}, \{b_j\}_1^q$, we get $\{\mathcal{X}^{(k)}\}_{k=0}^m$ such that

$$\sum_{k=0}^m A_k \mathcal{X}^{(k)} = \mathcal{Y}, \text{ and } R - \|a + X^{(j)}\| \geq b_j, \quad 1 \leq j \leq q. \tag{7.2.6.32}$$

Now let $\mathcal{G}_{n+1}$ be the function vector in $\mathbf{C}^{m+1}$ defined by

$$\mathcal{G}_{n+1}|_{I_j} = a + X^{(j)}, \quad j = 1, \cdots, q.$$

Since $\sum_{j=1}^{q} X^{(j)} = 0$ (noticing $X^{(j)} = (\mathcal{X}_j^{(0)}, \cdots, \mathcal{X}_j^{(m)})$), we see $E(\mathcal{G}_{n+1}|\mathcal{F}_n) = \mathcal{G}_n$. In addition, we have

$$\Big(\sum_{k=0}^{m} T_k g_{k,n+1}|_{I_j}\Big)_{j=1}^{q} = \Big(\sum_{k=0}^{m} T_k g_{k,n}|_{I_j}\Big)_{j=1}^{q} + \Big(\sum_{k=0}^{m} A_k \mathcal{X}_j^{(k)}\Big)_{j=1}^{q}$$
$$= f_n|_I + (\Delta_{n+1} f|_{I_j})_{j=1}^{q} = (f_{n+1}|_{I_j})_{j=1}^{q}.$$

Thus we get $\sum_{k=0}^{m} T_k g_{k,n+1} = f_{n+1}$ by induction. Finally, $\mathcal{G}_{n+1}$ satisfies (7.2.6.31) with n replaced by $n+1$,

$$R - ||\mathcal{G}_{n+1}|_{I_j}|| = R - ||a + X^{(j)}|| \geq b_j$$
$$= \Big\{E\Big(\sum_{l=n+2}^{\infty} |\Delta_l f|^2 |\mathcal{F}_{n+1}\Big)\Big|_{I_j}\Big\}^{\frac{1}{2}}. \tag{7.2.6.33}$$

This means

$$||\mathcal{G}_n||_\infty \leq R, \quad \forall n.$$

Since $\mathcal{G}_n$ converges to $\mathcal{G}$ in L^2 at least, we get

$$\sum_{k=0}^{m} T_k g_k = f, \quad \text{and } ||\mathcal{G}||_\infty \leq R = R||f||_*.$$

The proof of the theorem is finished. □

As a result of the operator characterization of H_1, we get the following Riesz brother's theorem. First we notice that any finite measure ν on $(\Omega, \mathcal{F}, \mu)$ generates a q-martingale $f = (f_n)_{n\geq 0}$ as follows,

$$f_n|_{I_{i_1,\cdots,i_n}} = q^n \nu(I_{i_1,\cdots,i_n}), \quad n \geq 0.$$

It is a martingale owing to the fact

$$E(f_n|\mathcal{F}_{n-1})|_{I_{i_1,\cdots,i_{n-1}}} = \frac{1}{q}\sum_{k=1}^{q} q^n \nu(I_{i_1,\cdots,i_{n-1},k})$$
$$= q^{n-1}\nu(I_{i_1,\cdots,i_{n-1}}) = f_{n-1}|_{I_{i_1,\cdots,i_{n-1}}}.$$

Theorem 7.2.6.8 *Let $\{A_k\}_0^m$ be a family of linear transforms in V which has no nonzero common eigenvector in $\mathbf{R}^q \bigcap V$. Let ν be a finite measure such that $T_k\nu$ are all finite measures, for all k. Then $T_k\nu$ are all absolutely continuous, $k = 0, \cdots, m$.*

Proof. Notice that { martingales generated by finite measures } $= L^1$ which follows from

$$\nu \text{ is a finite measure } \Rightarrow E(|f_n|) = \sum |\nu|(I_{i_1,\cdots,i_n}) = ||\nu||, \quad \forall n \Rightarrow f \in L^1;$$

Assume $f = (f_n)_{n\geq 0} \in L^1$. Then the measure ν defined first on {atoms} by $\nu(I_{i_1,\cdots,i_n}) = q^{-n} f_n|_{I_{i_1,\cdots,i_n}}$ is finite. (The additivity of ν comes from the martingale property of f, and the boundedness of ν comes from that for all n

$$\sum |\nu(I_{i_1,\cdots,i_n})| = \sum \int_{I_{i_1,\cdots,i_n}} |f_n| d\mu = \|f_n\|_1 \leq C.)$$

Now since $T_k\nu$ are all finite measures, $k = 0, \cdots, m$, from Theorem 7.2.6.5, we see that $T_k f \in L^1_u$, $k = 0, \cdots, m$. It is just what we want. The proof is finished. □

7.3 Weighted Φ-inequalities for regular martingales

In this section we want to discuss the weighted Φ-inequalities under the regularity introduced in §7.1. We will show that all inequalities we discussed before are general Φ-inequalities in this case. Let $(\Omega, \mathcal{F}, \mu, \{\mathcal{F}_n\})$ be usual but with the condition that "R" holds. Let z be a special weight, $d\hat{\mu} = z d\mu$. First we want to know if $(\Omega, \mathcal{F}, \hat{\mu}, \{\mathcal{F}_n\})$ possesses the regularity "$\widehat{\text{R}}$". The answer is conditionally positive.

Lemma 7.3.1 *Assume $z \in S$. Then conditions "R" and "$\widehat{\text{R}}$" are equivalent. Meanwhile, "R" implies $z \in S^+$, and "$\widehat{\text{R}}$" implies $z \in S^-$.*

Proof. Assume "R" holds, and $z \in S^-$. Then for all $F \in \mathcal{F}$,

$$\begin{aligned}\widehat{E}(\chi_F|\mathcal{F}_n) &= z_n^{-1} E(\chi_F z|\mathcal{F}_n) \leq d z_n^{-1} E(\chi_F z|\mathcal{F}_{n-1}) \\ &= d z_{n-1} z_n^{-1} \widehat{E}(\chi_F|\mathcal{F}_{n-1}) \leq dK\widehat{E}(\chi_F|\mathcal{F}_{n-1}), \quad \forall n \geq 1.\end{aligned}$$

This proves "$z \in S^-$ together with "R" implies "$\widehat{\text{R}}$" ". The reciprocal, i.e. "$z \in S^+$ together with "$\widehat{\text{R}}$" implies "R" " could be proved analogously. The remainders follow easily. For example, we want deduce $z \in S^-$ from "$\widehat{\text{R}}$". We have (Noticing $z^{-1} \in \widehat{L}^1$)

$$z_n^{-1} = \widehat{E}(z^{-1}|\mathcal{F}_n) \leq \hat{d}\widehat{E}(z^{-1}|\mathcal{F}_{n-1}) = \hat{d} z_{n-1}^{-1}.$$

The proof is finished. □

Lemma 7.3.2 *"$z \in A_\infty$ together with "R" " is equivalent to "$z^{-1} \in \widehat{A}_\infty$ together with "$\widehat{\text{R}}$" ".*

Proof. Assume "R" holds and $z \in A_\infty$, say $z \in A_p$. Then from Proposition 6.3.8, we see that for some $\varepsilon > 0$,

$$E(z^{1+\varepsilon}|\mathcal{F}_n) \leq C z_n^{1+\varepsilon}.$$

This means (see §6.1) that $z^{-1} \in \widehat{A}_q$ with $q = \frac{1+\varepsilon}{\varepsilon}$. In addition, from

$$1 \leq z_n E(z^{-\frac{1}{p-1}}|\mathcal{F}_n)^{p-1} \leq K, \quad E(z^{-\frac{1}{p-1}}|\mathcal{F}_n) \leq KE(z^{-\frac{1}{p-1}}|\mathcal{F}_{n-1}), \; n \geq 1,$$

we see $z_{n-1} \le Cz_n$, $n \ge 1$, i.e. $z \in S^-$. From Lemma 7.3.1, we see "$\widehat{\text{R}}$" holding. Thus we have proved " "R" together with $z \in A_\infty$ implies "$\widehat{\text{R}}$" together with $z^{-1} \in \widehat{A}_\infty$". The proof of the reciprocal is similar. The proof is finished. □

Now we study the weighted Φ-inequalities among $Mf, S(f)$ and $\sigma(f)$, where Φ is general one.

Theorem 7.3.3 *Assume that "R" holds and $z \in A_\infty$, Φ is general one. Then for all martingales $f = (f_n)_{n\ge0}$ with respect to $(\Omega, \mathcal{F}, \mu\{\mathcal{F}_n\}_{n\ge0})$, we have*

$$C\widehat{E}(\Phi(Mf)) \le \widehat{E}(\Phi(S(f))) \le C\widehat{E}(\Phi(Mf)). \tag{7.3.1}$$

Proof. We know that "$\widehat{\text{R}}$" holds and $z^{-1} \in \widehat{A}_\infty$. Without loss of generality, f_0 can be assumed to be 0. For the process $(|f_n|)_{n\ge0}$ and $\beta\lambda$ (for all $\lambda > 0$, some suitable $\beta > 0$ determined later) define a stopping time τ such that

$$\hat{\mu}(\{\tau < \infty\}) \le \hat{d}\hat{\mu}(\{Mf > \beta\lambda\}), \quad M_\tau f \le \beta\lambda. \tag{7.3.2}$$

Consider the stopped martingale $f^{(\tau)}$. For process $(S_n(f^{(\tau)}))_{n\ge0}$ and λ define T such that

$$S(f^{(\tau)}) \subset \{T < \infty\},\ \hat{\mu}(\{T < \infty\}) \le \hat{d}\hat{\mu}(\{S(f^{(\tau)}) > \lambda\}),\ S_T(f^{(\tau)}) \le \lambda. \tag{7.3.3}$$

For $\alpha > 1$, we have

$$\{S(f) > \alpha\lambda\} \subset \{S(f^{(\tau)}) > \alpha\lambda\} \bigcup \{\tau < \infty\}, \tag{7.3.4.i}$$

$$\{S(f^{(\tau)}) > \alpha\lambda\} \subset \left\{S^2(f^{(\tau)}) - S_T^2(f^{(\tau)}) > (\alpha^2 - 1)\lambda^2\right\}. \tag{7.3.4.ii}$$

Thus we get

$$\begin{aligned} E(\chi_{\{S(f^{(\tau)})>\alpha\lambda\}}|\mathcal{F}_T) &\le \frac{1}{(\alpha^2-1)\lambda^2} E(S^2(f^{(\tau)}) - S_T^2(f^{(\tau)})|\mathcal{F}_T) \\ &\le \frac{1}{(\alpha^2-1)\lambda^2} E\left(\left|f^{(\tau)} - f_T^{(\tau)}\right|^2 \Big| \mathcal{F}_T\right) \le \frac{4\beta^2}{\alpha^2 - 1}. \end{aligned}$$

Let p be such that $z^{-1} \in \widehat{A}_p$, we have

$$\widehat{E}(\chi_{\{S(f^{(\tau)})>\alpha\lambda\}}|\mathcal{F}_T) \le CE(\chi_{\{S(f^{(\tau)})>\alpha\lambda\}}|\mathcal{F}_T)^{\frac{1}{p}} \le \varepsilon_\beta.$$

Since $\{S(f^{(\tau)}) > \alpha\lambda\} \subset \{T < \infty\} \in \mathcal{F}_T$, we get

$$\begin{aligned} |\{S(f^{(\tau)}) > \lambda\}|_{\hat{\mu}} &= \int_{\{T<\infty\}} \widehat{E}(\chi_{\{S(f^{(\tau)})>\alpha\lambda\}}|\mathcal{F}_T) d\hat{\mu} \\ &\le \varepsilon_\beta |\{T < \infty\}|_{\hat{\mu}} \le \varepsilon_\beta \hat{d} |\{S(f^{(\tau)}) > \lambda\}|_{\hat{\mu}}. \end{aligned}$$

Thus we get the good λ-inequality between $S(f)$ and Mf with respect to $\hat{\mu}$, and hence the second inequality in (7.3.1). The other half of the assertion could be proved in same way. The proof is finished. □

Remark Under the same conditions "R" and $z \in A_\infty$, the rearrangement inequality between $S(f)$ and Mf with respect to $\hat{\mu}$ holds, too. In fact, define τ and T such that

$$\left\{Mf > F(t) = (Mf)^{\hat{*}}\left(\frac{t}{2\hat{d}}\right)\right\} \subset \{\tau < \infty\},$$

$$|\{\tau < \infty\}|_{\hat{\mu}} \le \hat{d}|\{Mf > F(t)\}|_{\hat{\mu}}, \quad M_\tau f \le F(t),$$

$$\{S(f^{(\tau)}) > G(t) = S(f)^{\hat{*}}(2t)\} \subset \{T < \infty\},$$

$$|\{T < \infty\}|_{\hat{\mu}} \le \hat{d}|\{S(f^{(\tau)}) > G(t)\}|_{\hat{\mu}}, \quad S_T(f^{(\tau)}) \le G(t).$$

Denoting $H(t) = \alpha^2 F^2(t) + G^2(t)$ with α determined later, then we have

$$\{S(f) > \alpha F(t) + G(t)\} \subset \{(S(f^{(\tau)}) > \alpha F(t) + G(t)\} \bigcup \{\tau < \infty\},$$

$$\begin{aligned}\{S(f^{(\tau)}) > \alpha F(t) + G(t)\} &\subset \{S^2(f^{(\tau)}) > H(t)\}\\ &\subset \left\{S^2(f^{(\tau)}) - S_T^2(f^{(\tau)}) > \alpha^2 F^2(t)\right\},\end{aligned}$$

$$\begin{aligned}E(\chi_{\{S(f^{(\tau)})>\alpha F(t)+G(t)\}}|\mathcal{F}_T) &\le E(\chi_{\{S^2(f^{(\tau)})-S_T^2(f^{(\tau)})>\alpha^2F^2(t)\}}|\mathcal{F}_T)\\ &\le \frac{1}{\alpha^2 F^2(t)} E(S^2(f^{(\tau)}) - S_T^2(f^{(\tau)})|\mathcal{F}_T)\\ &\le \frac{C}{\alpha^2 F^2(t)} E\left(\left|f^{(\tau)} - f_T^{(\tau)}\right|^2 \middle|\mathcal{F}_T\right) \le \frac{4C}{\alpha^2},\end{aligned}$$

$$\widehat{E}(\chi_{\{S(f^{(\tau)})>\alpha F(t)+G(t)\}}|\mathcal{F}_T) \le \frac{1}{4\hat{d}}, \text{ provided } \alpha \text{ big enough.}$$

Thus, we get

$$|\{S(f^{(\tau)}) > \alpha F(t) + G(t)\}|_{\hat{\mu}} \le \frac{1}{4\hat{d}}|\{T < \infty\}|_{\hat{\mu}} \le \frac{t}{2},$$

and hence

$$|\{S(f) > \alpha F(t) + G(t)\}|_{\hat{\mu}} \le |\{\tau < \infty\}|_{\hat{\mu}} + \frac{t}{2} \le \hat{d}|\{Mf > F(t)\}| + \frac{t}{2} \le t,$$

$$S(f)^{\hat{*}}(t) \le \alpha(Mf)^{\hat{*}}\left(\frac{t}{2\hat{d}}\right) + S(f)^{\hat{*}}(2t), \quad \forall t > 0. \tag{7.3.5}$$

Similarly, we also have

$$(Mf)^{\hat{*}}(t) \le \alpha S(f)^{\hat{*}}\left(\frac{t}{2\hat{d}}\right) + (Mf)^{\hat{*}}(2t), \quad \forall t > 0. \tag{7.3 6}$$

With slightly weaker conditions, we have a general Φ-inequality between $S(f)$ and $\sigma(f)$.

Theorem 7.3.4 *Assume that "R" holds and $z \in S$ (equivalently, "$\widehat{\text{R}}$" holds and $z \in S$). Then for all $f = (f_n)_{n\geq 0}$ with $f_0 = 0$ (martingale with respect to $(\Omega, \mathcal{F}, \mu, \{\mathcal{F}_n\}_{n\geq 0})$) we have*

$$C\widehat{E}(\Phi(\sigma(f))) \leq \widehat{E}(\Phi(S(f))) \leq \widehat{E}(\Phi(\sigma(f))). \tag{7.3.7}$$

Proof. For $\lambda > 0$, define stopping times $\tau = \inf\{n : \sigma_{n+1}(f) > \beta\lambda\}$, and T such that

$$\{S(f^{(\tau)}) > \lambda\} \subset \{T < \infty\}, \hat{\mu}(\{T < \infty\}) \leq \hat{d}\hat{\mu}(\{S(f^{(\tau)}) > \lambda\}),\ S_T(f^{(\tau)}) \leq \lambda.$$

Then for $\alpha > 1$, we have

$$\begin{aligned}
\widehat{E}(\chi_{\{S(f^{(\tau)})>\alpha\lambda\}}|\mathcal{F}_T) &\leq \frac{1}{(\alpha^2-1)\lambda^2}\widehat{E}(S^2(f^{(\tau)}) - S_T^2(f^{(\tau)})|\mathcal{F}_T) \\
&= \frac{1}{(\alpha^2-1)\lambda^2}\widehat{E}\Big(\sum_1^\infty \chi_{\{T\leq n-1\}}|\Delta_n f^{(\tau)}|^2\Big|\mathcal{F}_T\Big) \\
&= \frac{1}{(\alpha^2-1)\lambda^2}\widehat{E}\Big(\sum_1^\infty \widehat{E}(\chi_{\{T\leq n-1\}}|\Delta_n f^{(\tau)}|^2|\mathcal{F}_{n-1})\Big|\mathcal{F}_T\Big) \\
&= \frac{1}{(\alpha^2-1)\lambda^2}\widehat{E}\Big(\sum_1^\infty z_{n-1}^{-1}E(\chi_{\{T\leq n-1\}}|\Delta_n f^{(\tau)}|^2 z_n|\mathcal{F}_{n-1})\Big|\mathcal{F}_T\Big) \\
&\leq \frac{C}{(\alpha^2-1)\lambda^2}\widehat{E}\Big(\sigma^2(f^{(\tau)}) - \sigma_T^2(f^{(\tau)})|\mathcal{F}_T\Big) \leq \frac{C\beta^2}{\alpha^2-1} = \varepsilon_\beta.
\end{aligned} \tag{7.3.8}$$

$$|\{S(f) > \alpha\lambda\}|_{\hat{\mu}} \leq \hat{d}|\{\sigma(f) < \beta\lambda\}|_{\hat{\mu}} + \varepsilon_\beta\hat{d}|\{S(f) > \lambda\}|_{\hat{\mu}}.$$

Thus we obtain the good λ-inequality with respect to $\hat{\mu}$ directly. The remainder of the proof is unchanged. □

Remarks

1. The condition "R" together with $z \in S$ is weaker than the condition "R" together with $z \in A_\infty$.

2. Under the same conditions, i.e., "R" and $z \in S$, we have also the rearrangement inequalities

$$S(f)^{\hat{*}}(t) \leq \alpha\sigma(f)^{\hat{*}}\Big(\frac{t}{2\hat{d}}\Big) + S(f)^{\hat{*}}(2t), \quad \forall t > 0. \tag{7.3.9}$$

$$\sigma(f)^{\hat{*}}(t) \leq \alpha S(f)^{\hat{*}}\Big(\frac{t}{2\hat{d}}\Big) + \sigma(f)^{\hat{*}}(2t), \quad \forall t > 0. \tag{7.3.10}$$

We have also the general Φ-inequality between $Mf \vee S(f)$ and $Mf \wedge S(f)$.

Theorem 7.3.5 *Assume that "R" holds, $z \in A_\infty$ and Φ is general. Then for all $f = (f_n)_{n\geq 0}$ (martingale with respect to $\{\mathcal{F}_n\}_{n\geq 0}$), we have*

$$\widehat{E}(\Phi(Mf \vee S(f))) \leq C\widehat{E}(\Phi(Mf \wedge S(f))). \tag{7.3.11}$$

Proof. We need the conditional version of the simplest unweighted inequality between $Mf \vee S(f)$ and $Mf \wedge S(f)$, as follows (see §3.5)

$$E(Mf \vee S(f))|\mathcal{F}_0) \leq CE(Mf \wedge S(f)|\mathcal{F}_0). \tag{7.3.12}$$

Now assume $f_0 = 0$ without loss of generality. Denote $A(f) = (A_n(f))_{n\geq 0}$ $A_n(f) = M_nf \vee S_n(f)$, $B(f) = (B_n(f))_{n\geq 0}$, $B_n(f) = M_n(f) \wedge S_n(f)$. For $(B_n(f))_{n\geq 0}$ and $\beta\lambda$ define a stopping time τ such that

$$|\{\tau < \infty\}|_{\hat{\mu}} \leq \hat{d}|\{B_\infty(f) > \beta\lambda\}|_{\hat{\mu}}, \quad B_\tau(f) \leq \beta\lambda.$$

Consider the stopping martingale $f^{(\tau)}$, and for the process $A(f^{(\tau)}) = (A_n(f^{(\tau)}))_{n\geq 0}$ and λ define a stopping time T such that

$$\{A(f^{(\tau)}) > \lambda\} \subset \{T < \infty\}, \quad \{T < \infty\}|_{\hat{\mu}} \leq \hat{d}|\{A(f^{(\tau)}) > \lambda\}|_{\hat{\mu}}, \; A_T(f^{(\tau)}) \leq \lambda.$$

Then for $\alpha > 1$ determined later, we have

$$\{A(f) > \alpha\lambda\} \subset \{A(f^{(\tau)} > \alpha\lambda\} \bigcup \{\tau < \infty\},$$

$$\{A(f^{(\tau)}) > \alpha\lambda\} \subset \{A(f^{(\tau)}) - A_T(f^{(\tau)}) > (\alpha - 1)\lambda\}\}.$$

Consider a new σ-field family $\{\mathcal{F}'_m\}_{m\geq 0} = \{\mathcal{F}_{m+T}\}_{m\geq 0}$, and a new martingale $g' = (g'_m)_{m\geq 0}$ with respect to the new family, where

$$g'_m = f^{(\tau)}_{m+T} - f^{(\tau)}_T, \quad m \geq 0.$$

Noticing that

$$A(f^{(\tau)}) - A_T(f^{(\tau)}) \leq A(g'), \quad B(g') \leq 2B(f^{(\tau)}),$$

and applying (7.3.12) with respect to $(\Omega, \mathcal{F}, \mu, \{\mathcal{F}'_m\}_{m\geq 0})$,

$$E(A(g')|\mathcal{F}'_0) \leq CE(B(g')|\mathcal{F}'_0),$$

we get

$$\begin{aligned} E(\chi_{\{A(f^{(\tau)})>\alpha\lambda\}} \mid \mathcal{F}_T) &\leq \frac{1}{(\alpha-1)\lambda} E(A(f^{(\tau)}) - A_T(f^{(\tau)}) \mid \mathcal{F}_T) \\ &\leq \frac{C}{(\alpha-1)\lambda} E(B(f^{(\tau)}) \mid \mathcal{F}_T) \\ &= \frac{C}{(\alpha-1)\lambda} E(B_\tau(f) \mid \mathcal{F}_T) \leq \frac{C\beta}{\alpha-1}. \end{aligned} \tag{7.3.13}$$

The remaining steps to get (7.3.11) are unchanged. The proof is finished. □

Remark As same as above, under the conditions "R" and $z \in A_\infty$, we have also the weighted rearrangement inequality between $Af = Mf \vee S(f)$, $Bf = Mf \wedge S(f)$.

$$(A_\infty f)^*(t) \leq \alpha(B_\infty f)^*\left(\frac{t}{2\hat{d}}\right) + (A_\infty f)^*(2t), \quad \forall t > 0, \text{ for some } \alpha. \tag{7.3.14}$$

Notes to Chapter 7

§7.1. The condition "R" was well known before. For example, see Neveu [1] and Brossard [1]. As for the condition "R_ω" and other conditions formulated in Definition 7.1.1, it seems to occur first in Long [1,2]. In Theorem 7.1.2, the part (d) $\Rightarrow$ (e) is due to Gundy [1], (e) $\Rightarrow$ (a) is due to B-H-L [1], other implications are due to Long [1]. Theorem 7.1.2′ is also due to B-H-L [1].

§7.2.1. About atomic decomposition. As we pointed out, the idea of atomic decomposition was originated from Herz's works (Herz [1,2]), and formulated explicitly in the classical case by Coifman [1]. But it was B-M [1] who defined the atom concept in martingale setting, and gave H_1's atomic decomposition in the continuous martingale case or dyadic martingale case. Theorem 7.2.1 is due to Long [2].

§7.2.2. About duality. Making use of atomic decomposition to establish the duality H_1-BMO, is well known for specialists, such as R.Coifman, C. Fefferman, C. Herz, etc. Such discussion could be seen in B-M [1], there the case $p = 1$ was investigated in detail.

§7.2.3. About interpolation. The interpolation between H_p in martingale setting has not seen before. Lemma 7.2.3.3 is purely of real analysis, it is called Hardy's inequality in generalized form which is well known for real analysts. As for the operator interpolation concerning H_p, it was Igari [1] who started the topic in the classical case. Okada [1] discussed the topic in martingale setting, there the weak type operator interpolation concerning H_1 was discussed without any regularity. As for the addition of weak type estimates, the idea is due to E. M. Stein - G.Weiss. Lemma 7.2.3.6 except the case $p > 1$, is due to them. Theorems 7.2.3.7, 7.2.3.8 are well known, at least their idea. The kind of H_p's decomposition formulated in Theorem 7.2.3.9 is well known, for example, see Han [1] in the classical case. Its martingale version occurs first in the book.

§7.3.4. About the integrability $L\log^+ L$. Theorem 7.2.4.1 is well known, see for example Gundy [1], Garsia [1], Neveu [1] etc., its classical version see Zygmund [1], Stein [1].

§7.2.5. About the rearrangement of $f \in H_p \bigcap \operatorname{Re} L^1$. Davis [2] established theorem 7.2.5.1 in the classical version. There the reciprocal result has been established, too: let f be such that $\int_0^1 \left|\frac{1}{x}\int_{-x}^{x} f_d dt\right|^p dx < \infty$, then there is a rearrangement (on original space $\mathbf{T}^n$ or $\mathbf{R}^n$) of f, which is in H_p.

§7.2.6. About operator characterization of H_p. Lemmas 7.2.6.1, 7.2.6.2 are due to Janson [1], but the idea of the latter seems to be due to Chao-Taibleson [1]. Lemmas 7.2.6.5, 7.2.6.5′ are due to Janson [1]. A. Uchiyama [2] pointed out that unlike the classical case, we could not expect that Theorem 7.2.6.5 holds for all $p > 0$. Here is his proposition: Consider the q-martingale case. Let $m \geq 1$, and $A_1, \cdots, A_m$ be given arbitrarily. Then there exists $p_0(q) > 0$, such that for all $p \in (0, p_0(q)]$,

$$\inf\left\{\liminf_{n\to\infty} \|Mf\|_p^{-1}\left\{\|f_n\|_p + \sum_{j=1}^{m}\|(T_i f)_n\|_p\right\} : f \in H_p, f \not\equiv 0, f_0 = 0\right\} = 0.$$

The argument we have adopted in this section works only for $q \geq 3$, since $V = \{x \in$

$\mathbf{C}^2 : x_1 + x_2 = 0\}$ is of dimension one, all linear operators on it must be of the form $Af = \lambda f$, with $\lambda \in \mathbf{C}$, and hence all vectors are eigenvectors. Chao [1] gave methods to deal with the case $q = 2$. For the detail, see Chao [1].

§7.3. Main results are due to Long [1,4], some of them occur first in the book, for example, the inequality $\widehat{E}(\Phi(Mf \vee S(f))) \leq C\widehat{E}(\Phi(Mf \wedge S(f)))$. All the rearrangement versions of the inequalities cited in this section, occurred first in Long [7].

8 Some Applications of Martingale Techniques in Harmonic Analysis

As we mentioned in the preface of the book, Martingale Theory played a remarkable role in the development of Harmonic Analysis. It mainly means that Martingale Theory was one of the sources providing new ideas and methods to the latter. Many good examples showing this could not be put in the book because of the limitation from author's interests and book's space. Here we just take two examples to show such significance of Martingale Theory. The first example is concerned lacunary Fourier series. As well known, for any lacunary series $\sum_k c_{n_k} e^{in_k x}$ in the sense $n_{k+1} \geq qn_k > 0$, $q > 1$, for all k, we have

$$c\Big(\sum_k |c_{n_k}|^2\Big)^{\frac{1}{2}} \leq \Big(\int_0^{2\pi} |f(x)|^p \, dx\Big)^{\frac{1}{p}} \leq c\Big(\sum_k |c_{n_k}|^2\Big)^{\frac{1}{2}},$$

where $f(x)$ means the pointwise sum of the series, the existence of which is assumed (for the first inequality), or implied (for the second inequality in the preceding inequality chain). In the special case $n_k = r^k, k \geq 0, r > 1$, Gundy-Varopoulos [1] discovered a kind of martingales which could give a martingale explanation for this equivalence, even in a generalized version. §8.1 is devoted to this example. The second example is the martingale proof of the $T(b)$ Theorem, a very big theorem in Calderón-Zygmund's singular integral operator theory. The proof presented here is quite elementary, the simplicity of which comes mainly from a martingale construction of a nice frame of $L^2(\mathbf{R}^d, dx)$ with respect to a given $b(x)$. The involved martingales are not the classical ones, but with respect to the complex measure $b(x)dx$. Although the facts we need for the $T(b)$ Theorem from the Martingale Theory are only a few, for the sake of completeness, we would like still to devote some words to such martingales themselves. These materials are put in §8.2.

8.1 Gundy-Varopoulos' backward martingales and its application in lacunary Fourier series

First, let us introduce a kind of backward martingales which can be used to study the lacunary Fourier series. Let $\Omega = \mathbf{T} = [0,1)$, $\mathcal{F} = \mathcal{B}$ (Lebesgue family), $d\mu = dx$ (Lebesgue measure), and

$$\mathcal{F}_n = \left\{F \in \mathcal{B} : \ F + \frac{j}{r^n} = F \text{ (modulo measure zero)}, \ \forall j \in \mathbf{Z}\right\}, \ n \geq 0, \qquad (8.1.1)$$

where $r > 1$ is a given integer. Obviously, each $\mathcal{F}_n$ is a σ-field. $\{\mathcal{F}_n\}_{n\geq 0}$ is not nondecreasing any longer, but nonincreasing. Notice as well that, $\mathcal{F}_0 = \mathcal{F}$, $\bigcap_n \mathcal{F}_n =$ trivial one. The latter can be seen as follows. Let $F \in \mathcal{F}$ be such that $|F|\,|F^c| > 0$. We want to show that we can not have $F + \frac{j}{r^n} = F$, for all j, for all n. In fact,

$$\int_0^1 |(x+F)\bigcap F^c|dx = \int_0^1 \chi_{F^c}(t)\int_0^1 \chi_F(t-x)dxdt = |F||F^c| > 0.$$

Since $|(x+F)\bigcap F^c|$ is a continuous function of x, there exists $\frac{j}{r^n}$ such that $|(\frac{j}{r^n} + F)\bigcap F^c| > 0$. This means that F is not $\frac{1}{r^n}$-periodic. Thus $F \in \bigcap \mathcal{F}_n$ must be of measure 0 or 1.

Gundy-Varopoulos' $([0,1), \mathcal{B}, dx, \{\mathcal{F}_n\}_{n\geq 0})$ is a special example of $(\Omega, \mathcal{F}, \mu, \{\mathcal{F}_n\}_{n\geq 0})$, where

$$\{\mathcal{F}_n\}_{n\geq 0} \text{ is nonincreasing one}, \quad \mathcal{F}_0 = \mathcal{F}, \quad \bigcap \mathcal{F}_n = \text{ trivial one.}$$

In this case, an adapted process is called a martingale, if

$$E(f_n|\mathcal{F}_{n+1}) = f_{n+1}, \quad n = 0,1,2,\cdots. \qquad (8.1.2)$$

Denote f_0 by f, (8.1.2) means more, i.e.,

$$f_n = E(f_{n-1}|\mathcal{F}_n) = \cdots = E(f|\mathcal{F}_n), \quad n \geq 0.$$

Usually, we assume $f \in L^1$, so all martingales are in L^1_u. As for the convergences of martingales, it is a little complicated to get

$$\lim_{n\to\infty} f_n = E(f), \text{ a.e.}, \quad \forall \text{ martingale } f = (f_n)_{n\geq 0}, \qquad (8.1.3)$$

we take it as granted(see Dellacherie-Meyer [1]). The convergence is also in $L^p, 1 \leq p < \infty$, when $f = (f_n)_{n\geq 0} \in L^p$, owing to the uniform integrability and the pointwise convergence of $\{f_n\}_{n\geq 0}$. The operators M and S are defined by

$$M_n f = \sup_{k\geq n} |f_k|, \quad Mf = M_0 f, \qquad (8.1.4)$$

$$S_n(f) = \left(|E(f)|^2 + \sum_{k\geq n}^{\infty} |d_n|^2\right)^{\frac{1}{2}}, \quad S(f) = S_0(f), \qquad (8.1.5)$$

where

$$d_n = f_n - f_{n+1}, \quad n \geq 0. \tag{8.1.6}$$

A similar discussion as in §2.1 can show that M is L^p-bounded $1 < p < \infty$, and weak L^1-bounded. Take the weak L^1-boundedness of M as an example. Assume that $f = (f_n)_{n\geq 0} \in L^1$, and $E(f) = 0$. For $\lambda > 0$, define a stopping time

$$\tau = \sup\{n : \ |f_n| > \lambda\} \quad (\text{as usual, } \sup\{\emptyset\} = 0).$$

Then $\tau \neq \infty$, a.e., $\{M_1 f > \lambda\} \subset \{\tau > 0\}$, and on $\{\tau > 0\}$, we have $|f_\tau| > \lambda$. So

$$|\{M_1 f > \lambda\}| \leq \sum_1^\infty |\{\tau = n\}| \leq \frac{1}{\lambda}\sum_1^\infty \int_{\{\tau=n\}} |f_\tau| d\mu \leq \frac{1}{\lambda}\int_{\{\tau>0\}} |f| d\mu.$$

When $E(f) \neq 0$, consider $f - E(f)$ at first. Since $Mf \leq |f - E(f)| + M_1(f - E(f)) + |E(f)|$, we get

$$|\{Mf > \lambda\}| \leq \frac{c}{\lambda}||f||_1, \quad \forall\lambda > 0.$$

As for the boundedness of S, it should be better to establish them in the regular case, since the preceding $([0,1), \mathcal{B}, dx, \{\mathcal{F}_n\}_{n\geq 0})$ has some regularity. As in §7.1, $(\Omega, \mathcal{F}, \mu, \{\mathcal{F}_n\}_{n\geq 0})$ is called "R" regular, if

$$E(f|\mathcal{F}_n) \leq dE(f|\mathcal{F}_{n+1}), \quad \forall n \geq 0, \forall f \in L^1_+, \tag{8.1.7}$$

or equivalently

$$\forall F_n \in \mathcal{F}_n, \ \exists G_n \in \mathcal{F}_{n+1}, \text{ s.t. } F_n \subset G_n, \ |G_n| \leq d|F_n|, \ \forall n \geq 0. \tag{8.1.8}$$

As shown in Theorem 7.1.2, there are a lot of equivalent statements for this regularity. We just want to show that we have the following consequence of this regularity which is more convenient to use: for all nonnegative adapted processes $(\gamma_n)_{n\geq 0}$ with $\lim\limits_{n\to\infty} \gamma_n = 0$, a.e., for all $\lambda > 0$, there exists a stopping time τ_λ such that

$$|\{\tau_\lambda > 0\}| \leq d|\{M\gamma > \lambda\}|, \tag{8.1.9}$$

$$\{M\gamma > \lambda\} \subset \{\tau_\lambda > 0\}, \tag{8.1.10}$$

$$M_{\tau_\lambda}\gamma \leq \lambda. \tag{8.1.11}$$

To prove this assertion, we define $\tau_\lambda = \sup\{n : \ E(\chi_{\{\gamma_{n-1}>\lambda\}}|\mathcal{F}_n) \geq \frac{1}{d}\}$. Then $\tau_\lambda \neq \infty$, a.e. and (8.1.9)–(8.1.11) can be proved by the argument in (a) $\Rightarrow$ (f) of Theorem 7.1.2.

Thus, for regular $(\Omega, \mathcal{F}, \mu, \{\mathcal{F}_n\}_{n\geq 0})$, by making use of (8.1.9), (8.1.11), we can get (by considering $f - E(f) = (f_n - E(f))_{n\geq 0}$ instead of $f = (f_n)_{n\geq 0}$ at first)

$$||Mf||_p \approx ||S(f)||_p, \quad 0 < p < \infty. \tag{8.1.12}$$

In the regular case, we can get the weak L^1-boundedness of S easily. In fact, let $f = (f_n)_{n\geq 0} \in L^1$ with $E(f) = 0$. For $\lambda > 0$, and the process $(|f_n|)_{n\geq 0}$, we have the stopping time τ satisfying $|\{\tau > 0\}| \leq d|\{Mf > \lambda\}|$, and $M_\tau f \leq \lambda$. Thus

$$\{S(f) > \lambda\} \subset \{\tau > 0\} \bigcup \{\tau = 0, S(f^{(\tau)}) > \lambda\} \quad \text{with } f^{(\tau)} = (f_{\tau \vee n})_{n\geq 0},$$

$$\begin{aligned} |\{S(f) > \lambda\}| &\leq d|\{Mf > \lambda\}| + \frac{c}{\lambda^2}\int_\Omega |f_\tau|^2 d\mu \\ &= d|\{Mf > \lambda\}| + \frac{c}{\lambda}\int_\Omega |f_\tau| d\mu \leq \frac{c}{\lambda}\|f\|_1. \end{aligned}$$

Now we return to the Gundy-Varopoulos' martingales case. We will show that it is regular, and hence S is L^p-bounded, $1 < p < \infty$, and weak L^1-bounded. First, we want to know how the conditional expectation is formulated. We have

$$E_n(f)(x) = E(f|\mathcal{F}_n)(x) = \frac{1}{r^n}\sum_{j=0}^{r^n-1} f(x + j/r^n). \tag{8.1.13}$$

It is easy to verify. Since the right-hand side is $\mathcal{F}_n$-measurable, and has same integral value as f does on any $F \in \mathcal{F}_n$ (it follows from the $\frac{1}{r^n}$-periodicity of F), we get (8.1.13). We also need another expression of $E_n(f)$ by means of Fourier transform. Let $f \in L^1$ (denoting the Fourier transform of f by $\hat{f}$, i.e., $\hat{f}(n) = \int_0^1 f(x)e^{-2\pi i n x}dx$),

$$f \sim \sum_{-\infty}^{\infty} \hat{f}(n)e^{2\pi i n x},$$

then we have

$$f_k = E(f|\mathcal{F}_k) \sim \sum_{-\infty}^{\infty} \hat{f}(nr^k)e^{2\pi i n r^k x}, \tag{8.1.14}$$

owing to the fact

$$\frac{1}{r^k}\sum_{j=0}^{r^k-1} e^{2\pi i n j r^{-k}} = \begin{cases} 0, & n \neq mr^k, \quad \forall m \in \mathbf{Z}, \\ 1, & n = mr^k, \quad \text{some } m \in \mathbf{Z}. \end{cases}$$

And hence

$$d_k = f_k - f_{k+1} \sim {\sum_{-\infty}^{\infty}}' \hat{f}(nr^k)e^{2\pi i n r^k x}, \tag{8.1.15}$$

where $\sum'$ means that the summation is over all $n \neq mr$, for all $m \in \mathbf{Z}$. From (8.1.13), we see that for all $f \in L^1_+$, for all n,

$$f_n(x) \leq r f_{n+1}(x).$$

This means that the regularity condition "R" holds.

Now we use this kind of backward martingales to the lacunary Fourier series.

Proposition 8.1.1 *Let $r > 1$ be an integer, $f \in L^1(\mathbf{T})$ with its Fourier expansion $f \sim \sum_{n=0}^{\infty} \hat{f}(r^n)e^{2\pi i r^n x}$. Then for all $p > 1$,*

$$c\|f\|_p \leq \Big(\sum_0^{\infty} |\hat{f}(r^n)|^2\Big)^{\frac{1}{2}} \leq c\|f\|_1. \tag{8.1.16}$$

Proof. From (8.1.15) we see that f considered as a martingale, has

$$d_k = \hat{f}(r^k)e^{2\pi i r^k x}, \quad S(f) = \Big(\sum_1^{\infty} |\hat{f}(r^k)|^2\Big)^{\frac{1}{2}} = \text{ const. (since } E(f) = 0).$$

By making use of the weak type $(1,1)$ of the square operator S, we see (denoting $\lambda = \big(\sum_0^{\infty} |\hat{f}(r^k)|^2\big)^{\frac{1}{2}}$)

$$1 = \Big|\Big\{S(f) > \frac{\lambda}{2}\Big\}\Big| \leq \frac{c}{\lambda}\|f\|_1,$$

it is just the second inequality in (8.1.16). The first inequality in (8.1.16) follows from

$$\|f\|_p \leq c\|S(f)\|_p \leq c\Big(\sum_0^{\infty} |\hat{f}(r^k)|^2\Big)^{\frac{1}{2}}.$$

The proof is finished. □

For general Fourier series, we have a natural version of the second inequality in (8.1.16).

Theorem 8.1.2 *Let $r > 1$ be an integer, and $f \in L^1(\mathbf{T})$, with $\{\hat{f}(n)\}_{-\infty}^{\infty}$ as its Fourier coefficients, such that f considered as a martingale is in H_1. Then we have*

$$\Big(\sum_0^{\infty} |\hat{f}(r^k)|^2\Big)^{\frac{1}{2}} \leq c\|f\|_{H_1}. \tag{8.1.17}$$

Proof. By duality, the problem is reduced to one concerning BMO, the definition of which is (owing to the regularity)

$$BMO = \{g \in L^2 : \; E(|g - g_k|^2 | \mathcal{F}_k) \leq c\}. \tag{8.1.18}$$

For those $g \sim \sum_0^{\infty} \hat{g}(r^n)e^{2\pi i r^n x}$, we have

$$E(|g - g_k|^2|\mathcal{F}_k) = E\Big(\sum_0^{k-1} |d_n|^2 |\mathcal{F}_k\Big) = \sum_{n=0}^{k-1} |\hat{g}(r^n)|^2.$$

This means that we have

$$\|g\|_{BMO} \leq c\|g\|_2, \quad \forall g \sim \sum_0^{\infty} \hat{g}(r^n)e^{2\pi i r^n x}. \tag{8.1.19}$$

Now (8.1.17) follows from (the H_1–BMO duality is true, too)

$$\Big(\sum_0^\infty |\hat{f}(r^k)|^2\Big)^{\frac{1}{2}} = \sup\{|E(fg)| : g \text{ is as in (8.1.19), and } \|g\|_2 = 1\}$$
$$\le \sup_g C\|Mf\|_1\|g\|_{BMO} \le C\sup_g \|Mf\|_1\|g\|_2 \le C\|Mf\|_1.$$

The proof is finished. □

Our sake is to give a generalized version of (8.1.16). We need some elementary facts concerning the martingales with two indices which follow from the corresponding facts concerning the Hilbert space valued martingales with one index. Let us devote some words to the latter. Let f be H-valued with H a Hilbert space, and $\mathcal{F}_0$ a sub-σ-field satisfying the usual conditions. Then the conditional expectation can be defined in the same way, and it has the same properties as in **C**-valued case, for example, when $f, g \in L^2(H)$, g is $\mathcal{F}_0$ measurable,

$$E_{\mathcal{F}_0}(\langle f, g\rangle_H) = \langle E_{\mathcal{F}_0}(f), g\rangle_H, \quad E_{\mathcal{F}_0}(\langle g, f\rangle_H) = \langle g, E_{\mathcal{F}_0}(f)\rangle_H,$$

and when $\mathcal{F}_0 \subset \mathcal{F}_1$,

$$E_{\mathcal{F}_0}E_{\mathcal{F}_1} = E_{\mathcal{F}_1}E_{\mathcal{F}_0} = E_{\mathcal{F}_0},$$

and when $\{\mathcal{F}_n\}_{n\ge0}$ is a nondecreasing family, denoting $E_m = E_{\mathcal{F}_m}$, $\Delta_k = E_k - E_{k-1}$,

$$E_m(\langle\Delta_k f, \Delta_l g\rangle_H) = 0, \quad \text{when } k \ne l,\ k, l \ge m,\ f, g \in L^2(H).$$

(When $\{\mathcal{F}_n\}_{n\ge0}$ is nonincreasing, we have the similar orthogonality.) Thus, the arguments given in §2.1 work well in the present case, and hence, when $\{\mathcal{F}_n\}_{n\ge0}$ is nondecreasing, we have

$$\|S(f)\|_p \approx \|f\|_{L^p(H)}, \quad 1 < p < \infty,$$

$$|\{S(f) > \lambda\}| \le \frac{c}{\lambda}\|f\|_{L^1(H)}, \quad \forall\lambda > 0,$$

where

$$S(f) := \Big(\sum_0^\infty \langle d_k, d_k\rangle_H\Big)^{\frac{1}{2}}, \quad d_k = E(f|\mathcal{F}_k) - E(f|\mathcal{F}_{k-1}),\ k \ge 0.$$

When $\{\mathcal{F}_n\}_{n\ge0}$ is nonincreasing, we have the same inequalities with f replaced by $f - E(f)$, and the same definition of S (but $d_k = E(f|\mathcal{F}_k) - E(f|\mathcal{F}_{k+1})$, of course). Now consider the two indices case. Let $\{\mathcal{F}_n\}_{n\ge0}$, $\{\mathcal{G}_m\}_{m\ge0}$ be two nonincreasing families of sub-σ-fields satisfying the usual conditions. In addition assume that the conditional expectation is commutative in the following sense

$$f_{n,m} = E(E(f|\mathcal{F}_n)|\mathcal{G}_m) = E(E(f|\mathcal{G}_m)|\mathcal{F}_n), \quad \forall n, m. \tag{8.1.20}$$

We consider martingales $f = (f_{n,m})_{n,m\ge0}$ generalized by $f \in L^1$, i.e., $f_{n,m} = E(E(f|\mathcal{F}_n)|\mathcal{G}_m)$, for all n, m.

Lemma 8.1.3 *Let $f = (f_{n,m})_{n,m\geq 0}$ be a martingale as above, such that $f = (f_n)_{n\geq 0}$, $f_n = E(f|\mathcal{F}_n)$, is in H_1. Then we have*

$$\|\bar{S}(f)\|_p \approx \|f - E(f)\|_p, \qquad 1 < p < \infty, \tag{8.1.21}$$

$$|\{\bar{S}(f) > \lambda\}| \leq \frac{c}{\lambda}\|f - E(f)\|_{H_1}, \qquad \forall\lambda > 0, \tag{8.1.22}$$

where

$$\bar{S}(f) = \Big(\sum_{n,m} |d_{n,m}|^2\Big)^{\frac{1}{2}}, \; d_{n,m} = f_{n,m} - f_{n+1,m} - f_{n,m+1} + f_{n+1,m+1}. \tag{8.1.23}$$

Proof. For f given in the lemma, define a l^2-valued function G,

$$G = (d_0, d_1, \cdots), \quad d_k = E(f|\mathcal{F}_k) - E(f|\mathcal{F}_{k+1}), \quad k \geq 0,$$

and the martingale $(G_m)_{m\geq 0}$ generated by G,

$$G_m = E(G|\mathcal{G}_m) = (E(d_0|\mathcal{G}_m), E(d_1|\mathcal{G}_m), \cdots), \quad m \geq 0.$$

Notice that we have

$$\|G\|_{L^p(l^2)} = \Big\|\Big(\sum_0^\infty |d_k|^2\Big)^{\frac{1}{2}}\Big\|_p \approx \|f - E(f)\|_p, \qquad 1 < p < \infty,$$

and because of

$$\begin{aligned} d_{n,m} &= E(f_n|\mathcal{G}_m) - E(f_{n+1}|\mathcal{G}_m) - E(f_n|\mathcal{G}_{m+1}) + E(f_{n+1}|\mathcal{G}_{m+1}) \\ &= E(d_n|\mathcal{G}_m) - E(d_n|\mathcal{G}_{m+1}), \end{aligned}$$

we have

$$S(G) = \Big(\sum_{m=0}^\infty \|\{E(d_n|\mathcal{G}_m) - E(d_n|\mathcal{G}_{m+1})\}\|_{l^2}^2\Big)^{\frac{1}{2}} = \Big(\sum_{n,m} |d_{n,m}|^2\Big)^{\frac{1}{2}} = \bar{S}(f).$$

Thus, we get

$$\|\bar{S}(f)\|_p = \|S(G)\|_p \approx \|G\|_{L^p(l^2)} \approx \|f - E(f)\|_p, \quad 1 < p < \infty,$$

$$|\{\bar{S}(f) > \lambda\}| = |\{S(G) > \lambda\}| \leq \frac{c}{\lambda}\|G\|_{L^1(l^2)} \leq \frac{c}{\lambda}\|f - E(f)\|_{H_1}.$$

This proves the lemma. □

Now let p, q be two different primes. Consider two sub-σ-fields on $(\mathbf{T}, \mathcal{B}, dx)$ as above: $\mathcal{F}_n$ is consisting of all p^{-n}-periodic sets, and $\mathcal{G}_m$ is consisting of all q^{-m}-periodic sets. Obviously, such $\{\mathcal{F}_n\}$, $\{\mathcal{G}_m\}$ satisfy the conditions we need. Let $f \in L^1$, then

$$d_n \sim \sum_{k:p\nmid k} \hat{f}(kp^n)e^{2\pi ikp^n x} \quad (d_n = E(f|\mathcal{F}_n) - E(f|\mathcal{F}_{n+1})),$$

here $p \nmid k$ means $k \neq mp$, for all $m \in \mathbf{Z}$, and hence

$$E(d_n|\mathcal{G}_m) - E(d_n|\mathcal{G}_{m+1}) \sim \sum_{l:\ p\nmid l, q\nmid l} \hat{f}(lp^n q^m) e^{2\pi i l p^n q^m x}. \tag{8.1.24}$$

If f has the form

$$f \sim \sum_{n,m} \hat{f}(p^n q^m) e^{2\pi i p^n q^m x}, \tag{8.1.25}$$

then

$$d_{n,m} = \hat{f}(p^n q^m) e^{2\pi i p^n q^m x}, \quad \bar{S}(f) = \Big(\sum_{n,m} |\hat{f}(p^n q^m)|^2\Big)^{\frac{1}{2}}.$$

Thus we get

Theorem 8.1.4 *Let p, q be two different primes, $f \in L^1(\mathbf{T})$ such that f has the form* (8.1.25), *and the martingale $f = (f_n)_{n\geq 0} \in H_1$, $f_n = E(f|\mathcal{F}_n)$, $\mathcal{F}_n = \{$all p^{-n}-periodic sets$\}$. (Or $f = (g_m)_{m\geq 0} \in H_1, g_m = E(f|\mathcal{G}_m)$, $\mathcal{G}_m = \{$all q^{-m}-periodic sets$\}$.) Then*

$$c_p \|f\|_p \leq \Big(\sum_{n,m} |\hat{f}(p^n q^m)|^2\Big)^{\frac{1}{2}} \leq c\|f\|_{H_1}, \quad \forall p,\ 1 < p < \infty. \tag{8.1.26}$$

Proof. By making use of Lemma 8.1.3, and reasoning just like in the proof of Proposition 8.1.1, we get (8.1.26). □

For general Fourier series, we have

Theorem 8.1.5 *Let p, q be two different primes, $f \in L^r(\mathbf{T})$, $r > 1$. Then we have*

$$\Big(\sum_{n,m} |\hat{f}(p^n q^m)|^2\Big)^{\frac{1}{2}} \leq c\|f\|_r. \tag{8.1.27}$$

Proof. Consider those functions g having the form (8.1.25), and $\|g\|_2 = 1$. Noticing $\bar{S}(g) = \big(\sum_{n,m} |\hat{g}(p^n q^m)|^2\big)^{\frac{1}{2}} = \|g\|_2$, we have

$$\begin{aligned}\Big(\sum_{n,m} |\hat{f}(p^n q^m)|^2\Big)^{\frac{1}{2}} &= \sup_g |E(fg)| \leq \sup_g \|f\|_r \|g\|_{r'} \\ &\leq c \sup_g \|f\|_r \|\bar{S}(g)\|_{r'} = c \sup_g \|f\|_r \|g\|_2 = c\|f\|_r.\end{aligned}$$

This completes the proof. □

Corollary 8.1.6 *Let $f \in L^p(\mathbf{T}), 1 < p < \infty$, $a_1, \cdots, a_m \in \mathbf{Z}^+$. Then*

$$\Big(\sum_{n_1, \cdots, n_m} |\hat{f}(a_1^{n_1} \cdots a_m^{n_m})|^2 \Big)^{\frac{1}{2}} \leq c\|f\|_p. \tag{8.1.28}$$

Proof. It follows by factorizing each a_j. □

8.2 The martingale proof of the $T(b)$ Theorem and the related martingales

Let us first remind what the $T(b)$ Theorem is like. The so-called Hilbert transform is defined by (modulo a constant)

$$Hf(x) = \text{ p.v. } \int_{-\infty}^{\infty} \frac{f(y)}{x-y} dy = \lim_{\varepsilon \to 0} \int_{|x-y|>\varepsilon} \frac{f(y)}{x-y} dy, \tag{8 2.1}$$

here p.v. means "principal value". It is a topic in Complex Analysis, and can be studied by making use of the complex variable theory. In order to extend it to $\mathbf{R}^d$, Calderón-Zygmund introduced the so-called Real Method, and studied following operators systematically in the early 50's,

$$Tf(x) = \text{p.v.} \int_{\mathbf{R}^d} K(x-y) f(y) dy, \tag{8.2.2}$$

where $K(x)$ is a function defined on $\mathbf{R}^d - \{0\}$, satisfying the size and smoothness conditions as follows: for $r \in (0,1]$,

$$|K(x)| \leq c|x|^{-d}, \quad \forall x \neq 0, \tag{8.2.3}$$

$$|K(x) - K(x')| \leq c|x-x'|^r |x|^{-d-r}, \quad \text{when } |x| \geq 2|x-x'|. \tag{8.2.4}$$

Such T is called a singular integral operator, and $K(x)$, a Calderón-Zygmund kernel. When some other natural conditions are imposed on the kernel $K(x)$, for example

$$\int_{\alpha<|x|<\beta} K(x) dx = 0, \quad \forall \beta > \alpha > 0, \tag{8.2.5}$$

then it is easy to see that T is a L^2-bounded operator, i.e.,

$$\|Tf\|_2 \leq c\|f\|_2, \quad \forall f \in L^2, \tag{8.2.6}$$

by making use of Plancherel Theorem concerning the Fourier transform, since under the conditions imposed on K, we have

$$\|\widehat{K}(\xi)\|_\infty \leq c, \quad \text{with } \widehat{K}(\xi) = \lim_{\varepsilon \to 0} \int_{\varepsilon \leq |x| \leq \frac{1}{\varepsilon}} K(x) e^{-ix\cdot\xi} dx.$$

Calderón-Zygmund's Real Method tells us that T is also L^p-bounded, for $1 < p < \infty$, and weakly L^1-bounded. Such singular integral operators are of convolution type, so the powerful Fourier transform can be used. In the 80's, people are interested in the third generation of singular integral operators, which are not of convolution type. They are defined by

$$Tf(x) = \text{p.v.} \int_{\mathbf{R}^d} K(x,y)f(y)dy, \tag{8.2.7}$$

where $K(x,y)$ is defined on $\mathbf{R}^d \times \mathbf{R}^d - \{x = y\}$, and satisfying the size and smoothness conditions

$$|K(x,y)| \leq c|x-y|^{-d}, \quad \forall x \neq y. \tag{8.2.8}$$

$$|K(x,y) - K(x',y)| + |K(y,x) - K(y,x')|$$
$$\leq c|x-x'|^r |x-y|^{-d-r}, \quad |x-y| \geq 2|x-x'|. \tag{8.2.9}$$

A typical example of such singular integral operators is the Cauchy integral operator H_Γ defined on a Lipschitz curve Γ,

$$H_\Gamma f(x) = \text{p.v.} \int_{-\infty}^{\infty} \frac{f(y)}{z(x) - z(y)} dy, \tag{8.2.10}$$

where $z(x)$ is the arc-length parameterization of Γ, satisfying

$$|z(x) - z(y)| \approx |x-y|, \quad \forall x, y \in \mathbf{R}. \tag{8.2.11}$$

(Notice that a usual Cauchy integral is

$$\int_\Gamma \frac{f(\zeta)}{z - \zeta} d\zeta,$$

in terms of the parameterization expression, it should read

$$\int_{-\infty}^{\infty} \frac{f(y)z'(y)}{z(x) - z(y)} dy.$$

We omit the factor $z'(y)$ because of $|z'(y)| \approx 1$.) A natural question is that if H_Γ is L^2-bounded. The conjecture "H_Γ is L^2-bounded" kept to be open for a long time untill in 1982, Coifman-McIntosh-Meyer [1] gave an affirmative answer to it. Meanwhile, people tried to solve the general question: under what conditions, T defined in (8.2.7) is L^2-bounded. Once it is L^2-bounded, then Calderón-Zygmund's program works well to get T's other boundedness. In 1984, David-Journé [1] gave a beautiful answer, i.e., $T(1)$ Theorem, which says

$$T \text{ is } L^2\text{-bounded} \iff T \text{ has WBP, and } T(1),\ T^t(1) \in BMO,$$

here WBP means "weak boundedness property", and T^t is the transpose of T (i.e., the operator defined by the kernel $K(y,x)$), and BMO is the classical BMO space mentioned in the notes of Chapter 4. By making use of the $T(1)$ Theorem, many operators can be shown to be L^2-bounded except the Cauchy integral operator. The reason is that the function 1 is not a good testing function for H_Γ, but the function $z'(x)$ is, because $H_\Gamma(z') = 0$ by Cauchy Theorem. So the question is if the function 1 in $T(1)$ Theorem could be replaced by a suitable function class. In 1985, David-Journé-Semmes [1] obtained the $T(b)$ Theorem, the most important ingredient of which is following

Special $T(b)$ Theorem *Let $b(x)$ be a pseudoaccretive function in the sense*

$$b \in L^\infty, \text{ and } \left|\frac{1}{|I|}\int_I b(x)dx\right| \geq c > 0, \quad \forall I \in \mathcal{I}, \tag{8.2.12}$$

where $\mathcal{I}$ is the set of all quasi-dyadic-cubes constructed soon, such that $T(b) = 0 = T^t(b)$, and T has WBP with respect to b in the sense (8.2.1.8). Then T is a L^2-bounded operator.

Up to now, there have been many proofs for this theorem. Among them, a quite satisfactory one is: finding a quasi-basis $\{\alpha_i\}$ of $L^2(\mathbf{R}^d, dx)$ with respect to the given $b(x)$, such that T's matrix in this basis could be estimated easily. Of course, the approach to find such quasi-basis are numerous. The martingale approach we will introduce is elementary and simple enough.

8.2.1 The proof of the special $T(b)$ theorem

This subsection will be in analysis style, we do not want to go far away from the main topics of the book, so we will omit some calculations, although all of them would be elementary.

Denote $\mathcal{I}_0 = \{\text{all dyadic cubes of length } 1\}$. Divide each $I \in \mathcal{I}_0$ into two equal parts by hyperplanes perpendicular to the x_1-axis, and let $\mathcal{I}_1 = \{I : \text{so produced}\}$. Then continue this way along with the axes $x_2, x_3, \cdots, x_d, x_1, \cdots$, and get $\mathcal{I}_2, \mathcal{I}_3, \cdots, \mathcal{I}_d, \mathcal{I}_{d+1}, \cdots$ respectively. How about $\mathcal{I}_k$ for $k < 0$? They come from the procedure reverse to previous one. Let $\mathcal{I} = \bigcup \mathcal{I}_k$. Notice that for each $I \in \mathcal{I}_k$, we have $I = I_1 \bigcup I_2$ with $I_1, I_2 \in \mathcal{I}_{k+1}$, for all $k \in \mathbf{Z}$. This is the main feature of the construction. Now for all k, for all $I \in \mathcal{I}_k$, define

$$\begin{aligned} \alpha_I &= |I|^{-\frac{1}{2}}|I|_b^{-1}(|I_2|_b\chi_{I_1} - |I_1|_b\chi_{I_2}), \\ \beta_I &= |I|^{\frac{1}{2}}(|I_1|_b^{-1}\chi_{I_1} - |I_2|_b^{-1}\chi_{I_2}), \end{aligned} \tag{8.2.1.1}$$

where $|I|_b = \int_I b\,dx$, i.e. the b-measure of I. Obviously, $\{\alpha_I, \beta_I\}_{I\in\mathcal{I}}$ satisfies

$$\int b\alpha_I\,dx = 0 = \int \beta_I b\,dx, \quad \forall I \in \mathcal{I}, \tag{8.2.1.2}$$

$$\int |\alpha_I|^2\,dx \approx 1 \approx \int |\beta_I|^2\,dx, \quad \forall I \in \mathcal{I}, \tag{8.2.1.3}$$

$$\int \beta_J b\alpha_I\,dx = 0, \quad \forall I \neq J. \tag{8.2.1.4}$$

From these, it looks that $\{\alpha_I, \beta_I\}_{I\in\mathcal{I}}$ behaves like an orthonormal system. And that it behaves like a basis follows from some elementary facts, at least some basic ideas, of the related martingale theory. In the next subsection, we will show this. At the moment, assume that we have, for all $f \in L^2$,

$$f = \sum_I \alpha_I \langle \beta_I, f\rangle_b, \tag{8.2.1.5}$$

with the series converging in L^2, and

$$\|f\|_2 \approx \Big(\sum_I |\langle \beta_I, f\rangle_b|^2\Big)^{\frac{1}{2}}, \tag{8.2.1.6}$$

where

$$\langle \beta_I, f\rangle_b = \int \beta_I b f dx, \quad \forall I. \tag{8.2.1.7}$$

Let T be a singular integral operator defined by the kernel $K(x, y)$ satisfying (8.2.8), (8.2.9), and T has WBP with respect to b in the sense

$$|\langle \beta_I, T(b\alpha_I)\rangle_b| \le c, \quad \forall I, \tag{8.2.1.8}$$

and

$$T(b) = 0 = T^t(b). \tag{8.2.1.9}$$

We want to prove that T is L^2-bounded.

By making use of the quasi-basis $\{\alpha_I, \beta_I\}_{I\in\mathcal{I}}$, we have

$$\begin{aligned} T(bf) &= \sum_I T(b\alpha_I)\langle \beta_I, f\rangle_b \\ &= \sum_J \alpha_J \sum_I \langle \beta_J, T(b\alpha_I)\rangle_b \langle \beta_I, f\rangle_b, \end{aligned}$$

$$\|T(bf)\|_2 \approx \Big(\sum_J \Big|\sum_I \langle \beta_J, T(b\alpha_I)\rangle_b \langle \beta_I, f\rangle_b\Big|^2\Big)^{\frac{1}{2}}.$$

Thus we get a matrix operator

$$t: \ \{\langle \beta_I, f\rangle_b\}_I \to \Big\{\sum_I \langle \beta_J, T(b\alpha_I)\rangle_b \langle \beta_I, f\rangle_b\Big\}_J,$$

and T's L^2-boundedness is reduced to t's l^2-boundedness. There is a useful lemma to give sufficient conditons for the l^2-boundedness of a matrix operator.

Schur Lemma. *Let $(\varepsilon_{i,j})$ be a nonnegative infinite matrix, such that there exists a nonnegative sequence $\{w_j\}$ satisfying*

$$\sum_j (\varepsilon_{i,j} + \varepsilon_{j,i}) w_j \le c w_i, \quad \forall i,$$

then the matrix operator defined by $(\varepsilon_{i,j})$ *is* l^2*-bounded.*

Now we want to apply the lemma to our case: $\varepsilon_{i,j} = |\langle\beta_J, T(b\alpha_I)\rangle_b|$, $w_j = |J|^{\frac{1}{2}-\delta}$ with $\delta \in (0, \frac{r}{d})$. Noticing the symmetry between $K(x,y)$ and $K(y,x)$, and between $\{b\alpha_I\}$ and $\{\beta_J b\}$, we see that for the T's L^2-boundedness, it is enough to get

$$\sum_J |\langle\beta_J, T(b\alpha_I)\rangle_b||J|^{\frac{1}{2}-\delta} \leq c|I|^{\frac{1}{2}-\delta}, \tag{8.2.1.10}$$

where

$$\begin{aligned} c_{I,J} &= |\langle\beta_J, T(b\alpha_I)\rangle_b| \\ &= \left|\int_{\mathbf{R}^d}\int_{\mathbf{R}^d} \beta_J(x)b(x)K(x,y)b(y)\alpha_I(y)dydx\right|. \end{aligned} \tag{8.2.1.11}$$

To make the roles of each condition in the $T(b)$ Theorem clear, we summarize as follows.

(8.2.1.8) will be used to estimate $c_{I,J}$ when $I = J$. The size condition (8.2.8) implies

$$\int_I\int_J |K(x,y)|dx\,dy \leq c|I|\log\frac{c|J|}{|I|}, \quad \text{when } I\bigcap J = \emptyset,\ I \subset cJ, \tag{8.2.1.12}$$

where cJ denotes the c-time extension of J, which will be used to estimate $c_{I,J}$ when I, J are disjoint but near each other. From $T^t(b) = 0 = T(b)$, we get

$$\begin{aligned} &\iint \chi_{J_1}(x)b(x)K(x,y)b(y)\alpha_I(y)\,dy\,dx \\ &\quad = -\iint \chi_{J_1^c}(x)b(x)K(x,y)b(y)\alpha_I(y)\,dy\,dx, \end{aligned} \tag{8.2.1.13}$$

$$\begin{aligned} &\iint \beta_J(x)b(x)K(x,y)b(y)\chi_{I_1}(y)\,dy\,dx \\ &\quad = -\iint \beta_J(x)b(x)K(x,y)b(y)\chi_{I_1^c}(y)\,dy\,dx, \end{aligned} \tag{8.2.1.14}$$

where the superscript c denotes the complementary set, which will be used to estimate $c_{I,J}$ when $J = J_1 \bigcup J_2$, $I \subset J_1$, or $I = I_1 \bigcup I_2$, $J \subset I_1$. Finally the smoothness condition (8.2.9) will be used to estimate $c_{I,J}$ when I, J are far away from each other. More precisely, we have

$$|T(b\alpha_I)(x)| + |T^t(\beta_I b)(x)| \leq c|I|^{\frac{1}{2}+\frac{r}{d}}|x - y_I|^{-d-r}, \quad \forall x \notin c_0 I,\ \forall I, \tag{8.2.1.15}$$

where c_0 is a constant depending only on d, and y_I is the center of I. In fact, only see $T(b\alpha_I)(x)$, we have

$$\begin{aligned} |T(b\alpha_I)(x)| &= \left|\int_{\mathbf{R}^d} (K(x,y) - K(x,y_I))b(y)\alpha_I(y)\,dy\right| \\ &\leq c|I|^{-\frac{1}{2}+1+\frac{r}{d}}|x - y_I|^{-d-r} = c|I|^{\frac{1}{2}+\frac{r}{d}}|x - y_I|^{-d-r}. \end{aligned}$$

Now (8.2.1.8), (8.2.1.12)–(8.2.1.15) are enough to get (8.2.1.10) by considering six cases according to

$$|I| \le |J| : \; J \bigcap c_0 I = \emptyset; \; J \bigcap c_0 I \neq \emptyset, \text{ but } I \bigcap J = \emptyset; \; I \subset J,$$

$$|I| > |J| : J \bigcap c_0 I = \emptyset; \; J \subset 2c_0 I, \quad \text{dist} \le cl(J); \; J \subset 2c_0 I, \quad \text{dist} > cl(J),$$

where dist $=$ dist$(x_J, b(I))$ with x_J the center of J and $b(I) = \bigcup_1^2 \text{bd}(I_j)$ (bd $=$ "boundary", and $I = I_1 \bigcup I_2$). For the estimates of $c_{I,J}$, and hence of $\sum_J c_{I,J}|J|^{\frac{1}{2}-\delta}$, in all six cases, only elementary calculations are needed. For the details we refer to Long [8], for those readers who are interested in the topic. Anyway, the proof of the special $T(b)$ Theorem is finished, modulo (8.2.1.5), (8.2.1.6).

8.2.2 Martingales with respect to complex measures

For the completeness, we want to give some fundamental facts concerning the martingales with respect to complex measures in this subsection, which are beyond the need to verify (8.2.1.5), (8.2.1.6).

Let $(\Omega, \mathcal{F}, \nu)$ be a σ-finite, complete, nonnegative measure space with $|\Omega| = \infty$, $\psi(x)$ be a complex measurable function in L^∞, and $\{\mathcal{F}_n\}_{-\infty}^{\infty}$ be a nondecreasing family of sub-σ-fields satisfying

$$\mathcal{F} = \bigvee \mathcal{F}_n, \; \bigcap \mathcal{F}_n = \text{ trivial, each } (\Omega, \mathcal{F}_n, \nu) \text{ is complete, } \sigma\text{-finite.} \tag{8.2.2.1}$$

Consider the measure $d\mu = \psi d\nu$. The conditional expectation and martingale considered in this subsection will be with respect to $d\mu$, but the L^p-estimate will be with respect to $d\nu$. This is a new feature.

Let $\mathcal{G}$ be a sub-σ-field of $\mathcal{F}$ such that $(\Omega, \mathcal{G}, \nu)$ is complete and σ-finite. Denote the expectations with respect to $d\nu$ and $d\mu$ by $\widetilde{E}$ and E respectively. Since $(\Omega, \mathcal{G}, \nu)$ is σ-finite, $\Omega = \bigcup \Omega_k$, $\{\Omega_k\}$ is disjoint, $\Omega_k \in \mathcal{G}$, and $|\Omega_k| < \infty$ (as usual $|\cdot|$ denotes the measure in the underlying space, meanwhile $|\cdot|_\mu$ denotes the μ-measure). For any $f \in L^1_{\text{loc}}$ (means $f\chi_F \in L^1$, for all $F \in \mathcal{F}, |F| < \infty$), $\widetilde{E}_\mathcal{G}(f) = \widetilde{E}(f|\mathcal{G})$ has meaning on each Ω_k, so we can define

$$\widetilde{E}_\mathcal{G}(f) = \sum_k \widetilde{E}(f|\mathcal{G})\chi_{\Omega_k}.$$

Obviously the definition is independent of the partition of Ω. Assume that $\widetilde{E}(\psi|\mathcal{G}) \neq 0$, a.e. Then we can define

Definition 8.2.2.1 *Let $\mathcal{G}$ be as above, $\psi \in L^\infty$ such that $\widetilde{E}(\psi|\mathcal{G}) \neq 0$. Then the conditional expectation E with respect to $d\mu$ is defined by*

$$E_\mathcal{G}(f) = E(f|\mathcal{G}) = \widetilde{E}(\psi|\mathcal{G})^{-1}\widetilde{E}(\psi f|\mathcal{G}), \quad \forall f \in L^1_{\text{loc}}. \tag{8.2.2.2}$$

Proposition 8.2.2.2 *E has following elementary properties*

(a) $E(1) = 1$.

(b) *E is linear, furthermore when g is $\mathcal{G}$-measurable, we have $E(fg) = E(f)g$.*
(c) *$E(f)$ is $\mathcal{G}$-measurable, and*

$$\int_A E(f)d\mu = \int_A f d\mu, \quad \forall A \in \mathcal{G}, \quad \forall f \in L^1(A, \nu).$$

(d) *Let $\mathcal{G}_1 \subset \mathcal{G}_2$, then*

$$E(E(f|\mathcal{G}_2)|\mathcal{G}_1) = E(f|\mathcal{G}_1).$$

In particular, from it , we have

$$E(E(f|\mathcal{G}_2) - E(f|\mathcal{G}_1)|\mathcal{G}_1) = 0.$$

(e) *Let $\{\mathcal{F}_n\}_{-\infty}^{\infty}$ be as above, denote $\Delta_n = E_n - E_{n-1}$, $E_n = E_{\mathcal{F}_n}$. Then*

$$E(\Delta_n f \Delta_m g|\mathcal{F}_k) = 0, \quad n \neq m, \ n, m \geq k.$$

(f) *Assume $c_0^{-1} \leq |\psi| \leq c_0$, a.e. (It is natural, since the most interesting case is of the case when $d\mu$ is absolutely continuous with respect to $d\nu$). Then for $1 \leq p \leq \infty$, $E = E_{\mathcal{G}}$ is L^p-bounded, if and only if*

$$c^{-1}c_0^{-1} \leq |\widetilde{E}(\psi|\mathcal{G})| \leq cc_0, \quad a.e.$$

We omit the proofs, which are elementary.

Remark The property (c) is the characterizing property of the conditional expectation as in the classical case. That is to say if we define the operator E by (c), then other properties remain true. Now deduce (8.2.2.2) from (c). Let A be any set in $\mathcal{G}$ with finite measure, and $f \in L^1_{\text{loc}}$. Then we have

$$\int_A f d\mu = \int_A \widetilde{E}(\psi f) d\nu,$$

$$\int_A E(f) d\mu = \int_A \psi E(f) d\nu = \int_A \widetilde{E}(\psi E(f)) d\nu = \int_A \widetilde{E}(\psi) E(f) d\nu,$$

and hence (from(c))

$$\int_A \widetilde{E}(\psi f) d\nu = \int_A \widetilde{E}(\psi) E(f) d\nu,$$

$$\widetilde{E}(\psi f) = \widetilde{E}(\psi) E(f).$$

This proves the assertion. But in this definition, we should claim the uniquence of $E(f)$, and describe the domain of f. We still need to consider the measure $|\mu|$. That is why we consider $d\nu$ at first, then $d\mu$.

Definition 8.2.2.3 *The adapted process $f = (f_n)_{-\infty}^{\infty}$ is called a martingale (with respect to $d\mu$, except otherwise stated), if*

$$f_n = E(f_{n+1}|\mathcal{F}_n), \quad \text{a.e.}, \quad \forall n. \tag{8.2.2.3}$$

For any martingale $f = (f_n)_{-\infty}^{\infty}$ *(even any process), we define*

$$M_n f = \sup_{k \le n} |f_k|, \quad Mf = M_\infty f, \tag{8.2.2.4}$$

$$S_n(f) = \left(|f_{-\infty}|^2 + \sum_{-\infty}^{n} |\Delta_k f|^2\right)^{\frac{1}{2}}, \quad S(f) = S_\infty(f), \tag{8.2.2.5}$$

where $f_{-\infty} = \lim_{n \to -\infty} f_n$*, pointwise.* $f = (f_n)_{-\infty}^{\infty}$ *is called an* L^p*-bounded martingale,* $1 \le p \le \infty$*, if* $\|f\|_p = \sup_n \|f_n\|_p < \infty$.

In what follows, we always assume

$$c_0^{-1} \le |\widetilde{E}(\psi|\mathcal{F}_n)| \le c_0, \quad \text{a.e.} \quad \forall n. \tag{8.2.2.6}$$

We will show that M is L^p-bounded, $1 < p \le \infty$, and weakly L^1-bounded; for $1 < p \le \infty$, each L^p-bounded martingale $f = (f_n)_{-\infty}^{\infty}$ is generated by some $f \in L^p(\nu)$, in the sense

$$f_n = E(f|\mathcal{F}_n), \quad \forall n; \tag{8.2.2.7}$$

and for $1 \le p \le \infty$, each L^p-bounded martingale $f = (f_n)_{-\infty}^{\infty}$ has the pointwise limits $\lim_{n \to \infty} f_n$ and $\lim_{n \to -\infty} f_n$.

Proposition 8.2.2.4 *Let* $1 < p \le \infty$*, then* M *is an* L^p*-bounded operator, as well as a weakly* L^1*-bounded operator, i.e.,*

$$\|Mf\|_p \le c\|f\|_p, \quad \forall f = (f_n)_{-\infty}^{\infty}, \quad 1 < p \le \infty, \tag{8.2.2.8}$$

$$|\{Mf > \lambda\}| \le \frac{c}{\lambda}\|f\|_1, \quad \forall f = (f_n)_{-\infty}^{\infty}, \quad \forall \lambda > 0. \tag{8.2.2.9}$$

Finally, for $1 < p \le \infty$*, each* L^p*-bounded martingale* $f = (f_n)_{-\infty}^{\infty}$ *is generated by some* $f \in L^p(\nu)$*, and* $\|f\|_p \approx \sup_n \|f_n\|_p$.

Proof. Let $f = (f_n)_{-\infty}^{\infty}$ be a martingale. Then

$$f_n = E(f_{n+1}|\mathcal{F}_n) = \widetilde{E}(\psi|\mathcal{F}_n)^{-1}\widetilde{E}(\psi f_{n+1}|\mathcal{F}_n),$$

$$\begin{aligned} f_n = E(f_{n+2}|\mathcal{F}_n) &= \widetilde{E}(\psi|\mathcal{F}_n)^{-1}\widetilde{E}(\psi f_{n+2}|\mathcal{F}_n), \\ &= \widetilde{E}(\psi|\mathcal{F}_n)^{-1}\widetilde{E}(\widetilde{E}(\psi f_{n+2}|\mathcal{F}_{n+1})|\mathcal{F}_n). \end{aligned}$$

Since $\widetilde{E}(\psi|\mathcal{F}_n) \ne 0, \ne \infty$, a.e., we get

$$\widetilde{E}(\psi f_{n+1}|\mathcal{F}_n) = \widetilde{E}(\widetilde{E}(\psi f_{n+2}|\mathcal{F}_{n+1})|\mathcal{F}_n), \quad \forall n.$$

This means that $(\widetilde{E}(\psi f_{n+1}|\mathcal{F}_n))_{-\infty}^{\infty}$ is a martingale with respect to $(\Omega, \mathcal{F}, \nu, \{\mathcal{F}_n\}_{-\infty}^{\infty})$. It is an L^p-bounded martingale, since

$$\widetilde{E}(\psi f_{n+1}|\mathcal{F}_n) = \widetilde{E}(\psi|\mathcal{F}_n) f_n,$$

and (8.2.2.6) is assumed. In addition, we have

$$\sup_n \|f_n\|_p \approx \sup_n \|\widetilde{E}(\psi f_{n+1}|\mathcal{F}_n)\|_p,$$

$$Mf \approx \sup_n |\widetilde{E}(\psi f_{n+1}|\mathcal{F}_n)|.$$

From the corresponding results in the classical case, we get

$$\begin{aligned}\|Mf_n\|_p &\approx \|\sup_n |\widetilde{E}(\psi f_{n+1}|\mathcal{F}_n)|\,\|_p \\ &\le c\sup_n \|\widetilde{E}(\psi f_{n+1}|\mathcal{F}_n)\|_p \approx \sup_n \|f_n\|_p, \quad 1 < p \le \infty,\end{aligned}$$

$$\begin{aligned}|\{Mf > \lambda\}| &\le |\{\sup_n |\widetilde{E}(\psi f_{n+1}|\mathcal{F}_n)| > c\lambda\}| \\ &\le c\lambda^{-1}\sup_n \|\widetilde{E}(\psi f_{n+1}|\mathcal{F}_n)\|_1 \approx \lambda^{-1}\sup_n \|f_n\|_1, \quad \forall\lambda > 0.\end{aligned}$$

Now let $1 < p \le \infty$. For $M \in \mathbf{Z}^+$ arbitrarily large, we have $\Omega = \bigcup \Omega_k$, $\Omega_k \in \mathcal{F}_{-M}, |\Omega_k| < \infty$, for all k. Since $(\widetilde{E}(\psi f_{n+1}|\mathcal{F}_n)\chi_{\Omega_k})_{n\ge -M}$ is an L^p-bounded martingale on Ω_k, for all k, there exist $\psi f \in L^p(\Omega_k, \nu)$ such that on Ω_k, for all k,

$$\widetilde{E}(\psi f_{n+1}|\mathcal{F}_n) = \widetilde{E}(\psi f|\mathcal{F}_n), \quad n \ge -M. \tag{8.2.2.10}$$

Comparing with $f_n = \widetilde{E}(\psi|\mathcal{F}_n)^{-1}\widetilde{E}(\psi f_{n+1}|\mathcal{F}_n)$, and letting $M \to \infty$, we get (8.2.2.7). In addition, we have

$$\|\psi f\|_p \le c\sup_n \|\widetilde{E}(\psi f_{n+1}|\mathcal{F}_n)\|_p, \quad \|f\|_p \le c\sup_n \|f\|_p.$$

Obviously, we have also $\sup_n \|f_n\|_p \le c\|f\|_p$. This proves $\|f\|_p \approx \sup_n \|f_n\|_p$, which means the equivalence of the two explanations of $\|f\|_p$. The proof is finished. □

Proposition 8.2.2.5 *Let $1 \le p \le \infty$, and $f = (f_n)_{-\infty}^{\infty}$ be an L^p-bounded martingale. Then*

$$\lim_{n\to\infty} f_n = f, \quad 1 < p \le \infty; \quad \lim_{n\to\infty} f_n \text{ exists, a.e. for } p = 1, \tag{8.2.2.11}$$

$$\lim_{n\to-\infty} f_n = 0, \quad \text{a.e. for } 1 \le p < \infty. \tag{8.2.2.12}$$

Proof. Make a partition $\Omega = \bigcup \Omega_k$, $\Omega_k \in \mathcal{F}_0$, $|\Omega_k| < \infty$, for all k. Then both $(\widetilde{E}(\psi|\mathcal{F}_n)\chi_{\Omega_k})_{n\ge 0}$ and $(\widetilde{E}(\psi f_{n+1}|\mathcal{F}_n)\chi_{\Omega_k})_{n\ge 0}$ are L^p-bounded martingales with respect to $(\Omega_k, \mathcal{F}\bigcap\Omega_k, \nu|_{\Omega_k}, \{\mathcal{F}_n\bigcap\Omega_k\}_{n\ge 0})$, for each k. So

$$\lim_{n\to\infty} \widetilde{E}(\psi|\mathcal{F}_n) = \psi, \quad \text{a.e. on } \Omega_k, \quad \forall k,$$

$$\lim_{n\to\infty} \widetilde{E}(\psi f_{n+1}|\mathcal{F}_n) = \psi g, \quad \text{a.e. on } \Omega_k, \quad \forall k,$$

where $g \in L^1(\Omega_k, \nu)$ for $p = 1$, and $g = f$, for $1 < p \leq \infty$. Thus on each Ω_k,

$$\lim_{n\to\infty} f_n = \lim_{n\to\infty} \widetilde{E}(\psi|\mathcal{F}_n)^{-1}\widetilde{E}(\psi f_{n+1}|\mathcal{F}_n) = \begin{cases} g, & \text{for } p = 1; \\ f, & \text{for } 1 < p \leq \infty. \end{cases}$$

This is just (8.2.2.11). Now we prove (8.2.2.12). Denote $\theta(x) = \overline{\lim}_{n\to-\infty} |f_n|$. Then $\theta(x) \leq Mf(x)$. Since $\theta(x)$ is $\bigcap \mathcal{F}_n$ - measurable, $\theta(x) = a \geq 0$. Because of the weak L^p-boundedness, $1 \leq p < \infty$ (notice that L^p-boundedness is stronger than weak L^p-boundedness), we have

$$|\{\theta > \lambda\}| \leq |\{Mf > \lambda\}| \leq \left(\frac{c}{\lambda}\|f\|_p\right)^p, \quad \forall \lambda > 0.$$

Since $|\Omega| = \infty$, this implies $a = 0$. □

Remark When $1 < p < \infty$, we have also $\lim_{n\to\infty} f_n = f$ in L^p. It follows from the Dominated Convergence Theorem of Lebesgue.

Now we turn to the L^2-equivalence between Mf and $S(f)$, a crucial result in the field. In the classical case, this equivalence is trivial, since

$$\widetilde{E}(\widetilde{S}(f)^2) = \widetilde{E}\Big(\sum_{-\infty}^{\infty} |\widetilde{\Delta}_k f|^2\Big) = \widetilde{E}\Big(\sum_{-\infty}^{\infty} \widetilde{\Delta}_k f \sum_{-\infty}^{\infty} \overline{\widetilde{\Delta}_l f}\Big) = \widetilde{E}(|f|^2).$$

In the present case, the orthogonality is not with respect to the measure $d\nu$. But the difficulty is not so serious. In order to establish a more general equivalence between M and S later on, we want to establish the conditional L^2-equivalence between M and S at first. That is to say, we consider the underlying space $(\Omega, \mathcal{F}, \nu, \{\mathcal{F}_n\}_{n\geq 0})$, and want to obtain

$$c\widetilde{E}(S(f)^2|\mathcal{F}_0) \leq \widetilde{E}(|f|^2|\mathcal{F}_0) \leq c\widetilde{E}(S(f)^2|\mathcal{F}_0), \tag{8.2.2.13}$$

with the coefficient c independent of both $\{\mathcal{F}_n\}_{n\geq 0}$ and $f = (f_n)_{n\geq 0}$. Once it is established, we can get more general equivalence between M and S on the underlying space $(\Omega, \mathcal{F}, \nu, \{\mathcal{F}_n\}_{-\infty}^{\infty})$.

Theorem 8.2.2.6 *Let $(\Omega, \mathcal{F}, \nu, \{\mathcal{F}_n\}_{n\geq 0})$ be as usual, $d\mu = \psi d\nu$ be as above, $f = (f_n)_{n\geq 0}$ be a martingale with respect to $d\mu$. Then* (8.2.2.13) *holds with c depending only on c_0 in* (8.2.2.6).

Proof. Assume that $f = (f_n)_{n\geq 0}$ is an L^2-bounded martingale. We have

$$\begin{aligned} \Delta_n f &= \widetilde{E}(\psi|\mathcal{F}_n)^{-1}\widetilde{E}(\psi f|\mathcal{F}_n) - \widetilde{E}(\psi|\mathcal{F}_{n-1})^{-1}\widetilde{E}(\psi f|\mathcal{F}_{n-1}) \\ &= (\widetilde{E}(\psi|\mathcal{F}_n)^{-1} - \widetilde{E}(\psi|\mathcal{F}_{n-1})^{-1})\widetilde{E}(\psi f|\mathcal{F}_{n-1}) \\ &\quad + \widetilde{E}(\psi|\mathcal{F}_n)^{-1}(\widetilde{E}(\psi f|\mathcal{F}_n) - \widetilde{E}(\psi f|\mathcal{F}_{n-1})), \quad n \geq 0. \end{aligned}$$

(As usual, when $n = 0$, all terms with the subscript -1 disappear, so the identity holds.) Thus we get

$$|\Delta_n f|^2 \le c|\widetilde{\Delta}_n(\psi f)|^2 + c|\widetilde{E}_{n-1}(\psi f)|^2|\widetilde{\Delta}_n(\psi)|^2, \quad n \ge 0, \tag{8.2.2.14}$$

$$\begin{aligned}\widetilde{E}\Big(\sum_{n=0}^{\infty}|\Delta_n f|^2\Big|\mathcal{F}_0\Big) &\le c\widetilde{E}\Big(\sum_{n=0}^{\infty}|\widetilde{\Delta}_n(\psi f)|^2\Big|\mathcal{F}_0\Big)\\ &\quad + c\widetilde{E}\Big(\sum_{n=1}^{\infty}|\widetilde{E}_{n-1}(\psi f)|^2|\widetilde{\Delta}_n(\psi)|^2\Big|\mathcal{F}_0\Big)\\ &\le c\widetilde{E}(|f|^2|\mathcal{F}_0) + cJ,\end{aligned}$$

$$\begin{aligned}J &\le \widetilde{E}\Big(\sum_{n=1}^{\infty}M_{n-1}(\psi f)^2\Big(\sum_{k=n}^{\infty}|\widetilde{\Delta}_k(\psi)|^2 - \sum_{k=n+1}^{\infty}|\widetilde{\Delta}_k(\psi)|^2\Big)\Big|\mathcal{F}_0\Big)\\ &= \widetilde{E}\Big(\sum_{n=1}^{\infty}\widetilde{E}\Big(\sum_{k=n}^{\infty}|\widetilde{\Delta}_k(\psi)|^2\Big|\mathcal{F}_n\Big)(M_{n-1}(\psi f)^2 - M_{n-2}(\psi f)^2)\Big|\mathcal{F}_0\Big)\\ &\le c\|\psi\|_\infty^2\widetilde{E}\Big(\sum_{n=1}^{\infty}(M_{n-1}(\psi f)^2 - M_{n-2}(\psi f)^2)\Big|\mathcal{F}_0\Big) \le c\widetilde{E}(|f|^2|\mathcal{F}_0).\end{aligned}$$

This proves one half of (8.2.2.13) Now prove the another half of (8.2.2.13). Assume $S(f) \in L^2$. We have

$$\widetilde{E}(|f|^2|\mathcal{F}_0)^{\frac{1}{2}} \le c\widetilde{E}(|\psi f|^2|\mathcal{F}_0)^{\frac{1}{2}} = c \sup_{g:\widetilde{E}(|g|^2|\mathcal{F}_0)\le 1}|\widetilde{E}(g\psi f|\mathcal{F}_0)|, \tag{8.2.2.15}$$

where the equality can be seen as follows: by taking

$$g = \bar{f}(\widetilde{E}(|f|^2|\mathcal{F}_0) + \varepsilon)^{-\frac{1}{2}},$$

we have

$$\widetilde{E}(|g|^2|\mathcal{F}_0) < 1, \text{ and } \widetilde{E}(gf|\mathcal{F}_0) = \widetilde{E}(|f|^2|\mathcal{F}_0)(\widetilde{E}(|f|^2|\mathcal{F}_0) + \varepsilon)^{-\frac{1}{2}},$$

and hence

$$\widetilde{E}(|f|^2|\mathcal{F}_0)^{\frac{1}{2}} \le \sup_g|\widetilde{E}(gf|\mathcal{F}_0)| \le \widetilde{E}(|f|^2|\mathcal{F}_0)^{\frac{1}{2}}.$$

Noticing that $g = \sum_{l=0}^{\infty}\Delta_l g$ with the series converging in L^2, and considering $f_N = \sum_{k=0}^{N}\Delta_k f$, and applying (8.2.2.15), we get

$$\begin{aligned}\widetilde{E}(|f_N|^2|\mathcal{F}_0)^{\frac{1}{2}} &\le c\sup_g\Big|\widetilde{E}\Big(\sum_{l=0}^{\infty}\Delta_l g\cdot\psi\cdot\sum_{k=0}^{N}\Delta_k f\Big|\mathcal{F}_0\Big)\Big|\\ &= c\sup_g\Big|\widetilde{E}\Big(\sum_{k=0}^{N}\Delta_k g\cdot\psi\cdot\Delta_k f\Big|\mathcal{F}_0\Big)\Big|\\ &\le c\sup_g\widetilde{E}(S(g)S(f)|\mathcal{F}_0) \le c\widetilde{E}(S(f)^2|\mathcal{F}_0)^{\frac{1}{2}}\sup_g\widetilde{E}(S(g)^2|\mathcal{F}_0)^{\frac{1}{2}}\\ &\le c\widetilde{E}(S(f)^2|\mathcal{F}_0)^{\frac{1}{2}},\end{aligned}$$

here we have used the orthogonality

$$E(\Delta_k g \Delta_l f | \mathcal{F}_0) = 0, \quad k \neq l, \ k,l \geq 0,$$

another formulation of which is just

$$\widetilde{E}(\Delta_k g \cdot \psi \cdot \Delta_l f | \mathcal{F}_0) = 0, \quad k \neq l, \ k,l \geq 0.$$

Since $S(f) \in L^2$, the inequality $\widetilde{E}(|f_N|^2|\mathcal{F}_0) \leq c\widetilde{E}(S(f)^2|\mathcal{F}_0)$ implies the L^2-boundedness of the martingale $f = (f_n)_{n\geq 0}$, and hence

$$\widetilde{E}(|f|^2|\mathcal{F}_0) \leq c\widetilde{E}(S(f)^2|\mathcal{F}_0).$$

This proves the theorem. □

Remark As we pointed out, the result for $(\Omega, \mathcal{F}, \nu, \{\mathcal{F}_n\}_{-\infty}^{\infty})$ follows from limit passage. Say $f = (f_n)_{-\infty}^{\infty}$ is a martingale on $(\Omega, \mathcal{F}, \nu, \{\mathcal{F}_n\}_{-\infty}^{\infty})$. Consider $f = (f_n)_{n\geq -M}$ on $(\Omega, \mathcal{F}, \nu, \{\mathcal{F}_n\}_{n\geq -M})$ at first. We have

$$\int_\Omega \Big(|f_{-M}|^2 + \sum_{n=-M+1}^{\infty} |\Delta_n f|^2 \Big) d\nu \approx \int_\Omega \Big(\sup_{n\geq -M} |f_n|^2 \Big) d\nu,$$

with the equivalence coefficient independent of M. Letting $M \to \infty$, we get $\widetilde{E}(S(f)^2) \approx \widetilde{E}(|f|^2) \approx \widetilde{E}((Mf)^2)$.

Now consider the special case where $\Omega = \mathbf{R}^d$, $\mathcal{F} = \mathcal{B}$, $d\nu = dx$, $d\mu = b(x)dx$ with $b(x)$ being a pseudoaccretive function, and $\{\mathcal{F}_n\}_{-\infty}^{\infty}$ with $\mathcal{F}_n$ generated by $\mathcal{I}_n$ constructed in §8.2.1, for all n. Obviously, $(\mathbf{R}^d, \mathcal{B}, dx, \{\mathcal{F}_n\}_{-\infty}^{\infty})$ satisfies the usual conditions. So for all $f \in L^2$,

$$f = \sum_{-\infty}^{\infty} \Delta_n f,$$

with the series converging both pointwise and in L^2. Now we want to find a function pair family $\{\alpha_I, \beta_I\}_{I\in\mathcal{I}}$ such that

$$\Delta_n f = \sum_{I\in\mathcal{I}_{n-1}} \alpha_I \langle \beta_I, f\rangle_b, \quad \forall f, \ \forall n,$$

where b is the given pseudoaccretive function in the $T(b)$ Theorem, and

$$\langle \beta_I, f\rangle_b = \int_{\mathbf{R}^d} \beta_I(y) b(y) f(y)\, dx, \quad \forall I.$$

Define

$$\alpha_I = a_1\chi_{I_1} + a_2\chi_{I_2}, \quad \beta_I = b_1\chi_{I_1} + b_2\chi_{I_2}, \quad I = I_1 \bigcup I_2,$$

with $a_1, a_2, b_1, b_2 \in \mathbf{C}$ determined later. Comparing the two expressions below, say $I \in \mathcal{I}_{n-1}$, $I = I_1 \bigcup I_2$,

$$\begin{aligned}\Delta_n f|_I &= \frac{1}{|I_1|_b}\int_{I_1} bf\,dx\,\chi_{I_1} + \frac{1}{|I_2|_b}\int_{I_2} bf\,dx\,\chi_{I_2} - \frac{1}{|I|_b}\int_I bf\,dx\,\chi_I \\ &= \Big(\Big(\frac{1}{|I_1|_b} - \frac{1}{|I|_b}\Big)\int_{I_1} bf\,dx - \frac{1}{|I|_b}\int_{I_2} bf\,dx\Big)\chi_{I_1} \\ &\quad + \Big(\Big(\frac{1}{|I_2|_b} - \frac{1}{|I|_b}\Big)\int_{I_2} bf\,dx - \frac{1}{|I|_b}\int_{I_1} bf\,dx\Big)\chi_{I_2},\end{aligned}$$

and

$$\begin{aligned}\alpha_I\langle\beta_I, f\rangle_b &= \Big(a_1b_1\int_{I_1} bf\,dx + a_1b_2\int_{I_2} bf\,dx\Big)\chi_{I_1} \\ &\quad + \Big(a_2b_2\int_{I_2} bf\,dx + a_2b_1\int_{I_1} bf\,dx\Big)\chi_{I_2},\end{aligned}$$

we find that

$$a_1 = |I|^{-\frac{1}{2}}|I|_b^{-1}|I_2|_b, \quad a_2 = -|I|^{-\frac{1}{2}}|I|_b^{-1}|I_1|_b,$$

$$b_1 = |I|^{\frac{1}{2}}|I_1|_b^{-1}, \quad b_2 = -|I|^{\frac{1}{2}}|I_2|_b^{-1},$$

will make

$$\Delta_n f = \sum_{I\in\mathcal{I}_{n-1}} \alpha_I\langle\beta_I, f\rangle_b, \quad \forall n,\ \forall f \in L^2,$$

$$\int b\alpha_I\,dx = 0 = \int \beta_I b\,dx, \quad \forall I,$$

$$|\alpha_I| \approx |I|^{-\frac{1}{2}} \approx |\beta_I|, \quad \text{on } I, \quad \forall I.$$

Thus we will have

$$f = \sum_{n=-\infty}^{\infty} \Delta_n f = \sum_{-\infty}^{\infty}\sum_{I\in\mathcal{I}_{n-1}} \alpha_I\langle\beta_I, f\rangle_b = \sum_I \alpha_I\langle\beta_I, f\rangle_b,$$

and

$$\begin{aligned}\|f\|_2^2 \approx \|S(f)\|_2^2 &= \int\sum_{-\infty}^{\infty}\Big|\sum_{I\in\mathcal{I}_{n-1}} \alpha_I\langle\beta_I, f\rangle_b\Big|^2 dx \\ &= \sum_{-\infty}^{\infty}\sum_{I\in\mathcal{I}_{n-1}}\int_{\mathbf{R}^d} |\alpha_I|^2|\langle\beta_I, f\rangle_b|^2\,dx \approx \sum_I |\langle\beta_I, f\rangle_b|^2.\end{aligned}$$

This completes the verifications of (8.2.1.5) and (8.2.1.6). Up to now, the proof of the special $T(b)$ Theorem has been finished completely. Notice that $\{\alpha_I, \beta_I\}_{I\in\mathcal{I}}$ is like orthonormal in the sense

$$\int \beta_J b\alpha_I\,dx = \delta_{I,J}.$$

Now consider the convex Φ–equivalence between M and S. But first we consider the general Φ–equivalence between M and S for those martingales $f = (f_n)_{-\infty}^{\infty}$: there exists a nonnegative, nondecreasing, adapted process $D = (D_n)_{-\infty}^{\infty}$, such that

$$|\Delta_n f| \leq D_{n-1}, \qquad \forall n. \tag{8.2.2.16}$$

We still consider the case $(\Omega, \mathcal{F}, \nu, \{\mathcal{F}_n\}_{n\geq 0})$ at first. Notice that in this case, $n = 0$ in (8.2.2.16) implies $f_0 = 0$. For a general martingale $f = (f_n)_{n\geq 0}$, we should consider $f - f_0 = (f_n - f_0)_{n\geq 0}$ instead. It brings no essential influence on the problems we consider, since we have

$$S(f - f_0) \leq S(f) \leq S(f - f_0) + |f_0|,$$

$$M(f - f_0) - |f_0| \leq Mf \leq M(f - f_0) + |f_0|,$$

and the terms involving f_0 are easy to handle.

Theorem 8.2.2.7 *Let $f = (f_n)_{n\geq 0}$ be a martingale satisfying* (8.2.2.16). *Then for any general $\Phi(u)$ (in the sense in* §3.5), *we have*

$$\int_\Omega \Phi(S(f))\, d\nu \leq c \int_\Omega \Phi(Mf + D_\infty)\, d\nu, \tag{8.2.2.17}$$

$$\int_\Omega \Phi(Mf)\, d\nu \leq c \int_\Omega \Phi(S(f) + D_\infty)\, d\nu, \tag{8.2.2.18}$$

with the constant c independent both of $f = (f_n)_{n\geq 0}$ and $\{\mathcal{F}_n\}_{n\geq 0}$.

Proof. Let $\alpha > 1$, and $\beta > 0$ be determined suitably, $\lambda > 0$ be any level, $f = (f_n)_{n\geq 0}$ be any martingale with $f_0 = 0$. Notice

$$|f_n| \leq |f_{n-1}| + |\Delta_n f| \leq M_{n-1} f + D_{n-1} = \rho_{n-1}, \quad \forall n \geq 0.$$

Define a stopping time $\tau = \inf\{n : \rho_n > \beta\lambda\}$, and consider the stopped martingale

$$f^{(\tau)} = (f_n^{(\tau)})_{n\geq 0} = (f_{n\wedge\tau})_{n\geq 0}.$$

We have

$$\{\tau < \infty\} = \{\rho_\infty > \beta\lambda\}, \quad Mf^{(\tau)} = \sup_n |f_{n\wedge\tau}| \leq M_\tau f \leq \rho_{\tau-1} \leq \beta\lambda.$$

Furthermore, we have

$$\begin{aligned}\{S(f) > \alpha\lambda\} &\subset \{\tau < \infty\} \bigcup \{\tau = \infty,\ S_\tau(f)^2 > \alpha^2\lambda^2\} \\ &\subset \{\tau < \infty\} \bigcup \{S(f^{(\tau)})^2 > \alpha^2\lambda^2\}.\end{aligned}$$

Define another stopping time

$$T = \inf\{n : S_n(f^{(\tau)}) > \lambda\}.$$

then we have

$$\{\tau < \infty\} = \{S(f^{(\tau)}) > \lambda\}, \quad S_{T-1}(f^{(\tau)}) \leq \lambda.$$

Thus we get

$$\{S(f) > \alpha\lambda\} \subset \{\tau < \infty\} \bigcup \{S(f^{(\tau)})^2 - S_{T-1}(f^{(\tau)})^2 > (\alpha^2 - 1)\lambda^2\}. \quad (8.2.2.19)$$

We want to get the good λ-inequality of the pair $(S(f), \rho_\infty)$,

$$|\{S(f) > \alpha\lambda\}| \le c|\{\rho_\infty > \beta\lambda\}| + \varepsilon_\beta|\{S(f) > \lambda\}|, \quad \forall\lambda > 0, \quad (8.2.2.20)$$

with $\lim_{\beta\to 0} \varepsilon_\beta = 0$, from which (8.2.2.17), (8.2.2.18) follow immediately. (Here $|\Omega| = \infty$ brings no trouble owing to the σ-finiteness.) From (8.2.2.19), we see that for (8.2.2.20), it is enough to prove

$$|\{S(f^{(\tau)})^2 - S_{T-1}(f^{(\tau)})^2 > (\alpha^2 - 1)\lambda^2\}| \le \varepsilon_\beta|\{T < \infty\}|. \quad (8.2.2.21)$$

Just as in §3.5, after introducing a new martingale

$$g = (g_n)_{n\ge 0}, \; g_n = f^{(\tau)}_{T+n} - f^{(\tau)}_{T-1} \text{ (and hence } \Delta_n g = \Delta_{T+n}(f^{(\tau)}), \; \forall n \ge 0),$$

with respect to $d\mu$ and $\{\mathcal{G}_n\}_{n\ge 0} = \{\mathcal{F}_{T+n}\}_{n\ge 0}$, we get

$$S(g)^2 = \sum_{n=0}^{\infty} |\Delta_n g|^2 = \sum_{k=T}^{\infty} |\Delta_k f^{(\tau)}|^2 = S(f^{(\tau)})^2 - S_{T-1}(f^{(\tau)})^2,$$

and hence (by making use of the first inequality of (8.2.2.13))

$$\begin{aligned}\widetilde{E}(S(f^{(\tau)})^2 - S_{T-1}(f^{(\tau)})^2|\mathcal{F}_T) &= \widetilde{E}(S(g)^2|\mathcal{G}_0)\\ &\le c\widetilde{E}(|g|^2|\mathcal{G}_0) = c\widetilde{E}(|f^{(\tau)} - f^{(\tau)}_{T-1}|^2|\mathcal{F}_T) \le c\beta^2\lambda^2,\end{aligned}$$

$$\widetilde{E}(\chi_{\{S(f^{(\tau)})^2 - S_{T-1}(f^{(\tau)})^2 > (\alpha^2-1)\lambda^2\}}|\mathcal{F}_T) \le \frac{c\beta^2}{\alpha^2 - 1} = \varepsilon_\beta.$$

Thus (noticing $\{S(f^{(\tau)})^2 - S_{T-1}(f^{(\tau)})^2 > (\alpha^2 - 1)\lambda^2\} \subset \{T < \infty\}$)

$$\begin{aligned}&|\{S(f^{(\tau)})^2 - S_{T-1}(f^{(\tau)})^2 > (\alpha^2 - 1)\lambda^2\}|\\ &\quad\le \int_{\{T<\infty\}} \chi_{\{S(f^{(\tau)})^2 - S_{T-1}(f^{(\tau)})^2 > (\alpha^2-1)\lambda^2\}}\, d\nu\\ &\quad= \int_{\{T<\infty\}} \widetilde{E}(\chi_{\{S(f^{(\tau)})^2 - S_{T-1}(f^{(\tau)})^2 > (\alpha^2-1)\lambda^2\}}|\mathcal{F}_T)\, d\nu \le \varepsilon_\beta|\{T < \infty\}|.\end{aligned}$$

This proves (8.2.2.21), and hence (8.2.2.17). The proof of (8.2.2.18) is similar. Transferring to the case $\{\mathcal{F}_n\}_{-\infty}^{\infty}$ from the case $\{\mathcal{F}_n\}_{n\ge 0}$ is direct. The proof is finished. □

When $\Phi(u)$ is convex in addition, we can get rid of the factor D_∞ in (8.2.2.17) and (8.2.2.18), by making use of the Davis' decomposition, which holds in present case, too: For any martingale $f = (f_n)_{n\ge 0}$, with $d_n = \Delta_n f$, define

$$\Delta_n g = d_n \chi_{\{|d_n| \le 2M_{n-1}d\}} - E(d_n \chi_{\{|d_n| \le 2M_{n-1}d\}}|\mathcal{F}_{n-1}), \quad n \ge 0,$$

$$\Delta_n h = d_n \chi_{\{|d_n|>2M_{n-1}d\}} - E(d_n \chi_{\{|d_n|>2M_{n-1}d\}}|\mathcal{F}_{n-1}), \quad n \geq 0, \ h_0 = f_0,$$

then we have $f = g + h$, $g = (g_n)_{n\geq 0}$, $h = (h_n)_{n\geq 0}$, and

$$|\Delta_n g| \leq 4M_{n-1}d, \qquad n \geq 0, \tag{8.2.2.22}$$

$$\int_\Omega \Phi\Big(\sum_0^\infty |\Delta_n h|\Big)\, d\nu \leq c \int_\Omega \Phi(Md)\, d\nu, \tag{8.2.2.23}$$

where $\Phi(u)$ is a moderate convex function from $\mathbf{R}^+$ to $\mathbf{R}^+$ with $\Phi(0) = 0$.

Theorem 8.2.2.8 *Let $\Phi(u)$ be a moderate convex function as above, $f = (f_n)_{-\infty}^{\infty}$ be any martingale, then*

$$\int_\Omega \Phi(Mf)\, d\nu \approx \int_\Omega \Phi(S(f))\, d\nu, \tag{8.2.2.24}$$

with the equivalence constant depending only on c_0 and Φ.

Proof. First, consider the case $\{\mathcal{F}_n\}_{n\geq 0}$. Owing to Davis' decomposition, we get

$$\begin{aligned}\int_\Omega \Phi(S(f))\, d\nu &\leq c\int_\Omega \Phi(S(g))\, d\nu + c\int_\Omega \Phi(S(h))\, d\nu \\ &\leq c\int_\Omega \Phi(Mg)\, d\nu + c\int_\Omega \Phi(Md)\, d\nu + c\int_\Omega \Phi\Big(\sum_0^\infty |\Delta_n h|\Big)\, d\nu \\ &\leq c\int_\Omega \Phi(Mf)\, d\nu + c\int_\Omega \Phi(Mh)\, d\nu \leq c\int_\Omega \Phi(Mf)\, d\nu.\end{aligned}$$

This completes the proof of the theorem. □

Most of the problems studied in the preceding chapters can be discussed for this new kind of martingales, we omit the detail.

Notes to Chapter 8

§8.1 The kind of backward martingales and its application to lacunary Fourier series occur in this section are due to Gundy-Varopoulos [1].

§8.2.1. The idea to prove the $T(b)$ Theorem by martingale approach is due to Coifman-Jones-Semmes [1], where the proof of $T(b)$ Theorem in $\mathbf{R}$ case was given. For the $\mathbf{R}^d$ case, David [1] proved the existence of a quasi-basis of $L^2(\mathbf{R}^d, dx)$ adapted to given $b(x)$, and using it he proved the $T(b)$ Theorem. The explicit construction of the quasi-basis $\{\alpha_I, \beta_I\}_{I\in\mathcal{I}}$ given in this subsection is due to Long [8], there the $T(b)$ Theorem is Clifford algebra valued.

§8.2.2. The theory of martingales with respect to complex measures was motivated by the paper Coifman-Jones-Semmes [1]. The materials occur in this subsection are borrowed from Long-Qian [1].

References

C.R.A.S.P = C.R. Acad. Sc. Paris (Serie A)
I.U.M.J. = Indiana Univ. Math. Jour.
L.N.M. = Lect. Notes in Math. (Springer Verlag)
P.A.M.S. = Proc. Amer. Math. Soc.
P.S.P.M. = Proc. of Symp. in Pure Math.
T.A.M.S. = Trans. Amer. Math. Soc.

Azema, J., Gundy, R.F., Yor, M.
[1] Sur l'intégrabilité uniforme des martingales continues, Sem. Prob. XIV, L.N.M., **781**(1980), 53-61.
Bagby, R.J.
[1] Maximal functions and rearrangements, I.U.M.J., **32**(1983), 879–891.
Bagby, R.J., Kurtz, D.S.
[1] A rearranged good λ-inequality, T.A.M.S., **293**(1986), 71–81.
Bañuelos, R.
[1] Martingale transform and related singular integrals, T.A.M.S., **293**(1986), 547–564.
Bennett, A.G.
[1] Probabilistic square functions and a priori estimates, T.A.M.S., **291**(1985), 156–166.
Bennett, C., Sharpley, R.
[1] Weak type inequalities for H_p and BMO, P.S.P.M., **35**(1979), 201–229.
Bergh, J., Löfström, J.
[1] Interpolation spaces, An Introduction, Springer-Verlag, 1976.
Bernard, A
[1] Espaces H_1 de martingales à deux indices, Dualité avec les martingales de type "BMO" , Bull. Sc. Math. 2e serie, **103**(1979), 297–303.
Bernard, A., Maisonneuve, B.
[1] Decomposition atomique de martingales de la class H_1, Sem. Prob. XI, L.N.M., **581**(1977), 303–323.
Bonami, A., Lepingle, D.
[1] Fonction maximale et variation quadratique des martingales en presence d'un poids, Sem. Prob. XIII, L.N.M., **721**(1979), 294–306.
Brossard, J.
[1] Generalisation des inégalités de Burkholder et Gundy aux martingales régulières à deux indices, C.R.A.S.P., **288**(1979), 267–270.
Bru, B., Heinich, H., Lootgieter, J.C.
[1] Sur la régularité des filtrations, C.R.A.S.P., **294**(1982), 313–316.
Burkholder, D.L.
[1] Martingale transforms, Ann. Math. Sta., **37**(1966), 1494–1504.

[2] Distribution function inequalities for martingales, Ann. of Prob., **1**(1973), 19–42.
[3] One sided maximal functions and H_p, J. Func. Anal., **18**(1975), 429–454.
[4] Martingale theory and Harmonic Analysis in Euclidean spaces, P.S.P.M., **35** part 2(1979), 283–301.
[5] Sharp inequalities for martingales and stochastic integrals. Astérisque, Sem. Bourbaki, No. **157–158**(1988), 75–94.
[6] Differentional subordination of harmonic functions and martingales, Proc. of the Seminar on Harmonic Analysis and PDE (Spain 1987), L.N.M., **1384**(1988), 1–23.

Burkholder, D.L., Davis, B., Gundy, R.
[1] Integral inequalities for convex functions of operators on martingales, Proc. 6th Berkley. Symp., **2**. 1972.

Burkholder, D.L., Gundy, R.F.
[1] Extrapolation and interpolation of quasi-linear operators on martingales, Acta Math., **124**(1970), 249–304.

Burkholder, D.L., Gundy, R.F., Silverstein, M.L.
[1] A maximal function characterization of the class H_p, T.A.M.S., **157**(1971), 137–153.

Carleson, L.
[1] Two remarks on H_1 and BMO, Adv. in Math., **22**(1976), 269–277.

Chang, S.Y.A.
[1] Carleson measure on the bidisk, Ann. Math., **109**(1979), 613–620.

Chang, S.Y.A., Fefferman, R.
[1] The Calderón-Zygmund decomposition on product domain, Amer. J of Math., **104**(1982), 455–468.

Chao, J.A.
[1] Triadic and Dyadic conjugate martingales, P.S.P.M., **35**(1979).
[2] Hardy Spaces on regular martingales, L.N.M., **939**(1982), 18–28.

Chao, J.A., Janson, S.
[1] A note on H_1 q-martingales, Pacific J. of Math., **92**(1981), 307–317.

Chao, J.A., Long R.L.
[1] Martingale transforms with unbounded multipliers, P.A.M.S., **114**(1992), 831–838.
[2] Martingale transforms and Hardy spaces, Prob. Theory and Related Fields, **91**(1992), 399–404.

Chao, J.A., Taibleson, M.H.
[1] A sub-regularity inequality for conjugate systems on local fields, Studia Math., **46**(1973), 249–257.

Chow, C.S.
[1] Les inégalités surmartingales d'après, A.M.Garsia. Sem. Prob. IX, L.N.M., **465**(1975), 206–212.

Chow, Y.S.
[1] Martingales in a σ-finite measure spaces indexed by directed sets, T.A.M.S., **97**(1960), 254–285.

Coifman, R.R.
[1] A real variable characterization of H_p, Studia Math., **51**(1974), 269–274.
Coifman, R.R., Fefferman, C.
[1] Weighted norm inequalities for maximal functions and singular integrals, Studia Math., **51**(1974), 241-250.
Coifman, R.R., Jones, P., Semmes, S.
[1] Two elementary proofs of the L^2-boundedness of Cauchy integrals on Lipschitz curves, J. Amer. M.S., **2**(1989), 553-564.
Coifman, R.R., Rochberg, R.
[1] Another characterization of BMO, P.A.M.S., **79**(1980), 249–254.
Coifman, R.R., Weiss, G.
[1] Extensions of Hardy spaces and their use in analysis, Bull. A.M.S., **83**(1977), 596–645.
Coifman, R.R., Rochberg, R., Weiss, G.
[1] Factorization theorems for Hardy spaces in several variables, Ann. of Math., **103**(1976), 611–635.
David, G.
[1] Wavelets, Calderón-Zygmund operators and singular integrals on curves and surfaces, L.N.M., **1465**(1991).
David, G., Journé, J.L.
[1] A boundedness criterion for generalized Calderón-Zygmund operators, Ann. of Math., **120**(1984), 371–397.
David, G., Journé, J.L., Semmes, S.
[1] Operateur de Calderón-Zygmund, fonctions paraaccrétifes et interpolation, Rev. Math. Iberaamericana, **1**(1985), 1–55.
Davis, B
[1] On the integrability of the martingales square function, Israel J. Math., **8**(1970), 187–190.
[2] Brownian motion and analytic functions, The Ann. of Prob., **7**(1979), 913–932.
[3] Hardy spaces and rearrangements, T.A.M.S., **261**(1980), 211–233.
Dellacherie, C.
[1] Inégalités de convexité pour les processus croissants et les sousmartingales, Sem. Prob. XIII, L.N.M., **721**(1979), 371-377.
Dellacherie, C., Meyer, P.A.
[1] Probabilites et Potentiels, 2e edition, chapitres I-IV, chapitres V-VIII, Hermann, Paris, 1975, 1980.
Dellacherie, C., Meyer, P.A., Yor. M.
[1] Sur certains properties des espaces de Banach H_1 et BMO, Sem. Prob. XII, L.N.M., **649**(1978), 98–113.
Doleans-Dade, C., Meyer, P.A.
[1] Une characterisation de BMO, Sem. Prob. XI, L.N.M., **581**(1977), 383–389.
[2] Inégalités de norms avec poids, Sem. Prob. XIII, L.N.M., **721**(1979), 313–331.

Doob, J.L.
[1] Stochastic Processes, Wiley, New York, 1953.
Dunford, N., Schwartz, J.T.
[1] Linear Operators, New York, Interscience, 1958.
Duren, P.L.
[1] Theory of H_p spaces. Acad. Press. New York, 1970.
Duren, P.L., Romberg, B.W., Shields, A.L.
[1] Linear functionals on H_p spaces with $0< p <1$, J. reine angew. Math., **238**(1969), 32–60.
Edwards, R.E., Gaudry, G.I.
[1] Littlewood-Paley and Multiplier Theory, Erg. 90, Springer-Verlag, 1977.
Emery, M.
[1] Le theorems de Garnett-Jones, D'après Varopoulos, Sem. Prob. L.N.M., **850**(1981), 278–284.
Fefferman, C.
[1] Characterization of bounded mean oscillation, Bull. A.M.S., **77**(1971), 587–588.
Fefferman, C., Stein, E.M.
[1] H_p spaces of several variables, Acta Math., **129**(1972), 137–194.
Fefferman, R.
[1] Bounded Mean Oscillation on the polydisk, Ann. Math., **110**(1979), 395–406.
Frazier, A.P.
[1] The dual space of H_p of the polydisc for $0 < p < 1$, Duke Math. J., **39**(1972), 369–379.
Garnett. J.B., Jones, P.W.
[1] The distance in BMO to L^∞, Ann. of Math., **108**(1978), 373–396.
[2] BMO from dyadic BMO, Pacific J. of Math., **99**(1982), 351–371.
Garsia, A.
[1] Martingale Inequalities, Sem. Notes on Recent Progress, Benjamin, 1973.
Getoor, R.K., Sharpe, M.J.
[1] Conformal martingales, Invent. Math., **16**(1972), 271–308.
Gundy, R.F.
[1] A decomposition for L^1-bounded martingales, Ann. of Math., Statis., **39**(1968), 134–138.
[2] On the class $L\log^+ L$, martingales, and singular integrals, Studia Math., **33**(1969), 109–118.
[3] Inégalité pour martingales, à un et deux indices. L' espace H_p, Cours à l'Ecole d'été de Prob. de Saint-Flour, VIII (1978), L.N.M., **774**(1980).
Gundy, R.F., Varopoulos, N. Th.
[1] A martingale that occurs in Harmonic Analysis, Ark. for Mat., **14**(1976), 179–187.
Han Y.S.
[1] Decomposition of functions and its applications in operator intepolation, Sci. Sinica, **7**(1983).

Hanks, R.

[1] Interpolation by the real method between BMO, $L^a(0 < a < \infty)$ and $H_a(0 < a < \infty)$. I.U.M.J., **26**(1977), 679–689.

Herz, C.S.

[1] Bounded mean oscillation and regulated martingales, T.A.M.S., **193**(1974), 199–215.

[2] H_p-spaces of martingales, $0 < p \leq 1$, Zeit, Wahrscheinlichkeits theorie, **28**(1974), 189–265.

Igari, S.

[1] An extension of the interpolation theorem of Marcinkiewicz II, Tohoku Math. J., **15**(1963), 343–358.

Izumisawa, M., Kazamaki, N.

[1] Weighted norm inequalities for martingales, Tohoku Math. J., **29**(1977), 115–124.

Izumisawa, M., Sckiguchi, T., Shiota, Y.

[1] Remark on a characterization of BMO-martingales, Tohoku Math. J., **31**(1979), 281–284.

Jacod, J.

[1] Calcul stochastique et problèmes de martingales, L.N.M., **714**(1979).

Janson, S.

[1] Characterization of H_1 by singular integral transforms on martingales and R^n, Math. Scand., **41**(1974), 140–152.

Jawerth, B

[1] Weighted inequalities for maximal operators: linearization, localization and factorization, Amer. J. of Math., **108**(1986), 361–414.

John, F., Nirenberg, L.

[1] On functions of bounded mean oscillation, Comm. Pure Appl. Math., **14**(1961), 415–426.

Jones, P.W.

[1] Factorization of A_p weight, Ann. of Math., **111**(1980), 511–530.

Kazamaki, N.

[1] A characterization of BMO martingales, Sem. Prob. X, L.N.M., **511**(1976), 536–538.

[2] Changes of law, martingales, and the conditioned square function, Tohoku Math. J., **31**(1979), 549–552.

Khanh, B.D.

[1] Intégrales singulières, commutateurs, et la function $f^{\#}$, Bull. Sc. Math. 2e, Serie **103**(1979), 241–253.

Koosis, P.

[1] Introduction to H_p Space, Lond. M.S., Lect Note Serie, **40**(1980).

Latter, R.H.

[1] The atomic decomposition of Hardy spaces, P.S.P.M., **35**(1979), 275–279.

Lenglart, E., Lepingle, D., Pratelli, M.

[1] Presentation unifiée de certaines inégalités de la theorie des martingales, Sem. Prob. XIV, L.N.M., **781**(1980).

Lepingle, D.

[1] Sur certains commutateurs de la theorie des martingales, Sem. Prob. XIII, L.N.M., **721**(1979), 138–147.

Long R.L. =Long J.L. =Long, R. = Ruilin Long

[1] Martingale régulière et Φ-inégalités avec poids enter f, $S(f)$, $\sigma(f)$, C.R.A.S.P., **291**(1980), 31-34.

[2] Sur I'espace H_p de martingales régalières $(0 < p \leq 1)$, Ann. Inst. H. Poincare (B), XVII (1981), 123–142.

[3] Distance of $f \in BMO$ to L^∞, Acta Math. Sinica, **25**(1982), 189–201.

[4] Weighted Φ-inequalities on martingales, Acta Math. Sinica, **26**(1983), 173–178.

[5] A property of convex functions, Kexue Tongbao, **27**(1982), 641–642.

[6] Two classes of martingale spaces, Scientia Sinica A, **26**(1983), 362–375.

[7] Rearrangement techniques in martingale setting, Illinois J. of Math., **35**(1991), 506–521.

[8] Martingale proof of Clifford valued $T(b)$ Theorem on $\mathbf{R}^d$, to appear in Bull. des Sciences Mathématiques.

Long R.L., Peng L.Z.

[1] Decomposition of BMO functions and factorization of A_p weights in martingale setting, Chin. Ann. of Math., **4**B(1)(1983), 117–128.

[2] Two weighted maximal (p, q) inequalities in martingale setting, Acta Math. Sinica, **29**(1986), 253–258.

Long, R., Qian, T.

[1] Clifford martingales, Φ-equivalence between $S(f)$ and $M(f)$, preprint.

Meyer, P.A.

[1] Un cours sur les intégrales stochastiques, Sem. Prob. X, L.N.M., **511**(1976).

[2] Notes sur les intégrales stochastiques III, sur un theoreme de C. Herz et D. Lepingle, Sem. Prob., XI, L.N.M., **581**(1977), 465–469.

[3] –IV. Caracterisation de BMO pour un operateur maximal, ibid, 470–475.

Muckenhoupt, B.

[1] Weighted norm inequalities for the Hardy maximal function, T.A.M.S., **165**(1972), 207–226.

Neveu, J.

[1] Martingales à temps discret. Masson, Paris, (1972).

Okada, M

[1] An interpolation of operators in the martingale H_p-spaces. Proc. Japan. Acad., **52**(1976), 55–57.

Petersen, K.E.

[1] Brownian motion, Hardy spaces and Bounded Mean Oscillation, Cambridge Univ. Press, Cambridge, 1977.

Rao, K.M.

[1] Quasi-martingales, Math. Scand., **24**(1969), 79–92.

Rao, M.M.

[1] Interpolation, ergodicity and martingales, J. Math. Mech., **16**(1966), 543–568.

[2] Conjugate series, convergence and martingales, Rev. Romm. Math. Pures Appl., **22**(1977), 219–254.

[3] Stochastic processes and integration. Sijthoff & Noordhoff, 1979.

Reimann, H.M., Rychener, T.

[1] Functionen beschrankter mitterer oszillation, L.N.M., **487**(1975).

Sarason, D.

[1] Functions of vanishing mean oscillation, T.A.M.S., **207**(1975), 391–405.

Sawyer, E.

[1] A characterization of a two-weight inequality for maximal operators, Studia Math., **75**(1982), 1–11.

Stein, E.M.

[1] Note on the class $L \log^+ L$, Studia Math., (1969), 305–310.

[2] Singular integrals and differentiability properties of functions, Princeton Univ. Press, Princeton, N.J., 1970.

Stein, E.M., Weiss. G.

[1] Introduction to Fourier Analysis on Euclidean Spaces, Princeton, 1971.

Stroock, D.W.

[1] Applications of Fefferman-Stein type interpolation to probability theory and analysis, Comm. Pure. Appl. Math., **26**(1973).

Strömberg, J.O.

[1] Bounded Mean Oscillation with Orlicz norms and duality of Hardy spaces, I.U.M.J., **28**(1979), 511–544.

Uchiyama, A.

[1] A constructive proof of the Fefferman-Stein decompostion of BMO on simple martingales, Conference on Harmonic Analysis in Honor of A. Zygmund, (1981), 495–505.

[2] The singular integral characterization of H_p simple martingales, Proc. of A.M.S., **88**(1983), 617–623.

Varopoulos, N.Th.

[1] A probabilistic proof of the Garnett-Jones theorem on BMO, Pacific J M., **90**(1980), 200–221

[2] The Helson-Szego theorem and A_p functions for Brownian Motion of several variables, J. of Func. Anal., **39**(1980), 85–121.

[3] A theorem on weak type estimates for Riesz transforms and martingale transforms, Ann. de L' Inst. Fourier, XXXI(1981), 257–264.

Weiss, G.

[1] Some problems in the theory of Hardy spaces, P.S.P.M., **35**(1979), 189–200.

Woyczynski, W.A.

[1] Geometry and martingale in Banach spaces, L.N.M., **472**(1975).

Yor, M.

[1] Convergence de martingale dans L^1 et H_1, C.R.A.S.P., **286**(1978).

[2] Les inégalités de sous-martingales comme consequence de la relation de domination, Stochastics, **3**(1979).

Yen, K.A.

[1] Martingale and Stochastic Integral, Shanghai Sci & Tec. Press, 198[illegible].

Zygmund, A.
[1] Trigonometric Series I, II, 2nd ed. Cambridge, 1968.
Zaanen, A.C.
[1] Linear Analysis, North Holland, 1953.

Symbols

Index